SOME THEORY OF
SAMPLING

SOME THEORY OF

SAMPLING

By

WILLIAM EDWARDS DEMING

DOVER PUBLICATIONS, INC.
NEW YORK

This Dover edition, first published in 1966, is an
unabridged and unaltered republication of the work
originally published by John Wiley & Sons, Inc., in
1950.

International Standard Book Number: 0-486-64684-X
Library of Congress Catalog Card Number: 66-30538

Manufactured in the United States of America
Dover Publications, Inc.
180 Varick Street
New York, N.Y. 10014

PREFACE

The preface gives an author a chance to write his own review of the book. A review should state what the author tried to do, and why. It should also state whether he succeeded, but on this point only the judgment of the reader counts. Briefly, the aim here is to teach some theory of sampling as met in large-scale surveys in government and industry, and to develop in the student some power and desire for originality in dealing with problems of sampling.

This book is planned for two types of teaching. In the social sciences and commerce, teachers will find that Chapters 1 through 13 constitute a year's study. A first course in statistical methods is assumed. The day is past when students of the social sciences may hope to learn their subjects without thinking quantitatively with the aid of mathematics, yet it is a fact that most of the essential theory of these chapters goes not beyond the level of college algebra, although occasionally some forgotten calculus may need refreshment. In the natural sciences, engineering, and industrial management, students may start with Chapter 4 and work their way through to the end of the book, touching Chapters 11 and 12 only lightly for want of time. Such students, it is presupposed, will have done reading in the statistical control of quality and in the design of experiment.

Graduates in mathematical statistics, when taking up practice, discover yawning gaps between theory and practice: the better their theoretical training, the wider the gap. Chapters 1, 2, 4, 5, 11, and 12 are designed to help to bridge this gap.

One aim of the book has been directed toward the needs of the mature specialist in subject-matter who, like the author, must teach himself in the theory of statistics.

Copious exercises have been provided for the classroom and for the self-taught student. Almost every exercise illustrates some principle that has been found useful in the author's experience as a teacher and as a consultant in government, industry, and marketing.

It should be made clear that this book is not intended as a textbook in mathematical statistics. The reader is therefore advised to supplement his studies by pursuing mathematical works like the books by Fisher, Aitken, Neyman, E. S. Pearson, Cramér, Wilks, Kendall, Wald, Bose, the Statistical Research Group, and others, and by attendance at a statistical teaching center, if possible.

To the theoretical statistician of today the problem of sampling is the development and application of the theory of probability to the planning and interpretation of surveys, with the aim of acquiring and presenting to management and to other research workers quantitative information having maximum usefulness and maximum reliability per unit cost. Through proper planning with the aid of statistical theory, the reliability of the information obtained in a survey is controllable and demonstrable, so that the range of possible interpretations of the information is delimited.

It is interesting to note that sampling today is not confined to partial coverages: the censuses of population, agriculture, commerce, and industry in the United States and Canada now include concurrent and supplementary samples, not only to broaden the scope of the information, but to study and evaluate the errors and biases of the census in order that the data may be made more useful. It is significant that the recent excellent book on sampling by my friend F. Yates, D.Sc., F.R.S., is entitled *Sampling Methods for Censuses and Surveys* (Griffin, 1949).

Modern statistical practice requires knowledge of statistical theory, but knowledge exists only where there is research, however humble. Statistical research is therefore one of the most vital components of any statistical program. Statistical research is particularly necessary in the government service because of the high level of quality and economy that the public has a right to expect in government statistics. Moreover, there are many theoretical and practical problems that are encountered only in large-scale statistical surveys. The chief concern of any government statistical agency should be assurance of the necessity, the success, and the economy of any survey that is authorized at public expense. Such obligations can not be met in the absence of knowledge and research in statistical theory. Statistical research is constantly lowering costs and enhancing the reliability and usefulness of statistical information.

The intuition, like the conscience, must be trained. Gone are the days when the intuition, guided by expert knowledge of subject-matter, constituted sufficient qualifications for carrying out sample studies. Knowledge of the subject-matter of a survey is as essential as ever, but in modern practice this knowledge must be combined with the theory of sampling. Some of the most important contributions to research in population, vital statistics, economics, agricultural science, psychological testing, industrial relations, standardization, development and testing of product, manufacturing, and marketing, have come from theoretical statisticians.

The treatment given here omits the theory of several important developments in sampling that are now from three to six years old and already in use in government surveys. One important omission is an extended account of Hansen and Hurwitz's theories of sampling with varying probabilities. Another omission is an adequate account of new theories of estimation. Optimum allocation in stratified multistage sampling is barely mentioned. The very act of writing a book isolates its author for a time while the stream of research moves forward, leaving him behind. This stream is moving so rapidly that any book in statistical theory produced today will be out of date before it is printed, but this does not mean that none should be printed. Moreover, the particular deficiencies mentioned may be excused on the ground that they will be included by Mr. Hansen and his colleagues in a subsequent volume. At best, a book today on statistical theory can be only an exposure of some man's ignorance.

A legion of friends have helped in the writing of this book. First, there is my wife, Lola S. Deming, without whose dependable aid in calculation, preparation of manuscript, and proofreading, attempts to produce a book would long ago have been abandoned. A roster of the world's leading statisticians has assisted generously. Foremost, for continual inspiration and guidance, I record with pleasure my indebtedness to Morris H. Hansen, William N. Hurwitz, and P. C. Mahalanobis, F.R.S. Crushed with the load of work that government statistical consultants must bear, Mr. Hurwitz nevertheless spent many weary hours reading manuscript and deliberating with the author. Mr. Arnold Frank has kindly worked through many of the exercises and has given attention to a number of mathematical details. Miss Theresa Hoerner of the Forest Service has assisted me expertly in the proof stages. Mr. Jacob E. Lieberman, Dr. P. C. Tang, and Mr. Richard H. Blythe have read parts of the manuscript and made numerous suggestions. Contributions from Messrs. Frederick F. Stephan, Harold F. Dodge, H. G. Romig, Jerome Cornfield and Miss Mary N. Torrey, and others are partially recorded in appropriate places in the text. I am indebted to Professor Ronald A. Fisher and to Messrs. Oliver & Boyd Ltd., Edinburgh, for permission to reprint Table VI from their book *Statistical Methods for Research Workers*. For prominence, the last name to be mentioned will be that of Professor Philip M. Hauser, who in 1939 invited me out of the natural sciences and into the problems of the sampling of human populations as they are met in the Census, and who, from the vantage point of keen technical sense and professional prestige, coupled with high administrative responsibilities, cleared away the brush of administrative hierarchy and tradition that so often foil, bewilder, and discourage real technical ability, and provided room for the growth of modern sampling practice.

W. E. D.

Washington
1 January 1950

CONTENTS

PART I. THE SPECIFICATION OF THE RELIABILITY REQUIRED

PART II. SOME ELEMENTARY THEORY FOR DESIGN

PART IV. APPLICATIONS OF SOME OF THE FOREGOING THEORY

11. INVENTORIES BY SAMPLING

12. A POPULATION SAMPLE FOR GREECE

PART V. SOME FURTHER THEORY FOR DESIGN AND ANALYSIS

13. DETAILED STUDY OF SOME BINOMIAL AND RELATED DISTRIBUTIONS

A. THE BINOMIAL AND POISSON SERIES

SOME THEORY OF
SAMPLING

PART I. THE SPECIFICATION OF THE RELIABILITY REQUIRED

CHAPTER 1. THE PLANNING OF SURVEYS

In these days when so much emphasis is properly being placed on economy in government research operations, it is important to take advantage of the substantial savings that can be effected by substituting sound mathematical analysis for costly experimentation. In science as well as in business, it pays to stop and figure things out in advance.—Edward U. Condon, *The National Applied Mathematics Laboratories* (The National Bureau of Standards, February 1947).

The three major theoretical problems of survey-design. The statistician's speciality is *measurement:* more precisely, the *method and design of measurement.* Part of his problem is to decide the following: What is to be measured? What questions should be asked? What precision is needed? How can the survey best be carried out to provide the information desired with the desired precision and no more? What will the survey cost? What do the results mean? How can objective measures of the sampling errors and biases be obtained, so that the reliability and meaning of the results can be assessed and methods improved? To him, once the nature of the measurement has been decided upon, the theoretical aspects of the design of surveys consist of three major problems:

Specify the reliability to be aimed at, in view of the allowable cost (Step vi in the section "Steps in taking a survey," p. 6).

A suitable specification of reliability will consist of an aimed-at precision (such as a coefficient of variation of 1 percent, 5 percent, or 25 percent in some important characteristic). In addition, in order to achieve better interpretation of the data, the specifications may require measurements of the differences between various procedures (two or more different ways of stating the questions, different methods of training the interviewers, different definitions, etc.).

Design the survey or experiment so that it will produce the prescribed precision at the lowest possible cost and with the personnel and physical facilities likely to be available (Steps v–xi). The design must meet any irremovable administrative restrictions. It must provide an index of precision and comparisons between various alternative procedures.

If the expected cost appears to be too great, the problem must be reconsidered, the scope of the survey narrowed, the prescribed precision relaxed; perhaps the survey may be abandoned.

1

Appraise the precision actually attained in several important characteristics, and evaluate the differences between the various procedures specified for comparison. Also, compute the costs of various phases of the survey, and evaluate any remaining variances or other statistical measures that will be useful in cutting the costs or increasing the reliability of future surveys of this kind.

In modern statistical practice the three problems of sampling (viz., specification, design, and appraisal, as enumerated above) are not independent, but react on each other. Thus the specification of the precision to be aimed at is not finalized until some work has been done on the design to get some idea concerning the costs of various levels of precision. Preliminary calculations of costs and expected precision (Step vi) may show, for instance, that the survey that was contemplated will be too costly; whereupon the aims of the survey (Step i) must in part be reformulated on a smaller scale. Conversely, one can not start laying out the mathematical design of a survey without some preliminary specifications of what is wanted. Finally, Step xiv tells what precision was attained and how the cost of a similar survey in the future can be reduced (Ch. 10).

Sampling is not mere substitution of a partial coverage for a total coverage. Sampling is the science and art of controlling and measuring the reliability of useful statistical information through the theory of probability.

Before the theory of probability was used, the three problems listed above, *specification*, *design*, and *appraisal*, were independent. It was possible to specify that accuracy was desired, and it was possible to design surveys, but the two were not linked together or linked with cost. One could only hope that his plan would produce what was wanted, but he could never know whether it would or whether it did, or whether he had bought what he bought at a much higher figure than necessary, as without probability theory no objective calculations of precision were possible, either before or after a survey was carried out. There was no way of comparing the efficiency of one plan with the efficiency of another; in fact, there was no talk of efficiency. Little was learned from the experience of one survey that could be helpful in the design of another, as no appraisal of precision was possible or even contemplated. The only thing that could be done without probability theory toward appraising the results of a survey was to make comparisons with other surveys and census material. If they appeared to agree, the survey was pronounced good; if they appeared to disagree in some respects, there was no way of knowing whether the difference was attributable to sampling errors, or to "significant" differences between the questionnaires or the training of interviewers, or to other possible contributory causes. Without probability, "significance" may be unrecognizable.

Economic balance in design. The statistician's aim in designing surveys and experiments is to meet a desired degree of reliability at the lowest possible cost under the existing budgetary, administrative, and physical limitations within which the work must be conducted. In other words, the aim is *efficiency*—the most information (smallest error) for the money. These aims accord with Fisher's principles of modern design of experiment.[1]

A great deal is implied by good design. It would be possible to design a very efficient plan for obtaining too little precision for the purpose, or too much. The statistician, through the use of statistical theory, aims to ride between two types of error:

1. His plan may yield more precision than is needed and may thus be too costly, too slow, and (in the case of population-studies) excessively burdensome on the public.

2. It may yield insufficient precision, in which case significant results fail of attainment, and the efforts and expense of the survey are largely lost.

The statistician himself will occasionally slip into one of these errors, but through his specialized knowledge of theory he minimizes, in the long run, the net economic losses resulting therefrom.

A third type of error, much more deplorable than either of these two, is to design a beautiful plan that elicits irrelevant information or sets up protection where none is needed. The first step is therefore to find out what the problem is: *what is wanted?* With the aid of theory, and by finding out first of all what is wanted, the statistician aims to strike an economic balance between all these errors: he aims to obtain enough but no more precision or protection than necessary, and not to incur losses in carrying out meaningless surveys and experiments, however efficient.

The requirement of a plain statement of what is wanted (the *specification* of the survey) is perhaps one of the greatest contributions of modern theoretical statistics.

The relation of sampling to other characteristics of a statistical program. A good statistical program for a government or a corporation possesses six important characteristics:

i. Usefulness and comprehensiveness of content
ii. Reliability of results, sufficient for the purpose
iii. Intelligibility (classifications and definitions that are understood)

---

[1] No special reference is necessary: this aim permeates all his books and papers.

    iv. Speed

    v. Economy of operation

    vi. Accurate interpretation and presentation

In general, *statistical work consists first of all of determining what kinds of statistical information would be useful for the ends in view; of deciding whether the desired information can be obtained at all or at reasonable cost; and then of procuring this information at the lowest possible cost, and interpreting it in a form that assists rational decisions and adds to knowledge.* Statistical research itself is directed toward increasing the speed, reliability, and usefulness of whatever statistical information is deemed necessary; decreasing the cost of getting it; providing better measures of its reliability and more accurate interpretations.

A matured perspective for the mathematical statistician is best gained through several years' experience under competent leadership,[2] during which time his mathematical reading and research are expected to continue. For instance, in the planning of surveys it is important for the statistician not to insist on sampling refinements that are costly and unnecessary for the purpose intended, but without a matured perspective he can not see what levels of refinement are requisite and proper in any particular survey. Moreover, he must know how to estimate the costs and time required for various statistical operations in field and office.

Statistics are a basis for action, and *every survey therefore has a purpose,* namely, to get the answers to certain questions that will affect decisions or provide increased knowledge. Until the purpose is stated, there is no right or wrong way of going about the survey. To be specific, in the planning of a sample the statistician must know whether to aim for a standard error of 2 percent or 10 percent or 50 percent, and this requires that a decision be reached on what is wanted, and when, why, and how.

Steps in taking a survey. The following summary of the various steps that are passed through in taking a survey is presented here with no claim that it is complete or that the steps must take place in this or any other particular order, but with the hope that it provides some indication of the framework in which a statistician finds himself in practice. The student should note that most of the steps are the same whether the survey is a complete count or a sample. Preparations cease at any stage where it seems unwise or futile to proceed.

    [2] This is the only satisfactory way of learning how to estimate the costs and time required for various operations in field and office, how to write instructions for field and office, how to lay out training programs, how to provide suitable controls on the office-work, and how to carry out a host of other kinds of detailed work without which a statistical program breaks down.

i. Define the problem statistically. Decide what statistical information is really needed.

Several paragraphs have already appeared regarding this step. See if there is really a statistical problem at all. List the various decisions that are possible and determine whether these decisions will depend on the possible results of any proposed survey. If the possible results of a survey will not help in the decisions, no survey is needed. The best way to go about this is to draw up some specimen tabulations that might arise from a proposed survey, and observe whether any quantitative information will be helpful in leading to the right decision. The specimen tabulation plans will also be of inestimable help in defining the universe and drawing up the questionnaire.

The reader might now turn to the beginning of Chapter 12 to perceive how the problem of observing an election was translated into several statistical questions, capable of measurement.

ii. Define the universe to be studied.

The universe must be defined so clearly that the interviewers will have no difficulty deciding whether a particular farm or household or which member of the family belongs in the survey. A careful statement of the problem and consideration of the tabulation plans and how they will be used will assist greatly in defining the universe. Further notes occur in the next chapter; also in Chapter 4 wherein it will be seen that the definition of the universe is often necessarily limited by the frames that can be found or constructed.

It is important to note that the universe can only be defined operationally in terms of a real frame or a combination of frames *and* in the manner of using these frames. (See Chs. 2 and 4 for the definition of the frame.)

iii. Make a thorough investigation to see how much of the information that is needed is already available in published or unpublished reports. Aim to keep the survey, if any, as small as possible.

iv. Decide what type of survey, if any, could possibly provide the information that is desired, and do so at reasonable cost. Decide also the best frequency of coverage, and the best time of the year for the date of the survey. Should this be a single survey? Or would it be better and perhaps less expensive in the end to plan a series of surveys? If so, should a relatively large survey be taken at rare intervals, or would a series of smaller surveys at frequent intervals be better? (Ch. 7 has a bearing on this question.)

v. Lay plans for reducing the burdens of response, and for eliciting clear, intelligible information. Begin work on definitions and classifications, keeping the field-workers and the respondents in mind. Follow conventions where possible. Get started on the questionnaire, for which skilled consulting assistance may well be required. Start the hiring and training of supervisors.

Consider the difficulties of definition and interviewing and obtaining the information desired. There will be some nonresponse. How much? Enough to impair the survey seriously? How can it be reduced? Can the first wave of response be corrected by interviewing a sample (e.g., every third) of the people not responding at the first interview, or not returning their questionnaires by mail? Will the people have the information that is wanted? Small business firms are notorious for lack of records concerning sales, purchases, and costs of operation.

> But I keep no log of my daily grog,
> For what's the use o' being bothered?
> I drink a little more when the wind's off shore,
> And most when the wind's from the north'ard.

> —Arthur Macy, *The Indifferent Mariner.*

Even income taxes bring blessings: they have forced people to keep some records where none existed before. Unemployment compensation also necessitates records of wages where no records existed before.

If this is to be a postal survey, devise systematic procedures for sending one, two, or three letters to delinquents, to be followed by telegrams, telephone calls, and finally by personal calls on all or a definite sample of the hard core of resistance.

Consider the resources that are available with which to conduct the survey—the office-equipment and personnel, field-force, maps, lists, instructions, and experience in similar work. Decide whether it is worth while to go ahead.

vi. Lay out roughly several alternative sample designs, to show approximately what the costs will be for various degrees of precision. Aim to keep the survey small and well controlled in office and field.

Decide the maximum allowable sampling error (2, 5, 25, or 50 percent). Again, Steps i and ii may now need revision.

The decision on the allowable error is an administrative matter but must usually be solved by the statistician in deliberation with others. The administrator who needs the information to be supplied by the survey usually thinks of figures as being absolute and may be unaware of the difficulties in collecting or interpreting data. He may think of sampling as a game of chance, not appreciating the fact that sampling errors are under control. He is responsible for administering a program and can not take unnecessary chances. He needs facts and not errors or probabilities (pp. 298 and 562). It is therefore natural for him to demand too big a sample or even a complete count. The statistician must stand his ground firmly, weighing the requirements and balancing the loosest precision that will serve these requirements against the additional cost of obtaining greater precision.

It is sometimes possible to fix the precision objectively on the basis of expected net profit: cf. Richard H. Blythe, "The economics of sample-size applied to the scaling of sawlogs," *The Biometrics Bulletin*, Washington, vol. 1, 1945: pp. 67–70.

vii. Provide by proper design interpenetrating, concurrent, and supplementary "illuminating" samples that will measure (*a*) the completeness of coverage and the possible effects on the data attributable to incomplete coverage; (b) the possible effects of errors arising from response and nonresponse, and from differences between interviewers; (c) differences arising from various admissible procedures of collection, interviewing, or training; (d) differences in cost between various admissible procedures.

These "illuminating" samples are as important or more important for a complete census than they are for a sample. In fact, no census can be regarded as finished unless it is accompanied by properly designed tests of the quality of the job, and a frank discussion of the deficiencies so discovered. A splendid example is furnished by the "complete" census of the commerce of France in 1946, which was tested by a subsequent areal sample of re-interviews and found to be too deficient to warrant publication. (G. Chevry, "Control of a general census by means of an areal sampling method," *J. Amer. Stat. Assoc.*, vol. 44, 1949: pp. 373–9.) Other examples are furnished by the censuses of population, agriculture, and housing in the United States in 1950, and by the census of Canada in 1951.

viii. Draw up instructions for the field-workers.

The field-workers must first of all be hired and told when and where to report for training, how long the survey will occupy them, etc. A set of instructions must be prepared for every step that takes place: listing, drawing the sample from the list, interviewing (definitions, staying within bounds), cleaning up, sending the completed forms to headquarters. Serious bias of almost any kind may occur through faulty instructions. The supervisors will require special additional instructions. Examples of instructions for listing, drawing the sample, and interviewing will be found in Appendix A to *A Chapter in Population Sampling* (full reference on p. 79).

Schools of instruction are to be conducted, and they will require materials for teaching (the instructions just mentioned, records of interviews, review of the sampling procedure, examinations).

Remember that instructions are for the benefit of the people that are going to do the work—not for the writer. The aim of the survey, and the reasons for each instruction must be made clear.

ix. Get started in earnest on the tabulation plans and eventually finalize them. Settle upon the areas of tabulation, detail, sizes of classes, table-captions, headings, stubs, etc. (Permissible minimum size of class is a problem in statistical theory, whether the survey be a complete count or a sample: see Ch. 7.)

x. Pretest the questionnaire and instructions for the field-workers, which have supposedly been in preparation and pretty well developed by now. Several pretests may be a good investment. If pretests show that the refusal rate is high, or that the quality of the information is poor, the survey may well be abandoned or modified at this point.

A pilot survey offers an opportunity not only to pretest and compare different versions of the questionnaire and instructions so as to improve the accuracy and amount of response, but also to obtain advance estimates of some of the variances, proportions, and correlations that appear in the formulas for the sampling errors, so that the cost of the sample may be trimmed to a minimum.

A pilot survey should not be used merely as a device to be doing something. No procedure should be tested unless it has a reasonable chance of acceptance.

This is a good time to draw up the instructions for coding: the results of the pretest will illustrate most of the different kinds of problems that will be met. The coding should do more than merely classify the answers: it should be carried out so that the tabulations will convey information on the subject of the survey, and the uses intended.

xi. Revise the questionnaire and instructions.

xii. Finalize the sampling procedure.

This is an attempt to meet the specifications finally decided upon.

Provide maps, lists, and controls for the interviewers and the supervisors. Draw up the tabulation plans and other office-procedures for forming estimates and estimates of reliability; they are part of the sample-design.

xiii. Carry out the survey and the tabulations.

It is difficult to lay enough emphasis on the need for careful execution of any survey, complete or sample. A good sample-design is lost if it is not carried out according to plans. A statistician's responsibility is not confined to plans: he must also seek assurance of cooperation in field and office, and maintain constant touch with the work, also with the interpretation of the results. Many so-called "complete counts" have been badly in error because of careless and incompetent work in field and office. Neither an incompleted "complete count" nor an incompleted sample is a sample, but rather a form of *chunk* (p. 14).

On the other hand, it is easy to become too finical about many of the operations in the collecting and processing of data. Too much time laboring over some fine points in the coding of unusual cases may not be worth while. One hundred percent verification of every step may be wasteful.[3] Editing by machine is gaining favor for searching out gross errors.

Adjustments must be made and rules established for handling difficulties and departures from instructions. For example, during the interviewing, certain problems will be discovered in the listing. Some dwelling units (d.us.) listed will not be found by the interviewers. For some of these the explanation will be obvious; in others, not so. Some d.us. will be missed by the listers. Some d.us. will be burned or removed, and here and there new d.us. will have been constructed in the interim between listing and interviewing. Some listings will not be d.us., but will turn out to be warehouses, garages, or offices. Some d.us. will be found unoccupied. In others the inhabitants will have gone to Florida during the interval of interviewing. There will be refusals and incomplete returns. These must all be accounted

[3] W. Edwards Deming and Leon Geoffrey, "On sample inspection in the processing of census returns," *J. Amer. Stat. Assoc.*, vol. 36, 1941: pp. 351–60; W. Edwards Deming, Benjamin J. Tepping, and Leon Geoffrey, "Errors in card punching," *ibid.*, vol. 37, 1942: pp. 525–36.

for and reported, and their possible effects calculated. (Cf. Appendix C to *A Chapter in Population Sampling*.)

xiv. From a subsample of the returns compute the sampling errors of some of the important figures obtained in the survey (Ch. 10).

xv. Interpret and publish the results.

Study the sampling errors and any comparisons that were made between interviewers and different versions of the questionnaire. Compare the results of this survey with the results of other surveys; point out differences in definitions and procedure.

First comes a report for administrative use, setting forth the conclusions reached as a result of the survey. This report is expected to throw light on the question that prompted the survey in the first place, and thus to assist the administrator to come to a rational decision on his problem. This report may well be the chief aim of the survey. The findings therein will be based on first-hand knowledge of the field-work and on calculations made with the aid of the theory of probability, as well as on knowledge of the risks and gains that appear to be associated with the various alternative decisions. However, this report should not contain any talk of probabilities or errors: it is for action only (pp. 298, 562). It should be brief, confined if possible to one telegraph blank.

Next it may be worth while to prepare a research paper for the information of colleagues. This paper should contain detailed results and the conclusions reached. It should describe the sampling plan and the necessary theory or references thereto. It should likewise include the main definitions and a copy of the questionnaire if space permits. It should show the calculated sampling errors for the chief results, and the conclusions derived from any supplementary tests. It should include a careful statement of any difficulties encountered in the field or in the coding, and their possible effects on the data. Enough detail must be given so that the reader can verify the formulas that were used for the standard errors, and form his own opinions concerning the precision and the accuracy and the conclusions. Comparison of the results and procedures with those of other surveys, along with reasons for differences, may be worth while. The observed variances and correlations and any unusual experience encountered will be of aid to statisticians elsewhere in the planning of surveys dealing with similar material. Distinguish between sampling errors and errors of forecasting (p. 18).

The reader is directed to a memorandum entitled "Recommendations concerning the preparation of reports on sampling surveys," which was written by the United Nations Subcommission on Statistical Sampling in 1948. This memorandum is obtainable from the United Nations Statistical Office and will also be found on pp. 141 ff in F. Yates's *Sampling Methods for Censuses and Surveys* (Griffin, 1949).

Probability-samples and judgment-samples. In his daily practice the statistician must constantly be aware of two different types of samples, probability-samples and judgment-samples.[4]

[4] These terms and definitions were first put forward by the author in an article, "Some criteria for judging the quality of surveys," *J. of Marketing*, vol. xii, 1947: pp. 145–57. This article is a revision of a chapter bearing the same title in the book *Measurement of Consumer Interest* (University of Pennsylvania, 1947), edited by C. West Churchman, Russell L. Ackoff, and Murray Wax.

Probability-samples, for which the sampling errors can be calculated, and for which the biases of selection, nonresponse, and estimation are virtually eliminated or contained within known limits.

Judgment-samples, for which the biases and sampling errors can not be calculated from the sample but instead must be settled by judgment.

The two types of surveys are not distinguished by the questionnaire and instructions, but by the procedures for selecting the sample, for calculating the estimates, and for appraising the precisions of these estimates. A probability-survey is carried out according to a statistical plan embodying automatic selection of the elements (people, farms, manufactured material) concerning which information is to be obtained. In a probability-sample neither the interviewer nor the elements of the sample have any choice about who is in the sample. If a sample of individuals is desired, the design of a probability-sample must give rules for finding these individuals; it is not sufficient that it give rules that lead to a random selection of households, leaving the selection of the individuals in these households to the judgment of the interviewer. A probability-sample demands a competent field-force and careful execution of the instructions at all stages of the work. It is also to be noted that in a probability-sample the procedure for forming the estimates is automatic, being laid down beforehand as part of the sample-design. Unless these conditions are met, probability theory can not be used to appraise the precision of the results, and a survey can not be characterized as a probability-sample.

A probability-sample will send the interviewer through mud and cold, over long distances, up decrepit stairs, to people who do not welcome an interviewer; but such cases occur only in their correct proportions. Substitutions are not permitted: the rules are ruthless.

Actually, a pure probability-sample with complete response is a rarity. In practice there will usually be some nonresponse and some departure from instructions. An upper limit to the biases so created may often be assigned, nevertheless, through knowledge of the subject matter, in which case the survey will still satisfy the definition of a probability-sample, viz., *a calculable error.* Thus, suppose that in a survey of 1000 households, 500 are found to be users of a certain product, 450 are found to be nonusers, and 50 were never found at home. By assigning the 50 nonresponses first to the users and then to the nonusers, upper and lower limits to the mean square error of the results may be calculated (for the definition of mean square error; see p. 129).

In contrast, the results from a judgment-sample are obtained by procedures which depend to some appreciable part on i. a judgment-selection of "typical" or "representative" counties, cities, road-segments, blocks, individual people, households, firms, farms, articles, or packages concerning which information is to be obtained; or on ii. weighting factors that are prescribed arbitrarily or by expert judgment to make allowances for certain sizable segments of the population whose magnitudes and characteristics are unknown and not determined by the sample. The following examples may be noted in this respect: the assumption that nonresponding groups are similar to responding groups; that homes without telephones are similar to homes with telephones; that packages that are difficult to get at are similar to packages on the outside of a pile. There are many problems in which the survey itself, through (e.g.) failure of proper design, failure of the questionnaire, or for lack of sufficient response, fails to elicit certain information that is needed in calculating the final estimates: in such cases the survey is of the judgment type, whether originally intended thus or not.

The "quota" method is one type of judgment-sample. In this method an interviewer is assigned to procure (e.g.) 10 interviews with people conforming to certain sociological and economic characteristics within a prescribed area, such as housewives who do not work full time for pay, who own their homes, who belong in a certain economic level, a particular age-class, and live in a particular block, tract, or precinct. The quota method is subject to the biases of selectivity and availability, besides the errors of incorrect assignment of weights to the various classes of the population. This assertion, however, is not intended to cast doubts on the quota method, but to acquaint the reader with some of the problems.

This book will deal entirely with probability-samples; in other words, this is a book on statistical theory, not subject-matter or manipulation of data. Judgment-samples, so far as I know, are not amenable to statistical analysis. I know of no way to remove the biases of selectivity, availability, nonresponse, and incorrect assignment of weights. Moreover, I know of no way in which to calculate the standard errors of data from a quota sample, the reason being that a particular man or house has no assignable probability of coming into the sample; hence probability does not apply. It is more important to learn something about the *biases* of a judgment-sample than about its sampling errors. The usefulness of data from judgment-samples is judged by expert knowledge of the subject-matter and comparisons with the results of previous surveys, not from knowledge of probability. A skilled statistical theorist would be helpless in the analysis of a judgment-sample if he were to depend on his knowledge of theory. It is a fact, though, that some of the lessons regarding economy in the design (not analysis) of probability-samples are equally applicable to judgment-samples. For example, theory can assist judgment-samples in the choice of sampling unit, allocation of the sample to economic levels and to urban and rural areas, and in the number of survey points.

Such remarks are not meant to imply that judgment-samples can not and do not deliver useful results, but rather that the reasons why they do when they do are not well understood. Indeed, quota and other types of judgment-samples will undoubtedly continue to play an important role in research, and they will become more and more useful as their strong points and weak points are more generally understood.

Pilot surveys are usually judgment-samples. In trying out a questionnaire or set of instructions, or for getting a rough idea of how much a certain operation is going to cost,[5] or what the refusal rate is likely to be, it may not be necessary or desirable to carry out a probability-survey; it will often be sufficient to conduct a trial in a particular county or city or even in a few blocks, chosen by judgment. Examples abound. The proposed instructions and questionnaire for the decennial census of population in 1940 were put to a test in St. Joseph and Marshall Counties in Indiana in August 1939. These counties were not selected as a probability-sample, but because they contained an abundance of "typical" situations. They served the purpose well, as they focused attention on weak points of the instructions and the questionnaire. Moreover, a large operation in two adjoining counties provided a dress-rehearsal for the big census eight months later, as a widely dispersed probability-sample would not have done. Much of the experimental work in the planning of the 1950 censuses of population and agriculture is being conducted in areas chosen by judgment.

As for comparisons of costs between probability- and judgment-samples, no satisfactory basis for comparison is possible because the two types of survey are different commodities and are not interchangeable. Price without knowledge of quality is meaningless, and it is impossible to compare the costs of two proposed methods of conducting a study unless the precision and biases of the results of *both* methods are known and controllable. In many of the surveys on characteristics of the population, of farms, of agricultural production that are carried out by the government, a controllable and measurable error of sampling and freedom from the biases of selection and nonresponse are considered indispensable and cheaper than a wrong decision based on biased results. Moreover, business, industry, and private research demand quality in government statistics. For similar reasons there is a decided trend in private research in marketing toward the use of probability-samples.

[5] It should be emphasized that careful studies of cost require probability-samples and other skilled statistical techniques not treated here.

A relatively inefficient but unbiased design for a single (nonrecurring) probability-sample need not be costly to lay out. An inexpensive map and a visit to the library to look at Census figures will often provide sufficient information for the delineation of large roughly equal sampling units for single- or double-stage sampling. The inefficiency of the design is then to be counterbalanced by taking a sufficiently large sample. On the other hand, for a recurring survey, it usually pays to make more elaborate preparations by providing several years' supply of small efficient sampling units and listings so that smaller samples may be used month after month.

Either way, a probability-sample demands careful field-work, constantly reviewed by a competent statistician, with records and callbacks, proper training and supervision. These safeguards cost money, but there is no alternative if demonstrable precision is required. To say that the job can be done cheaper without them is to confuse the issue, as there can be no talk of price without a simultaneous measure of quality.

A judgment-sample can often be devised quickly without *benefit of skilled statistical assistance*, which is sometimes very hard to find.

Remark 1. As already stated, strictly, there is hardly ever a pure probability-sample. The purest examples are the simple ones in which the universe to be sampled is by definition a file of cards: there are then no refusals or nonresponses unless some entries are illegible. However, as there were refusals, nonresponses, and inevitable errors of response in the original collection of the information on the cards, these imperfections will be carried over into any sample, even 100 percent, that is drawn from the cards.

In the collection of original data, either by interview or mail, unless response is mandatory (e.g., income tax, electric bills, and the like), some incompleteness of returns is inevitable, and the reliability of the returns is then to some though perhaps negligibly small extent dependent on expert judgment. Under particularly unfortunate circumstances the unwillingness or inability to provide the information may be so serious that little semblance of a probability-sample remains, even though the selection of respondents was originally designed on a probability-basis: under such circumstances one might just as well have started off with a judgment-sample (e.g., an assignment of quotas). Careful pretests should avoid such difficulties.

A sample that is 95 or 98 percent a probability-sample and the other 5 or 2 percent a judgment-selection or judgment-adjustment for refusals, for people not at home, etc., may still be an excellent sample, although it is important to investigate the remaining 5 or even 2 percent as soon as possible. There have been instances in our experience when a nonresponse rate as low as 5 percent was found later to be seriously affecting the results. (Further remarks regarding the biases arising from nonresponse will be found in Ch. 2, pp. 33–6.)

It is sometimes supposed that there are no troubles with nonresponse when inanimate materials are sampled for testing. Curiously enough, this is not always so. In the author's experience, a sample of 208 manholes belonging to a large utility company appealing to the court for higher rates was drawn for the purpose of inspecting the manholes themselves, the ducts, and the cables therein. It was then discovered that 22 of the manholes designated for the sample had been paved over. Civic authorities would not let them be uncovered. Substitutions could not be permitted, nor a new sample. The difficulty was settled unfavorably to the company by declaring them to be one grade lower than average. The sample was still a probability-sample because a lower limit to the value of the property could be assigned.

Remark 2. Statistical research has disclosed and explained several amazing facts about sampling. It is entirely possible to build up a "sample" of people by adding a few names here and subtracting a few there, so that the list finally agrees almost perfectly with the last census and any additional information in regard to the proper proportions by area, age-groups, sex, color, education, economic level, ownership of home, telephone, and in fact with respect to almost any conceivably complex pattern.[6] This is what in lay language is sometimes described as "a perfect cross-section." In fact, however, this kind of "sample" is extremely dangerous, as it may fail miserably to correspond with the population of the country, city, or county that it was intended to represent *in regard to the characteristics that the survey is expected to measure* (e.g., the number of people intending to buy certain books or holding certain political opinions). Such hazards are avoided in probability-samples.

Remark 3. Judgment is indispensable in any survey. It would be decidedly incorrect to say that knowledge of the universe is not utilized in a probability-sample, and blind chance substituted. In modern sampling, judgment and all possible knowledge of the subject-matter under study are put to the best possible use. Knowledge and judgment come into play in many ways in the design of probability-samples; for instance, in defining the kind and size of sampling units, in delineating homogeneous or heterogeneous areas, and in classifying the households into strata in ways that will be contributory toward reduction of sampling error. There is no limitation to the amount of judgment or knowledge of the subject that can be used, but this kind of knowledge is not allowed to influence the final selection of the particular cities, counties, blocks, roads, households, or business establishments that are to be in the sample; this final selection must be automatic, for it is only then that the bias of selection is eliminated, and the sampling tolerance is measurable and controllable.

Definition of a chunk. A chunk [7] is a convenient slice of a population. A judgment-sample is planned with expert judgment; a chunk is dictated by convenience. The following examples may be classed as chunks:

A certain city, selected mainly because the surveying organization has a field-force there, not yet disbanded from a previous survey.

The first 1000 returns of any form, compulsory or voluntary.

Any group of people who happen to be handy (a class of students, for example).

---

[6] This device is often confused with stratification.

[7] The term "chunk" was first used in this connexion by Dr. Philip M. Hauser in an effort to distinguish between a sample and a chunk.

Interviews of "average people" on street-corners.

An investigation that is carried out by somehow finding people who fit the descriptions of varied classes of the population and will answer the questions.

A list of names, however large, unless selected by a random procedure from the entire universe.

A study of the economic status and religious affiliation of the principals in weddings held in June (cf. Werner J. Cahnman, *Amer. Sociol. Review*, vol. 13, February 1948: pp. 96–7).

A "flying questionnaire" is a familiar form of chunk which is occasionally used by restaurants, air-lines, and department stores which provide their patrons with simple questionnaires enquiring whether this or that item or service is satisfactory, why they bought what they bought, and why they bought it here. The returns show only the extremes in satisfaction and vexation, and at that only from the articulate. Such devices do serve some useful if limited purposes, but if interpreted as representative the results may be disastrous. An investment in even a very small sample, taken at a much higher cost per schedule, may be much wiser.

It would be wrong to imply that no good has ever been done by studying chunks. On the contrary, some very useful results have been accomplished in this way. Moreover, some of the best research that has been done in the arts of questioning and interviewing has been carried out on chunks of the population.

Definition of a preferred procedure. Definition of a procedural bias.
The definition of any characteristic, whether it be age, employment status, income from interest, change in liquid assets, yield per acre, quality of being defective, or anything else, must be given in terms of an *operation* or *procedure* for the *measurement* of this characteristic. For some characteristics it will be agreed by the experts that there is a *preferred procedure*, even if in practice some preferred procedures are never or hardly ever used. A preferred procedure is distinguished by the fact that it supposedly gives or would give results nearest to what are needed for a particular end; and also by the fact that it is more expensive, or more time-consuming, or even impossible to carry out.

A preferred definition for a person's *age* might be specified as the result of subtracting the date of his birth from the present date. In turn, date of birth might be defined as the figure obtained by examining the person's birth certificate; or if there is no birth certificate, his horoscope, family album, or sworn evidence, or by tying his past with some important event, preference being in the order given. This definition for age would of course be unworkable in the ordinary survey. In practice, an unpreferred procedure is substituted—for example, merely ask the man his age at his last birthday, or his date of birth, and write down the answer.

If the man himself is not at home, enquire of any responsible member of the household.

For the characteristic *seeking work*, there is no definition that is useful for all purposes. Being registered with an unemployment agency may be taken as one definition of seeking work, but there are many circumstances where supplementary sufficient conditions must be specified, and these conditions vary much with the intended uses of the data. An application for unemployment compensation might be taken as another definition of seeking work. Educational level, occupation, industry, income, expenditures, prices, yield, and in fact practically every other characteristic, are difficult to agree upon. People have devoted their careers to definitions and concepts in the labor force, income, expenditures, etc. An appreciation for such efforts is fundamental to good statistical work.

Besides preferred and unpreferred definitions and ways of asking questions, there are preferred and unpreferred ways of hiring, training, and supervising the field- and office-workers, and of conducting the field-work. Three call-backs might be preferred over two, one, or none; etc.

The results of two surveys, one using a preferred definition, the other an unpreferred definition, will be different—often greatly different. The result of a preferred procedure is sometimes called a *true value*.

A preferred procedure, if one exists, may be *random* or *stable* in the Shewhart sense and hence may possess a mathematically "expected" value, but again, it may not; and the same thing may be said about an unpreferred procedure. It is usually assumed without justification in the design of surveys that "expected" values do exist for the procedures contemplated, and such assumptions will be made in this book in spite of the fact that too little is known about the validity of these assumptions and about what to do when they are violated. It is to be hoped that the next decade will see much research into such questions.

A preferred procedure is always subject to modification and obsolescence. What is preferred today may not be preferred tomorrow.

The *bias* of any *unpreferred* or *biased procedure* is the difference between the results that it produces and the results that would be produced had a preferred procedure been used instead. If both the preferred and unpreferred procedures possess mathematically "expected" values, the definition of bias is very simple: it is then merely the difference between the two "expected" values.

The results of a biased or unpreferred procedure may safely be interpreted only when its relation to a preferred procedure is pretty thoroughly understood. Hence one of the most fruitful kinds of statistical research for the near future lies in the study of differences between various procedures, preferred and unpreferred. So-called "complete" censuses are in dire need of this kind of research. Field trials by means of interpenetrating or supplementary samples are required for such studies. Sampling and the design of experiment are the essential statistical tools.

As a preferred procedure is always subject to modification or obsolescence, we are forced to conclude that *neither the accuracy nor the bias of any procedure can ever be known in a logical sense.* The *precision* of a random or stable procedure, however, *may be measured and known.* For our purposes, the precision of some particular random procedure for measuring a particular characteristic of the universe may be defined as the inverse of the standard deviation of the distribution of the estimates obtained in repeated applications of the procedure.

In what follows, the bias of some proposed procedure will be the difference between its "expected" results and the "expected" results of a procedure that is by definition preferred as of today.

Besides procedural biases, there are also the small and usually negligible biases of sampling (next chapter).

Biases and accidental variation. There are two main kinds of errors in surveys—biases and accidental variation. The insidious thing about biases is their constancy and the consequent difficulty of detecting them. Tests conducted to demonstrate the absence of bias are ofttimes only experimental demonstrations of remarkable ability to repeat the same mistake. To be specific, if the results of a large survey are divided into ten piles at random, or are divided according to the geographic locations of the regions whence they originate, intercomparisons are incapable of detecting a bias in the overall procedure because the results in each pile may *all be wrong by the same amount.* Similarly, agreement year after year does not demonstrate the absence of a bias. It should also be remarked that most biases are not removed or diminished simply by increasing the size of the sample.

Accidental variations are disclosed by a visible scattering of results when a survey is repeated. Variability exists whether the biases are appreciable or negligible. When a survey is repeated, the counties, cities, blocks, road-segments, households, farms, business establishments, people interviewed, interviewers, time of day when particular people are interviewed, and results obtained, will vary. It may come as a shock to some readers to learn that accidental variability is not all wiped out by a complete canvass. The reasons lie in the changing picture of a multitude of circumstances that lead to the results that are actually obtained. In any complete canvass (and of course the same remarks hold for a sample) a particular set of interviewers is engaged to carry out the work. That these particular ones were engaged is to some extent a matter of chance. Certain people happened to see the advertisement for interviewers; others might have applied for the jobs had they not failed to see it. A new selection of interviewers will bring forth a new set of results, perhaps inappreciably different when tabulated in classes, but nevertheless different. A new selection of supervisors who are to train the interviewers will bring still another new set of results. Even the weather has an effect; an interviewer on a household survey finds a particular woman at home merely because a thundershower is in the offing and she decides not to go shopping just now: the replies that she gives to the questionnaire will be different in some respects from the replies that would have been obtained from her daughter, who would have given the responses had the thundershower not come up just then. A lawn-sprinkler sends an interviewer around her assigned area in a different direction than she would otherwise have taken, and she finds certain people at home to give responses who otherwise, a few minutes later or a few minutes earlier, would not have been found at home. The time of day and a multitude of accidental circumstances affect responses.

In sampling, a part of the accidental variations arises from the chance selection of the areas, households, people, farms, manufactured articles, or other elements which are drawn into the sample. In a probability-sample the accidental variability arising from selection may be usefully treated mathematically as a "random variable."

Definition of sampling error. Thus, as we shall see, a sample can be designed so that it will estimate within 2 percent, or closer if desired, what *would have been* the result of applying *the same procedure* to *every member* of the universe. That is, through sampling, we can discover to within (e.g.) 2 percent what would have been the result of asking *every person* in some region, by use of the same procedures and care as were exercised on the sample, how many families in an area own the homes that they are living in, how many homes are mortgaged, how much certain types of families spent for various items of food and clothing, how much wheat they raised last year, and how many acres were in wheat, and how many families intend to purchase certain items of house-hold equipment next year. Yet the sampling error of such a survey is no measure of the reliability of someone's prediction of how many families will own their homes five years hence, what they will spend for various items of food, how many men will be employed or unemployed a year hence, how much wheat they will raise next year, or what they will purchase. Too many people have confused sampling error with a meas-ure of the validity of a prediction of the condition of the universe a year hence. The distinction is important.

Good sampling is essential for good prediction but it is no guarantee of good prediction. Anyone can easily misuse good data. Prediction of what people will do, even on the basis of a complete and perfect census, can fail for many reasons—unreliable methods of predicting, failure to understand the questionnaire and the information that was obtained, or because of unforeseen events, such as people changing their minds, and for many other reasons.

An empirical test of sampling theory may be made as follows: (a) carry out a complete census, and then (b) designate a sample of areas, people, firms, or farms; (c) tabulate the returns for the sample and for the whole; (d) by subtraction, compute the actual sampling errors. By designating a long series of independent samples of the one complete count, a distribution of sampling errors can be constructed. This distribution will follow closely the predictions of the theory of sampling. In this way, comparability of procedures between sample and complete count is assured, and the test is valid. This definition for the measurement of sampling error studiously avoids confusion between a prediction and the actual error of sampling. It likewise avoids confusion between the real errors of sampling and the biases of various kinds which are never exactly the same in two different surveys, sample or complete.[8]

Illustration of sampling error and bias. One of the simplest illustrations of bias and sampling variability is found in shots aimed at a target. The target might be the vertical line seen in Fig. 1. Under the assumption of randomness there will be an "expected" center of gravity when any number of shots is fired. If the "expected" center of gravity of the shots falls to the right, as in the two top panels, there is a bias which can be corrected by changing the setting of the sights. In surveys, a bias can be corrected by revising the questionnaire, by changing the method of survey (such as by changing from a judgment-selection to a random selection of the households that are to be interviewed), or by removing the bias of nonresponse by calling back again and again if necessary on people not at home at first call (instead of omitting such people or making substitutions or adjustments by judgment).

An important characteristic of repeated shots, even under ideal conditions, is that they do not all fall at the same spot; there is a scatter, even with a fixed setting of the sights. Under conditions of randomness, the sampling variability has the property of possessing a range or *tolerance* with *an associated degree of probability*. In Chapter 9 it will be seen that the 3-sigma sampling tolerance is a band outside of which practically no shots ever fall. A 2-sigma band is narrower and may be

---

[8] This definition of sampling error is taken from a paper by Morris H. Hansen and W. Edwards Deming entitled "On an important limitation to the use of data from samples," which was read at the meeting of the International Statistical Institute in Berne, 5–10 September 1949.

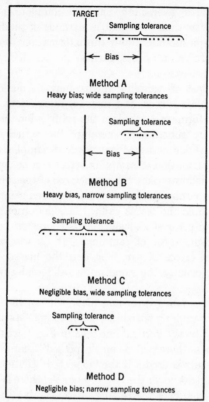

FIG. 1.   Illustrating the meaning of bias and sampling error.   The dots in any panel represent repetitions of a particular sampling and estimating procedure and of a particular operation by which the surveys are carried out.   If the sampling procedure is "random," a large number of repetitions will cluster about an "expected" value, and a prescribed percentage of them will fall within a "sampling tolerance" of some prescribed size such as 95 percent or 99 percent.   The vertical line represents the "expected" value of a preferred method, which might be a combination of better training, better supervision, improved questionnaire, etc.   The "bias" of any method is defined as the distance between its "expected" value and the "expected" value of some particular preferred method.

expected to contain about 95 percent of the shots.   It is impossible to predict the exact position of an individual shot, but under conditions of randomness it is possible to predict and control the *probability* or *proportion of hits* that will fall inside a particular band of error, and such is the aim of this book.[9]

[9] A magnificent set of comparisons between predictions of sampling tolerances and actual subsequent results from the complete Census of Agriculture of 1945 was published by Morris H. Hansen and William N. Hurwitz, "On dependable samples

In Fig. 1 are four panels illustrating different degrees of bias and sampling variability that are produced by different types of sample-design.[10] In the top panel heavy biases are present. One contributory bias might be the bias of selection—as, for example, exists when the interviewer's judgment is allowed to come into play (mentioned earlier), or when nothing is done to elicit responses from people who are not at home at first call or who mislay their questionnaires. The sampling tolerance in the top panel is wide, but in a probability-sample it is under control and can be made narrower (as in Panel B) by increasing the size of the sample or changing the procedure of selection or using more efficient methods of estimation (cf. Ch. 5). In Panel C the biases have practically all been removed, as by using automatic selection and energetic follow-up of nonresponse, or corrections by other devices, but the sampling tolerance is still wide. In Panel D the sampling tolerance of Panel C above has been narrowed, again possibly by increasing the size of sample or making other suitable modifications in procedure.

Remark. Like most illustrations the chart is oversimplified. Almost every sample actually consists of several samples, as many as there are questions to be tabulated, whereas the chart applies to any one question, but to only one at a time.

In designing a sample it will usually be found that there is one essential characteristic whose desired precision determines the type and size of sample. Thus, in a sample that is intended to produce a population count as well as population characteristics, the design will ordinarily be laid out along lines that will produce the required reliability in the population count (such as a standard error of 1 percent), because a sample that will do this will yield more reliability than is usually required for characteristics such as age distribution in 5-year age-classes, the sex ratio, school attendance, classes of employment, marital status, and other characteristics for which ordinarily only proportions in broad classes are desired.

Number of field-workers for greatest precision. An important consideration in deciding the number of interviewers to be employed arises from a desire to minimize the effect of differences between interviewers.

for market surveys," *J. Marketing*, vol. xiv, 1949: pp. 363–72. "On the basis of tabulations from a sample of the returns, the Bureau of the Census in July 1946 published national estimates for 61 agricultural items, together with a statement of the precision of each estimate. Corresponding figures from the complete Census of Agriculture became available about a year later. The estimates and their sampling errors as originally published, together with the relative differences between the sample estimates and the complete census returns, appear in the table below. It is seen that the complete census was in reasonable agreement with the advance statements of the precision of the original estimates. Three (5 percent) of the 61 differences between sample estimate and census exceeded 2 standard deviations, and none exceeded 3 standard deviations."

[10] This figure and several accompanying paragraphs in the text appeared in the article and book cited in footnote 4.

Given a job to do of a certain size, save for limitations of time it would be conceptually possible for a single interviewer to perform the entire task. At the other extreme, a large corps of interviewers might be hired and trained and the job finished up in a hurry. Of course, the latter procedure is more costly because of the higher proportionate cost of training the large corps of interviewers. It might seem that the more interviewers the better, in order to give differences between interviewers a chance to cancel out. This would be so if each member of a large corps of interviewers could be hired, trained, and supervised as well as if he were a member of a small corps, but in many situations this is not so—in fact, it is hardly ever so.

The following simple model serves as a suggestion in many actual situations. Assume that if the size of the corps of interviewers exceeds some easily manageable number (perhaps 10, 15, or 20 at a single administrative center; the exact number is unimportant now), the effectiveness of the interviewing falls off, and may fall off rapidly.

Suppose that the variance $\sigma_F{}^2$ between interviewers is crudely but effectively represented by the function

$$\sigma_F{}^2 = a^2 + b^2 F^2 \tag{1}$$

wherein a^2 is roughly the average variance between interviewers in ideally small groups, and b^2 is a coefficient that is introduced to take care of the poorer work done by larger numbers. F is the number of interviewers. Drawing upon a general theorem from Chapter 4, we may say that the variance of the results ($\bar{x}$) of the survey may be represented by

$$\sigma_{\bar{x}}{}^2 = \frac{\sigma_F{}^2}{F} + f(n)$$

$$= \frac{a^2}{F} + b^2 F + f(n) \tag{2}$$

wherein $f(n)$ is some function of the size n of the sample and the manner of selecting it. By differentiating $\sigma_{\bar{x}}{}^2$ with respect to F it is seen that $\sigma_{\bar{x}}{}^2$ is a minimum when

$$F = \frac{a}{b} \tag{3}$$

This is the optimum number of interviewers from the standpoint of minimizing the effects of differences between interviewers, regardless of cost.

It may be noted that when this number of interviewers is employed, $\sigma_F{}^2 = 2a^2$; hence to evaluate b we need only discover at what number (F) $\sigma_F{}^2$ is double its ideal value, a^2. To date, no such measurements exist, so far as I know; but in a particular survey which had as its aim the evaluation of the plant and equipment of a public utility company, σ_F^2 was measured for several types of equipment. As the corps of observers were all small (usually from 2 to 10), the values of a^2 may be assumed equal to σ_F^2. As a surmise, and only that, we might suppose that σ_F^2 would have been doubled if F had been 25. Then 25 would have been the optimum number of observers, had cost not been a consideration.

It is important to note that the optimum number of interviewers or observers apparently does not depend on the sample-size n or on $f(n)$; hence the number obtained as optimum is *still optimum for all sizes of sample, even for a complete count.* Of course, not all functional relations between σ_F^2 and F lead to a like conclusion (e.g., $\sigma_F^2 = a^2$ and $\sigma_F^2 = a^2 + bF$ do not), but any function σ_F^2 possessing a minimum will do so, and the conclusions just arrived at are probably valid under most circumstances.

Exercise 1. If you were required to pick out a county to represent the whole country for studies of a price index, or unemployment, or income, or consumption of various foods, or crop yield, or a combination of such measures, would you prefer to pick out the county by judgment or to pick it out at random?

Personally, I should refuse to do it, but if it must be done, I should prefer to see a judgment-selection made. As the number of sampling units increases, the reliability of a judgment-selection remains about constant, whereas the reliability of a probability-selection increases. For a single unit, or even six, I should prefer to see judgment-selections.

Exercise 2. A publisher of technical books wishes to determine whether his public prefer their books in bright or dead colors. If you were asked to carry out a study to get an answer to this question, how would you define the universe and what sort of questionnaire would you draw up?

CHAPTER 2. THE VARIOUS ERRORS OF A SURVEY

The Bureau of Standards has worked hard to learn the accuracy of its measurements and it supplies each weight with a certificate indicating how much the weight may differ from one pound. The calibration of the weight is valuable just because its possible error is known. When the Bureau of the Census makes an enumeration, there are errors, which they acknowledge. They know the extent of the errors from many sources and they try to learn more about them from others. . . . It is far easier to put out a figure than to accompany it with a wise and reasoned account of its liability to systematic and fluctuating errors. Yet if the figure is . . . to serve as the basis of an important decision, the accompanying account may be more important than the figure itself.—John W. Tukey in *Amer. Statistician*, vol. 3, 1949, p. 9.

A. LIST OF ERRORS AND BIASES

Imperfections in all surveys. In any survey there are many sources of error (cf. the list on pp. 26–30). A perfect survey is a myth. Some surveys are of course better than others, but even the best surveys contain imperfections. It must not be supposed, however, that all surveys are worthless because all have errors.[1] There are varying degrees and kinds of error, and some types are less disturbing than others. Errors possessing something of a random character may partially cancel each other. Other errors, even though nonrandom and hence persistent, may be harmless. Sometimes this persistence (bias) can be measured and subtracted out; at least this is the aim in research on biases. Much depends on how the data are to be used. For instance, when trends or proportions rather than absolute numbers are to be considered, no harm is done if the figures that are to be compared are all in error by the same percentage.[2] Moreover, in some problems, an error of 100 percent or even much more will not affect the decision one way or another.

The figures (tabulations) that are produced by a survey are the results of carrying out certain operations; a change in any of the operations, even a change in the color of the paper on which the questionnaire is printed, may be expected to produce different results. Improvement in survey-design comes about as a result of better understanding of

---

[1] Discussions on errors in surveys, limitations of data, and possible improvements in methods sometimes lead some people to suppose that all statistical data are worthless. This effect was actually observed in a meeting of the Washington Statistical Society in 1948. It was also observed on one occasion in a hearing when a certain government statistical agency requested additional funds with which to test the quality of the data collected, and to measure the completeness of coverage: the committee were amazed to learn that the agency had not been doing a perfect job.

[2] Too often, however, people forget that the assumption of equal percentages of error requires demonstration.

what information is wanted and how to get it more accurately and cheaper.

The statistician can not, in practice, overspecialize in sampling theory. He must be on the alert for all kinds of errors and biases. For what profiteth a statistician to design a beautiful sample when the questionnaire will not elicit the information desired, or if the universe has not been satisfactorily defined, or the field-force is so badly organized that the results will not be worth tabulating? And again, what is accomplished if a well-designed questionnaire and well-disciplined field-force are used with a biased sampling procedure?

A sampling procedure is to be judged adequate or over- or under-refined only relative to the other errors that will afflict any survey. It thus falls to the statistician to act as a court of equalization. He is error-conscious by training, and he must keep his eye on all the sources of error that may afflict a survey if he is to come out with a proper balance and anything useful. In order to minimize costs he must be careful not to over-refine his sampling error or any other kind of error.

To the administrator who must rely on information from a survey in making a decision affecting the pocketbooks or the health of the people or stockholders whom he serves, it makes little difference where an error in a figure came from—badly designed questionnaire, sampling error, unsatisfactory definition of the universe, sloppy field-work, or mistakes in interpretation; it is all the same to him; it is still a wrong figure. The statistician who designed the sample is guilty along with his colleagues who took part in the planning if the figures obtained are not useful. Occasionally, owing to insurmountable difficulties in timing, lack of facilities (maps, lists, personnel), impossible problems in the definition of the universe or in the questionnaire, or lack of funds, the best survey-design that can possibly be devised within the limitations may not be good enough to attain the reliability desired. In such a situation the statistician owes a responsibility to his employer and his profession to issue a firm disavowal of participation. A better plan is to find ways of overcoming the apparently insurmountable obstacles.

Economic balancing of errors. For many purposes the vector or right-angled addition of the biases and sampling error of a survey procedure is a useful concept, as depicted in Fig. 9, page 129. The statistician should keep in mind the hypotenuse and strive to reduce it. The hypotenuse is the root-mean-square error, which is the resultant of the two legs—the sampling errors on one leg, and the biases from all causes on the other leg. When the biases are large, it is uneconomical and ineffective to spend a lot of money to keep the sampling errors small. As a matter of fact, ofttimes the most effective way to shorten the hypotenuse is to decrease the bias while increasing the sampling error.

That is to say, the overall *usefulness and reliability of a survey may actually be enhanced by cutting down on the size of sample* and using the money so saved to hire better interviewers and to provide better training and supervision in the field, thus trimming the biases of interviewing, response, and nonresponse more than enough to counterbalance the increased sampling error.

Detailed list of sources of error. As has already been stated several times, some sources of error are present in both samples and complete counts; others owe their origin only to sampling. Some of the first group are in practice not entirely independent of the size of sample and may actually increase as the size of sample is increased. A list of some of the sources of error, along with some brief remarks, is now to be given.[3] Most of them can be controlled or measured by careful planning, following the steps that are outlined in the preceding chapter. Not all of them are independent; e.g., mistakes of the interviewer can sometimes hardly be distinguished from careless and disorganized field-procedure, or even from bias arising from the interviewer. In turn, bias arising from the interviewer is sometimes indistinguishable from errors in response. Failure to design the questionnaire properly can hardly be distinguished from a wrong interpretation of the results. Nevertheless, the list hereunder has been most useful in the form given.

The various errors and biases will rank differently in importance for different surveys. Thus, in a survey asking farmers their intentions to plant a certain crop next spring, the date of the survey would perhaps be more important than the auspices. On the other hand, we know from experience that in times of economic depression, the results of a survey of unemployment are greatly influenced by the auspices of the survey, whereas in better times the auspices may not be so important.

PROCEDURAL BIASES, OR
ERRORS COMMON TO BOTH COMPLETE COUNTS AND SAMPLES [4]

1. Failure to state the problem carefully and to decide just what statistical information is needed (Step i on p. 5).

Failure to grasp the aims of the survey: failure to perceive what information will be needed and in what form.

[3] This list is similar to one given in a paper by the author entitled, "On errors in surveys," *Amer. Sociol. Review*, vol. 9, 1944: pp. 359–69. Some paragraphs in this chapter are copied with little change.

[4] The student of engineering, physics, or chemistry may wish to formulate a parallel list of errors for physical measurements.

2. Failure of the questionnaire.

Failure to recognize the difficulties of acquiring certain types of information, through unwillingness of respondents to give it or their sheer inability to provide it or understand what is wanted.

Lack of clarity in definitions; ambiguity; varying meanings of same word to different groups of people; eliciting an answer liable to misinterpretation.

Omitting questions that would be illuminating to the interpretation of other questions.

Use of emotional words; leading questions; patterned response.

Fostering errors and nonresponse of a questionnaire through formidable appearance and lack of clarity in printing and layout (so that the respondent writes his answer in the wrong place; checks the wrong block; lays it away; etc.)

As stated in Chapter 1, page 5, the design of a questionnaire requires professional skill, which should be engaged early in the planning.

3. Failure to recognize differences between various kinds and degrees of canvass, and to fit the questions to the type of canvass.

Mail, telephone, telegraph, direct interview.
Intensive vs. extensive interviews.
Long vs. short schedules.
Check-block plan vs. free response.
Correspondence panels.
Key reporters.

4. Failure to define the universe with enough precision and to provide a satisfactory frame therefor.[5]

5. Faulty instructions and definitions.

Too many words and too many instructions, without disclosure of the reasons underlying the procedures, is a common fault. An interviewer may become as confused as a waitress with too much detail and not enough purpose.

6. Bias arising from nonresponse.   Respondent—

Not found at home, even after repeated calls.
Refuses to give the information.
Merely fails to take the trouble to return the questionnaire, or to keep an appointment with the interviewer.
Is unable to furnish the information.

7. Bias arising from late returns.

The bias of late returns is not as serious as the bias of nonresponse. Moreover, in a periodic series, late returns can be added to produce revised figures in subsequent reports.

[5] The *frame* is a description of the sampling units that compose the universe. This term was adopted by the United Nations Subcommission on Statistical Sampling in 1948. It was suggested by Fisher, who in the deliberations of the Subcommission also suggested the term *substrate*, for which I now have a slight preference.

8. Errors in response, voluntary and involuntary.

Plain honest accidental mistakes in responding.
Illegible entries.
Failure of memory.
Memory-bias.
Guessing, made necessary through lack of records.
Unwillingness to give the right answers.
Refusal to give any answer.
Failure to understand the question: answering the wrong question.
Wrong answers arising from pride, called "prestige-bias," by which the respondent upgrades his education, occupation, income, expenditures, literature, or purchases, or downgrades her age.
Wrong answers arising from protection of self-interest: understatement of income and production, overstatement of expenses.
Response-conditioning—the impact on the respondent of being invited to participate, and the effect of being asked or interviewed a number of times in repeated surveys. Repeated interviews have advantages and disadvantages. My friend Sam Barton of Industrial Surveys Company, New York, tells me that in his experience about 13 weekly questionnaires are required before response on consumption of foods and drugs becomes reliable enough for use.

9. Accidental variations in response.

10. Bias (in response) arising from the interviewer.

Answers guided by suggestions from the interviewer or by check blocks or other suggestions in the questions, any of which may channel the respondent into grooves.
Injection of the interviewer's beliefs and prejudices into his interpretations of the answers and in the way he records them (cf. the reference to Rice, p. 42).
Failure of the interviewer to establish good relations, with the result that many questions are incorrectly or incompletely answered.
The opposite failure, establishing relations that are entirely too cordial, with the result that the respondent tries to please the interviewer.

11. Bias (in response) arising from the auspices.

People will report differently in different surveys, because they can not separate information from its ultimate uses. They will unintentionally report differently in censuses of agriculture and population, and still differently in a census of unemployment or to the tax-collector. They will unintentionally report differently in surveys carried out by the government and in surveys carried out by private research organizations.

12. Blunders of the interviewer.

Failure to understand the definition of the universe (thereby including certain people or business establishments that should not be included; failing to include some that should be included).
Failure to understand the questions, definitions, and instructions, and thereby doing things incorrectly.

Failure to heed the questions, definitions, and instructions.
Failure to stay within the bounds of the area assigned to him.
Failure to cover his area or assignment completely.
Failure to state his business and the questions clearly.
Failure to persist in finding someone at home.
Getting responses from some member of the family who does not know the answers.
Omitting some questions and guessing at the answers.
Fatigue bias, or dilution bias, whereby an interviewer, inspector, or coder fails to note a characteristic that appears infrequently, such as a rare race, occupation, industry, or defective article.

13. Careless and disorganized field-procedure.

Faulty instructions and definitions, faulty methods of selecting supervisors and interviewers, faulty or insufficient training, negligence through which some areas and classes of the universe are missed, areas and classes being covered that should not be covered, areas covered in duplicate (hardly possible except in sampling), or just plain misunderstanding.

14. Bias arising from an unrepresentative date for the survey or of the period covered.

This might be classed as an error of sampling, even though the canvass is complete.

15. Ineffectual tabulation plans.

Failure to portray the results effectively for the uses intended (poor selection of characteristics, class intervals, too many or too few cross-tabulations).

16. Errors in processing (coding, editing, tabulating, calculating, tallying, posting, and consolidating).

17. Faulty publication and interpretation of the results.

Failure to understand the definitions.
Failure to calculate the sampling errors and to give information on the other errors in the data.
Failure to understand the difficulties of the field-work or of the coding and tabulating.
Failure to recognize secular changes that take place in the universe before the results are written up and recommendations made.
Personal bias.
Bias arising from bad curve-fitting; wrong weighting; incorrect adjustment.
Placing too much confidence in small numbers.
Making unwarranted forecasts. (A sample, however good, can only describe the past, not the future; see p. 18.)
Applying the interpretations and forecasts to domains or other universes not covered by the survey.

ERRORS HAVING THEIR ORIGIN IN SAMPLING

Probability-samples

18. Random sampling errors.

The aim of this book is the evaluation and control of these errors.

19. Sampling biases.

   i. Human failures.
   a. Getting into the wrong areas, going beyond bounds, partially or wholly omitting an occasional sample area or household that presents difficulties.
   b. Preferential treatment of sample areas and households.
   ii. Biases of the estimating procedure.
   a. Use of the wrong formula of estimation; usually a failure to tailor the estimating procedure to the probabilities of selection (cf. the exercises commencing on p. 87).
   b. Bias of the base. Usually the proportion f contains an unbiased estimate of B in its denominator, in which case the product fB is unaffected by error in B. This is the usual case with a ratio-estimate (Ch. 5). Sometimes, however, the sample gives an estimated proportion f independently of error in the base, B. The error in B is then carried into the product fB.
   iii. Bias arising from failure to randomize the starting points in a systematic selection (p. 89).

Judgment-samples
(Including the quota method)

18′. Random sampling errors.

19′. Sampling biases.

   i. Biases of selection, conveniently described as the biases of judgment, selectivity, and availability.

   ii. Bias of the estimating procedure. In a quota sample, a base must be used in the estimating procedure because the sample can only give proportions and can not stand on its own feet.

B. REMARKS ON THE VARIOUS ERRORS AND BIASES

Numerous vigorous researches on the biases of response, interviewing, format, response-conditioning, prestige-bias, effect of nonresponse, and biases of both probability- and quota-samples under various conditions are in progress. Results of such studies will help to improve the quality and lower the costs of obtaining information on population, agriculture, and commerce, in regular censuses and occasional or monthly surveys.

In spite of the fact that the subject is in a state of flux owing to the large amount of research now in progress, a few notes will be attempted in this chapter regarding some of the errors and biases in surveys.

Failure of the questionnaire. A questionnaire is never perfect: some are simply better than others. All questionnaires are difficult. As has been said already, the first step in planning a survey is to find out what is wanted. The questionnaire is the channel through which the needed information is elicited. Faulty design usually arises from lack of knowledge of the subject-matter and failure to grasp the problem, but a questionnaire can fail in many ways. It can fail to ask relevant questions: it may fail to elicit sufficient ancillary information to permit correct interpretation. It may be too detailed, too tedious, resulting in nonresponse. Instructions and format are exceedingly important.[6]

Failure to define the universe with enough precision and to provide a satisfactory frame. Only a few remarks will be made here on this topic, in spite of its importance. The definition of the universe follows almost automatically from a careful statement of the problem. However, this is only the start: there is still the mechanical labor of finding ways of making the definition intelligible and workable, and to provide a frame. In the actual operation of drawing a sample of areas, a good rule to follow is this: exclude from the universe any area that would cause embarrassment if it were to fall into the sample. Swamp-land, railroad yards, rivers, lakes, parks, and other uninhabited territory should be excluded in a survey of the population of a city. Mountain tops, bad-lands, forests, and any uninhabited areas should be excluded. If a fully unbiased procedure is intended, one must, before excluding any area on such grounds, make certain that it is really uninhabited. Likewise, in a survey of poultry farms, any areas where there are no poultry farms should be excluded.

One further remark must be made in this connexion. Although it is not necessary to attempt to exclude uninhabited areas from the frame in a population-survey, it is certainly desirable to do so, for two reasons. First, as will be seen later, the variance of the sample is reduced if the areal units are made as nearly alike as possible: in other words, the presence of 0-areas (no population) increases the variance between areas ($\sigma_b{}^2$ in Ch. 5) and raises the cost per unit amount of information. It should be carefully noted, however, that 0-areas *do not cause bias.* Second, an interviewer feels as if he had been sent on a wild-goose chase when he follows his map to an area where no one lives, particularly if he

[6] John A. Clausen and Robert N. Ford, "Controlling bias in mail questionnaires," *J. Amer. Stat. Assoc.*, vol. 42, 1947: pp. 497–511.

traveled a long way to get there: moreover, the interviewer, unless he has been given some instruction in the theory of sampling, imagines that the sample is seriously impaired, and he may easily develop a detrimental lack of confidence in the organization. What is really bad, in his zeal to improve the sample, he may substitute an inhabited area for an uninhabited area, and thus impair the sample instead of improving it.

Sometimes it is well to exclude from the universe sparsely populated areas whose total population is *known* to be not more than 2 or 3 percent of the type of population to be covered in the universe, particularly if the costs of canvassing such sparsely populated areas is excessive. Another plan, which is free from bias, is to sample such areas very lightly, a suggestion which is in line with the mathematical theory of Chapter 6 on the most efficient allocation of the sample.

The problem of providing an adequate frame [5] for the universe is often serious and costly. It is easy to define a universe as "all concerns that export or import petroleum-products." But where *are* all the concerns that export or import petroleum-products? A few big companies are known, but there must be a myriad of smaller ones whose aggregate business would perhaps be a sizable portion of the total. And how about companies that export only for their own use abroad? Are they to be included?

The reader might pause further to think of the difficulties in providing a frame for a census of transportation. The Class I railroads and Class I trucking concerns already report monthly to the Interstate Commerce Commission, and their names and addresses are on file: census information from such concerns could be obtained through the Interstate Commerce Commission in a special enquiry, either sample or complete. But there remain literally scores of thousands of surface and water transportation companies not doing interstate business or not large enough to be registered with the Interstate Commerce Commission or anywhere else: how can they be covered? How can they be found and identified? There are telephone and business directories, but they are not complete, and they do not definitely identify all the transportation companies as transportation companies. Even if such concerns are discovered, they often do not have records by which the desired information may be obtained. Then there are numerous unknown and unreachable concerns that derive some sizable part of their incomes from transportation. Should they be included? Besides, there are as many more concerns (no one knows how many) that have their own transportation departments, serving only themselves (e.g., department stores). Some of these concerns even own small (i.e., short) railroads and ships. Should a census of transportation cover these? How? The answers to such questions lie in determining what is wanted and why. What will it cost? Where is the money coming from? These are some of the problems that the mathematical statistician must help solve, and solve satisfactorily, before he has a mathematical problem to work on.

The following actual examples illustrate a few more kinds of universes and some of the difficulties.

All farm operators in a certain area. (The definition of a farm is extremely difficult and must be carefully learned and followed out if useful results are to be compiled.)

School children. (Public or private or both? How about children not in school because of sickness?)

Costs of construction in a certain county. (Illustration of the kind of care that must be exercised: a site for a new building has been cleared of timber at a cost of $1000, and the timber sold for $2000. Is this $1000 to be charged as part of the cost of construction?)

In a survey of radio-ownership, is a radio to be included if it is not in working order? if in the repair shop? or if an order has been placed for its repair?

Differences between various kinds and degrees of canvass. Too little is known in regard to the differences in results obtained from mail, telephone, telegraph, and interview canvasses, or the results obtained from different plans of questionnaire.[7] The problem is not whether differences exist, but how great are the differences, and why do they exist, and what effect will they have on the uses that are made of the data? Theory and more extensive empirical evaluations are needed so that comparability can be obtained between different methods, and so that the cheaper methods may have greater utility.

Bias arising from nonresponse. As already indicated (p. 13) nonresponse in sufficient quantity may seriously impair the usefulness of a survey and undo the work that went into the sample and other tasks of preparation. The root of the difficulty lies in the fact that the people who do not respond are in some ways and to varying degrees different from them that do.[8] As was indicated earlier, nonresponse arises from *i.* people not found at home, even after repeated calls; *ii.* refusal to give the information; *iii.* mere failure to take the trouble to return the questionnaire, or to keep an appointment with the interviewer; *iv.* sheer inability to furnish the information. These points can not all be discussed adequately here. Some remarks on these points have been made already: for example, the use of pilot studies for improving the question-

---

[7] It is important not to confuse (a) the differences in response elicited from different kinds and degrees of canvass, with (b) the different proportions of response that will be obtained.

[8] At the time of writing many studies are being made of this problem, but few of them have so far been published. One in particular will be mentioned—an illuminating article by E. H. Hilgard and Stanley L. Payne, "Those not at home: riddle for pollsters," *Pub. Opin. Quart.*, vol. viii, 1944: pp. 254–61.

naire and instructions so as to get more accurate responses and more responses, and the failure of small business concerns to keep accurate records on purchases, sales, and costs of operation have been mentioned (Ch. 1). Many very important studies, for example, studies on family-income and expenditures, are rendered difficult through failure of families to have the required information on record.

The proportion of nonresponse in any voluntary survey depends strongly on several factors: *i.* the questionnaire; *ii.* the interviewer; *iii.* the training and supervision of the interviewer; *iv.* advance notice by letter; *v.* publicity by newspaper or radio; endorsement, as by the local chamber of commerce. The design of the questionnaire is important, as it must incite and maintain the interest of the respondent, even if it is long and tedious or requires daily records of purchases of food and clothing. Extreme skill is required as the interviewer introduces himself and his problem to the intended respondent. The variation in performance between interviewers is amazing. Some interviewers have persistently good luck at inducing people to give answers: others have persistently bad luck. Some interviewers actually find significantly fewer or more than an average proportion of people "at home" at first call. Much depends on the interviewer's self-reliance, persistence, and belief in the usefulness of the survey; also probably on a host of intangibles, such as clothing and speech. Sometimes a second trial at the door brings success: sometimes it only draws forth a more emphatic and dramatic refusal. Often a second interviewer will succeed where the first one failed.

The net amount of nonresponse in an area will depend greatly on the supervisor in charge, who may cleverly choose the interviewers, train them, and fire them with crusading enthusiasm; who may know his territory and match certain interviewers to certain areas; who may cleverly handle refusals and sense whether it will be advantageous to return the first interviewer or send another one to a household or business establishment that has refused, and whether to send a letter in advance.

Advance publicity by press and radio is very helpful, even if the publicity is unfavorable. A whispering campaign, however, may be disastrous. One interviewer that I talked with recently had sat three hours in the kitchen of a respondent listening to family history and reasons why she was not going to give answers to the interviewer's questions, all as a result of coaching by neighbors who perhaps justifiably thought that their area was being oversurveyed.

It is important to bear in mind that the problem of *nonresponse is not solved by starting off with an excess of cases to allow for shrinkage.* There is no substitute for response.

A sample is no longer a probability-sample if it is ruined by non-response or any other difficulty of execution. The amount of nonresponse that may be tolerated in a probability-sample depends on the aims of the survey. As illustrated in Chapter 1, on the basis of substantive knowledge, outside limits may sometimes be placed on the bias arising from nonresponse, in which case the sample is still a probability-sample, as it permits calculation of the errors. However, in some surveys, for example on income, sales, inventories, purchases of raw materials, such limits may be so broad that they are useless, and the survey is no longer a probability-sample.

Supplementation by a judgment-sample of families to take the place of those that have dropped out and ruined a probability-sample is prescribed by some statisticians. However, the reliability of a ruined probability-sample must be evaluated as if it were a judgment-sample.

A mailed-canvass or postal survey presents special hazards from nonresponse. Too often a postal survey is a makeshift, hastily and incompletely devised, trusting to luck. Too often it is not realized that a postal survey, if it is to be a probability-sample, requires careful and explicit preparation. First, the sample for a postal survey is more exacting than a sample for interviews (cf. Ch. 4 on this point). Second, provision must be made for keeping records of nonreturns, so that letters may be sent out pleading for cooperation. Finally, pressure in the form of telegrams, telephone calls, and personal interviews on a fraction (1 in 2 or 1 in 3) of the delinquents must be exerted on the hard core of resistance.[9]

Through perseverance in a recurring survey, mail response may sometimes be boosted to 95 percent or better: in fact, in the monthly report on retail business activity conducted by the Bureau of the Census, 95, 98, and 100 percent response is being maintained in some cities.

Important economies may often be effected by making the fullest possible use of postal surveys combined with direct interviews in the non-responding segment.

The consulting statistician must be explicit at the outset regarding support for battling nonresponse; otherwise his mathematical skill in design may go for naught. From the standpoint of sheer cost, provision for whittling the nonresponse at the first call or at the first receipt of a mailed questionnaire is of vital concern in the planning because of

[9] This device was suggested by Frank Yates in "Methods and purposes of agricultural surveys," *J. Royal Soc. Arts*, vol. xci, 1943: pp. 367–79. The optimum proportion of the delinquents to be interviewed was computed by Hansen and Hurwitz in "The problem of nonresponse in sample surveys," *J. Amer. Stat. Assoc.*, vol. 41, 1946: pp. 517–29. A sample of 1 interview in 3 of the nonreturns is a good rule, to be increased to 1 in 2 if interviewing costs are not excessive.

the terrific expense of second and third calls and (in the case of a postal survey) the relatively high cost of telegrams and personal interviews compared with the cost of mailing a questionnaire. Pilot studies in which the questionnaire is refined and improved, and variances and costs are studied, may thus be a wise investment. From 5 to 25 per cent of the total cost of a survey may well be invested in pilot studies.

Occasionally an unsuspecting statistician encounters a postal survey in which there is no interest in correction for nonresponse for the simple reason that the questionnaire is intended mainly as an advertisement. Its utility as an advertisement is hardly diminished by the absence of follow-up. Such uses of questionnaires should not be confused with statistics.

Postal surveys are not feasible for studies of the general population in some parts of the world where there is preponderant illiteracy.

Nonresponse is an annoyance peculiar to probability-sampling. Non response does not appear in a quota-sample because interviewers fill their quotas with people who are willing or eager to talk, and they do not come in contact with people who are timid or do not like to be inter viewed. Nonresponse is rarely encountered when the sample is drawn from a "chunk" consisting of a list of "representative" respectable, literate people of the community who are known or thought to be willing to cooperate in various studies. Peace of mind under such circumstances is unfortunately a delusion because such methods are biased to start with; 100 percent response may still be badly in error. To do good sampling one must face the problem of nonresponse and not bury it.

Small gifts are often given out to people as an inducement to keep records of consumption or purchases and thus to build up response. Undoubtedly a bias is introduced, but it is probably small in most surveys. It could be measured by comparing two interpenetrating net works of samples, one with a gift and one without.

Remark. Alfred Politz of New York has kept the author informed of a series of experiments in which correction for the nonresponse of people not at home is accomplished without calling back.[10] The plan depends on the fact that, of people who are at home half the days at a given hour, half are actually at home at that hour, and a sample run over all days of the week will find half of them. When the responses of these people are multiplied by 2, the result is unbiased. Similar corrections are made for people who are customarily at home a quarter of the time or three-quarters of the time at a given hour. This suggestion was made independently by H. O. Hartley, *J. Royal Stat. Soc.*, vol. cix, 1946: p. 37.

[10] Alfred Politz and Willard Simmons, "An attempt to get the not-at-homes into the sample without callbacks," *J. Amer. Stat. Assoc.*, vol. 44, 1949: pp. 9–31.

Errors in response. There are two kinds of variability in response, different descriptions of the same situation i. given by the same person at two different times; ii. given by different persons. Both kinds of error are often much greater than is ordinarily supposed, and both can be controlled to some extent by the drafting of the questionnaire and the training of the interviewers. In a continuing survey the cooperation and education of the respondent may often be fostered so as to decrease the first type. However, it must be recognized that respondents under repeated questioning often change their characteristics.

It might be thought that factual data such as age could be collected with little error, and that only data with looser definition such as employment status and education are subject to wide variation. An extensive study carried out by Gladys L. Palmer,[11] however, shows that variation in response is indeed large in all these characteristics, and that age is certainly no exception. Yet what property could be more objective? In a recanvass of 8500 people in Philadelphia, after an interval of only 8 to 10 days, 10 percent of the ages were different by 1 year or more when reported by the same respondent in both canvasses (an example of the first kind of variability), and 17 percent of the ages were different by 1 year or more when reported by different respondents (an example of the second kind of variability).

The author once saw a diagram in a government office on which were plotted the self-reported ages of 300,000 men. On the x-axis were the ages as reported on a certain date; on the y-axis were the ages as reported just 2 years later by the same men. Perfectly consistent reports would have given a 45-degree line with an intercept of 2 years. Actually, 88 percent of the points lay on this line. About 9 percent lay on parallel lines a unit above and below. Distinct paths were also traced out 5 and 10 years above and below, with many points scattered indiscriminately.

Another example of the second kind of variability in response is furnished by Katherine D. Wood,[12] who exhibited tables showing the discrepancies between duplicate reports of the occupations of 4500 workers, one report coming from the worker himself or some member of the houshold, and the other report coming from the worker's employer. Table 1 in her article shows that when the occupations are classified into only 9 major occupational groups, 21.7 percent of the total number of duplicate reports are in disagreement—i.e., fall in a different one of the 9 broad groups. Her Table 2 shows that when the occupations are classified into 233 groups, the difference jumps to 35.5 percent.

[11] Gladys L. Palmer, "Factors in the variability of response in enumerative studies," *J. Amer. Stat. Assoc.*, vol. 38, 1943: pp. 143–52.

[12] Katherine D. Wood, "The statistical adequacy of employers' occupational records," *Social Sec. Bull.*, vol. 2, May 1939: pp. 21–4.

Apparent variations from place to place and from time to time in the incidence of disease and crime are often only variations in definitions and in the thoroughness and accuracy of reporting.

It should be pointed out that the net effect of variability in reporting is not always as bad as might be surmised. One reason is that many errors can be caught in a careful job of editing. For instance, in processing the reports on the annual production of lumber which are sent into the Census from sawmills, every effort is made to diminish the net effect of variability and carelessness in response. Each report is carefully compared with the previous annual report from that mill. To expert editors who know the lumber business, the respondent's difficulty and the consequent correction of an erroneous report are often obvious. When not obvious, the case may be turned over to the Forest Service, which in turn may initiate correspondence or send a local representative to the mill to discover what difficulty if any exists. A second reason is that the poorest reporting on production and sales often occurs in the small establishments, which all told contribute only a small fraction of the total of the annual production or sales. The larger establishments keep records and can make better reports. For a third reason there is an element of randomness in reporting dictated by the accident of circumstance. The weather, time of day, the particular person providing the information, the route followed by the interviewer, and many other factors are accidental in nature and affect the results. As a result, some reports (of age, number of board-feet of wood cut, sales, and stocks) are accidentally higher and others are accidentally lower than they might have been under other circumstances. This random element is compensating on a probability basis, the net effect being that the final tabulations may portray distributions that are reasonably independent of the random element of variability and able to serve many useful purposes. Random errors have less chance of canceling each other if the tabulations are made in fine classes.

It is a mistake, however, to take refuge in the assumption that errors in response are going to cancel each other and thus to excuse poorly designed questionnaires and inexpert interviewing. The random element may wash out, but a bias is different; it is not necessarily partially or wholly compensated by another bias in the opposite direction. For instance, in spite of variability in the reporting of age, frequencies showing characteristics of the population by age will usually turn out to be remarkably independent of the random errors in reporting, but will clearly show the downward and upward heaping toward the fives and tens. Likewise, the random errors that occur in taking inventories of canned peas in a number of grocery stores may pretty well cancel each other, leaving only the effect of (a) the downward bias that arises from failures to look in the basement or out-of-the-way places for peas, and (b) the downward or upward bias that arises from the natural tendency to undercount or overcount, whichever it may be.

In view of errors in response, not to speak of the other factors that affect the usefulness of a survey, it is obvious that a complete coverage can not give absolute accuracy. As a matter of fact, absolute accuracy is nondefinable.

In this connexion I am reminded of a conversation with Frederick F. Stephan. He was once asked how big a sample would be required to measure within 5 percent the extent of unemployment in the country. This was in 1934, when plans for a sample census of unemployment were being considered. His reply was that even a 100 percent sample could not give 5 percent accuracy because of differing ideas regarding definitions of unemployment and the interpretation of the questions. Even with the elimination of sampling errors, there would remain unsettled differences between various alternative definitions of unemployment. There would remain, moreover, errors of enumeration (variability in response; housewife doesn't know the answer but answers anyhow; some families missed; some refuse; etc.). Before it is profitable to talk of reducing sampling errors to 5 percent, it would be necessary to reduce both the variability in response (by sharpening the definition) and the error of enumeration to magnitudes comparable with 5 percent accuracy.

A magnificent series of experiments on errors of response has been carried on by Mahalanobis in connexion with his surveys of jute and other crops in Bengal. Such surveys show the limitations of data collected on a 100 percent basis, and give guidance on methods of training interviewers. They show the necessity of building up a reliable human agency, whether one is to do sampling or a complete canvass. The following paragraphs are quoted from Mahalanobis,[13] beginning on page 408.

210. Several things became clear in the course of these studies. The number of discrepancies at the stage of the field survey was very high. The absolute sum of both positive and negative discrepancies gives a convenient picture of the accuracy of the field work. In 1937, for example, it was found that for a group of villages taken together the absolute discrepancy was as high as 58 percent of the actual area under jute. The positive and negative discrepancies, however, occurred to a large extent in equal proportions, so that they tended to cancel out. The algebraic sum of discrepancies was thus much smaller; and in the case of the same group of villages considered above the total algebraic discrepancy was of the order

[13] P. C. Mahalanobis, "On large-scale sample surveys," *Phil. Trans. Royal Soc.*, vol. 231B, 1944: pp. 329–451; "Recent experiments in statistical sampling in the Indian Statistical Institute," *J. Royal Stat. Soc.*, vol. cix, 1946: pp. 325–78.

of only 5 percent.   A good proportion of the recording mistakes at this stage were thus amenable to statistical treatment.

211. This is satisfactory, but clear evidence was also found of inaccuracies which could not have arisen excepting from false entries or gross negligence. The magnitude of the discrepancy, both algebraic and absolute, also varied widely from one investigator to another.   A part of this no doubt may be ascribed to differences in the "personal equation" of the individual workers, but detailed comparison and scrutiny of the material left little doubt that some of the investigators were dishonest in their field work.

212. *Crop record.*   In the next stage of the work, namely, preparation of crop records, a similar detailed comparison was carried out.   Here the absolute discrepancy was something of the order of 9 or 10 percent, while the algebraic discrepancy was less than 2 percent.   On the whole inaccuracies at this stage were far smaller in magnitude than the mistakes which occurred at the stage of field survey.   This is, of course, just what may be expected in view of the fact that the field survey has had to be carried out under far more difficult conditions.

213. *Area measurement.*   A detailed study was also made of errors occurring at the stage of copying the area of individual plots from revenue records which were kept in the district headquarters and were thus scattered all over the province.   This arrangement was difficult to supervise, and large mistakes were detected.   From 1940, therefore, the practice was adopted of measuring the area of individual plots directly in the Laboratory with the help of photographic scales.   The absolute discrepancy by this method is of the order of 2 percent, and the algebraic discrepancy appreciably below 1 percent.

214. *Border effect.*   In a sample survey on a large scale there were naturally many other sources of error, some of which were studied experimentally.   For example, there was the question of the border effect.   It was found that there was persistent overestimating in working with units of very small size.   In the case of field survey the obvious explanation is that the investigator has a tendency to include rather than to exclude plants or land which stand near the boundary line or perimeter of the grid.   This boundary effect naturally becomes less and less important as the size of the grid is increased.   In crop-cutting work on jute it was found, for example, that mean values for all the characters studied (such as number of plants per acre, weight of green plants, weight of dry fibre) were much higher for sample units of small size, so that it was not at all safe to work with cuts of a size less than say 25 sq. ft.   In the case of the area survey it was generally found inadvisable to work with grids of size less than about 1 acre.

215. The above studies revealed the great importance of controlling and eliminating as far as possible the mistakes which occurred at the stage of the field survey.   This is why from the very beginning special attention was given to the need of building up a reliable human agency.   In 1937 there was not a single trained field worker, and only about half a dozen computers.   Whatever training was possible was given in the very short time at the disposition of the Laboratory, and this had to be repeated every year, as the scheme was sanctioned from year to year.   The whole of the field staff was recruited for only three or four months, and continuity of employment could not be guaranteed.   A large number, especially the abler men, left after one season and did not come back, so that work had to be carried on with a large proportion of untrained men each year.   On the statistical side, however, it became possible to train up and give more or less con-

tinuous employment to a good proportion of computers by employing them on other projects.

216. Various attempts were made to improve the efficiency of the survey by proper selection of workers. With this purpose in view, a study was made of variations in output (and in certain instances also of mistakes) of individual workers. Without entering into details I may mention one or two typical results. The average output of all workers for any particular type of work was adopted as the basis for comparison, and the index of output of each individual worker was found by dividing his actual output by the adopted standard and multiplying by 100. Individual variations were enormous. For example, in the field survey the index number in 1937 varied from 48 for a particular worker to 146 for another investigator; the output of the quickest worker was three times as large as the output of the slowest. The coefficient of variation fluctuated roughly between 25 and 40 percent, depending on the particular type of work in the case of the field survey.

217. The position was much the same in the statistical portion of the work. The coefficient of variation in output among individual workers was roughly about 20 or 25 percent in the case of simple operations like listing and comparison of entries, and of the order of 30 or 35 percent in the case of work involving computations. The question of accuracy was also studied to some extent by comparing the proportion of mistakes made by different workers for different types of work. Here also large variations were found.

218. In 1940 and 1941 arrangements were made from the Indian Statistical Institute to hold examinations for the award of certificates for computing work and field survey. I am making a passing reference to these things to indicate the kind of methods which were adopted from time to time for selecting suitable workers with a view to improving the general efficiency of the survey.

Examples of errors in response in a census of business.[14] The Census mailing list shows 24 business establishments in an area. A questionnaire is mailed to each establishment (and this might be either a sample or a complete count). Some of the questionnaires come back correctly filled out to the best of the respondents' knowledge and understanding, but the following errors can and do occur. They ought not to occur, but they do.

1. In spite of care taken to compile lists of businesses from previous Census records and from Social Security records, some businesses exist in the area that were not on the list. No list has ever been completely correct and up to date. Unlisted businesses have no chance of being in the sample (unless the proprietors hear about it and enquire at the nearest Census office), and bias of underenumeration is the result.

2. Some of the 24 businesses have gone bankrupt, but there is still a receptacle for mail, and the envelope is delivered by the Post Office. No questionnaire is returned from such people, and a bias is introduced by any attempt to adjust the results.

3. Some of the 24 businesses have moved, and the questionnaire is forwarded to them. Some questionnaires come back filled out with no indication of the new address, and the results are biased by overenumeration.

[14] The examples are typical, but do not refer to any actual area.

4. Some of the names are duplicates of the same business—Smith's Cycle Shop and the Green Meadow Sport Mart being the same establishment. Only one report comes back, and the other is charged to nonresponse, an adjustment made, and the result biased. Or, both reports come back, filled out, and the results are biased from overenumeration.

5. A business is owned jointly by two partners who do not keep in touch with each other. One fills out the report and sends it in. The other's conscience hurts him when he hears of the survey; he obtains a form or writes the information in a letter and sends it in. The names and figures are sometimes so different that it is difficult to recognize the second report as a duplicate. This is another form of overenumeration.

6. A man has two businesses at the same address but only one may be listed. He makes ice cream and retails it in his own ice-cream parlor, but he also sells it to several ice-cream parlors located elsewhere. He fills out the form for his retail business, skipping his more important business of manufacturing and wholesaling ice cream, and the results are biased by underenumeration.

7. A man changes his type of business: the form that he receives does not now apply, and he fails to send it in or to heed later pleas for its return. This is one form of nonresponse.

8. A business changes hands and name. The questionnaire is forwarded by the Post Office, and that is the end of it. The business that is now carried on in the old premises is not reported. This is another form of underenumeration.

9. Plain nonresponse through inadvertence or refusal.

10. Errors in response.

These last two errors are listed at the end, not because they are least important, but to emphasize the fact that they are *only two* of the errors of response in a business survey. It should be noted that none of the errors mentioned is a sampling error.

Skilled workers in the Census office in Washington are able to uncover most of these errors and correct them with uncanny ability. They of course occasionally require the assistance of a local official. Some errors, however, can be discovered only by a personal canvass.

Bias arising from the interviewer. In 1914 Rice [15] in a social study of 2000 destitute men found that the reasons given by them for being down and out carried a strong flavor of the interviewer. Results recorded by a prohibitionist showed a strong tendency for the men that he interviewed to ascribe their sorry existence to drink; those interviewed by a man with socialist leanings showed a strong tendency to blame their plight on industrial causes. Quantitative measures of the interviewer bias in this particular survey turned out to be amazingly large. The men may have been glad to please anyone that showed an interest in them.

Variation attributable to the interviewer arises from many factors: the political, religious, and social beliefs of the interviewer; his economic

[15] Stuart A. Rice, "Contagious bias in the interview," *Amer. J. Sociology*, vol. 35, 1929: pp. 420–23. See also C. C. Lienau, "Selection, training, and performance," *Amer. J. Hygiene*, vol. 34, 1941: pp. 110–32.

A sample study or other partial coverage possesses a distinct advantage
[i]n the processing for the same reason that it does in the interviewing,
[v]iz., the smaller force required to do the work, and the consequent better
[c]ontrol that is possible.

Faulty interpretation of the results. In any study made for the analy-
[si]s of causes, preliminary to formulating a course of action for the future,
[th]ere must be inferences drawn from empirical data. These inferences,
[if] they are to be useful, must often take the form of predictions—pre-
[di]ctions of future populations, where they will live and what they will
[ea]t and wear, and how they will react to a particular product or service.
[U]nfortunately, even the best survey is a story of the past, not of the
[fu]ture; and unless the underlying cause system is expressible analytically
[or] in probability form, there may be no statistical method of predicting
[th]e future.

[E]ven with the best of intentions there may be a personal and profes-
[sio]nal bias in interpretation. This fact is so well known that it would
[be] superfluous to go into the subject here or to point out the magnitude
[of] the differences that can exist purely on the grounds of personal dif-
[fer]ences in education, experience, and environment. A familiar example
[is t]he picture of a labor situation presented by management, as opposed
[to t]he picture presented by labor organizations.

[E]rrors and differences in interpretation sometimes arise from mis-
[und]erstanding the questionnaire or failure to take into consideration
[the] form of the questions as written on it or as actually used in the
[inte]rview. Without some recognition of the problems involved in carry-
[ing] out the survey, from the standpoint of both the collecting agency
[and] the respondent, sizable errors in interpretation are almost sure to
[aris]e. The more important the survey, the more important are the
[erro]rs of interpretation. For careful interpretation it is necessary to be
[acqu]ainted with the field-work; not just with the instructions which tell
[how] the field-work should have been carried out, but with the procedure as
[actu]ally followed (see the quotation from *The Production of Lumber*,
[p. 9]9).

[T]he conditions that are described by a survey may have changed by
[the t]ime the tabulations are ready for processing. These changes detract
[from] the utility of the survey, and if they are ignored, serious errors may
[go] into the interpretations.

[A] complete count requires a longer time than a sample for processing—
[so m]uch longer, in fact, that often because of changes in conditions, it is
[only] a historical record by the time it is ready. As a basis for action
[(the] only excuse for taking a survey) a sample will therefore often be
[more] reliable because of the shorter time required for collecting and
[proce]ssing the data.

status, environment, and education. Also, perhaps most interviewers
can not help being swayed in the direction of their employers' interests.
But how much? What is the effect on the tabulations? Different interviewers will record different descriptions of the same situation and different interpretations to identical statements from a respondent.

One source of bias and variability arising from the interviewer has its
roots in lack of understanding of the subject and purpose under investigation, without which the interviewer can not evaluate a situation or
properly record the respondent's statements.

Part of the variation attributable to the interviewer arises from the
different moods into which different interviewers cast their respondents.
The interviewer may make the respondent gay or despairing, garrulous
or clammish. Some interviewers unconsciously cause respondents to
take sides with them, some against them. This kind of variability is
difficult to distinguish from the error in response.

A small corps of interviewers can be trained to a high level of homogeneity; hence in sample surveys and other partial coverages it is possible
to diminish differences between interviewers to a degree not attainable
in large-scale surveys. In particular, partial coverages repeated at
intervals may possess an enhanced degree of comparability from one
survey to another.

Training will sometimes introduce biases in a corps of interviewers,
depending on how they are trained. A corps of enumerators with less
training and greater variability might come nearer to finding out what a
social scientist really wishes to know about. Bias produced by training
partakes of bias of the auspices (see below), and it is sometimes difficult
to make the distinction.

Bias of the auspices. Any change in the method of collecting or
processing data can be expected to show a change in results. A shift
in the sponsoring organization is no exception. Bias of the auspices
probably stems from a conscious or unconscious desire on the part of
the respondent to take sides for or against the organization sponsoring the
survey, but perhaps more to protect his own interests, which may vary
with the sponsoring agency. Everyone supposes, for instance, that the
replies elicited by an agent of a relief organization concerning income and
work status are different, on the whole, from those elicited by a government agency such as the Census. The Census and the WPA (Works
Progress Administration) both collected information on the work status
of the people the week containing the 1st April 1940. The WPA found
more people working for the WPA, and more females seeking work, than
the Census found.

**Bias arising from an unrepresentative date for the survey or the period
covered.** The measurement of total annual sales, total annual postal
traffic in various classes, telephone, telegraph, freight, passenger, or air

traffic, or movement of some particular commodity, consumption of foods of various kinds, or the pattern of consumption or service rendered, and a host of other problems which require totals or averages over a year or some other period, present difficult problems because of heavy weekly or seasonal variability. Actually, in such problems it is necessary to recognize the fact that a good sample of time is as imperative as a good sample of areas, business establishments, families, or anything else. In many cases it is possible and advisable to conduct the survey on a random sample of days scattered throughout the year in sufficient number to give a good total or average. Often the problem is but a collection or transcription of records, such as waybills, sales slips, orders, toll tickets, railway tickets, air tickets, in which case a continuous sample of (e.g.) every 100th waybill or toll ticket from every station or from a good sample of stations may be the answer. The collection and tabulation then proceed on an orderly basis; they can be run efficiently and accurately, and can be adjusted to the requirements. More important, they provide speedy information so that shifts in distribution, rates, and services can be made at the most opportune time. A number of business firms and government agencies have adopted such methods.

Too often in the past, a huge survey or collection of transactions has been conducted during a selected week (or during two or four selected weeks of the year), the particular week or weeks having been chosen by judgment as "average" or "representative." No matter how good be the sample of areas, such a sample is still a judgment-sample in time. The reader may agree that the tabulation of more than a sample (e.g., every 10th or 100th) of all the waybills, toll tickets, or sales slips that were issued during "representative" periods would in most cases be an unjustifiable expenditure. Not even a complete tabulation, nor any amount of wishful thinking, can alter the fact that such a sample is inadequate for measuring either an annual total or the pattern of services rendered.

On more than one occasion the author has seen complete tabulations of "representative" weeks bogged down with sheer bulk, months after the date of collection. The speedy tabulation of every 100th ticket would have displayed most of the information contained in the original collection, in time to be useful.

A continuous sample of transactions, with its speed, simplicity, and flexibility, provides the answer to many problems where current information on totals, averages, patterns, and changes is needed for intelligent administration in government and private business. It would be wrong to leave the impression, however, that continuous samples are always possible. A census of population, agriculture, or commerce, for example, must give full and complete details for a particular date, and this date must be selected by judgment. A date that is satisfactory for a

census of population may not be so good for a census of agri for economy and comparability the two censuses must be tak at a time when the vast army of interviewers will not be snow or floods. Intermediate samples, monthly, quarter nually, then provide current and continuous information while these changes are taking place.[16]

Errors in processing. A review of the codes assigned o is oftentimes not a matter of correcting wrong codes, but me of honest differences of opinion between coder and reviewer. will often find themselves in disagreement on the correct c to a response. Two coders working on the same set of going to turn out two different sets of results; likewise tw coders working on the same set of schedules are going to different sets of results. *A fortiori*, two sections of coders slightly different instructions will show still greater dif though the two sets of instructions supposedly say the different words. The two sets of results may, however, pr tions so nearly alike that in most problems they would le action, and that is what counts. Research needs to be show the extent of the differences to be expected from va wording of instructions for coding, editing, and field-w clusion seems inevitable that unless it is merely a matter (such as 1 for male and 2 for female) it is impossible to job of coding except in terms of the distributions produce is no way of determining whether the individual codes ha correctly. One can say only that two different sets o two different sets of coders produced substantially th tions. In view of this fact it seems to follow that wh coder or editor or punch operator is uniformly good er errors are relatively insignificant compared with the o as variability of response) it is necessary only to perfor of his work (preferably by sampling methods) to be as tinuity of control.[17] Workers who can not qualify f should be transferred.

Machine and tally errors are often supposed to be existent, but the actual situation is otherwise. These at a reasonable minimum, however, by machine c checks, especially with a force of workers in which th people with seasoned experience.

[16] In Chapters 4 and 5 we shall see ways in which data from to enhance the efficiency of these intermediate samples.

[17] W. Edwards Deming and Leon Geoffrey, *J. Amer. Stat* pp. 351–60; W. Edwards Deming, Benjamin J. Tepping, and *Stat. Assoc.*, vol. 37, 1942: pp. 525–36.

Sampling errors. One often hears objections to sampling because of sampling errors. Such objections can be sustained only if, after consideration of the other inaccuracies, the elimination or reduction of the sampling errors seems to be a wise investment. Sampling errors have the favorable characteristics of being controllable through the size and design of the sample, which is the purpose of this book. It is now possible to lay out sample-designs in many types of surveys whereby one can state in advance the width of a band that will contain 99 percent or any other percent of the sampling errors. Sampling errors, even for small samples, are often the least of the errors present.

The next step in the direction of greater reliability of surveys must lie along the line of further research in other types of errors.

At present, sampling errors are the only errors that are in satisfactory condition so far as theoretical and experimental knowledge is concerned.

Sampling biases. Complete counts and samples may be expected to show persistent differences arising from psychological factors associated with fatigue and differences in the training and procedures. Intermissions occupied in travel between sample-areas and sample-households relieve fatigue. Knowledge that one sample-area or household represents many others apparently calls forth special efforts on the part of the interviewers. A similar difference between complete and sample-inspection has long been noted in industry; and it has been noted also that, in the sample-inspection of office work like coding and punching,[17] sample-inspection usually discovers about 25 percent more errors than complete inspection.

When a sample is taken along with a complete count in a population study, the sample-areas or names are designated on maps and lists by heavy boundaries and other distinguishing marks, and these designations apparently lead some interviewers to give the sample slight preferential treatment.[18, 19] A deleterious effect arises when an interviewer, waxing overzealous in his efforts to produce a good census, substitutes what he regards as an average home, average farm, or average person, when the rules for the selection of the sample have led him to an unusual home, farm, or person.

[18] The definition of sampling error given on page 18 is not afflicted with preferential treatment of the sample.

[19] In the United States Census of Population in 1940, in spite of instructions and training, several hundred of the 110,000 interviewers substituted an adult for a child in the sample, apparently because the "supplementary questions" (asked only of the people in the sample) on usual occupation and fertility would not apply to a child. It is to be noted that this kind of substitution required extra effort on the part of the interviewer, at no extra pay.

Incidentally, in this instance, little loss of census information resulted from these substitutions because the supplemental information required for a child could be largely inferred from information regarding an adult member of the family.

It may be of interest to census workers to note that, as a sample taken by itself is a smaller operation, a higher grade of worker can be recruited and trained more satisfactorily than the larger corps of workers that is required for a complete canvass. It would be wrong, however, to give the impression that the differences are always large or that the sample always gives better results. The purpose of this section is merely to say that differences are to be expected.

Other types of sampling biases, specifically the bias that arises from failure to randomize the starting points in a systematic selection, or the tiny bias that arises from a ratio-estimate, are treated elsewhere in the text.

A word on sample studies of complete returns. Samples of completed reports, which might be waybills, tax forms, census returns, wage reports, hospital records, relief records, consumers' accounts, or the like, constitute one of the chief uses of sampling.[20] In fact, several censuses have been salvaged through sampling, even after the field-work was completed. Examples are the 1:1000 sample of the Japanese census in 1923, interrupted in processing by the earthquake, and the 1:50 sample of the Indian census of 1941. Many of the publications of the census of the United States in 1940 were tabulated by a 1:20 sample. There is sometimes reluctance to adopt sampling methods because of a commendable pride in traditional accuracy. But let us look at the problem in its entirety and see just how far this accuracy goes. If the study were purely for accounting purposes, a complete count with an attempt at perfect processing would be justified or even demanded. It should be borne in mind that the purpose is not accounting, however, unless the action to be taken is with respect to each respondent by reason of the data on his response; an income tax report is an example. Most studies are for purposes of analysis, wherein the ultimate aim is policy and action for the *future*, not the past. For purposes of analysis, even a complete count, however perfect, is still a sample and must be interpreted as such (Ch. 7).

Presentation of data requires description of errors and difficulties.[21] In the presentation of data the omission of an adequate discussion of all the errors present and the difficulties encountered constitutes a serious

---

[20] The reader may wish to consult Walter M. Perkins, *Simple Methods for Representative Sampling in Studies of Public Assistance Case-loads* (Social Security Board, Public Assistance Research Memorandum No. 6, Washington, March 1944). At the time of writing, copies of this publication are on hand for distribution.

Some exercises on the sampling of records occur in Chapter 4 of this book, beginning on page 87.

[21] Chapter III in Shewhart's *Statistical Method from the Viewpoint of Quality Control* (Graduate School, Department of Agriculture, Washington, 1939) should be read in connexion with these remarks.

defect in the data and is sure to lead to misinterpretation and misuse. It is common in a sample study to point out the sampling errors, as should always be done. One of the main things to keep in mind is that the figures obtained in any survey are useful to the careful research worker only if the operations by which they were obtained are carefully described, and all weaknesses reported. There are several ways of doing this. The paragraph below appears in many of the reports published from the *Sixteenth Census* (1940) on the basis of the 5 percent sample.

> The statistics based on the sample tabulations are expected to differ somewhat from those which would have been obtained from a complete count of the population. An analysis of the statistics based on the tabulations of the 5 percent sample of the population for items that were obtained also for the total population indicates that in 95 percent of the cases the sample statistics differ from the complete census statistics by less than 5 percent of all numbers of 10,000 or more, by less than 10 percent for numbers between 5000 and 10,000, and by less than 20 percent for numbers between 2000 and 5000. Somewhat larger variation may be expected in numbers below 2000. Even for these small numbers, however, the majority of the differences between the sample and the complete census statistics are less than 10 percent, although much larger differences occasionally occur.

The statement of a standard deviation or band of variation in the form of a plus and minus (e.g., 1123 ± 42), along with *the number of independent sampling units* on which the calculation is based, is a common way of calling attention to the sampling errors. The formula for the calculation must also be given.

Unfortunately there is no simple way of indicating the possible magnitudes of the other errors, but it can be done in one way or another. As an example it is a pleasure to cite a few lines from *The Production of Lumber, by States and by Species; 1942* (Bureau of the Census, November 1943), published under the direction of Mr. Maxwell R. Conklin, Chief of the Industry Division.

> These statistics are based on a mail canvass, supplemented by a field enumeration conducted by the U. S. Forest Service and the Tennessee Valley Authority. In the field enumeration, Forest Service and TVA representatives interviewed mills that did not respond to the mail canvass, and, in addition, conducted an intensive search for mills. . . . Among the smaller mills, bookkeeping is generally inadequate. Even the total cut for a mill may be an estimate, and the species breakdown for such a mill, particularly in areas of diversified growth, must frequently be estimated by the mill operator or by the enumerator. . . . Difficulties in enumeration because of lack of adequate mill records were overcome in many cases where the mill disposed of its total cut through a concentration yard. In such instances enumerators were able to obtain information for individual mills from the yard operator, particularly in the South and Southeast, where concentration yards are an important factor in the distribution of lumber.

This approach was not satisfactory, however, when an operator sold his lumber to several different yards in the course of the year, and where the records at the concentration yard did not indicate clearly whether the cut was for 1942 or 1941. . . . Mills engaged solely in remanufacturing, finishing, or otherwise processing lumber were excluded. . . . In a number of cases, the mill reports were in terms of dressed or processed lumber, since many integrated mills, i.e., those both sawing and dressing, were able to report only on a finished basis. The discrepancy, which is of unknown magnitude, is equivalent to the amount of waste in processing. In canvassing integrated mills, however, the cut was counted at only one point in the processing operation, so that no duplication occurred. . . . An ever-present complicating factor in the canvass was the extreme mobility of the smaller mills. . . .

Exercise 1.   Comment on the following excerpts.   Keep in mind the possible biases in nonresponse, and the aims of the surveys.   Think through the following questions before you comment.

In what ways could the samples of respondents be criticized?

How important was it in each case to obtain interviews of a sample of nonresponses?

Describe the additional uses that could be made of data on such subjects if a probability-sample had been taken and the field-work well controlled, including rigid following of nonresponses or a sample thereof.

Are the conclusions justified?

Could you devise better procedures at the same cost?   Are higher costs warranted?

a. (From the *New York World-Telegram*, 17 Nov. 1947, p. 17.)

To date—two weeks after the questions were sent out—40 percent of the replies have been received. This is regarded by professional poll-takers as a good return and a fair sample. On the basis of answers thus far, these main trends can already be observed:

1. Further American aid to Europe is supported seven to one.
2. A two-to-one majority say they would support a $15-billion-to-$30-billion, four-year plan to put Europe on its feet.

b. (Leaflet distributed by the Central Railroad of New Jersey, 1948.)

On November 12, 1947, the questionnaire "Now . . . YOU tell us!" was distributed on regularly scheduled Jersey Central trains. Here are the results of that survey, which covered Jersey Central schedules and air-conditioned coaches for "Commuter Clubs."

We received 4024 replies to the questionnaire, which we consider sufficient to give us an accurate cross-section of opinion. . . . In regard to changes in schedules, however, majority preference indicated satisfaction with present arrival and departure times. In many cases, requests for changes in schedules were not only in the minority, but, for all practical purposes, actually canceled themselves. . . . *Overcrowding of trains*: Our records show that less than 1 percent of the passengers are forced to stand, and these only for short distances. . . . *Poor lighting:* The lighting on some of our battery-equipped cars is not always satisfactory and this is particularly noticeable during severe cold weather.   To correct this, we are equipping our cars with "head-end" lighting which eliminates the dim lights at slow speeds and while standing at stations.   Two hundred of our cars have been so equipped and we are continuing the program.   We even are making some changes in the "head-end" lighted equipment which will further improve the illumination, including the elimination of the "flicker" you may have noticed on our long trains when hauled by the Diesel locomotives. . . . *Poor coffee on ferries:* Yes, one person complained about this—and we've spoken to the chef!

c. (*The Evening Star*, Washington, 1 July 1946, p. 1.)

The White House reported that telegrams received there since the President's veto message Saturday, and his radio address to the country that night, have been running 50 to 1 in support of the President's action on the basis of the first 2500 communications examined.

d. Criticize this description of a proposed method of study (an actual example).   Bear in mind that this is all of the description.

The questionnaire will be used on a sampling basis, with the size of sample determined by local conditions.   The number of interviews will not exceed 10 percent of the prospective and present consumers in any area and never less than 30.

Exercise 2.   Criticize the following plans for a proposed survey of social workers.   (This is an actual example; practically every statement represents dangerous technology.)

Since it has not been possible to provide as much publicity for the study as is desirable, it is believed necessary to send out questionnaires to all social workers in the state [rather than to a sample] so as to insure an adequate number of replies within a relatively short period of time.   Moreover, it will be necessary to make a number of relatively fine classifications on the basis of the returns that would not normally be needed for a single state.   No field recalls are to be made but we plan to send out a short flyer about a week after the original mailing to all those to whom the original questionnaire was sent, to remind them to send in their replies.

Exercise 3.   The publishers of a certain magazine desire to learn something about the characteristics of its subscribers.   It is a magazine devoted mostly to commerce, and the publishers wish to learn (e.g.) how many of their subscribers control the purchasing of typewriters and other divers types of office equipment within their organizations.   A random sample of subscribers is to be drawn and interviewed, and amongst a battery of questions, two of them are these:

How many typewriters has your company on hand now, under your control?

Do you exercise control over the purchase of typewriters in your company?

The reason for including these particular questions is to have some basis for selling advertising space to manufacturers of typewriters. Criticize the above questions. (Hint: how would you carry out the tabulations to estimate how many typewriters are controlled by all the subscribers, and how many subscribers control the purchase of typewriters? Remember that two or more executives in a company may subscribe to the magazine, and that both or all, because of divided and group authority, may think that they have jurisdiction over all typewriters: each typewriter then has more than one chance of getting into the sample. It would be impossible to untangle the probabilities and derive an unbiased estimate. An entirely new approach was adopted. Note that increasing the size of the sample will not help.)

Exercise 4. (Suggested by a remark from Frederick F. Stephan.) A random sample of individuals or of households is drawn, and a short screening questionnaire is used with the aim of identifying which of them belong to a particular class (age group, home owners, farm operators), which it is desired to study with a longer questionnaire a few weeks later. Describe a reason why the results of this study will differ from the results that would be obtained by using the detailed questionnaire on the original and larger sample, and throwing away the questionnaires that apparently do not come from the class that it is desired to study. (Hint: the short screening questionnaire does not do its work perfectly: it admits some to the main study which should not be admitted, and it fails to admit some that should be admitted. The longer questionnaire does a better job of screening, but is more expensive to administer.[22])

Exercise 5. In a certain survey a total of 3000 interviews was desired. A refusal rate of 25 percent was expected in one class of areas, and a rate of 10 percent in another class. The sample-sizes in these areas were increased by $\frac{1}{3}$ and $\frac{1}{9}$ respectively, to allow for shrinkage. Criticize this procedure. (It does not correct the nonresponse, but it does help to hold the precision to desired levels.)

Exercise 6. A sample-survey is to be carried out by the quota method of selection: if the size of the survey is doubled, are your chances of being interviewed doubled? In the notation of Chapter 4, if $n = N$ are you *certain* to be interviewed? (If the survey were a probability-sample, and $n = N$, you would be interviewed with certainty.)

---

[22] An article of interest here is Joseph Berkson, "Cost-utility as a function measure of the efficiency of a test," *J. Amer. Stat. Assoc.*, vol. 42, 1947: pp. 246–55.

PART II. SOME ELEMENTARY THEORY FOR DESIGN

CHAPTER 3. MOMENTS AND EXPECTED VALUES

"Yes, books are all right." Winslow gave a little sigh. "Though it's remarkable how little help they offer in some of the more curious problems of life."—From James Hilton's novel *So Well Remembered* (Little, Brown & Co., 1945).

A. MOMENTS OF DISTRIBUTIONS

The moment coefficients of a set of numbers. The moment coefficients of a set of numbers are of great importance in the study of statistics. Any set of finite numbers possesses moment coefficients. The set of numbers may be grouped into a distribution (Fig. 2), and this distribution may describe a universe or a sample or a number of samples thrown together. It may in fact be a universe whose shape (proportions p_i) is determined by calculations in probabilities, rather than by an actual set of observations.

The source of the numbers is of no importance so far as the formal definitions of moment coefficients are concerned. So let

$$x_1, x_2, \cdots, x_N$$

be a set of N numbers, some or all of which may be replicates. Then their kth moment coefficient about the origin ($x = 0$) is defined by the equation

$$\mu_k = \frac{1}{N} \Sigma\, x_i{}^k \tag{1}$$

in which the summation is to cover all N numbers. The suffix k may take on the values 0, 1, 2, 3, etc.—0 for the 0-th moment coefficient, 1 for the 1st, 2 for the 2d, etc.

In Eq. 1 the moments are said to be calculated about the origin ($x = 0$), but it is often useful to calculate moments about some point other than 0. To calculate a moment about the point $x = a$, Eq. 1 is rewritten in the form

$$\mu_k = \frac{1}{N} \Sigma\, (x_i - a)^k \tag{1a}$$

in which each x_i is measured from a. The *mean* of the set of numbers is that value of a about which μ_1 has the value 0; in other words, μ_1 by definition satisfies the equation

$$\Sigma\, (x_i - \mu_1) = 0 \qquad \begin{array}{l}\text{[Put } k = 1 \text{ and}\\ a = \mu_1 \text{ in Eq. 1}a]\end{array}$$

which gives $\Sigma \, x_i - N\mu_1 = 0$, or

$$\mu_1 = \frac{1}{N} \Sigma \, x_i = \text{the mean} \tag{2}$$

Otherwise said, the mean μ_1 (or simply μ) is the 1st moment coefficient of the N numbers about 0.

The *central* moment coefficients are calculated about the mean. Sometimes it will be desirable to distinguish the symbols for moments about the mean from the symbols for moments about some other axis, in which case the notation of Eqs. 21 and 22 will be used. The 2d moment coefficient about the mean has a special name, the *variance*, designated by σ^2 or s^2 (a distinction between which will be made later), and it should be noted that

$$\sigma^2 = \frac{1}{N} \Sigma \, (x_i - \mu)^2 \qquad \begin{array}{l}\text{[Put } k = 2 \text{ and} \\ a = \mu \text{ in Eq. } 1a\text{]}\end{array}$$

$$= \frac{1}{N} \Sigma \, x_i^2 - \mu^2 \tag{3}$$

The square root of the variance is called the *standard deviation* and is designated by σ or s, whichever is applicable (cf. the Remark, p. 70).

Remark 1. The 0-th moment coefficient of any set of numbers is equal to 1.

Remark 2. The kth moment coefficient μ_k is an average value of x^k, and the kth root of μ_k is an average value of x. This can be seen best, perhaps, by rewriting Eq. 1 in the form

$$\frac{1}{N} \, (\mu_k + \mu_k + \cdots \text{ to } N \text{ terms}) = \frac{1}{N} \, (x_1^k + x_2^k + \cdots + x_N^k)$$

The constant μ_k on the left gives the same result as the variable x_i^k on the right. Each value of k gives a different kind of average (cf. Exs. 4 and 7 on p. 61).

Exercise. Work out the first four moment coefficients about 0 of the set of five numbers 1, 2, 3, 4, 5; also their variance and standard deviation.

Answer: $\mu_0 = 1, \quad \mu = 3, \quad \mu_2 = 11, \quad \mu_3 = 45,$

$\mu_4 = 195.8, \quad \sigma^2 = 11 - 3^2 = 2;$

$\sigma = \sqrt{2} = 1.41.$

The moment coefficients of a distribution. The N numbers may not all be different; in fact there may be only $M < N$ different numbers, in which case let the N numbers be sorted into M classes, described by the accompanying table and depicted in Fig. 2, for which it may be supposed that each of the N numbers is written on a poker chip, and the chips sorted and stacked at the proper positions along the horizontal axis.

COMPOSITION OF THE N NUMBERS

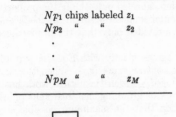

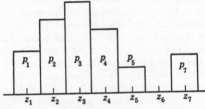

FIG. 2. A distribution of N numbers. They have been grouped by size into cells centered at $z_1, z_2, \cdots, z_6, z_7$, the proportions in these cells being $p_1, p_2, \cdots, p_6, p_7$. In the figure Cell 6 is empty; i.e., $p_6 = 0$. The centers need not be equally spaced.

The proportions in the various cells will be denoted by $p_1, p_2, \cdots, p_M$. The whole set of N numbers is composed of its M parts, wherefore

$$p_1 + p_2 + \cdots + p_M = 1 \tag{4}$$

Some of the proportions (p_i) may be equal, and in a rectangular distribution, all are equal. According to Eq. 1, the kth moment coefficient of the distribution described in the accompanying table (or by Fig. 2) will be calculated as

$$\mu_k = \frac{1}{N}(Np_1z_1{}^k + Np_2z_2{}^k + \cdots + Np_Mz_M{}^k)$$
$$= p_1z_1{}^k + p_2z_2{}^k + \cdots + p_Mz_M{}^k \tag{5}$$

Its mean will be gotten by setting $k = 1$, whereupon (writing μ for μ_1)

$$\mu = p_1z_1 + p_2z_2 + \cdots + p_Mz_M = \Sigma\, p_iz_i \tag{6}$$

The variance σ^2 of the distribution is gotten by putting $k = 2$ and measuring each number from the mean; thus

$$\sigma^2 = \frac{1}{N} \Sigma\, Np_i(z_i - \mu)^2 \tag{7}$$

which reduces to

$$\sigma^2 = \Sigma \, p_i z_i^2 - \mu^2 = \mu_2 - \mu^2 \tag{8}$$

as in Eq. 3, μ_2 being the 2d moment coefficient about 0.

Remark 1. It should be noted that the number N does not appear in these equations, having canceled out. Only the proportions $p_1, p_2, \cdots$ are left. Thus, a set of numbers possesses moment coefficients even if the number N is unspecified, provided only that the proportions p_i in the various classes are known.

Remark 2. Eq. 5 agrees strictly with Eq. 1 only if all the numbers in a cell are equal to the central value (z_i). When this is not so, "errors of grouping" are introduced. Much theoretical work has been presented in the literature on errors of grouping, no treatment of which will be given here except to mention that the maximum possible error in the mean is $\frac{1}{2}h$, where h is the (constant) class-interval. Errors of grouping are relatively more pronounced for the higher moments, but for any moment they diminish with the class-interval. On the other hand, the relative standard error or coefficient of variation of an estimate of the frequency within any cell increases as the class-interval is diminished, causing the "expected" frequency to diminish also. For most distributions met in practice, the errors of grouping are largely under control. Thus, in a distribution of incomes in a population in which most incomes fall at discrete levels such as $1000, $1440, $1860, it might be advisable to choose intervals that would contain these principal levels approximately in the middle. Similar decisions must be made in tabulating hours worked per week, or weeks unemployed. When nothing much is known beforehand about the material being tabulated, it is advisable to specify too many intervals rather than too few. There are two reasons: first, classes can be consolidated afterward; second, with most distributions met in social and economic surveys, the relative errors of grouping are almost certain to be negligible in the first four moments if there are from 10 to 15 class-intervals. Costs of tabulating and printing must be kept in mind; they rise rapidly as the number of intervals increases.

Exercise. The standard deviation of two numbers is half the distance between them.

The definition of a moment. The kth moment M_k of a set of numbers is defined as the numerator of Eq. 1. In accordance with this definition it follows that the kth moment coefficient is that quantity which, when multiplied by the total frequency N, gives the kth moment. In symbols

$$M_k = N\mu_k \tag{9}$$

When the scales along the axes of a distribution are so chosen that the area under the distribution curve is equal to unity, instead of the total frequency N, the curve is said to be *normalized*. The moment coefficient μ_k and the moment M_k are then numerically equal. In statistical work, distribution curves are nearly always normalized, and it is easy to fall into the habit of speaking of the moments of a curve instead of its moment coefficients.

Remark. The kth moment coefficient is dimensionally a distance raised to the kth power, but the kth moment is a distance raised to the kth power multiplied by an area. So, although for a normalized curve the kth moment coefficient and the kth moment are numerically equal, they are physically different things.

Moment coefficients of combined distributions. Sometimes in statistical work, after the mean and variance and perhaps other moments of a distribution have been worked out, supplemental tabulations bring in additional data requiring that new values for the moment coefficients be obtained. It is not necessary to recompute the work that has already been done.

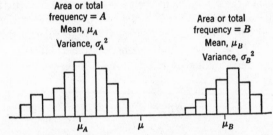

Fig. 3. The means and variances of two distributions have been calculated. The problem is to find the means and variances of the combination.

It will be useful to go into the matter in a little detail. In Fig. 3 are shown two distributions. For convenience they are shown disjointed, but it is not necessary that they be. They will be referred to as the A-distribution and the B-distribution. The letters A and B will also be used to denote the separate total frequencies or areas of the two distributions. A and B are often usefully measured in terms of their ratio to the total frequency $A + B$, for which the letters P and Q may be introduced and defined by the equations

$$P = \frac{A}{A + B}, \quad Q = \frac{B}{A + B} \tag{10}$$

It follows from the definition contained in Eq. 1 that the kth moment coefficient about 0 for the combined distribution is

$$\mu_k = \frac{\Sigma \, x^k}{A + B} \tag{11}$$

The summation in the numerator is to be taken over both parts, A and B. It can be broken up into two terms, one for a summation of x^k over distribution A, and another term for the summation of x^k over distribution B. Thus

$$\mu_k = \frac{\sum_A x^k + \sum_B x^k}{A + B} = \frac{A\mu_{kA} + B\mu_{kB}}{A + B}$$
$$= P\mu_{kA} + Q\mu_{kB} \tag{12}$$

wherein μ_k refers to the combined distribution, and μ_{kA} and μ_{kB} to the two component distributions. This equation can easily be extended to any number of component distributions in the form

$$\mu_k = \Sigma P_i \mu_{ki} \qquad \begin{array}{l} \text{[The subscript } i \text{ runs} \\ \text{over the number of} \\ \text{component distri-} \\ \text{butions]} \end{array} \tag{13}$$

which says that the kth moment coefficient of a combined distribution is equal to the weighted average of the kth moment coefficients of the component distributions. P_i is the relative frequency in the ith component, and $\Sigma P_i = 1$; i.e., the frequencies are normalized.

By multiplying through by N, the total frequency, the last equation gives

$$N\mu_k = \Sigma NP_i \mu_{ki} \qquad \begin{array}{l} \text{[} NP_i \text{ is the frequency in the} \\ i\text{th component, } \mu_{ki} \text{ its } k\text{th} \\ \text{moment coefficient]} \end{array}$$

or

$$M_k = \Sigma M_{ki} \qquad \begin{array}{l} \text{[} M_{ki} \text{ is the } k\text{th mo-} \\ \text{ment of the } i\text{th} \\ \text{component]} \end{array} \tag{14}$$

Thus *moments are additive*.

Remark 1. The distribution shown in Fig. 3 may be regarded as a combined distribution made up of M component distributions, the ith component having a moment of $NP_i x_i{}^k$. Eq. 5 is then only a simplified form of Eq. 13 in which each component consists of NP_i numbers all equal to x_i.

Exercise 1. Let the means of the two distributions A and B in Fig. 3 be denoted by μ_A and μ_B, and their standard deviations by σ_A and σ_B. Prove that for the combined distribution

$$\mu = P\mu_A + Q\mu_B$$
$$\sigma^2 = P\sigma_A{}^2 + Q\sigma_B{}^2 + PQ(\mu_B - \mu_A)^2$$
$$= P\sigma_A{}^2 + Q\sigma_B{}^2 + P(\mu_A - \mu)^2 + Q(\mu_B - \mu)^2$$

Remark 2. If the frequencies A and B are equal, the mean μ is midway between the component means. Whether equal or not, the mean μ is always between them. Only exceptionally $\mu = \mu_A$ if $Q = 0$, and $\mu = \mu_B$ if $P = 0$.

Remark 3. These equations have application in statistical offices when the means and variances are computed for the returns that have arrived up to a certain date. Later, when more returns come in, and there is need of taking account of all of them to make final estimates, it is not necessary to recompute *ab initio* the means and variances or any other moment coefficient, but only to add correction terms to the earlier computations, as called for in the equations just written. For example, in the spring of 1944, when sample censuses of certain areas were taken, the mean and standard deviation for the number of rooms in occupied dwelling places were computed immediately, but the unoccupied dwelling places were not counted till later. When the figures on unoccupied dwelling places were counted, the corrected means and standard deviations were quickly computed, and the sampling errors for the censuses made ready without delay.

Exercise 2. If $\mu_{(i)}$ is the mean of the ith component, prove that

$$\Sigma \, P_i(\mu_{(i)}{}^2 - \mu^2) = \Sigma \, P_i(\mu_{(i)} - \mu)^2$$

hence, for the mean and the variance of a combined distribution made up of any number of component distributions,

$$\mu = \Sigma \, P_i \, \mu_{(i)}$$

and

$$\sigma^2 = \Sigma \, P_i \, \sigma_i{}^2 + \Sigma \, P_i(\mu_{(i)} - \mu)^2$$
$$= \Sigma \, P_i \, \sigma_i{}^2 + \Sigma \, P_i(\mu_{(i)}{}^2 - \mu^2)$$

Remark 4. The second term on the right is the external variance of the separate means $\mu_1, \mu_2, \cdots$. Thus the variance σ^2 of the combined distribution is equal to the sum of the "internal variances" $(\sigma_i{}^2)$, properly weighted, added to the "external variance."

Moment coefficients of the area under a continuous curve.[1] For mathematical convenience it is customary to deal with continuous distribution curves instead of discrete summations. Actual distributions of universes and samples are always discrete, but the probability distributions that are derived by using continuous curves as approximations are almost always thoroughly satisfactory and calculated at considerably less mathematical labor than if a discrete distribution were rigidly adhered to. The moment coefficients of the area under a curve, about the y-axis, are defined

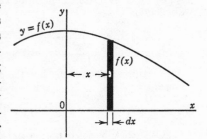

Fig. 4. Illustrating the area under the curve $y = f(x)$, for which moments are to be computed.

in precisely the same way as the moment coefficients of a discrete distribution, except that integrals replace summations as the limits of the sums that are obtained by taking finer and finer intervals along the approximating curve. The kth moment coefficient about 0 of the curve $y = f(x)$ between $x = a$ and $x = b$ is then

$$\mu_k = \frac{\displaystyle\int_a^b x^k f(x)\, dx}{\displaystyle\int_a^b f(x)\, dx} \tag{15}$$

The limits of integration are to cover the entire curve or whatever portion is under consideration. The limits will be $-\infty$ and $+\infty$ when the curve has no definite cut-off and the entire curve is to be covered. The normal curve is a familiar example. Sometimes the limits 0 and $+\infty$ will cover the entire curve, an example being the Type III curve (Fig. 57, p. 468) or half of a normal curve (see Ex. 13, p. 64).

Ordinarily, a frequency curve is *normalized*—i.e., the area under it is made equal to unity by proper choice of scales. $y = f(x)$ is the *distribution curve* of x or the *density function* of x if $f(x)\, dx$ is the *probability* of x in the interval $x \pm \frac{1}{2}\, dx$. $\displaystyle\int_{-\infty}^x f(x)\, dx$ is the *cumulative distribution function* (*c.d.f.*) of x. The moment coefficients μ_k and moments M_k are obviously equal numerically when a curve is normalized.

Remark 1. The odd moment coefficients about the mean of a symmetric distribution are all 0. A difficulty may seem to appear in certain cases when the limits of integration are $-\infty$ and $+\infty$. For example, the student might try to evaluate the mean or $3d$ moment coefficient of the Cauchy distribution

$$f(x) = \frac{2}{\pi} \frac{1}{1 + x^2} \tag{16}$$

as the integral $\displaystyle\int_{-\infty}^{\infty} \frac{x^k}{1 + x^2}\, dx$ does not exist for $k = 1, 3, 5, \cdots$. This difficulty disappears, however, when the principal value of the integral in the Cauchy sense is specified. This is done by writing

$$\mu_k = \operatorname*{Lim}_{b \to \infty} \int_{-b}^b x^k f(x)\, dx \qquad [k \text{ odd}] \quad (17)$$

A positive contribution to the integral for a symmetric curve is then annulled by an equal negative contribution, and the integral is 0 for all values of b if k is odd.

Remark 2. In contrast, an even moment coefficient of a symmetrical distribution can not be negative.

Remark 3. The Cauchy distribution does not possess a standard deviation because

$$\int_0^b \frac{x^2}{1 + x^2}\, dx \to \infty \quad \text{as} \quad b \to \infty$$

Accordingly the distributions of Student's z and Fisher's t do not possess standard deviations for sample-sizes $n = 3$.

Exercise 1. *a.* Prove that the curve $y = e^{-x}$ between $x = 0$ and $x = \infty$ is normalized.

b. Its kth moment coefficient is equal to $k!$ (HINT: Prove, by integrating by parts, that $\int_0^\infty x^k e^{-x} dx = k \int_0^\infty x^{k-1} e^{-x} dx$, which by continued reduction gives $k!$)

c. By the same integral, the kth moment of the Type III curve $y = x^{n-1} e^{-x}$ is $(k + n - 1)!$, and its kth moment coefficient is

$$(k + n - 1)! / \int_0^\infty x^{n-1} e^{-x} dx = (k + n - 1)!/(n - 1)!$$

Exercise 2. The moments of the curve $y = C f(x)$ about the y-axis depend on the value of C, but the moment coefficients do not. Hence, though the ordinates of a curve be doubled or in fact be multiplied by any constant factor whatever, the moment coefficients of the curve remain unaltered.

Exercise 3. The 0-th moment of any figure about any axis is the area of the figure.

Exercise 4. As the y-axis recedes further and further to the left from a given figure, $\mu_1, \mu_2^{\frac{1}{2}}, \mu_3^{\frac{1}{3}}, \mu_4^{\frac{1}{4}}, \cdots$ finally become more and more nearly equal. Illustrate this principle with the two sets of numbers

$$(a) \ 1, 2, 3, 4, 5; \quad (b) \ 11, 12, 13, 14, 15.$$

The numerical calculations give the accompanying table, whence it is obvious

	a	b
μ	3	13
$\sqrt{\mu_2}$	3.316	13.077
$\sqrt[3]{\mu_3}$	3.557	13.152

that the variation in Col. a is much greater than the variation in Col. b, although the variances of the original sets of numbers are identical.

Exercise 5. If a figure lies wholly to the positive side of the axis of moments, then $\mu_1 < \mu_2^{\frac{1}{2}} < \mu_3^{\frac{1}{3}} < \mu_4^{\frac{1}{4}} < \cdots$. When the axis recedes indefinitely, the inequalities tend to equalities.

Exercise 6. No matter where the y-axis is placed, the even moment coefficients stand in the following order: $\mu_2^{\frac{1}{2}} < \mu_4^{\frac{1}{4}} < \mu_6^{\frac{1}{6}} < \cdots$. As the y-axis recedes indefinitely, the inequalities tend to equalities.

Exercise 7. The harmonic mean H of the numbers $x_1, x_2, \cdots, x_n$ is defined by the equation

$$\frac{1}{H} = \frac{1}{n}\left(\frac{1}{x_1} + \frac{1}{x_2} + \cdots + \frac{1}{x_n}\right)$$

The geometric mean G is defined by the equation

$$G = \sqrt[n]{x_1 x_2 \cdots x_n}$$

Find the harmonic and geometric means of the two sets of numbers in Exercise 4 and compare them with the arithmetic mean and standard deviation.

Exercise 8. Show that the arithmetic mean of the sum of the logarithms of n numbers is the logarithm of the geometric mean of the n numbers.

Exercise 9. Derive the variances shown for the accompanying panels, which might represent various possible shapes of universes. Without knowing the answers, could you have arranged the panels in descending order of variance?

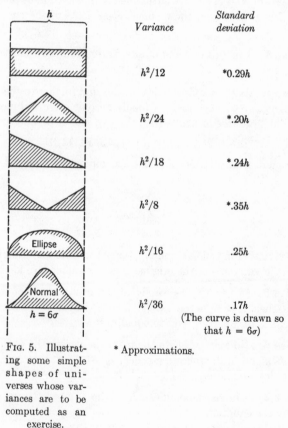

	Variance	Standard deviation
	$h^2/12$	*0.29h
	$h^2/24$	*.20h
	$h^2/18$	*.24h
	$h^2/8$	*.35h
Ellipse	$h^2/16$	.25h
Normal ($h = 6\sigma$)	$h^2/36$	.17h (The curve is drawn so that $h = 6\sigma$)

Fig. 5. Illustrating some simple shapes of universes whose variances are to be computed as an exercise.

* Approximations.

The results of this exercise have been very useful to the author in the planning of surveys (cf. p. 108).

Exercise 10. *a.* Reckoning moments about the y-axis, find the 1st, 2d, and 3d moment coefficients of the positive quadrant of the ellipse $x^2/a^2 + y^2/b^2 = 1$. *Answers:* $4a/3\pi$, $a^2/4$, $8a^3/15\pi$. (HINT: Use $x = a \cos \theta$ and $y = b \sin \theta$ for the ellipse.)

b. Do the same for the positive quadrant of the hypocycloid $(x/a)^{2/3} + (y/b)^{2/3} = 1$. *Answers:* $256a/315\pi$, $7a^2/64$, $8192a^3/45045\pi$. ($x = a \cos^3 \theta$, $y = b \sin^3 \theta$.)

c. From considerations of symmetry, write down the three moment coefficients of the same areas about the x-axis.

d. Without further calculation write down the 1st, 2d, and 3d moment coefficients of a semicircle of radius r about its diameter.

e. For Parts a and b, take the square root of the 2d moment coefficient and the cube root of the 3d moment coefficient, and note that $\mu_1 < \mu_2^{1/2} < \mu_3^{1/3}$. (See also Ex. 4 on p. 61.)

Suggestion. The integrals that occur in this exercise can all be thrown into the form $\int_0^{\frac{1}{2}\pi} \sin^m \theta \cos^n \theta \, d\theta$ by trigonometric substitution. The value of this integral is simply

$$\frac{(m - 1)(m - 3)(m - 5)\cdots 2 \text{ or } 1 \quad (n - 1)(n - 3)(n - 5)\cdots 2 \text{ or } 1}{(m + n)(m + n - 2)(m + n - 4)\cdots 2 \text{ or } 1} \alpha$$

$$[m \text{ and } n \text{ integral and positive}]$$

wherein α is unity unless m and n are *both even*, in which event α is to be replaced by $\frac{1}{2}\pi$. This extremely convenient formula was given by Wallis. Integrals such as $\int_0^{\frac{1}{2}\pi} \sin^m \theta \, d\theta$ also yield to this rule; the missing index is simply ignored, save that in determining the value of α it is to be classed as even. Clearly then, $\int_0^{\frac{1}{2}\pi} \sin \theta \, d\theta = \int_0^{\frac{1}{2}\pi} \cos \theta \, d\theta = 1$.

Exercise 11. Let $\phi(t) = \left[\dfrac{x_1{}^t + x_2{}^t + \cdots + x_n{}^t}{n} \right]^{1/t}$. $\phi(t)$ is an average value of the positive numbers $x_1, x_2, \cdots, x_n$: thus, if $t = -1$, $\phi(t)$ is the harmonic mean; if $t = +1$, $\phi(t)$ is the ordinary arithmetic mean; if $t = 2$, $\phi(t)$ is the root-mean-square; etc. Prove that, if $t \to 0$, $\phi(t)$ becomes the geometric mean of $x_1, x_2, \cdots, x_n$. (The case $t = 0$ gives an indeterminate form.)

A simple demonstration of the monotonic character of $\phi(t)$ was given by Nilan Norris, *Annals Math. Stat.*, vol. 6, 1935: pp. 27–9. As t increases, $\phi(t)$ also increases. Hence the geometric mean is larger than the harmonic mean but smaller than the arithmetic mean, which in turn is less than the root-mean-square value. This was illustrated in Exercises 4 and 5.

Exercise 12. If the mass of a body be divided into particles that are equal and sufficiently small to be treated as points, the sum of the squares of the distances from the particles to the center of gravity will be smaller than if measured to any other point. (Due to Legendre; see David Eugene Smith's *Source Book in Mathematics*, McGraw-Hill, 1929.)

Exercise 13. Deduce the formula of reduction

$$\int_0^\infty x^m e^{-\frac{1}{2}x^2/\sigma^2}\, dx = \sigma^2(m-1)\int_0^\infty x^{m-2} e^{-\frac{1}{2}x^2/\sigma^2}\, dx, \quad m > 0$$

and thence show that (a) the 2nth moment coefficient of the normal curve

$$y = \frac{1}{\sqrt{2\pi}\,\sigma}\, e^{-\frac{1}{2}x^2/\sigma^2}$$

about the central axis is $(\frac{1}{2}\sigma^2)^n (2n)!/n!$; and (b) that the $(2n+1)$-th moment coefficient of one-half of the normal curve is $2^{\frac{1}{2}(2n+1)}\sigma^{2n+1}$

$$\times\, n!/\sqrt{\pi} = \frac{n!}{\sqrt{\pi}}\, (\sigma\sqrt{2})^{2n+1}.$$

Given:
$$\frac{1}{\sqrt{2\pi}\,\sigma}\int_{-\infty}^\infty e^{-\frac{1}{2}x^2/\sigma^2}\, dx = 1$$

and

$$\frac{1}{\sqrt{2\pi}\,\sigma}\int_0^\infty x e^{-\frac{1}{2}x^2/\sigma^2}\, dx = \frac{\sigma}{\sqrt{2\pi}}\int_0^\infty e^{-v}\, dv = \frac{\sigma}{\sqrt{2\pi}}$$

(These results give the even and odd moment coefficients of half the normal curve.)

c. In particular, for the normal curve $\mu_4 = 3\mu_2{}^2 = 3\sigma^4$ and the Pearson measure

$$\beta_2 = \frac{\mu_4}{\mu_2{}^2} = 3$$

Exercise 14. Find the limiting average length of the parallel chords of a circle when the chords are distributed uniformly along (a) a line perpendicular to the chords; (b) the circumference. *Answers:* $\frac{1}{2}\pi r$ and $4r/\pi$.

Relations between moment coefficients about the mean and about some other axis. Suppose that a set of numbers x_1, x_2, etc., are all measured from the mean of their distribution, and that their moment coefficients μ_1, μ_2, μ_3, etc., about the mean are known. It is frequently necessary to know what their moment coefficients would be about some other axis, distant μ' to the left (negative side) of the mean. It is fortunately easy to derive a set of relations between the two sets of moment coefficients, and the work may proceed as follows. Let primes be attached to symbols that are reckoned from the new axis. Then

$$x_i' = x_i + u' \tag{18}$$

and

$$\mu_k' = \frac{1}{N} \Sigma\, x_i'^{k} = \frac{1}{N} \Sigma\, (x_i + u')^{k} \tag{19}$$

The parenthetical term $(x_i + \mu')^{k}$ may be expanded by the binomial theorem, and the relations between the two sets of moment coefficients written down mentally. In particular, with $k = 1$ the last equation gives

$$\mu' = \mu + \mu'$$

requiring μ to be zero, as of course it was to start with. If $k = 2$, Eq. 19 gives

$$\mu_2' = \frac{1}{N} \Sigma\, (x_i + \mu')^2$$

$$= \frac{1}{N} \Sigma\, (x_i{}^2 + 2x_i\mu' + \mu'^2)$$

$$= \mu_2 + 2\mu\mu' + \mu'^2$$

$$= \mu_2 + \mu'^2 \qquad \text{[Because } \mu = 0] \tag{20}$$

which is equivalent to Eq. 8 because μ_2 is the variance, σ^2.

Exercise 15. Derive the following relations between the two sets of moment coefficients. On the left are the moment coefficients about the new axis, which is distant μ' to the left of the mean.

$$\left.\begin{aligned}
\mu_0' &= 1 \\
\mu' &= 0 + \mu' \\
\mu_2' &= \mu_2 + 0 + \mu'^2 \\
\mu_3' &= \mu_3 + 3\mu'\mu_2 + 0 + \mu'^3 \\
\mu_4' &= \mu_4 + 4\mu'\mu_3 + 6\mu'^2\mu_2 + 0 + \mu'^4 \\
&\text{Etc.} \\
\mu_k' &= \sum_{r=0}^{k} \binom{k}{r} \mu'^r \mu_{k-r}
\end{aligned}\right\} \tag{21}$$

where $\binom{k}{r}$ is the binomial coefficient $\dfrac{k!}{r!(k-r)!}$.

The converse problem can be answered by solving these equations for the unstroked symbols in terms of the stroked ones, or afresh by putting $x = x' - \mu'$. Either way, the results will be the equations shown below. On the left are the moment coefficients about the mean, as is obvious from the fact that $\mu = 0$.

$$\left.\begin{aligned}
\mu_0 &= 1 \\
\mu &= 0 \\
\mu_2 &= \mu_2' - \mu'^2 \\
\mu_3 &= \mu_3' - 3\mu'\mu_2' + 2\mu'^3 \\
\mu_4 &= \mu_4' - 4\mu'\mu_3' + 6\mu'^2\mu_2' - 3\mu'^4 \\
&\text{Etc.} \\
\mu_k &= \sum_{r=0}^{k} \binom{k}{r} \mu'_{k-r} \, (-\mu')^r
\end{aligned}\right\} \quad (22)$$

Remark. The student should note that in every row of these equations the sum of the subscripts, with due account of the exponents, is a constant. (Remember that the subscript 1 is omitted for the mean.)

Exercise 16. Find by direct integration the first five moment coefficients of a circle of radius r about a diameter and then about a tangent. See if your results satisfy Eqs. 21 and 22.

Answer:

$$\left.\begin{aligned}
\text{About a diameter, } \mu_k &= \frac{k!\,r^k}{2^{k-1}(k+2)(\frac{1}{2}k!)^2} \quad [k \text{ even}] \\
&= 0 \quad\quad\quad\quad\quad\quad [k \text{ odd}]
\end{aligned}\right\}$$

$$\text{About a tangent, } \mu_k' = \frac{(2k+1)!\,r^k}{2^{k-1}(k+2)!\,k!} \quad \begin{bmatrix} k \text{ even or} \\ \text{odd} \end{bmatrix}$$

Short-cuts in the calculation of the mean and variance. It can be seen that the following equations for the mean and variance of a sample or any set of numbers are identities:

$$\bar{x} = \frac{1}{n} \sum_{1}^{n} (x_i - a) + a \qquad \begin{bmatrix} a \text{ being any} \\ \text{constant} \\ \text{whatever} \end{bmatrix} \quad (23)$$

$$\sigma^2 = \frac{1}{n} \sum_{1}^{n} (x_i - a)^2 - \left[\frac{1}{n} \Sigma \, (x_i - a) \right]^2 \quad (24)$$

With $a = 0$ this reduces to

$$\sigma^2 = \frac{1}{n} \Sigma \, x_i^2 - \bar{x}^2 \quad (25)$$

The last three equations are widely used for computational convenience. Usually the use of Eq. 23, with a as a convenient approximation to $\bar{x}$, will save a considerable amount of labor.

The accompanying tabular form will be found convenient for simultaneous computation of mean and variance. Cols. 1 and 2 show the original data. Cols. 3 and 4 can be filled in quickly by choosing a con-

COMPUTATIONAL FORM FOR FINDING THE MEAN AND VARIANCE OF A SET OF FREQUENCIES

(a is any convenient datum)

(1) Abscissa or x-value x_i	(2) Frequency or weight * y_i	(3) Weighted deviation from a $y_i(x_i - a)$	(4) Weighted square of the deviation from a $y_i(x_i - a)^2$
x_1	y_1	$y_1(x_1 - a)$	$y_1(x_1 - a)^2$
x_2	y_2	$y_2(x_2 - a)$	$y_2(x_2 - a)^2$
x_3	y_3	$y_3(x_3 - a)$	$y_3(x_3 - a)^2$
.	.	.	.
.	.	.	.
.	.	.	.
x_M	y_M	$y_M(x_M - a)$	$y_M(x_M - a)^2$
Sum	N	B	C
Weighted averages		B/N	C/N

* NOTE. If the frequencies or weights are relative to the total frequency or total weight, then y_i is to be replaced by p_i in conformity with the previous notation. The symbol N in this table is then to be replaced by 1.

veniently. The sums N, B, and C are then formed as indicated. The next step is to find the quotients B/N and C/N. Then the mean and variance are computed by the equations

$$\bar{x} = \frac{B}{N} + a \qquad \text{[Equivalent to Eq. 23]}$$

$$\sigma^2 = \frac{C}{N} - \left(\frac{B}{N}\right)^2 \qquad \text{[Equivalent to Eq. 24]}$$

Exercise 1. Ten readings of a micrometer screw are listed in Col. 1 below. Use the computational form of the last section to compute their mean and standard deviation. a. Take $a = 1.075$ for an arbitrary datum. The results follow, as the student should verify.

(1)	(2)	(3)	(4)
		Deviation from	Square of the devia-
Observation	Weight	1.075	tion from 1.075
1.078	1	3×10^{-3}	9×10^{-6}
.080	1	5	25
.071	1	−4	16
.076	1	1	1
.081	1	6	36
.082	1	7	49
.077	1	2	4
.073	1	−2	4
.079	1	4	16
.070	1	−5	25
Sum	$N = 10$	17×10^{-3}	185×10^{-6}
Average		1.7×10^{-3}	18.5×10^{-6}
		$= B/N$	$= C/N$

Then
$$\bar{x} = \frac{B}{N} + a$$
$$= 0.0017 + 1.075 = 1.0767$$
$$\sigma^2 = \frac{C}{N} - \left(\frac{B}{N}\right)^2$$
$$= 18.5 \times 10^{-6} - 1.7^2 \times 10^{-6}$$
$$= 15.61 \times 10^{-6}$$
$$\sigma = 0.00395$$

b. Repeat, using 1.070 as datum.

Exercise 2. Find the mean and standard deviation of the frequency distribution shown below. Let x be the midpoint of the wage or salary

THE SOUTH: FAMILY WAGE OR SALARY INCOME IN 1939

(Characteristics of Families, *Sixteenth Census:* 1940; Table 17, p. 47)

Wage or salary income (dollars) x	Number of families y
0	3,397,500
1–199	858,380
200–499	1,535,080
500–999	1,627,260
1000–1499	1,020,460
1500–1999	709,520
2000–2499	421,060
2500–2999	213,780
3000–4999	300,320
5000 and over	94,220

Suggestion. Cut off the last three figures and deal in thousands of families.

interval (0, 100, 350, 750, etc.).   Use $6000 for the mean of the group $5000 and over.

B. EXPECTED VALUES

Definitions: random variable, probability or theoretical distribution, "expected" value.   A *random variable* or random number is the result of applying a random operation.   A function of a random variable or variables is itself a random variable.   To the statistician there are two essential features of a random variable: *i*. as the operation is repeated, it generates a sequence of values of the random variable whose limiting distribution (called a *probability-distribution* or *theoretical distribution*) can be derived by sufficient mathematical prowess; *ii*. a particular random operation corresponding to a particular set of premises (size of sample, size and distribution of universe, technique of estimation) may be simulated in the real world by one device or another, such as the ideal bowl (next chapter) or by selecting the sample by the use of tables of random numbers (see the exercises in the next chapter, pp. 87–99).

A repetition of the random operation n times produces n values of the random variable, called a *random sample* of size n.   In conformity with one of the statements made above, any function of these n numbers is itself a random variable.   Some usual functions are listed below:

> The mean
> The sum
> Any multiple of the mean or sum ⎫ of the n numbers
> The median   ⎬ constituting a random
> Any percentile   ⎪ sample of size n
> The standard deviation
> The range   ⎭
> The first member of the sample, or the second, or the third, etc.
> The mean of the first two members
> The mean of three successive random samples of size n
> The difference between two random means
> The difference between the mean and any constant
> Any number calculated from a random sample as an estimate of some characteristic of the universe whence the sample was drawn

In application, a probability-distribution represents a prediction concerning the shape of the distribution that will be obtained for some random variable such as the mean, standard deviation, percentile, or range, of 1000 samples of n.   Comparison between a probability-distribution and a real distribution is one of the main problems of statistical analysis, because such comparisons often enable the scientist to understand nature better, which is to say that he is better able to predict the results of future samples, or in other circumstances to make the equally useful statement that no valid prediction is possible. Tests to determine which variety of wheat does best under given conditions, or which insecticide is best, tests to determine the effect of any medicinal treatment or social program, are but examples of the scientist's attempts to predict. The design of samples and experiments is the science of acquiring information at the least cost for making such predictions possible.

An *"expected" value Ex* is by definition the mean of the *probability-distribution* of a random variable x. The standard deviation of the probability-distribution of an estimate x is the *standard error* of x (i.e., of the *procedure* or *operation* by which the estimate x is produced). The fundamental requirement for the existence of an "expected" value Ex is that the distribution of x be produced by a random operation.

Remark 1. As a matter of notation, the Greek letters μ and σ^2 will usually be used for the mean and variance of a set of numbers that represent a universe. For contrast, the symbols $\bar{x}$ and s^2 will be used when the set of numbers has been obtained as a sample. So far as the formal computation is concerned, it makes no difference, but there is good reason on other grounds to distinguish between a universe and the samples that are drawn from it—in fact, this distinction is precisely the impetus to modern statistical research. A further differentiation will be needed in dealing with probability-distributions. For such distributions the symbol $E\bar{x}$ will be used for the mean of the probability-distribution of $\bar{x}$; Var $\bar{x}$ or $\sigma_{\bar{x}}^2$ for the variance $E(\bar{x} - E\bar{x})^2$ of the probability-distribution of $\bar{x}$; $E(\bar{x} - E\bar{x})^3$ will denote the 3d moment coefficient about the mean, etc. As a matter of fact, it will often be possible to calculate $E\bar{x}$ and Var $\bar{x}$ without calculating the shape of the probability-distribution of $\bar{x}$ (as in the next chapter).

If by theory it is calculated that a probability-distribution of x-values contains z_1 in the proportion p_1, z_2 in the proportion p_2, etc., then the "expected" value of x, denoted by Ex, is merely the mean of this probability-distribution, wherefore Eq. 6 may be rewritten as

$$Ex = p_1 z_1 + p_2 z_2 + \cdots + p_M z_M \tag{26}$$

Hence the rule: to calculate the "expected" value of a random variable x take the sum of all the values that x can assume, each value (z_i) of x first being multiplied (weighted) by the probability or proportion (p_i) in which that value occurs in the distribution of x.

Now, if p_i is the proportion of x-values that has the value z_i, then if a distribution of x^2 were drawn up, its mean would be

$$Ex^2 = p_1 z_1^2 + p_2 z_2^2 + \cdots + p_M z_M^2 \tag{27}$$

A similar expression with exponent 3 would give Ex^3; the exponent 4 would give Ex^4; etc. In fact, the "expected" value of any function $f(x)$ will be

$$E f(x) = p_1 f(z_1) + p_2 f(z_2) + \cdots + p_M f(z_M) \tag{28}$$

The variance of the probability-distribution of x will be

$$\text{Var } x = E(x - Ex)^2 \qquad \begin{matrix}\text{[Definition cor-}\\ \text{responding to}\end{matrix}$$
$$= Ex^2 - (Ex)^2 \qquad \begin{matrix}\text{Eq. 3]}\end{matrix} \tag{29}$$

The two terms on the right can be calculated from Eqs. 6 and 8 when the proportions p_1, p_2, $\cdots$ are known. Turned around, this last equation is also useful in the form

$$Ex^2 = \text{Var } x + (Ex)^2 = \sigma_x{}^2 + (Ex)^2 \tag{30}$$

These two forms of the same equation are of extraordinary importance and will be applied many times in succeeding pages. Sometimes Ex^2 can be evaluated first (as by Eq. 27), and the Var x found from Eq. 29. At other times the Var x will be known from some source, and Ex^2 will be calculated by Eq. 30.

Remark 2. It is instructive to regard the application of the f-operator as creating a new distribution (called a *derived* distribution) out of the original one. The process can be repeated any number of times: i.e., one might create a first derived distribution with an f_1-operator; out of that a second derived distribution with an f_2-operator.

Remark 3. The word "expected" and the symbol E signify a mathematical average calculated for a probability-distribution. They do *not* signify that any particular value of x or anything else is really to be expected in the ordinary sense of the word, either in any one sample or as an average of several samples. In sampling, as in any science, one takes what he gets. For this reason I use quotation marks and speak of the "expected" value of a random variable, even though I do not really expect to get it when sampling.

Commutation of E and Σ. It is often necessary to calculate the "expected" value of a sum. This can often be accomplished by noting that the "expected" value of a sum is the sum of the "expected" values. Symbolically,

$$E \Sigma = \Sigma E \tag{31}$$

For example,

$$E(x + y) = Ex + Ey \tag{32}$$

The commutative property of E and Σ, just expressed, is very useful and holds under all conditions, whether the two-way table of x and y shows independence or not. A proof will follow further on. Meanwhile

we pause to note that in contrast,

$$Exy = Ex\,Ey \qquad (33)$$

only if x and y are uncorrelated.   Otherwise,

$$Exy = Ex\,Ey + \rho\sigma_x\sigma_y \qquad (34)$$

where ρ is by definition the Pearson correlation coefficient between x and y.

Exercise 1.   Prove that

a. $$E\,a = a$$

b. $$E\,ax = a\,Ex$$

c. $$\text{Var}\,(x + a) = \text{Var}\,x \quad \text{or} \quad \sigma_x^2$$

d. $$\text{Var}\,ax = a^2\,\text{Var}\,x, \quad \text{or} \quad \sigma_{ax}^2 = a^2\sigma_x^2$$

In particular, $\text{Var}\,n\bar{x} = n^2\,\text{Var}\,\bar{x}$.

e. $$E(x - a)^2 = E(x - Ex)^2 + (Ex - a)^2$$
$$= \sigma_x^2 + (Ex - a)^2$$
$$= \sigma_x^2 \text{ only if } a = Ex$$

Thus $E(x - a)^2$ has its minimum value, σ_x^2, when $a = Ex$.   It may be helpful to see $E(x - a)^2$ plotted as a function of a, as in Fig. 6.   The curve is a parabola with intercept σ_x^2.

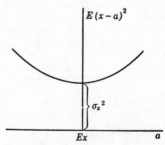

Fig. 6.   Showing $E(x - a)^2$ plotted as a function of a. $E(x - a)^2$ has its minimum, σ_x^2, when $a = Ex$.

Exercise 2.   Prove that, if $Ex_1 = Ex_2 = \cdots = Ex_n = \mu$, then

$$E\sum_1^n x_i = \Sigma\,Ex_i \qquad \text{[Interchange } E \text{ and } \Sigma\text{]}$$
$$= n\mu$$

Exercise 3. Prove that if $Ex_i = 0$, then

$$E[x_1 + x_2 + \cdots + x_n]^2 = E\sum_1^n x_i^2 + E\sum_{i=1}^n \sum_{j=1}^n x_i x_j$$

$$= \sum_1^n Ex_i^2 + E\sum_{i=1}^n \sum_{j=1}^n x_i x_j \qquad [j \neq i]$$

If x_i represents the ith random drawing from a universe, and if any two drawings (i and j) are uncorrelated, the last term is 0. If $Ex_i^2 = \sigma^2$ and if the drawings are independent, then $E[x_1 + x_2 + \cdots + x_n]^2 = n\sigma^2$. This result will be used in later chapters.

Exercise 4. Prove that, for any random variable x and any number A,

$$E\{x - Ex\}^3 = E(x - A)^3 - 3[E(x - A)^2](Ex - A) + 2(Ex - A)^3$$

Exercise 5. Prove that, for any random variable x, the standardized variable

$$t = \frac{x - Ex}{\sigma_x}$$

has the mathematical expectation 0 and unit variance. σ_x stands for the standard deviation of the distribution of x.

Exercise 6. Prove that

$$\Sigma (x_i - \bar{x})(y_i - \bar{y}) = \Sigma x_i y_i - n\bar{x}\bar{y}$$

where x_i, y_i is a pair of numbers, and i runs from 1 to n. $\bar{x}$ and $\bar{y}$ denote the averages of the x- and y-series.

Exercise 7. Given a universe

$$a_1, a_2, \cdots, a_N \qquad \text{Mean } \mu \qquad \text{Standard deviation } \sigma$$

Let a sample be drawn and recorded as

$$x_1, x_2, \cdots, x_n \qquad \text{Mean } \bar{x} \qquad \text{Standard deviation } s$$

x_1 being the 1st chip drawn, x_2 the 2d, etc. Let all the members of the universe have equal probabilities. Prove that

$$E x_1 = \mu$$

$$E x_2 = \mu$$

$$E x_3 = \mu$$

whether the sample be drawn with or without replacement.

a. First do this by putting $N = 5$ and writing down with equal frequencies all the possible values of x_1 and taking their average to find that $E x_1 = \mu$. Then for each possible value of x_1, write down all the possible values of x_2, and take their average to find that $E x_2 = \mu$. Similarly prove $E x_3 = \mu$. It is simplest to assume that the drawings are made with replacement, then to star those values which are to be excluded when the sample is drawn without replacement and to take the averages in the two cases. It follows that $E \bar{x} = \mu$ whether the sample is drawn with or without replacement. This type of illustration has been very helpful to the author.

b. Now derive a proof. Prove $E x_2 = \mu$ by noting that when the drawing is done without replacement, the probability of any chip being drawn at the second draw is $\dfrac{N-1}{N} \dfrac{1}{N-1} = \dfrac{1}{N}$, whence

$$E x_2 = \frac{1}{N}(a_1 + a_2 + \cdots + a_N) = \mu$$

Similarly prove that $E x_3 = \mu$, $E x_4 = \mu$, etc.

Exercise 8. Show that if a_1 has twice the probability of the other chips, then $E x_1 \neq \mu$, $E x_2 \neq \mu$, etc., and $E \bar{x} \neq \mu$. Under such conditions $\bar{x}$ is a biased estimate of μ. (This bias can be removed by weighting; see Ex. 6 on p. 91.)

Exercise 9. Put $N = 5$ and write down all the possible combinations of $x_i x_j$, and illustrate the fact that

$$\begin{aligned} &E \underset{(j \neq i)}{x_i x_j} = E x_i \, E x_i = \mu^2 && \text{with} \quad \text{replacement} \\ &E \underset{(j \neq i)}{x_i x_j} = \mu^2 - \frac{\sigma^2}{N-1} \neq E x_i \, E x_i \;\; \text{without} \quad \text{``} \end{aligned} \left. \vphantom{\begin{aligned} \\ \\ \end{aligned}} \right]$$

Exercise 10 (Proof that $E \Sigma = \Sigma E$). Let there be two universes, i.e., two bowls, designated A and B, containing chips marked in the following manner:

$$A : a_1, a_2, \cdots, a_N$$
$$B : b_1, b_2, \cdots, b_M$$

M and N may or may not be equal. Let p_{ij} be the probability that a_i and b_j will be the result of the ith draw from A and the jth draw from B. Make up a two-way table of all the possible values of $p_{ij}(a_i + b_j)$. For illustration this may be done with $N = 5$ and $M = 3$. By summation show that

$$\sum_i \sum_j p_{ij}(a_i + b_j) = \Sigma p_i a_i + \Sigma p_j b_j$$

where

$$\sum_j p_{ij} = p_i \quad \text{and} \quad \sum_i p_{ij} = p_j$$

The result so obtained may be rewritten as $E(x+y) = Ex + Ey$, which may be generalized to any number of random variables in the form

$$E \Sigma x = \Sigma E x$$

which is Eq. 31, q. e. d.

Note that this result holds under any values of p_{ij}. If the two drawings come out of the same bowl, we merely recognize that $p_{11} = p_{22} = p_{NN} = 0$. The results of the above exercises will be needed in the next chapter.

The coefficient of variation. One of the most useful statistical measures in sampling is the coefficient of variation of a distribution. The coefficient of variation of a distribution is defined as its standard deviation divided by the mean. In symbols, the coefficient of variation may be written

$$
\left.
\begin{aligned}
\gamma &= \frac{\sigma}{\mu} && \text{for a universe} \\[2ex]
c &= \frac{s}{\bar{x}} && \text{for a sample} \\[2ex]
C_x \text{ or C.V. } x &= \frac{\sqrt{\text{Var } x}}{Ex} = \frac{\sigma_x}{Ex} && \text{for a probability-distribution}
\end{aligned}
\right\} \quad (35)
$$

It will not be practicable to try to maintain a strict notation for the coefficient of variation.

A statement to the effect that an estimate of the total population of a particular area or class has a standard error of 1.5 percent is really only a statement of the coefficient of variation of the probability-distribution of estimates made according to prescribed plans. Thus, a sample estimate of a population might be written

$$500,000 \pm 7500 \qquad \text{[7500 being the standard error]}$$

or

$$500,000 \, (1 \pm 0.015)$$

The figure 0.015 or 1.5 percent is the estimated coefficient of variation of the sampling distribution of the estimated population.

The student should prove that a coefficient of variation is independent of the units in which the measurements are made—pounds, kilograms, feet, centimeters, hours, seconds, etc. A coefficient of variation is not used to describe a distribution whose mean is extremely small compared with its standard deviation.

CHAPTER 4. SOME VARIANCES IN RANDOM SAMPLING

Before the inherent variability of the test-animals was appreciated, assays were sometimes carried out on as few as three rabbits: as one pharmacologist put it, those were the happy days.—E. C. Fieller, Suppl., *J. Royal Stat. Soc.*, vol. vii, 1940–41: p. 3.

A. SOME PRINCIPLES OF PROCEDURE

Some definitions: universe, frame, ideal bowl, random sampling. This chapter will contain some basic theory for simple designs and for further development in later chapters. There will be a *frame*, which is to be thought of as a list of the N sampling units which constitute the *universe*. A list is one kind of frame, but the frame is often a file of cards, a map or set of maps, or verbal descriptions—any device by which the N sampling units are definitely identifiable one by one. In sampling a file of cards, the cards themselves constitute the frame.

For ease in classroom demonstrations, and for simplicity in discourse, the identification (e.g., name and address) of every sampling unit will be written on a poker chip, and the N physically similar poker chips will be placed in a bowl, the *ideal bowl*. One or more of the chips is to be drawn out of the bowl at random; the corresponding sampling units constitute the sample. The identifying information is necessary so that if a particular chip is drawn into the sample, an interviewer may be sent to the corresponding sampling unit to determine its population. In industrial sampling, the unit (manufactured article) is usually brought to the inspector who determines its *quality*, which in this text will be called the population of the unit (see Table 1 on p. 84).

Associated with each sampling unit (each chip) is a certain P-value or probability of being drawn into the sample. The P-values will correspond to some specific procedure of sampling. In the theory to be developed in this book, the P-values will all be equal within any bowl.[1] It is not to be inferred, however, that P-values must always be equal: some of the recent advances made by Neyman [2] and Hansen and Hurwitz [3,4] have involved unequal probabilities (e.g., sampling with proba-

[1] An exception will be seen in the sample of Greece in Chapter 12, in which the probabilities are sometimes proportionate to size.

[2] J. Neyman, "On the two different aspects of the representative method: the method of stratified sampling and the method of purposive selection," *J. Roy. Stat. Soc.*, vol. 97, 1934: pp. 558–625.

[3] Morris H. Hansen and William N. Hurwitz, "On the theory of sampling from finite populations," *Annals Math. Stat.*, vol. xiv, 1943: pp. 333–62.

[4] Morris H. Hansen and William N. Hurwitz, "A new sample of the population" (Bureau of the Census, Sept. 1944). "On the determination of the optimum probabilities in sampling," *Annals Math. Stat.*, vol. xx, 1949: pp. 429–32.

bility proportionate to the standard deviation of the stratum, or proportionate to some measure of size of the sampling unit). Although the P-values need not be equal, they must always be known: otherwise the development of a theory of sampling can not proceed.

In stratified sampling there will be several bowls (*strata*), and our theory, when developed, will apply to each bowl separately. The P-values may be constant within any bowl, yet different from bowl to bowl (cf. Ch. 6 on allocation of the sample in stratified sampling).

Random sampling with equal probabilities (equal P-values) is simulated in practice by making the chips physically alike, as nearly as possible, by shuffling them thoroughly (much more thoroughly than might be supposed necessary), and then drawing a chip after being blindfolded. Following Shewhart, we shall speak of this procedure as the *ideal-bowl experiment*. Sampling theory will give rational predictions of the distributions obtained by sampling, provided the sampling is carried out along the line of the ideal-bowl experiment or some procedure equivalent to it. In practice, random numbers are used (pp. 86 ff).

Some respect for the word "random" may be acquired by studying Shewhart's *Statistical Method from the Viewpoint of Quality Control* (The Graduate School, Department of Agriculture, Washington, 1939). There is grave danger in assuming that a procedure is random when it is only haphazard.

The frame and the sampling unit. As has been mentioned in Chapter 1, one difficult problem commonly encountered in sampling is to provide a suitable frame. Statistical information is desired concerning individuals, households, farms, firms, areas, manufactured articles, or some other entity, animate or inanimate. These entities may be called the *ultimate units* of the universe. The frame need not show these ultimate units: often the best frame obtainable merely divides the universe into an exhaustive set of *sampling units* (e.g., areas) each of which may contain ultimate units, none, one, or more than one. It is required that the totality of sampling units comprise the entire universe of ultimate units, and that each ultimate unit must be tied to one and only one sampling unit: otherwise the probabilities with which the various ultimate units may come into the sample can not be stated in advance, and probability theory can not apply.

A sampling unit (chip in the ideal bowl) might be intended to contain an individual person; a single household; 2, 3, 4, or more successive households (a *cluster;* cf. Ch. 5); a farm; a group of farms; a business firm; 1 bag or bale; 1 manufactured article; 4 or 5 successive articles; a whole case of articles, or something else definitely identifiable. Once a key map or list is made, showing definable units, a random sample of ultimate units can be drawn, with known probabilities (*vide infra*).

In the choice of sampling unit a great deal will depend on the possible frames that are possessed or can somehow be acquired. Sometimes the frames on hand permit little choice in the definition of the sampling unit. Sometimes they permit wide latitude, in which case certain special requirements can perhaps be met, such as near equality of size as measured by number of inhabitants or acreage or other content.

A sampling unit that is good for one purpose may not be so good for another, yet often a set of units, once defined, must be left as they are for some time to serve several surveys because of the time and expense required for drawing up a new frame. It should also be noted that any one survey is multipurpose to some degree, as no survey elicits information on only one question. Every survey is thus several surveys in one. If any latitude of choice is possible in the shapes or sizes of the sampling units, they should be fitted to the main purpose of the survey.

In areal sampling there is the added problem of striving not only for a frame of some kind but a frame that describes areal units that are fairly uniform in size as measured by the number of people, acreage, inventory, or something else. It will be seen later that if the populations a_1, a_2, $\cdots$, a_N of the sampling units are fairly uniform, the errors of sampling for characteristics closely correlated with these populations will be small without requiring a large sample.

In terms of the theory soon to be encountered, what is wanted is a small coefficient of variation, which is often sought by defining areal units that are small and as uniform as reasonably possible. The detailed labor of delineating small units provides good insurance against large ones, which do the damage (p. 344).

In many instances in census work a frame of small and efficient units is provided automatically by some inevitable routine operation not chargeable directly to the sample. The best example is a complete census, which happily provides a perfect frame consisting of units of any description, down to the individual person.

In taking a complete census, a listing of the entire population is made, and it is possible to draw from the listings, either concurrently or subsequently, a sample of (e.g.) every 20th individual, household, farm, male or female head, wife of head, oldest child, or of people of any characteristic. Sampling was so used in the Census of the United States in 1940 to obtain supplementary information from a 5 percent sample of the population, thus considerably broadening the scope of the census with great economy; and in fact many of the tabulations, not only of the supplementary information but of the regular information as well, were carried out only for the people in the sample. In the census of 1950, still heavier dependence is being placed on sampling; and I understand that similar dependence is in the plans for the census of Canada in 1951.

The extent, accuracy, and detail of maps showing recognizable boundaries, along with pertinent census information, will be limiting factors in the choice of areal unit. In the sampling of business and manufacturing firms, lists of large and medium-sized firms are necessary: fortunately such lists are in many countries kept fairly well current. Lists of the smaller firms by anything but name and address are incomplete, more or less: sometimes nonexistent. Fortunately again, lists of the smaller firms are not so necessary, as incomplete lists of small firms can be supplemented and corrected by drawing a sample of areas. Lists of the smaller firms, even though incomplete, are nevertheless useful also for control of the field-work in the sample-areas.

If a city is not too large every address may be prelisted over the whole city and every nth address selected for the sample, n and hence the sampling interval N/n being calculated after the manner to be expounded (cf. the exercise on p. 106). This plan was used in several cities in America in 1944 for samples designed to give estimates of the number of inhabitants—an extremely exacting problem in sampling.[5]

In using a plan of prelisting it is important to stress the need for thoroughness in listing in order that the sampling units (addresses in this case) be kept as uniform in size as is reasonably possible, and—even more important—that no areas be missed. In the particular surveys just mentioned, boarding houses and institutions were listed and sampled separately because they often contain large numbers of people and are highly variable.

In the United States small areal sampling units are now used in many government surveys, each unit containing a small group of households, farms, or business firms, or even a single household or business or manufacturing concern. In rural areas there is the Master Sample of Agriculture,[6] and in urban and suburban areas the Sanborn and other maps on which very small and uniform segments of area have been delineated through the vision of Morris H. Hansen, by whom also lists of large and medium-sized business and manufacturing concerns have been developed. These facilities are useful not only as aids to sampling, but

[5] In some of the nine cities, the sample censuses were taken in one stage of sampling, and in the remaining cities the sampling was done in two stages. The theory and procedure are described in a small volume by Hansen, Hurwitz, Tepping, and Deming, entitled *A Chapter in Population Sampling*, which the reader may find useful, as frequent references will be made to it. Purchase may be made from the Superintendent of Documents, Washington 25, at the price of one dollar. In future references the source and date will be listed as Bureau of the Census, 1947.

[6] R. J. Jessen, "The master sample project and its use in agricultural economics," *J. Farm Economics*, vol. xxix, 1947: pp. 531–40.

also as aids to the interpretation of data obtained in complete coverages of population, agriculture, business, and manufacturing, and for calibration of the completeness of coverage.

For recurring sample surveys the cost of elaborate maps and lists and of their upkeep is more than offset by savings in current surveys and improvement in the quality of the data.

It is often remarkable to observe how fruitful honest efforts can be when directed toward the procurement of a frame. Ration lists (may they soon be no more!) have provided frames in more than one instance. In one experience the electric light company kindly furnished a list of meters classified by business and residence, which provided the base for an excellent frame.

Howard Whipple Green's Real Property Inventory of Metropolitan Cleveland [7] and surrounding urban area provides an almost perfect list of addresses for sampling in Cleveland, either for population or business. Each address is classified by type of business or as a dwelling unit, and the whole list is brought up to date in the fall of every year. Similar inventories if developed for other cities would doubtless save their cost many times over by improving the quality of the information obtained in market and opinion surveys and other studies of the characteristics of business and population.

The author had the privilege of working in Japan on samples for studies of the population in 1947 and was pleased to find remarkable lists and maps.

In Japan the *shikuchōson* offices formerly contained complete lists of everybody in the *shi*, *ku*, or *chōson*. Furthermore, a separate list existed for the households in each *tonari-gumi* or neighborhood association which usually consisted of 10 to 15 families. However, these *tonari-gumi* were officially abolished two years ago for rationing and other administrative purposes. While it is still true that in a great many cases, perhaps more than half, nearly all the households in the *shiku-chōson* are shown on lists in the local offices, in some cases the lists are not kept up to date and therefore can not be used directly for sampling purposes. The reliance to be placed on the records varies considerably from one *shikuchōson* to another. This is true because there is now no compulsion to keep up-to-date records of the population. However, the tradition of record-keeping is an old one in Japan. For this and other reasons the records in many cases are still complete or almost complete on a fairly current basis. Furthermore, in the majority of cases, detailed maps show the location of every household within the *shi*, *ku*, or *chōson*. There is also some variation in the degree to which these maps are kept

[7] Address, Real Property Inventory of Metropolitan Cleveland, 1001 Huron Road, Cleveland 15.

up to date. Thus, while it is true that the *tonari-gumi* have been abolished and are no longer suitable sampling units, the maps still provide the basic information that is essential for efficient sampling. They are particularly good for the formation of small areal sampling units. The following excerpts from a letter from the author's colleague William J. Cobb at the present writing in Japan, represent the tentative opinions of Mr. Cobb and Dr. Kinichiro Saito of the Cabinet Bureau of Statistics, Tokyo, who made a joint investigation to obtain information on facilities available for the formation of small areal sampling units and for other sampling purposes.

All Japan is divided for administrative purposes into *shi* and *chōson*. The *shi* are cities of 30,000 or more population. *Ku* are divisions within the *shi* corresponding roughly to our wards. All Japan outside the *shi* is divided into *chōson*, which is a generic name for *machi* and *mura* (small towns and villages). In general the *machi* tend to be somewhat more urban and of somewhat larger population than the *mura*, although this is not universally true. The term *shikuchōson* is used to mean all administrative subdivisions of the country, including *shi*, *ku* within the *shi*, and *chōson*.

The expedients which may best be used vary from place to place and even within places. In general, the job is easy. This is because of the fact that very detailed maps exist, including plot maps, and, in most cases, inventories of all households in each enumeration district (*chōsa-ku*) by plot numbers. The inventories, being in most cases fairly recent, provide a measurement of the number of households in a given small area which is highly correlated with the actual number, and therefore extremely useful for the formation of basic small areal sampling units. Sometimes we use roads, sometimes streams or other natural features, occasionally plot lines, which we have verified, are known and can be found. In any case, it is surprisingly easy to set up small areal sampling units with the basic requisites—determinable boundaries, average number of households sufficiently small, and very small variation in number of households—in fact, much easier than in the U.S.A.

For the revision of the wage-survey, we wish to set up a monthly index of changes of employment. There is of course the possibility of very large sampling errors in estimates of this kind arising from great changes in one or a few establishments in a group from which only a small proportion is selected. I therefore explored the problem of finding other sources of information from which, at least, some indication of such changes might be obtained, so that such manufacturing plants could be sampled separately, with the assurance that we would have about the right proportion of rapidly changing (in number of employees) establishments. In nearby prefectures, I asked about this problem and was gratified to find that monthly changes are reported to the prefectural social security departments for all establishments having 5 or more workers. In some cases these reports may not be complete in coverage, but preliminary checks indicate that they are probably adequate to control this source of sampling variation.

The author has had the pleasure of talking with a number of Chinese scholars who convey the information that China both in urban and rural areas is organized on very nearly the same basis, the *chia* being a neighborhood of approximately 10 families, and the *pao* consisting of an aggregate of approximately 10 *chia*.

Detailed maps showing the plots of land by crop (jute, rice, maize, etc.) are to be found over almost every province in India, and the accuracy is such that Mahalanobis, in sampling for crop-acreages in Bengal, uses a square of area (grid or quad), stamped randomly on the map, for the sampling unit (cf. the quotation, p. 39).

National registers in some countries such as Sweden, Nederland, Belgium, and the United Kingdom provide reasonably accurate frames for studies of the population on a national basis, either directly from the register or extended to include direct interviews. For local frames, however, a national register could hardly be adequate, owing to lag in keeping any national register posted for local areas.

The subsampling of small areas through the use of rough detailed maps that are produced and used on the spot opens up many possibilities. An illustration is given in Fig. 10, page 139.

Remark 1. A frequent error made by beginners is to put in too much effort attempting to equalize the sizes of sampling units. The main requirement is to avoid relatively large units. Small units do relatively little harm (Ch. 10). If a ratio-estimate is to be used (*vide infra*), the attempts to equalize the units may be greatly relaxed.

Remark 2. As a matter of fact, even though one were completely successful in delineating sampling units of equal numbers of d.us. as measured on the basis of previous information, the sizes at the time of sampling will be different because of changes that take place meanwhile—building, demolition, and natural changes within d.us. (D.u. = dwelling units.)

Remark 3. It should be borne in mind, too, that sampling units that possess equal populations for one characteristic (e.g., total number of inhabitants or number of d.us.) will usually possess unequal populations for some other characteristic (e.g., number of unemployed or number of people with annual incomes above $4000). However, the populations of units as measured by one characteristic will often be related fairly closely to the several populations as measured by many other characteristics. For characteristics for which this is not so, a successful design for one characteristic may be much less successful for some other. One of the problems of sample-design is to compromise on such difficulties, not sacrificing one characteristic too far in order to achieve high efficiency for another.

Remark 4. It is to be remembered that lack of facilities for providing a frame for a sample will also make a respectable complete count equally difficult and inaccurate.

Remark 5. When a frame is a list of d.us. in geographical order (e.g., a city directory) it can and should be used in an unbiased manner so that omissions do not impair the final estimates. This can be done by defining the sampling unit as a half-open interval which extends (e.g.) from and including No. 17 Varick Street up to but not including No. 19. Dwellings for any reason not listed are thus given the same chance as others of coming into the sample. Omissions of course increase σ and the size of sample required to reach a desired degree of precision (invented by Stephan,[11] 1936).

Drawing the sample from the frame. The frame in some manner or other provides an identifying list of the N sampling units that compose the universe. A random sample of n units is desired.

Once the frame is established it is usually not difficult to devise instructions that can be followed by alert workmen for drawing a sample of n units with equal probabilities from the frame. One way is to use random numbers (see the exercises, p. 87), but in our work here in Washington we usually use *patterned* or *systematic sampling*, whereby the sample is designated on the frame by a random start and a constant sampling interval, usually a convenient integer not far from the calculated value of N/n (*vide infra*). One procedure for drawing a patterned or systematic sample of areas with equal probabilities is to number them on the key map in serpentine fashion, as in Fig. 11, page 141, commencing (e.g.) at the northeast corner, and then with a random start to designate for the sample every 5th or every 10th, or to use whatever other interval is conveniently near to the calculated magnitude of N/n.

Patterned or systematic sampling is simple and foolproof. It gives excellent geographic stratification, but under certain definable conditions there may be hidden difficulties.[8] Particular care is necessary in industrial sampling. But in the sampling of human populations, inventories, farms, and the like, experience shows that systematic sampling with a random start may be expected to yield small gains over random sampling, possibly from 2 to 10 percent smaller variance than the results of a sample of the same size selected entirely at random from the whole universe.

Much greater efficiency may be encountered under certain conditions. For example, in forest surveys Osborne [9] showed that systematic sampling may decrease the sampling variance tremendously—from 5 to 17 fold. In this book I shall recommend random selections, stratified or unstratified, for drawing samples of human populations, which include farms and business establishments. In stratified sampling, also studied by Osborne, and called by him *randomized blocks*, groups of con-

---

[8] *a.* William G. Madow and Lillian H. Madow, "On the theory of systematic sampling, I," *Annals Math. Stat.*, vol. xv, 1944: pp. 1–24. *b.* Lillian H. Madow, "Systematic sampling and its relation to other sampling designs," *J. Amer. Stat. Assoc.*, vol. 41, 1946: pp. 204–17. *c.* William G. Madow, "On the theory of systematic sampling, II," *Annals Math. Stat.*, vol. xx, 1949: pp. 333–54. *d.* W. G. Cochran, "Relative accuracy of systematic and stratified random samples," *ibid.*, vol. xvii, 1946: pp. 164–77.

[9] James G. Osborne, "On the precision of estimates from systematic versus random samples," *Science*, vol. 94: pp. 584–5; "Sampling errors of systematic and random surveys of cover-type areas," *J. Amer. Stat. Assoc.*, vol. 37, 1942: pp. 256–64.

tiguous sampling units are blocked off into contiguous or consecutive groups or strata, and one or two or more units drawn from each group by the use of random numbers. In a 10 percent sample, for example, 1 name in every consecutive 10 might be chosen at random. This plan preserves the advantage of stratification by geographic position or order, yet it avoids the risk of running into possible losses arising from patterned or systematic sampling.

In Chapter 10, where the appraisal of precision is treated, it is recommended that the subsample by which the appraisal is made be taken so as to reflect the gains and losses of the particular sampling procedure that was actually used.

Definition of population. Notation for universe and sample. The population of a sampling unit is determined by measurement, which in social and economic studies involves answers to questions, perhaps written at leisure or spoken in direct response to an interview. In industrial sampling and in surveys of equipment and inventories, the population of a unit is determined by some sort of visual test, or some mechanical or electrical test or measurement, which may perhaps be recorded automatically.

For any sampling unit, whatever it is, the population (a_i) depends on the particular characteristic being measured *and on the procedure for measuring this population.* For a given procedure there are in any one household as many values of a_i as there are questions in the enquiry. Some possible sampling units and values of population are listed in Table 1.

TABLE 1. SOME EXAMPLES OF POPULATIONS

Sampling unit or chip	Characteristic	Population of the unit, a_i
Household	Total number of inhabitants	5
Same household	Number of males 20–29 employed	0
" "	Number of girls under 15 in school	2
Adults in household	Satisfaction with laundry service	0 if satisfied
		1 if one is not satisfied
		2 if both are not satisfied
Manufactured article	Defective	0 if good
		1 if bad
" "	Diameter	1.061 inches

In this chapter the population a_i of sampling unit i is to be determined by a complete coverage of all the ultimate units therein. In Chapter 5 the population a_i will be estimated by a second stage of sampling.

It is brazenly assumed in the development of this theory that all measurements are exactly reproducible; in other words, any two interviewers will elicit the same information for the population a_i of sampling unit i. The sampling problem will thus be purely one of *selection*. This course is adopted here, not because errors of measurement and interviewing are any less important than the problems of sampling, but rather because this book treats only the random errors of selection.

Numerically the universe consists of a set of N finite numbers, the N populations $a_1, a_2, \cdots, a_N$. The universe therefore possesses a mean, a standard deviation, a range, and other statistical measures. A sample consists of a set of n numbers drawn from the universe. A sample will thus also possess a mean, a standard deviation, a range, and other statistical measures.

The N populations a_i of the universe need not all be different. If some of them are equal, the N populations of the universe may be grouped into M classes, assumed to be centered at $z_1, z_2, \cdots, z_M$, the respective proportions in these classes being initially [10] $p_1, p_2, \cdots, p_M$ as depicted in Fig. 2 on page 55. According to what has just preceded, the population of the universe (for any particular characteristic) is

$$A = a_1 + a_2 + \cdots + a_N \tag{1}$$

The mean population per unit is

$$\mu = \frac{A}{N} \tag{2}$$

and the variance of the universe is

$$\sigma^2 = \frac{1}{N} \sum_1^N (a_i - \mu)^2$$

$$= \frac{1}{N} \sum_1^N a_i{}^2 - \mu^2 \tag{3}$$

Now to return to the problem of sampling, it was stated earlier in this chapter that every sampling unit in the bowl must have associated with it a certain P-value or "probability" of being drawn. It is the mathematical manipulations of the P-values that build up probability-distributions for estimates computed from random samples.

As was also stated earlier, the P-values of the chips in the bowl may all be equal, or they may not be, but in either case it is absolutely essential that the probabilities be known. When the theory corresponds (or is intended to correspond) to drawings that are made with replacement, the P-value of each chip remains constant. An element may then appear twice or thrice or more times in one sample of n draws. But

[10] If the drawings are carried out with replacement, the adverb "initially" may be omitted.

when the drawings are made without replacement, an element, once drawn, has a P-value of 0 until replaced; hence it can not appear more than once in one sample.

It will suffice for the purposes of this book to identify a calculated probability-distribution with the theoretical or predicted distribution of a particular procedure of sampling, interviewing, and estimating. The mean of the probability-distribution of x is called the "expected" value of the estimate x, and the variance of this probability-distribution is the predicted variance of x.

The proper P-values to assign to the chips depend on the method of drawing. Experience has shown that when carefully manufactured physically similar poker chips are placed in a smooth bowl and very thoroughly shuffled, blindfolded drawings therefrom produce distributions in good accord with the probability-theory in which the P-values of all the chips in the bowl at any one time are equal. But if the weight or the texture of the chips affects their frequency of appearance in samples, as when the finish of the red chips is rougher than that of the white, the probability-distributions calculated on the basis of equal P-values will almost surely depart considerably from the actual results of real samples, and bias will result.

If certain people, such as families of managers of business establishments, fail to respond to a questionnaire or refuse to be interviewed, they have P-values of 0, and the distributions calculated on the basis of equal P-values for everyone in the sample may be considerably in error. These failures to respond or to be interviewed cause bias and alter the equations or at least the constants (such as certain variances of the universe) for calculating the sampling variances, because the responses constitute a sample only from that class of family or business establishment which will respond: the other class, the hard core of nonresponse, is left out unless special treatment is given to them (intensive follow-up or face-to-face interview). On the other hand, experience has shown that—

when employees, households, firms, areas, or any other sampling units are drawn by the use of random numbers or some equivalent device, with P-values properly assigned (not necessarily equal) in the various strata;

if no discretion is left to the interviewer as to who is in or out of the sample;

if the response from the elements selected for the sample is near 100 percent, or if adequate procedures are carried out to break up and elicit answers from a sample of the hard core of resistance;

then the bias of selection (Ch. 2) is practically eliminated, and the *sampling tolerances of selection are closely predictable by sampling theory.*

It is now necessary to introduce a notation for the n populations of a sample. The populations of the chips in the universe can be listed on N lines or cards, as stated earlier; but in recording the population of a sample on n lines, a change in the symbol for population is required because the first chip drawn into the sample can come from *any one* of the N lines on which the universe is listed, and thus it may or may not be equal to a_1. The distinction between the members of the universe and of a sample will be assisted by the use of two sets of letters, a_i for the population listed on the ith line of the universe, and x_i for the population recorded on the ith line of the sample.

In conformity with comments in Chapter 1, it is important to recognize the fact that a_i is the result of applying a particular operation (interview, test, or weighing procedure) on a sampling unit. If the operation is altered, intentionally or otherwise, a new set of values of a_i may be obtained. The theory of sampling can only predict probability limits for the result of expanding a particular operation of measurement to every member of the universe. It can not predict the result of some other operation (p. 18).

EXERCISES ON THE USE OF RANDOM NUMBERS AND ON THE BIASES OF CERTAIN SAMPLING PROCEDURES

At this point the student should acquire and learn to use a table of random numbers. Some of the following exercises will assist. An excellent table of 100,000 random numbers is contained in Fisher and Yates's *Statistical Tables for Biological, Agricultural and Medical Research* (Oliver & Boyd, 1938), which is indispensable not only as a set of tables, but also as a book on statistical methods. Another table of 100,000 random numbers was published by M. G. Kendall and B. Babington Smith, *Tracts for Computers*, No. 24 (Cambridge, 1939). A new table of 105,000 numbers in groups of 5 digits has been issued by H. Burke Horton (Interstate Commerce Commission, Washington, 1949). Tippett's famous table of 1927 is now out of print. A sample page of random numbers appears in the Appendix for assistance in the following exercises.

Exercise 1. *a.* Explain how to read out of a table of random numbers a number lying between 1 and 72 in such manner that all numbers between 1 and 72 have equal probabilities of being drawn.

b. A sample of 100 cards is to be drawn from a file of 759 cards all of which are to have equal probabilities. Explain how 100 random drawings may be made with random numbers.

c. Show how to use two columns of a table of random numbers to provide 10 random starts between 1 and 100. (Hint: $1 = 01$, $100 = 00$.)

d. An interviewer is told to interview the housewife in every seventh dwelling unit in a sample of small areas ("blocks"). Show how he may randomize his starting points in the areas, and thus avoid "corner bias," without the use of random numbers.

Exercise 2. *a.* A sampling interval of 10 is to be applied with a random start between 1 and 10 inclusive to a universe of 106 cards. Show that every card has the same probability of being drawn. This procedure gives a *systematic* sample.

b. An *unbiased* procedure is one whose "expected" result agrees with the total population of the universe, which in turn is defined as the sum of the populations of all N sampling units when each is measured with the same care as is exercised in the sample. Illustrate this definition by writing any 12 numbers on 12 cards. With 5 for a sampling interval, write down the results of all 5 possible samples (starting points 1, 2, 3, 4, 5). Introduce these estimating procedures:

$$X_1 = \frac{M}{m} \sum_1^m x_i$$

and

$$X_2 = k \sum_1^m x_i \qquad [k = 5, \text{ the sampling interval}]$$

Here $M = 12$, m is the number of cards in the sample (2 or 3), and x_i is the number on the ith card drawn into the sample. Compute X_1 and X_2 for all 5 samples, and find the average values of X_1 and X_2— in other words, their "expected" values.

Suggestion. As an illustration of what is wanted here, let the 12 cards be identified and sampled as shown in the accompanying tables.

Card $i = 1,\ 2,\ 3,\ 4,\ 5,\ 6,\ 7,\ 8,\ 9,\ 10,\ 11,\ 12$
Number $a_i = 2,\ 3,\ 5,\ 6,\ 6,\ 7,\ 9,\ 10,\ 14,\ 20,\ 25,\ 30$

$$\text{Sum} = \sum_1^M a_i = 137$$

Sample	Σx_i	$X_1 = \dfrac{M}{m}\Sigma x_i$	$X_2 = 5\Sigma x_i$
$i = 1, 6, 11$	34	136	170
2, 7, 12	42	168	210
3, 8	15	90	75
4, 9	20	120	100
5, 10	26	156	130
Average		134	137
Standard error		...	...

Note that—
  i. X_1 is biased. How much? (*Answer:* $134 - 137 = -3$.)
  ii. X_2 is unbiased, but apparently fluctuates over a greater range than X_1.

c. Calculate the standard errors of the two estimates X_1 and X_2, and enter them in the table.

d. Which estimate would you prefer in practice?

Exercise 3. Explain nonmathematically why you would expect the variance of any estimate derived from a systematic sample to change when the order of the members of the universe is shuffled.

Exercise 4. Often the files or lists that are to be sampled are located in several places, some here, some there, miles apart.

a. Devise a plan by which a uniform sampling interval, giving a proportionate sample (Ch. 6), may be applied to all the files simultaneously by as many workers as there are files. (*Hint:* Provide a list of random starts, as many random starts as there are files to be sampled. The author has done this with as many as 90 random starts. It may of course happen by chance that two random starts are the same, and if the sampling interval is less than the number of files, some of the starts must be the same.)

b. Why not deliberately give all workers the same random start? (In some types of work it would make little or no difference whether the starts were randomized, or the same random number used for all. Both procedures are unbiased, but there may be a gain in precision if the starts are randomized.)

c. Would you be worried if a particular file contains only 98 cards and the random start assigned for that file was 121? What would you do about it? (Nothing: this file simply does not contribute to the sample. Some other file, as small or smaller, might and occasionally will receive a low number as a random start and will contribute to the sample. Moreover, if the sample were to be redrawn, over and over, a file that does not contribute one time will contribute another time and in the right proportion of times. All this should be explained to the workers; otherwise they may suppose that the sampling procedure is defective through oversight, and they may in good faith and with the best of intentions draw some cards out of a file that should be skipped, to give it representation. Why would this impair the sampling procedure?)

d. Suppose that in some if not all the files the practice has been to put the rural cards at the front of the file. Would you modify your plan?

Remark. Perhaps as a preliminary step the rural cards might be counted and totalled, so that urban and rural cards could be sampled separately with prescribed random starts, just as if the number of files were doubled. The urban and rural intervals need not be the same and may in fact be adjusted so that approximately 1000 cards will be drawn from the urban universe and 1000 cards from the rural universe; separate estimates may then be prepared if desired. But if separate estimates are not desired, and if the urban and rural accounts are not greatly different, I would not go to the trouble of sampling them separately. Exact expressions for the gains and losses of stratified sampling will be found in Chapter 6.

Exercise 5. A list contains the names of all the people in an area, along with certain information concerning each, such as sex, age, relationship to head, school attendance, employment information, and perhaps other details as might be obtained by taking a complete census. The heads of the families are designated H. Some lines on the list do not contain names but contain notes regarding the description of the area, special difficulties encountered, explanations for the continuation of a particular family on another sheet, and the like. To save time and expense, the characteristics of families will be tabulated by a 5 percent sample instead of by a complete tabulation. The procedure will be to draw a 5 percent sample of heads, and to carry out the tabulations for the families whose heads are in the sample. The question is then how to draw a sample of heads so as to get a good 5 percent sample of families.

a. Show that any of the 8 plans described below will give an "expected" 5 percent sample of heads, the first four of which are systematic.

   i. With a random start take every 20th line.

   ii. With a random start take every 20th name.

   iii. With a random start take every 20th head.

   iv. With a random start take all the heads on every 20th sheet.

   v. From every successive 20 lines, draw one at random.

   vi. From every successive 20 names, draw one at random.

   vii. From every successive 20 heads, draw one at random.

   viii. From every successive 20 sheets, draw one at random, and take all the heads on these sample sheets.

b. Which plan do you think will give the smallest variance? (*Answer: iii* or *vii,* under conditions usually met in practice.)

c. Would the extra expense of using *iii* or *vii* be justified? (The answer depends on many considerations of cost, which vary depending on circumstances. In my own experience, I should be inclined toward plans *iii* or *vii,* especially as it is possible for alert workers to use these plans without making a separate list of heads.)

d. Suppose that the listing has been carried out by name, by family— head, wife, oldest child, etc., as in the regular census procedure, 30, 40, or 50 lines per page. The top lines of all pages will contain a preponderance of heads, and most heads are male. The bottom lines, on the other hand, will frequently be vacant: instructions to leave no lines vacant will not be wholly successful. The first line of the first page will in fact be the head of a family in a corner house; the second line will be his wife; etc. Then follows a second family, and a third, etc. Show that to the extent that lines are left vacant at the bottoms of pages, the systematic procedure i will be subject to sampling biases (loading in favor of heads, or wives, oldest children, etc.) unless the precaution is taken to divide the schedules into groups roughly equal in size (enumeration districts will do) and to randomize the starting point from group to group, or to rotate it systematically from line 1 to line 20. (Cf. a paper by Deming, Stephan, and Hansen, "The sampling procedures of the 1940 population census," *J. Amer. Stat. Assoc.*, vol. 35, 1940: pp. 615–30.)

Exercise 6. A file of cards contains one card for each address in an area. At some addresses there is 1 family, at other addresses there are 2 families, at others 3 families, at some addresses 10 families. An estimate of A, the total savings of all the people in the area, is desired.

a. If the cards are sampled with equal probabilities, and if *all* the families listed on a sample card are interviewed, then the estimate

$$X = \frac{M}{m} \sum_{1}^{m} X_i$$

[Unbiased. M is the total number of cards; m the number of cards in the sample; X_i the *total* savings of all the families on the ith card of the sample]

is an unbiased estimate of A, the total savings, no matter how distributed amongst the families of the area. (Recall Exercise 7, p. 73.)

b. Suppose that a sample of families is drawn by taking a family at random from each of m sample cards which have been drawn with equal probabilities. Show that the estimate

$$X = \frac{N}{m} \sum_{1}^{m} x_i$$

[Biased. m is the number of cards in the sample; x_i the savings of the family drawn from the ith card; N the total number of families.]

is biased. (However, the bias may not be serious; see Exs. 9 and 11.)

c. Suppose again that a sample is drawn by taking a family at random from each of the m sample cards which have been drawn with equal probabilities. Let there be N_i families on card i ($i = 1, 2, \cdots, M$), and let x_i be the savings of the family drawn at random from the card that was drawn at the ith draw ($i = 1, 2, \cdots, m$). Show that the estimate

$$X = \frac{M}{m} \sum_{1}^{m} N_i x_i \qquad \text{[Unbiased]}$$

is an unbiased estimate of A. This device, although it yields an unbiased estimate and does not require counting all the families, does require weighting by the factor N_i, which will raise costs and complicate the estimating procedure. The variance of this estimate will be calculated in Chapter 5 where sampling in two stages is treated.

d. Show that the estimate

$$X = \frac{M}{m^2} \left(\sum_1^m N_i \right) \left(\sum_1^m x_i \right) \qquad \text{[Biased]}$$

is biased unless N_i and x_i are uncorrelated.

Remark. A very important lesson is contained in the four estimates occurring in this exercise: two estimates are biased; the other two are not. A similar circumstance was met in Exercise 2. The bias or lack of bias in an estimate must thus be attributed not only to the manner of drawing the sample but to the manner of performing the calculations as well. Moreover, as will be seen, the variance of an estimate will also depend on the manner of performing the calculations. **The procedure of estimation must therefore always be considered as part of the sample design.**

Exercise 7. a. Suppose that probabilities P_i are assigned to the M cards ($i = 1, 2, \cdots, M$) and that a sample of m cards is drawn at random with these probabilities. Let A_i be the total savings shown for all the families listed on card i. Also let

$$A = \sum_1^M A_i = \text{the grand total}$$

Let X_i be the savings of all the families on the card that is drawn at the ith draw ($i = 1, 2, \cdots, m$) and redefine

$$X = \frac{1}{m} \sum_1^m \frac{X_i}{P_i} \qquad \text{[Unbiased]}$$

Prove that

$$EX = A$$

hence that X is an unbiased estimate of A for any set of probabilities P_i, none 0.

An important question is how in practice to assign these probabilities to give the smallest possible variance to X. The next part of the exercise shows that it is desirable that P_i be assigned as nearly proportional to A_i as possible with the knowledge at hand.

b. A sample of one card ($m = 1$) is to be drawn. Show that if P_i is proportional to A_i, which denotes the savings shown on card i ($i = 1, 2, \cdots, M$), then the variance of the sample is 0.

Solution

$$\text{Var } X = E(X - EX)^2$$

$$= \sum_1^M P_i \left[\frac{A_i}{P_i} - A \right]^2 \qquad \text{[Because } EX = A\text{]}$$

$$= \sum_1^M P_i \left(\frac{A_i}{P_i} \right)^2 - A^2 \qquad \left[\begin{array}{l} \text{Put } P_i = \frac{1}{k} A_i; \\ k = \sum_1^M A_i = A \end{array} \right]$$

$$= A \sum_1^M A_i - A^2$$

$$= 0$$

Note: The student may wish to show that the variance of X, although not 0, is very small if P_i is nearly but not quite proportional to A_i for every i. Of course, probabilities cannot be assigned in exact proportion to the sizes A_i, because the A_i are not known exactly. In practice approximations afforded by previous census or other information are used for assigning probabilities.

This result may be extended to samples of any size m by dividing the universe into m strata and then drawing one card at random from each stratum.

Exercise 8. In a survey of earnings a random sample of households was drawn with equal probabilities, and when several members of a sample household had income, one of them was drawn at random and an estimate of the average earnings per worker was calculated by the formula

$$\bar{x}_1 = \frac{1}{m} \sum_1^m x_i$$

[Biased. Here m is the number of households in the sample, and also the number of earners; x_i is the earnings of the earner selected from the ith household of the sample]

Show that this estimate is biased but that the bias can be removed by weighting the sample earner in a multiearner family by the factor N_i, where N_i is the number of earners in the ith household of the sample. An unbiased estimate is thus

$$\bar{x}_2 = \frac{1}{m\bar{N}} \sum_1^m N_i x_i$$

[Unbiased. The symbols m and x_i are as defined in the preceding equation, $\bar{N} = N/M$, N being the total number of earners in the universe, and M the total number of households therein. $\bar{N}$ is the average number of earners per household.]

The bias of the estimate $\bar{x}_1$ will be derived in Exercise 11. If N and $\bar{N}$ are not easily determined, as by counting, the ratio-estimate of Exercise 10f may be used.

Exercise 9. A card-file has been built up by commencing the history of each client on a card. With the passage of time some people's histories (accounts) run over on to two or more cards. A sample of accounts is to be drawn. If the cards have equal probabilities of being drawn, the individual accounts do not, as accounts covering several cards have higher probabilities than accounts that cover only one card.

Examine the theoretical and practical advantages and disadvantages of the two following plans for drawing a 1:50 sample of accounts: (a) with a random start, draw every 50th account; (b) with a random start, use a caliper to mark off intervals of 50 cards; include an account in the sample only if its initial card is at the commencement of one of these intervals.

By a clever device first used by Mildred Parten, the overexposure of accounts that cover several cards may be partially avoided by drawing cards systematically or at random with equal probabilities but taking into the sample the account next following any card that is drawn.[11] Show that although the individual accounts do not have equal probabilities under this plan either, the estimate

$$\bar{x} = \frac{1}{n} \sum_1^n x_i$$

[n is the number of accounts in the sample; x_i the ith account drawn into the sample]

will have small bias provided there is little correlation between A_i and N_{i-1}, wherein A_i is the ith account in the universe ($i = 1, 2, \cdots, M$) and N_{i-1} is the number of cards in the preceding account. N_0 is to be counted as N_M.

Solution

Let A_i be the amount of the ith account in the universe ($i = 1, 2, \cdots, M$).

$$\text{Bias} = E\bar{x} - \mu = \frac{1}{n} \sum_1^n Ex_i - \mu \qquad \left[\mu = \frac{1}{M} \sum_1^M A_i \right]$$

$$= \frac{1}{n} \sum_1^n \sum_1^M \frac{N_{i-1}}{N} A_i - \mu \quad [N = \sum_1^M N_i; \; \bar{N} = N/M]$$

$$= \frac{1}{N} \sum_1^M [N_{i-1} - \bar{N}] A_i$$

See also Cornfield's correction for overexposure on page 126.

[11] See Frederick F. Stephan, "Practical problems of sampling procedure," *Amer Soc. Rev.*, vol. 1, 1936: pp. 569–80.

Exercise 10. An electric company has 9800 pages of maps, bound into volumes, showing the locations of poles. On some pages there is only 1 pole; on others, 2, 3, or more poles—sometimes as many as 20 or 25. It is desired to draw a sample of poles in order to examine their condition and the condition of the cables and wires and cross-arms that the poles carry. The total number of poles is known from the accounting department to be about 105,000, and it has been decided that the sample should consist of approximately 1000 poles drawn systematically. Every pole and likewise every other piece of equipment supposedly possesses a "percent condition" which upon examination will be found to lie between 0 and 100, depending on its age, amount of rot, treatment (whether creosoted or not), or amount of rust, injury, and other factors such as loading, soil, and weather conditions. A crew of 2 inspectors will be sent to examine every pole in the sample and the equipment on these poles. An average "percent condition" for the universes of poles and for every other type of equipment will be calculated from the sample thus examined.

As there are 9800 pages of maps, and 105,000 poles, there is an average of about 10.7 poles per page. A sample that averages 1 pole for every 10 pages will therefore be about the desired size.

a. Show that the following plan does *not* give equal probabilities to the poles. "With a random start between 1 and 10, stop at every 10th page, and from these sample pages select a pole at random."

b. The estimate

$$\bar{x}_1 = \frac{1}{m} \sum_1^m x_i$$

[Biased. m is the number of pages in the sample; x_i the "percent condition" of the pole drawn from the ith sample page]

for the average percent condition of the universe of poles is therefore biased when the sample is drawn in the manner just described. (However, the bias may not be serious; see Exs. 9 and 11.)

c. Show that a sample consisting of all the poles on every 105th page with a random start will also give a sample of about the right size and will give equal probabilities to all poles. The estimate

$$\bar{x}_2 = \frac{1}{n} \sum_1^n x_i$$

[Slightly biased. n, the number of poles in the sample, is a random variable, hence this is a ratio-estimate. x_i is the condition of the ith pole of the sample]

will, however, be slightly biased, and this plan would probably have high variance because of the likelihood of similarity between adjacent poles.

d. Another plan, giving equal probabilities to all poles, is simply to count poles on the maps, beginning with pole No. 1 on page 1, and with a random start between 1 and 105, count off every 105th pole for the sample. Show that the estimate just written in part *c* is now unbiased.

Remark 1. Because of the likelihood of high serial correlation between adjacent poles, this plan may be expected to give much smaller variance than the previous plan of drawing all the poles on every 105th page.

Remark 2. Tukey's independent subsamples. By increasing the interval to 1050 and using 10 random starts between 1 and 1050 it is simple to specify 10 independent subsamples of 100 poles each. Moreover, the teams of inspectors (1 or 2 men to a team) might be designated *A*, *B*, *C*, *D*, *E*, *F* (if there were 6 teams), and they might be assigned to these subsamples equally at random or in some other balanced design. The variability of the inspectors will thus cancel out, leaving a pure measure of the sampling error along with separate estimates of the variability in performance of the teams of inspectors and the variance between teams. This plan has been used by the author at the suggestion of Professor John W. Tukey and Dr. W. J. Youden. (Cf. p. 353 for the reference.)

e. Show that the same size of sample would be obtained with two stages of sampling (next chapter) by setting

$$\frac{M}{m}\frac{N_i}{n_i} = \frac{105,000}{1000} = 105$$

wherein M/m is the counting interval for pages and N_i/n_i is the counting interval for poles. For example, let $M/m = 5$ and $N_i/n_i = 21$. Then *i.* with a random start draw every 5th page into the sample, numbering these pages $i = 1, 2, 3, \cdots$; *ii.* count the poles on these pages only, beginning with the first pole on the first sample page, and continue serially to the end; *iii.* with a random start mark every 21st pole for the sample (on the sample pages only, of course).

Remark 3. In practice the precision of this plan will be found close to the single-stage plan suggested in the preceding part of this exercise.

Remark 4. To introduce Tukey's plan of independent subsamples, increase the counting interval for poles to $N_i/n_i = 210$, and take 10 random starts between 1 and 210.

If desired, the pages may be drawn with probabilities closely proportional to their sizes as measured by number of poles, as is suggested in Exercise 12.

f. Suppose that the sample consists of one pole selected at random from every 10th page. If x_i is the condition of this pole, and if there are N_i poles on the page whence it came, the estimate

$$\bar{x} = \frac{\sum_1^m N_i x_i}{\sum_1^m N_i}$$

[Slightly biased; a ratio-estimate, to be encountered in Ch. 5] may be used. This is a ratio-estimate, so-called because the denominator as well as the numerator is a random variable. With this estimate, the total number of poles need not be counted. However, weighting by the factor N_i is required and this may be burdensome (as in Ex. 6c). Moreover, the variance of this estimate in this particular application would probably be pretty high and it would be far preferable to go to the trouble of using the sampling procedure of part d or e.

Exercise 11. Compute the bias in the biased estimate

$$\bar{x} = \frac{1}{m} \sum_1^m x_i$$

[In the sample, m pages, m poles. The pages have equal probabilities; the poles do not]

which was encountered in the preceding exercise, part b.

Solution

Let a_k be the condition of pole k in the universe ($k = 1, 2, \cdots, N$) and put

$$\mu = \frac{1}{N} \sum_1^N a_k$$

μ as thus defined is the true average condition of the N poles of the universe. Also let $\bar{a}_i$ be the average condition of the N_i poles on page i. Then

Bias in $\bar{x} = E\bar{x} - \mu$

$$= \frac{1}{m} \sum_1^m E x_i - \mu$$

$$= \frac{1}{m} \sum_1^m \frac{1}{M} \{\bar{a}_1 + \bar{a}_2 + \cdots + \bar{a}_M\} - \mu$$

$$= \frac{1}{M} \sum_1^M \bar{a}_i - \frac{1}{N} \sum_1^M N_i \bar{a}_i$$

[M is the total number of pages; $N = \bar{N}M$ the total number of poles]

$$= \frac{1}{M\bar{N}} \sum_1^M (\bar{N} - N_i)\bar{a}_i$$

Thus the bias in the given formula may be estimated in advance, provided something is known about the correlation between N_i and $\bar{a}_i$. In particular, if N_i and $\bar{a}_i$ are uncorrelated, the bias is 0.

Exercise 12. In the property of a large public utility company there are approximately 235,000 electrical multiple terminals, these being a mixture of cross-connecting terminals and distributing terminals, located in hundreds of buildings, scattered over a large city and several counties. An estimate of the physical condition of this property is desired. The detailed maps of the company show where each of these 235,000 terminals is located. The maps consist of 18,000 sheets, bound into volumes. The number of terminals per sheet ranges from 5 up to possibly 150, although some sheets contain only 1 terminal and some may contain more than 150. Discuss and criticise the following plan of sampling (an actual plan used by the author).

i. Assemble all the maps and freeze them until the sample is completed.

ii. Number the volumes 1, 2, 3, $\cdots$ and also number the sheets consecutively through all volumes.

iii. Make a quick eye-estimate T_i of the number of terminals on Sheet i ($i = 1, 2, 3, \cdots$). If this number appears to be more than 30, allow 2 or 3 more seconds for the estimate.

iv. Record this estimate T_i and the cumulative total, in the manner shown in Table 10, page 394.

v. Let $T = \Sigma\, T_i$. Suppose that a sample of about 300 terminals is desired. Take

$$C = \frac{T}{30}$$

for the counting interval, which is to be applied successively with 10 random starts to draw 10 subsamples of sheets.

vi. For the sample of terminals from Sheet i, first number the terminals on this sheet from 1 to $T_i{}'$. ($T_i{}'$ is an exact count and will be more or less than T_i, depending on how close the estimate T_i happened to be.)   Draw a random number Z_i between 1 and T_i. Terminal Z_i will be in the sample; also terminal $Z_i + T_i$ if it exists. If the eye-estimate was less than half the correct value $T_i{}'$, there may be a 3d terminal $Z_i + 2T_i$ in the sample. But if $T_i > T_i{}'$, it may be that $Z_i > T_i{}'$, in which case there is no sample from this sheet.

vii. Show that every terminal has the same probability as any other one of coming into any subsample; and that a good estimate of the average physical condition of the universe of terminals will be the simple average

$$\bar{x}^{(j)} = \frac{1}{n_j} \sum_1^{n_j} x_{jk} \qquad \begin{array}{l}\text{[Slightly biased; a ratio esti-}\\ \text{mate, because } n_j \text{ is a ran-}\\ \text{dom variable]}\end{array}$$

x_{jk} being the condition of the kth terminal in subsample j. The superscript (j) runs from 1 to 10 for the 10 subsamples.   Actually, the sample-sizes n_1, n_2, $\cdots$, n_{10} will all be practically equal. The grand average

$$\bar{x} = \frac{1}{10} \sum_{1}^{10} \bar{x}^{(j)}$$

would be used as the final estimate of the overall condition of the universe of terminals, and

$$t^2 = \frac{\left| \bar{x} - \mu \right|^2}{\dfrac{1}{9} \sum_{1}^{10} [\bar{x}^{(j)} - \bar{x}]^2}$$

would give an upper confidence limit to μ by the use of the Nekras-soff nomogram on page 555. As each $\bar{x}^{(j)}$ is very closely normally distributed because of the large number of terminals in each sub-sample, the conditions underlying the development of the t-test are in this case well satisfied.

Additional exercises on sampling biases will be found on pages 242 ff.

B. SOME GENERAL THEOREMS ON VARIANCE

Calculation of the mean and variance of the probability-distribution of the estimated sum and mean. The procedure to be followed here is a general one, applicable to any universe that has a variance. If at any time all the chips in the bowl have the same "probability" or P-value, then the P-value to be assigned to any one of them is

$$\frac{1}{\text{Number of chips in the bowl}}$$

Imagine that every chip in the universe bears a label typified by a_i, the population of the ith sampling unit. The first chip x_1 drawn in any sample might bear the label a_1, a_2, $\cdots$, or a_N. It will be assumed that x_1 is a random variable; that it varies by chance from one sample to another; that it has a theoretical distribution with a mean and variance which can be computed by the theory now to be developed.

Imagine that a large but finite number of samples has been drawn.[12] There will be an x_1 for the first sample, an x_1 for the second sample, etc.;

---

[12] If the chips of any one sample are drawn without replacement, each sample of n chips is to be returned to the bowl before the next sample is drawn.

likewise an x_2 for the first sample, an x_2 for the second sample, etc. So there will exist—

A sequence of x_1 values
"   "   " x_2 "
.   .
.   .
.   .
"   "   " x_n "
"   "   " x "
"   "   " X "
"   "   " $\bar{x}$ "

Each of these sequences is a set of numbers and has a distribution. The shapes and means and variances of these distributions change more or less as they are extended by taking more samples, but to each sequence there corresponds a probability-distribution, and the present problem is to calculate the means and variances of these probability-distributions. Later on, the exact shapes of these probability-distributions and meaningful ways of comparing them with the distributions of actual drawings may become a problem of interest (Ch. 9). For any one sample of the sequence compute

The sum of the sample, $\qquad x = x_1 + x_2 + \cdots + x_n \qquad$ (4)

The estimated population of the universe, $\qquad X = \dfrac{N}{n} x = N\bar{x} \qquad$ (5)

The estimated mean population per sampling unit in the universe, $\qquad \bar{x} = \dfrac{x}{n} \qquad$ (6)

In a moment it will be seen that X is an unbiased estimate of the total population A in the universe, and that $\bar{x}$ is an unbiased estimate of μ, the average population per unit.

The problem of finding the "expected" value of the sum x for samples of n is the problem of finding the mean of the probability-distribution of x. To proceed, it is noted first that

$$Ex = E(x_1 + x_2 + \cdots + x_n)$$

$$= Ex_1 + Ex_2 + \cdots + Ex_n \qquad \begin{array}{l}\text{[Theorem} \\ E \Sigma x = \Sigma Ex, \quad (7) \\ \text{Ch. 3]}\end{array}$$

Now $Ex_i = \mu$ for all drawings, whether with or without replacement.

This is so because any chip left in the bowl must be as good a sample as one already drawn out.   As a matter of fact, the probability-distributions of x_1, x_2, $\cdots$ are all identical to each other and to the original distribution of the a_i.   (See also Ex. 7 on p. 73.)   It follows that

$$Ex = n\mu \tag{8}$$

whence

$$\left. \begin{aligned} E\bar{x} &= \frac{1}{n} Ex = \mu \\ EX &= N\mu = A \end{aligned} \right\} \tag{9}$$

An estimate is *unbiased* if its "expected" value agrees with a particular characteristic of the universe.   Thus $\bar{x}$, according to Eq. 9, is an unbiased estimate of μ, which is by definition the average population per unit. Likewise $N\bar{x}$ and X are unbiased estimates of the total population A of the universe.

Next, the problem is to find the variances of the distributions of X and $\bar{x}$.

$$\begin{aligned} \operatorname{Var} x &= E(x - Ex)^2 && \text{[Definition; last chapter]} \\ &= Ex^2 - (Ex)^2 && \text{[An algebraic identity]} \\ &= E\left[\sum_{i=1}^{n} x_i\right]^2 - (n\mu)^2 \\ &= E\left[\sum_{1}^{n} x_i^2 + \sum_{i=1}^{n}\sum_{\substack{j=1 \\ (j \neq i)}}^{n} x_i x_j\right] - (n\mu)^2 \\ &= \left[n\sum_{1}^{N} \frac{1}{N} a_i^2 + E\sum_{i=1}^{n}\sum_{j=1}^{n} x_i x_j\right] - (n\mu)^2 \\ &\qquad \left[\text{Because } E\sum_{1}^{n} x_i^2 = \sum_{1}^{n} Ex_i^2 = \sum_{1}^{n} (\sigma^2 + \mu^2) \right. \\ &\qquad\qquad\qquad\qquad\qquad \left. = n(\sigma^2 + \mu^2)\right] \\ &= n(\sigma^2 + \mu^2) - (n\mu)^2 + E\sum_{i=1}^{n}\sum_{\substack{j=1 \\ (j \neq i)}}^{n} x_i x_j \tag{10} \end{aligned}$$

To evaluate the double summation on the right it is necessary to go separate paths, depending on whether the sampling is done with or without replacement.[13]   The derivation will be continued in Table 2.

[13] This order of presentation was suggested to me by my friend William N. Hurwitz of the Bureau of the Census.

TABLE 2. VARIANCES VALID FOR ANY UNIVERSE

With replacement	Without replacement

$$E\sum_{\substack{i=1 \\ (j \neq i)}}^{n}\sum_{\substack{j=1}}^{n} x_i x_j = \sum_{\substack{i=1 \\ (j \neq i)}}^{n}\sum_{\substack{j=1}}^{n} Ex_i x_j$$

$$= n(n-1)Ex_i x_j \qquad [j \neq i]$$

$$= n(n-1)\sum_{i=1}^{N}\sum_{j=1}^{N}\frac{1}{N^2}a_i a_j$$

$$= \frac{n(n-1)}{N^2}\left(\sum_{1}^{N} a_i\right)^2$$

$$= \frac{n(n-1)}{N^2}(N\mu)^2 \qquad (11)$$

$$E\sum_{\substack{i=1 \\ (j \neq i)}}^{n}\sum_{\substack{j=1}}^{n} x_i x_j = \sum_{\substack{i=1 \\ (j \neq i)}}^{n}\sum_{\substack{j=1}}^{n} Ex_i x_j$$

$$= n(n-1)Ex_i x_j \qquad [j \neq i]$$

$$= n(n-1)\sum_{\substack{i=1 \\ (j \neq i)}}^{N}\sum_{\substack{j=1}}^{N}\frac{1}{N(N-1)}a_i a_j$$

$$= \frac{n}{N}\frac{n-1}{N-1}\left[\left(\sum_{1}^{N} a_i\right)^2 - \sum_{1}^{N} a_i^2\right]$$

$$= \frac{n}{N}\frac{n-1}{N-1}[(N\mu)^2 - N(\sigma^2 + \mu^2)] \qquad (11')$$

Line by line, the explanation is this. First apply the theorem $E\Sigma = \Sigma E$ (p. 71). Next, by symmetry, each product $x_i x_j$ has the same "expected" value regardless of i and j, and there are $n(n-1)$ cross-products in a sample of n. Next, the P-value of any one cross-product is $1/N^2$ because there are N^2 possible products, including the admissible values $a_1 a_1$, $a_2 a_2$, etc. It follows that

$$\sigma_x{}^2 = n\sigma^2 + n\mu^2 - (n\mu)^2$$
$$+ \frac{n(n-1)}{N^2}(N\mu)^2$$

$$= n\sigma^2 \qquad (12)$$

Line by line, the explanation is this. First apply the theorem $E\Sigma = \Sigma E$ (p. 71). Next, by symmetry, each product $x_i x_j$ has the same "expected" value regardless of i and j, and there are $n(n-1)$ cross-products in a sample of n. Next, the P-value of any cross-product is $1/N(N-1)$ because there are $N(N-1)$ possible products, excluding the inadmissible values $a_1 a_1$, $a_2 a_2$, etc. It follows that

$$\sigma_x{}^2 = n\sigma^2 + n\mu^2 - (n\mu)^2$$
$$+ \frac{n(n-1)}{N(N-1)}[(N\mu)^2 - N(\sigma^2 + \mu^2)]$$

$$= \frac{N-n}{N-1}n\sigma^2 \qquad (12')$$

$$\doteq \left(1 - \frac{n}{N}\right)n\sigma^2 \qquad (12'')$$

Then, because $\bar{x} = x/n = X/N$, it follows from Exercise 1d on page 72 that

$$\sigma_{\bar{x}}{}^2 = \frac{\sigma^2}{n} \qquad (13)$$

Then, because $\bar{x} = x/n = X/N$, it follows from Exercise 1d on page 72 that

$$\sigma_{\bar{x}}{}^2 = \frac{N-n}{N-1}\frac{\sigma^2}{n} \qquad (13')$$

$$\doteq \left(1 - \frac{n}{N}\right)\frac{\sigma^2}{n} \qquad (13'')$$

TABLE 2. VARIANCES VALID FOR ANY UNIVERSE (*Continued*)

With replacement	*Without replacement*
The square root of each member gives	The square root of each member gives

$$\sigma_{\bar{x}} = \frac{\sigma}{\sqrt{n}} \qquad (14)$$

$$\sigma_{\bar{x}} = \sqrt{\frac{N-n}{N-1}} \frac{\sigma}{\sqrt{n}} \qquad (14')$$

$$\doteq \left(1 - \frac{n}{2N}\right) \frac{\sigma}{\sqrt{n}} \qquad (14'')$$

Then as $X = N\bar{x}$ (Eq. 5) it follows that

$$\sigma_X = N\sigma_{\bar{x}} \qquad (14a)$$

Then as $X = N\bar{x}$ (Eq. 5) it follows that

$$\sigma_X = N\sigma_{\bar{x}} \qquad (14a')$$

By dividing both sides of Eq. 14 by $E\bar{x}$ or μ, and by recognizing the equalities $\bar{x}/E\bar{x} = x/Ex = X/EX$, it is seen that

$$\text{C.V. } \bar{x} = \text{C.V. } x = \text{C.V. } X \qquad (15)$$

By dividing both sides of Eq. 14' by $E\bar{x}$ or μ, and by recognizing the equalities $\bar{x}/E\bar{x} = x/Ex = X/EX$, it is seen that

$$\text{C.V. } \bar{x} = \text{C.V. } x = \text{C.V. } X \qquad (15')$$

$$= \frac{\gamma}{\sqrt{n}} \qquad (16)$$

$$= \sqrt{\frac{N-n}{N-1}} \frac{\gamma}{\sqrt{n}} \qquad (16')$$

$$\doteq \left(1 - \frac{n}{2N}\right) \frac{\gamma}{\sqrt{n}} \qquad (16'')$$

where $\gamma = \sigma/\mu$, being the coefficient of variation of the universe (p. 75).

where $\gamma = \sigma/\mu$, being the coefficient of variation of the universe (p. 75).

Remark 1. It is to be noted that the two columns correspond to two different sampling procedures. The results are illustrative of the fact that if two sampling procedures are different, their variances are also different.

Remark 2. As $N \to \infty$ while the sample-size n remains finite, the results on the right approach those on the left. Moreover, if $n = 1$, the results in the two columns are identical, as should be so.

Remark 3. The factor $(N-n)/(N-1)$ in Eqs. 12' and 13' is known as the *finite multiplier* because it corrects for the finite size of the universe when the sampling is done without replacement. It is approximately equal to $1 - n/N$. Thus, if the sample decimates the universe ($n = 0.1N$), the finite multiplier is closely $1 - 0.1$ or 0.9, wherefore nonreplacement reduces the variances of X and $\bar{x}$ about a tenth and their standard errors about half as much, or 5 percent (note the term $n/2N$ in Eqs. 14'' and 16'').

It may be seen from Eqs. 16 and 16' that the effect of the finite multiplier is very closely this: if f_0 be defined as the sampling fraction $\frac{n}{N} = \frac{1}{N}\left(\frac{\gamma}{C}\right)^2$ calculated *without* the finite multiplier, then $f = f_0/(1 + f_0)$ will be the corrected sampling fraction [Yates, *Sampling Methods* (Griffin, 1949): p. 246].

Remark 4. Other derivations of Eqs. 12 and 12' will be found in the exercises on pages 119 and 131.

Further characteristics of the probability-distribution of the mean. As an exercise, the student is asked to find the 3d and 4th moment coefficients of the distribution of the means of samples of n drawn with replacement from any universe possessing 1st, 2d, 3d, and 4th moment coefficients. For convenience let the mean μ of the universe be 0. σ^2 will be its variance, and $\beta_2 = \mu_4/\mu_2{}^2 = \mu_4/\sigma^4$. The collected results are shown below.[14]

$$i. \qquad E\bar{x} = \mu = 0 \qquad\qquad\qquad \text{[Assumption]}$$

$$ii. \qquad E\bar{x}^2 = \text{Var } \bar{x} + (E\bar{x})^2 = \frac{\sigma^2}{n} \qquad \text{[Because } \mu = 0\text{]}$$

$$iii. \qquad E\bar{x}^3 = \frac{\mu_3}{n^2}$$

$$iv. \qquad E\bar{x}^4 = \frac{3\sigma^4}{n^2} + \frac{1}{n^3}(\mu_4 - 3\sigma^4) = \frac{3\sigma^4}{n^2} + \frac{\sigma^4}{n^3}(\beta_2 - 3)$$

whence

$$v. \quad \beta_2(\bar{x}) - 3 = \frac{1}{n}(\beta_2 - 3)$$

wherein $\beta_2(\bar{x})$ denotes the ratio $E\bar{x}^4 : (\text{Var } \bar{x})^2$.

These results illustrate how the distribution of the mean approaches the normal curve of standard deviation $\sigma/\sqrt{n}$. The skewness disappears rapidly with increasing sample-size because of the n^2 in the denominator of $E\bar{x}^3$. $\beta_2 - 3$ for a normal curve is 0 (p. 64); hence the last relation shows that $\beta_2(\bar{x})$ approaches the normal value 3 with increasing sample-size.

Remark. The reader should note that no finite number of relations like those shown above really proves that the distribution of $\bar{x}$ approaches the normal; they only show *how* the approach is made. A rigorous proof of the approach to normality involves the use of the characteristic function, and the assumption that all the moment coefficients of the universe exist.

Solution for Part iii

$$E\bar{x}^3 = \frac{1}{n^3} E\left(\sum_i^n x_i\right)^3 \qquad\qquad \begin{array}{l} [x_i \text{ denotes the } i\text{th} \\ \text{element of the} \\ \text{sample}] \end{array}$$

$$= \frac{1}{n^3}\left\{E\sum_1^n x_i{}^3 + E\sum_1^n x_i{}^2 x_j + E\sum_{i,j,k}^n x_i x_j x_k\right\} \quad \begin{array}{l} [i, j, k \text{ denote dif-} \\ \text{ferent elements} \\ \text{of the sample}] \end{array}$$

$$= \frac{1}{n^3}\left\{n\mu_3 + \sum_{i,j}^n Ex_i{}^2\, Ex_j + \sum_{i,j,k}^n Ex_i\, Ex_j\, E_k\right\}$$

$$= \frac{1}{n^3}\{n\mu_3 + 0 + 0\}$$

$$= \frac{\mu_3}{n^2} \text{ as required}$$

---

[14] The relation between β_2 for $\bar{x}$ and β_2 for the universe was given by L. Isserlis, "On the value of a mean as calculated from a sample," *J. Royal Stat. Soc.*, vol. lxxxi, 1918: pp. 75–81.

These results follow because x_i and x_j and x_k are independent, being different elements of the sample (though perhaps the same element of the universe); wherefore $Ex_i{}^2x_j = Ex_i{}^2 \, Ex_j = 0$ because $Ex_j = \mu = 0$. The result for $E\bar{x}^4$ is left to the student.

Use of the above theory. The importance of the theorems of variance that have just been derived is difficult to exaggerate. They are true for any universe that possesses a standard deviation. They are applicable to single-stage problems in which the sampling units are drawn independently and at random (i.e., not in clusters, Ch. 5).

Once the frame is provided, the theoretical problem from then on is to determine *in advance* the standard error of an estimated total (X) or mean ($\bar{x}$), and to calculate in advance the required size and cost of a sample that will yield a desired precision, such as a coefficient of variation of 5 percent. The essential quantity is the variance σ^2 or the coefficient of variation γ of the universe which is to be sampled: once these are known, the required size of sample can be calculated. Unfortunately σ and γ may not be known very closely before the sample is taken; nevertheless with some intelligent effort, supplemented by some numerical investigations of previously acquired data or by a pilot study,[15] adequately good prior estimates can almost always be made. Once a fairly good estimate of σ is made, the required size of sample (n) can be calculated. The smaller σ is, or γ, the smaller the sample required to give a sample estimate within some prescribed range of error, such as a standard error of 5 percent.

Experience along with good mathematical training is needed for the numerical guidance which is absolutely essential in the planning of a sample. Theoretically, the distribution of a universe can have almost any variance whatever, but there seem to be observable upper limits to σ and γ in nature, imposed by physical circumstances. Some grocery stores are bigger than others, some farms bigger than others, some trains longer than others, and some families bigger than others, but there seem to be natural limitations to many things. Sizes of banks and corporations, whether measured by demand-deposits or capital, vary more than inventories of grocers; and when sampling (e.g.) financial institutions to determine their total cash on hand, Series-G bonds, or net income before taxes, these institutions should, if at all possible, first be grouped or stratified by size as of some previous date, to cut down the coefficient of

---

[15] Early examples of the use of pilot studies and recurring surveys, in which each survey furnishes improved estimates of the variances and other characteristics of the universe, by which subsequent surveys can be carried out at decreased cost or increased precision, are furnished by P. V. Sukhatme, "Contribution to the theory of the representative method," Suppl., *J. Royal Stat. Soc.*, vol. ii, 1935: pp. 253–68, p. 255 in particular; and by Mahalanobis, references to whose work will be found on page 39. This idea is common practice now in several statistical centers.

variation within classes (Ch. 6). In sampling for the total inventory of grain, sugar, or tires held by retail dealers, it is useful to know that such universes almost never have coefficients of variation greater than 3. If they do, there is something wrong with the planning, because with some care the big dealers or warehouses (which are responsible for the high coefficient of variation) can usually be discovered and isolated and sampled separately, perhaps 100 percent. The inventories of the smaller dealers should then have a much reduced coefficient of variation. With some care the coefficient of variation in a stratum of inventories can often be made as low as 1.5 or 2. Any reduction in the variance of the universe cuts down the size of sample required for attaining a prescribed or aimed-at sampling error (such as a prescribed standard error of 5 percent). With a coefficient of variation of 3 for the universe, the sample-size required for a 5 percent coefficient of variation in the total estimated inventory would be calculated in the following manner. From Eq. 16'

$$n = \frac{N - n}{N - 1}\left(\frac{\gamma}{C}\right)^2 = \left\{1 - \frac{n - 1}{N - 1}\right\}\left(\frac{\gamma}{C}\right)^2 \qquad (17)$$

wherein C stands for the coefficient of variation of the estimates X, x, or $\bar{x}$. For a rough calculation, the finite multiplier might be neglected, giving

$$n = \left(\frac{\gamma}{C}\right)^2 \qquad (18)$$

as would be obtained from Eq. 16. Now, if $\gamma = 3$ and C is desired at 5 percent, then

$$n = \left(\frac{3}{.05}\right)^2 = 60^2 = 3600 \qquad (19)$$

If γ could be decreased to 2 by any device, or a 7 percent coefficient of variation tolerated in the sample-estimate, the sample-size could be greatly reduced. The finite multiplier should not be neglected if the sample is bigger than 10 percent of the universe. The effect of the finite multiplier is to reduce the required sample by the factor $f_0/(1 + f_0)$ as already mentioned.

Sampling for total acreage in wheat or some other crop is ordinarily difficult in the United States when a farm or group of farms forms the sampling unit; this is because of nonuniformity in size of farm. This difficulty is being reduced by constructing lists of extra-big farms and specialty farms, which may be set off into a separate stratum for separate treatment. The coefficient of variation of the remainder will be much reduced. In India, where holdings are much smaller and more uniform, the coefficient of variation of the area in rice or jute will run about unity, so I am told, after the large holdings are set off into a separate stratum—a relatively easy job there, because large farms are few and well known. Mahalanobis uses a geometrical grid for a sampling unit, which is stamped at random on a map: his unit, being of constant size, is independent of the size of farm (cf. the references and quotation on pp. 39 ff).

In sampling for crop-yield (bushels per acre), Sukhatme in Delhi tells me that the coefficient of variation of the number of bushels per acre of rice, wheat, and likewise the yield for many other crops, measured between fields within a village is usually somewhere between $\frac{1}{4}$ and $\frac{1}{2}$. The coefficient of variation measured between villages is somewhat smaller, $\frac{1}{5}$ to $\frac{2}{5}$. Abnormal areas or abnormal rainfall may raise or lower this figure, but usually the required sample (number of cuts) for a 5 percent coefficient of variation in estimated crop-yield can be calculated in advance with considerable confidence.[16]

Exercise 1. For many cities in the United States the coefficient of variation in the number of inhabitants per dwelling place is somewhat less than unity, often perhaps from 0.7 to 0.9. Assume it to be unity, for safety, and that a certain city contains about 40,000 dwelling places, a frame for which has been provided by a complete listing of dwelling places. The estimate of the number of inhabitants is to have a coefficient of variation of 1 percent. Find the required size of sample of dwelling places.

Solution

Turn back to Eq. 17 and write

$$n = \frac{N-n}{N-1}\left(\frac{\gamma}{C}\right)^2 = \left\{1 - \frac{n-1}{N-1}\right\}\left(\frac{\gamma}{C}\right)^2$$

wherein C stands for the desired coefficient of variation of the estimated number of inhabitants. With $\gamma = 1$ and $C = 0.01$ the required value of n would be 10,000 if N were infinite. With $N = 40,000$, it may be seen that $n = 8000$.

Exercise 2. Some pieces of a particular type of property (such as meters, switches, relays) belonging to a public utility company vary in

---

[16] Some useful numerical variances and theory will be found in the following papers by P. V. Sukhatme: "Reports on crop-estimating surveys: outturn of wheat in the Punjab, 1943–44," "Outturn of wheat in the Northwest Frontier Provinces, 1944–45," "Outturn of paddy in Madras, 1945–46," "Outturn of paddy in the Central Provinces and Berar, 1945–46," all published by the Indian Council of Agricultural Research (New Delhi); "Random sampling for estimating rice yield in Madras Province," *Indian J. Agric. Sci.*, vol. xv, 1945: pp. 308–17; "Random sampling for estimating rice yield in Kolaba District," *Proc. Indian Acad. Sci.*, vol. xxiii, 1946: pp. 194–209; "The problem of plot-size in large scale yield surveys," *J. Amer. Stat. Assoc.*, vol. 42, 1947: pp. 297–310.

condition only between 80 percent and 100 percent of their value when new. (The range of variability is small because the apparatus must be kept in excellent condition, else customers complain.)

a. Turn to Fig. 5 on page 62. Assume a rectangular distribution between the two given limits and show that a sample of only 133 pieces selected at random will provide an estimate of the total value of this type of property with a 3-sigma error band of $1\frac{1}{2}$ percent, which is well below the limits of observation. Thus, if the value of a particular type of property is \$200,000, its value would be determined by this sample within a 3-sigma band of \$3000, which is well within any rational basis for argument.

b. If the distribution were right-triangular, with the base at either end, the required size of sample would be 89.

c. If the distribution were normal, the required size of sample would be only 45.

Remark. This exercise illustrates the use of Fig. 5 on page 62 which has been helpful to the author in circumstances in which previous sampling experience is lacking, but in which enough substantive knowledge can be mustered for identifying the universe (or some stratum thereof that is to be sampled) as approximately rectangular, right-triangular, or some other shape suggested by the figure, and definitely contained within certain limits (80 and 100 in the above exercise). Under the circumstances of this exercise, for example, a sample of no more than 133 pieces would be required. Any shape, once assumed, fixes σ and γ, upon which the calculation of the required sample-size may proceed. After the sample is finished, the precision of the results will not be a matter of conjecture, but will be calculated independently of any assumptions that entered the planning stages (Ch. 8).

In the design of a sampling plan for measuring a new material, I prefer to make use of previous experience with similar materials or operations, along the lines just described, in preference to placing complete reliance on the use of some small number of measurements of the new material. No new problem is ever entirely new. Someone can always be found whose knowledge and experience can be translated, with the aid of theory, into a guide to a sampling plan not far from optimum. As experience accumulates with a "new" material there may be a chance to revise the plans.

Exercise 3. Bales of tobacco are being unloaded from a ship and from previous experience it is known that the coefficient of variation $\sigma : \mu$ of the weights of the bales from this particular source is not more than 5 percent. A load of $N = 1267$ bales is to be unloaded, and its total weight is desired with a precision of 1.5 percent for the 3-sigma limits. Solve Eq. 17 for n and observe that a random sample of $n = 93$ bales will be sufficient. In practice, a systematic sample of 1 bale in every 12 would be a good plan. Then the estimated total weight would be $X = N\bar{x}$, where $\bar{x}$ is the average weight of the n bales in the sample. σ and μ and the standard error of the estimate X should be computed after the sample of bales is weighed.

Special results for a universe of two cells. Sometimes the tabulation plans of a survey call for a dichotomy or the classification of the elements of the sample into two and only two mutually exclusive categories. Sampling from a universe of two categories is sometimes called "sampling for attributes." A few examples of dichotomies are written in the accompanying Table 3. It has been assumed all along and will continue

TABLE 3. EXAMPLES OF TWO-CELLED UNIVERSES

Object	*Dichotomy*	
Chip	0	1
"	White	Red
Area	Rural	Urban
House	Occupied	Vacant
Person	At work	Not at work
"	Under 14	14 and over
Family	With 1 child or none	With two or more children
Farm	Under 10 acres	10 or more acres
Industrial product	Accepted	Rejected
" "	Good	Defective
Chip or household	a_1	a_2

to be assumed that the elements have been unequivocally defined so that when drawn into the sample there will be no difficulties about classifying them into one pile or the other. Let the universe contain initially N chips, classifiable into two categories as follows—

Nq chips colored white, each counting x_1

Np " " red, " " x_2

q and p are the relative proportions; $q + p = 1$. The mean of this universe is

$$\mu = qa_1 + pa_2 = a_1 + p(a_2 - a_1) \tag{20}$$

and its variance is

$$\sigma^2 = qa_1{}^2 + pa_2{}^2 - \mu^2 \qquad \text{[Eq. 8, p. 56]}$$

$$= pq(a_2 - a_1)^2 \tag{21}$$

A special case of great practical utility is provided by setting $a_1 = 0$ and $a_2 = 1$ (called "the 0, 1 basis"). This system of numbering is ordinarily tacitly adopted in counting defectives in a lot or in a sample, in counting vacancies in a housing survey, or in counting the number of boys under 10 in a community of children. On the 0, 1 basis

$$\mu = p \qquad \text{[Put } a_1 = 0 \text{ and } a_2 = 1 \text{ in Eq. 20]} \tag{22}$$

$$\sigma^2 = pq \qquad [\text{ " \quad " \quad " \quad " \quad " Eq. 21]} \tag{23}$$

Suppose now that n chips be drawn to make a sample. The drawings are to be made independently and at random, which only means that we know the rules for assigning the P-values to the chips. The rule will be that, at every draw, every chip has the same probability of being drawn as any other chip still in the bowl.

Let several samples be drawn. If the n chips of a sample are drawn without replacement, they are to be returned before the next sample is

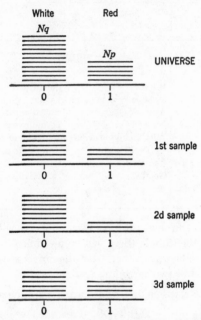

FIG. 7. A bowl-universe has in it initially N chips, Nq being white and counting 0, and Np being red and counting 1. Several successive samples of the same size n are drawn, with or without replacement after each draw. The chips in each sample are stacked as shown. The number of red chips varies from one sample to another. The problem is to find the theoretical distribution of r, the number of red chips in a sample of n.

drawn. The white (0) and red (1) chips of each sample are to be stacked, as shown in Fig. 7, and compared with the universe shown at the top. The number of red chips (r) varies from sample to sample; it is a random variable. In some samples there may be no red chips at all ($r = 0$). In other samples there may be 1 red chip, in others 2, etc.; and, finally, in some samples there may be n red chips. After some large number of samples is drawn, r being recorded for each, a distribution can be made for r; also for r/n.

Some interesting and useful results can be set down at once by rewriting the equations of Table 2 in terms of a two-celled universe, replacing μ by $a_1 + p(a_2 - a_1)$ and σ^2 by $pq(a_2 - a_1)^2$. On the 0, 1 basis, $\bar{x}$ is $r{:}n$, and $n\bar{x}$ is r. It should not be difficult for the student to derive from Table 2 the equations exhibited in Table 4.

TABLE 4.   VARIANCES FOR SAMPLES FROM A TWO-CELLED UNIVERSE

With replacement		*Without replacement*	

For any two-celled universe

$Ex = n\mu = na_1 + np(a_2 - a_1)$	(24)	$Ex = n\mu = na_1 + np(a_2 - a_1)$	(24')
$\text{Var } x = npq(a_2 - a_1)^2$	(25)	$\text{Var } x = \dfrac{N-n}{N-1}npq(a_2 - a_1)^2$	(25')
		$\doteq \left(1 - \dfrac{n}{N}\right)npq(a_2 - a_1)^2$	(25'')

It is also important to note that the C.V. x is the same as the C.V. r and C.V. r/n if $a_1 = 0$.

For a 0, 1 universe

$Er = np$	(26)	$Er = np$	(26')
$\text{Var } r = npq$	(27)	$\text{Var } r = \dfrac{N-n}{N-1}npq$	(27')
		$\doteq \left(1 - \dfrac{n}{N}\right)npq$	(27'')
$E\dfrac{r}{n} = p$	(28)	$E\dfrac{r}{n} = p$	(28')
$\text{Var }\dfrac{r}{n} = \dfrac{pq}{n}$	(29)	$\text{Var }\dfrac{r}{n} = \dfrac{N-n}{N-1}\dfrac{pq}{n}$	(29')
		$\doteq \left(1 - \dfrac{n}{N}\right)\dfrac{pq}{n}$	(29'')
$\text{C.V. } r = \text{C.V. }\dfrac{r}{n} = \sqrt{\dfrac{q}{np}}$	(30)	$\text{C.V. } r = \text{C.V. }\dfrac{r}{n} = \sqrt{\dfrac{N-n}{N-1}}\sqrt{\dfrac{q}{np}}$	(30')
		$\doteq \left(1 - \dfrac{n}{2N}\right)\sqrt{\dfrac{q}{np}}$	(30'')

The theoretical distribution of r and r/n in samples from a two-celled universe. So far, we have found the mean and variance of r and r/n, these symbols denoting the number of red chips, and the proportion red, respectively. This is a considerable amount of information, to be sure, but one might well proceed to enquire into the shapes of the probability-distributions of r and r/n. This is the problem of finding a series of probabilities for various values of r for a given sample-size n. The symbol $P(r, n, N)$ will be used for this probability, N being the initial size of the universe. As we shall see, the letter N can be dropped when sampling with replacement, because it disappears from the final result (cf. the last line of Table 5).

TABLE 5. DERIVATION OF THE PROBABILITY THAT A SAMPLE OF SIZE n WILL CONTAIN $n - r$ WHITE AND r RED CHIPS

With replacement	*Without replacement*
The number of combinations that can be formed with $n - r$ white chips and r red ones is $$\binom{n}{r}$$	The number of combinations that can be formed with $n - r$ white chips and r red ones is $$\binom{n}{r}$$
The number of ways of drawing a sample of size $n - r$ from Nq distinguishable white chips is $$(Nq)^{n-r}$$	The number of ways of drawing a sample of size $n - r$ from Nq distinguishable white chips is $$Nq \cdot Nq - 1 \cdot Nq - 2 \cdots Nq - \overline{n - r - 1}$$
The number of ways of drawing a sample of size r from Np distinguishable red chips is $$(Np)^{r}$$	The number of ways of drawing a sample of size r from Np distinguishable red chips is $$Np \cdot Np - 1 \cdot Np - 2 \cdots Np - \overline{r - 1}$$
The total number of possible samples containing $n - r$ white and r red chips drawn from a universe containing Nq white and Np red chips is therefore $$\binom{n}{r}(Nq)^{n-r}(Np)^{r}$$	The total number of possible samples containing $n - r$ white and r red chips drawn from a universe initially containing Nq white and Np red chips is therefore $$\binom{n}{r}\overbrace{Nq \cdot Nq - 1 \cdot Nq - 2 \cdots Nq - \overline{n-r-1}}^{n-r \text{ factors}}$$ $$\times \underbrace{Np \cdot Np - 1 \cdot Np - 2 \cdots Np - \overline{r-1}}_{r \text{ factors}}$$
The total number of possible samples of n chips that can be drawn from N distinguishable chips is $$N^{n}$$	The total number of possible samples of n chips that can be drawn from N distinguishable chips is $$N \cdot N - 1 \cdot N - 2 \cdots N - \overline{n - 1}$$
The reciprocal of this number is the P-value of any one sample. Therefore the sum of the P-values of all the samples	The reciprocal of this number is the P-value of any one sample. Therefore the sum of the P-values of all the samples

TABLE 5.   DERIVATION OF THE PROBABILITY THAT A SAMPLE OF SIZE n WILL CONTAIN $n - r$ WHITE AND r RED CHIPS (*Continued*)

With replacement	*Without replacement*
containing $n - r$ white and r red chips is	containing $n - r$ white and r red chips is

With replacement:

$$P(r, n) = \frac{1}{N^n} \binom{n}{r} (Nq)^{n-r}(Np)^r$$

$$= \binom{n}{r} q^{n-r} p^r \qquad (31)$$

[This is known as the rth term of the *binomial* or *Bernoulli* [17] series]

Note that the letter N has disappeared. Only the constant proportions q and p appear.

Without replacement:

$$P(r, n, N)$$
$$= \frac{1}{N \cdot N - 1 \cdot N - 2 \cdots N - \overline{n - 1}}$$
$$\times \binom{n}{r}$$
$$\times Nq \cdot Nq - 1 \cdots Nq - \overline{n - r - 1}$$
$$\times Np \cdot Np - 1 \cdots Np - \overline{r - 1}$$

$$= \frac{\binom{Nq}{n - r}\binom{Np}{r}}{\binom{N}{n}} \qquad (31')$$

[This is known as the rth term of the *hypergeometric series*]

Note that the letter N appears in the final result, along with the initial proportions q and p.

The symbol $\binom{n}{r}$ denotes the number of different combinations that can be made with r red chips and $n - r$ white ones. Another symbol C^n_r, which means the same thing, is sometimes used.

Thus, samples drawn with replacement give the *binomial* or *Bernoulli* series, whereas samples drawn without replacement give the *hypergeometric* series.[18] The two terms, binomial and hypergeometric, when calculated and compared numerically for particular values of n and r, will be found in near agreement when N is large in comparison with n, and the agreement improves as N increases relative to n (see Ex. 5, p. 122).

It is next to be noted that

$$x = (n - r)a_1 + ra_2 = na_1 + (a_2 - a_1)r \qquad (32)$$

$$\bar{x} = \frac{x}{n} = a_1 + \frac{1}{n}(a_2 - a_1)r \qquad (33)$$

[17] After Jacques Bernoulli (1654–1705), who studied this series with ability and zeal.

[18] The Bernoulli or binomial series is also hypergeometric in form as will be shown in Chapter 13, page 431.

Now the probability of getting a particular value of r in a sample of n is also the probability of getting the corresponding values of r/n, x, and $\bar{x}$. Except for appropriate changes in scale, the distributions of r, r/n, x, and $\bar{x}$ will therefore all be alike. The results so reached are summarized in Table 6.

Special difficulties encountered in sampling for small proportions. Some very important general principles of sampling are evident from Table 4, in spite of the fact that the theory thus far developed applies only to samples of n units drawn independently from a single bowl. In the bottom row of Table 4 it is seen that the coefficient of variation C of both r and r/n contains p in the denominator. Turned around and solved for n, Eq. 30' gives

$$n = \frac{N-n}{N-1} \frac{q}{C^2 p} \tag{34}$$

$$\doteq \frac{1}{C^2 p} \quad \begin{array}{l} \text{[An approximation,} \\ \text{useful if } p \text{ is very} \\ \text{small and } n/N \text{ not} \\ \text{large]} \end{array} \tag{35}$$

According to this equation, the smaller be p, the larger the sample required to give a desired precision C; or, alternatively, the lower the precision obtainable for a given size of sample. An example is a housing survey, aimed at discovering the number and characteristics of vacant dwelling units for rent within a particular city, when there are extremely few vacancies. Under such conditions a very large sample would be required if the proportion p of vacancies were desired with great precision.

Fortunately no great precision of sample is usually required when the proportion p is extremely small. In a housing survey difficulties with the definition of a habitable dwelling unit and a dwelling unit for rent will mask the sampling error. Thus a coefficient of variation of less than 30 percent may be considered wasteful when p is 1 percent or less. Any decision on the question of the need for new housing would be in the affirmative whether the vacancy ratio were $\frac{1}{4}$ percent or $\frac{3}{4}$ percent. If the sampling were binomial and hence described by Table 4, the required size of sample for a 30 percent coefficient of variation would be

$$n = \frac{1}{0.3^2 \times .01} = 1111 \text{ d.us.} \tag{36}$$

In practice, binomial sampling would not usually be used: instead, for economy, blocks of areas would perhaps be drawn into the sample as primary units, to be subsampled in various ways. By such a procedure the actual size of sample would be more like 3000 d.us., depending on the variances between blocks and within blocks for the particular city being sampled. Such calculations will form the subject of Chapter 5.

TABLE 6. PROBABILITIES, EXPECTED VALUES, AND VARIANCES FOR SAMPLES DRAWN FROM A TWO-CELLED UNIVERSE

Variable	Range of variable	With replacement			Without replacement		
		Probability	Expected value	Variance	Probability	Expected value	Variance
r	0 to n by steps of 1	$\binom{n}{r} q^{n-r} p^r$	np	npq	$\dfrac{\binom{Nq}{n-r}\binom{Np}{r}}{\binom{N}{n}}$	np	$\dfrac{N-n}{N-1} npq$
$\dfrac{r}{n}$	0 to 1 by steps of $1/n$	"	p	$\dfrac{pq}{n}$	"	p	$\dfrac{N-n}{N-1}\dfrac{pq}{n}$
x	nx_1 to nx_2 by steps of $x_2 - x_1$	"	$nx_1 + np(x_2 - x_1)$	$npq(x_2 - x_1)^2$	"	$nx_1 + np(x_2 - x_1)$	$\dfrac{N-n}{N-1} npq(x_2 - x_1)^2$
$\bar{x}$	x_1 to x_2 by steps of $\dfrac{1}{n}(x_2 - x_1)$	"	$x_1 + p(x_2 - x_1)$	$\dfrac{pq}{n}(x_2 - x_1)^2$	"	$x_1 + p(x_2 - x_1)$	$\dfrac{N-n}{N-1}\dfrac{pq}{n}(x_2 - x_1)^2$

When p is small, something like 99 dwelling units must be visited by the interviewers to find 1 vacancy. In a sense, visiting these 99 dwelling units represents wasted effort. If one only knew in advance where the vacant dwelling units were, so that he could go and count them and record their characteristics (whether single dwelling, apartment, duplex, in a building with business, etc.), the cost of a housing survey would be cut to a fractional part of the actual necessary cost. The *fraction* vacant over the entire city would of course not be known, however, unless the total number of dwelling units were known from external sources, as in fact it usually is from Census data and records of building and construction.

If the total number of dwelling units over the city, vacant plus occupied, is an unknown quantity, then the aim of the sample might be two-fold, not only to learn the proportion vacant and their characteristics, but also to determine the total number of all dwelling units. Visitation of the 99 occupied for every 1 vacant is then not to be regarded as a total loss, but rather as necessary to estimate the total number.

A similar problem arises in consumer research where the opinions of a rare group of people are desired, such as the owners of a particularly expensive radio set or high-class automobile. If all households are given equal probabilities, a score or more of them must be visited to find one that owns a particular type of radio set or automobile.

Such difficulties have led to the invention and attempted justifications of biased procedures for estimating proportions. One common plan is to attempt to isolate particular areas of the city in which most of certain classes of people live and to confine the survey to these areas. Surveys on the consumption and spending-habits of wage-earners are too often carried out in this manner. Bias is introduced in such a plan because it is usually impossible to isolate areas that will contain all or practically all of any particular class of people. Without covering the excluded areas with a sample, there is no telling how many people of any particular class are thus excluded, nor any way of knowing what their characteristics are in the particular question of investigation. The results from the delineated areas are thus *biased*, and the sample is a judgment-sample because the bias must be evaluated by judgment. As was mentioned in Chapter 1, and as will be proved in Chapter 6, an unbiased sample-design would call for a light sample from any areas that are lightly inhabited by the class of people that are to be studied. The added cost may be trifling compared with the enhanced usefulness of the data and peace of mind.

Another type of biased procedure for isolating particular classes of people is often obtained through the use of selected mailing lists. Thus the automobile registrations of a city will disclose the owners of high-priced automobiles. Interviews in a sample of such homes constitute a cheap but biased procedure for exacting information from an upper economic group. This is one kind of *chunk* (Ch. 1).

Here I have only attempted to state the basic difficulty in estimating a small proportion by sampling. There is no magic way out that I know of. The most economical solution appears to be through a combination of methods. The cheap method is exploited for all it is worth, and the more expensive method is used sparingly for finishing out the job. Mail responses from those who will respond, supplemented by face-to-face interviews of a sample of perhaps 1 in 3 of the people who failed to respond, is one application of this idea (Chs. 1 and 2). A plan was described two paragraphs above whereby a relatively heavy sample would be drawn from areas where most of a particular class of people are thought to live, and a light sample would be drawn from the other areas, instead of omitting them and running the risk of bias. Other ideas along these lines will readily suggest themselves when an unbiased procedure is really desired.

Remark. The requirement of large samples to reach high precision when the proportion p to be estimated is small is only one of the difficulties. There is also the *dilution bias* arising from the interviewer's or inspector's fatigue. After inspecting 999 pieces on the average to find 1 defective piece, an inspector is partially blind to defectives and misses some proportion of them. Similar remarks apply to the interviewing, coding, card-punching, and other operations in the preparation of statistical information concerning elements of a sample (individuals, houses, farms, etc.) that exist in great dilution. Just what proportions are lost by the dilution bias under various circumstances has not been carefully studied.

Chart for reading coefficients of variation in binomial sampling. Fig. 8 shows a chart that was in daily use at the Census during the planning and analysis of the $2\frac{1}{2}$ percent and 5 percent samples of population taken in 1940, for which the binomial theory applied acceptably for all population characteristics tabulated. If p represents the proportion of people in a population of N people having a particular characteristic, then a table made up from a perfect complete count would show Np people in the cell corresponding to that characteristic. A sample of n would have an "expected" value $Er = np$ in that cell. A single sample shows r, which when inflated by the factor N/n gives Nr/n for the cell. The following question often arises in sample censuses. A cell shows r people (or families or dwellings or farms) having a particular characteristic: what is the coefficient of variation associated with this cell-frequency? The chart in Fig. 8 was designed to answer this question. The coefficient of variation is read off in percent on the vertical scale. The lower solid line is used to find the coefficient of variation of a cell-frequency r as given by a sample, and the upper solid line is used to find the coefficient of variation of 5 percent samples, in terms of a cell-frequency in the universe, as explained further on.

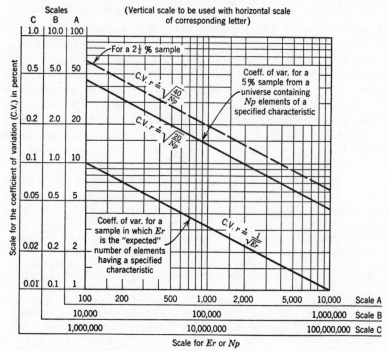

Fig. 8.　Chart for reading coefficients of variation in a 5 percent sample, calculated with binomial probabilities.　(Drawn for the author by his pupil Leonard Brickham.)

The chart is to be used for random sampling from a universe so large that the correction for nonreplacement may be ignored, in which case the binomial theory applies, giving

$$\text{C.V. } r = \sqrt{\frac{q}{np}} = \sqrt{\frac{q}{Er}} \tag{37}$$

$$= 100 \frac{1}{\sqrt{Er}} \quad \begin{matrix}\text{[In percent, provided } q \\ \text{is practically unity]}\end{matrix} \tag{38}$$

$$= 100 \sqrt{\frac{20}{Np}} \quad \begin{matrix}\text{[For a 5 percent sam-}\\ \text{ple from a universe}\\ \text{of which the propor-}\\ \text{tion } p \text{ has the speci-}\\ \text{fied characteristic]}\end{matrix} \tag{39}$$

These relations are linear on log-log paper.　With samples that are fairly large, coefficients of variation may be found closely enough by assuming that the observed r is equal to Er, and this is what must be done in practice. For example, suppose that in the sample for the state of Rhode Island (April 1, 1940) there were 413 families living in their own homes but having no member of the family in the labor force.[19] The coefficient of variation of this sample is seen from the chart to be about 5 percent.

Working the other way around, in planning a census one might wonder how reliable is a 5 percent random sample for determining the number of families having the aforesaid characteristic. On a guess that there are between 5000 and 10,000 such families in the state of Rhode Island, one could quickly see from the upper solid line of the chart that the coefficient of variation of the frequency of this class as given by a 5 percent sample would be about $5\frac{1}{2}$ percent. It is customary to accept 3 coefficients of variation (Ch. 9) as the extreme error of sampling, only rarely exceeded. A 5 percent sample for this characteristic would thus have a possible range of about 15 percent above and below Np. For the customary business-uses that are made of such figures, this margin of error is not too great, and a 5 percent sample should be considered adequate.

The dashed line was used for consideration of a $2\frac{1}{2}$ percent sample, which was taken in some areas instead of a 5 percent sample for certain family characteristics. Other parallel lines corresponding to other sizes of sample may be drawn as desired.

SOME EXERCISES ON THE BINOMIAL

Exercise 1. Find the mean and variance of the point binomial by evaluating the fundamental definitions

$$Er = \sum_{r=0}^{n} r\, P(r, n)$$

$$\sigma_r^2 = \Sigma\, (r - Er)^2\, P(r, n)$$

Solution (classical)

$$Er = \Sigma\, \frac{rn!}{(n-r)!\,r!}\, q^{n-r}p^r \qquad \left[\text{Because } P(r, n) = \frac{n!}{(n-r)!\,r!}\, q^{n-r}p^r\right]$$

$$= np\, \Sigma\, \frac{(n-1)!}{[(n-1)-(r-1)]!\,(r-1)!}\, q^{(n-1)-(r-1)}p^{r-1} \qquad \text{[This summation is unity; why?]}$$

$$= np(q+p)^{n-1} = np \quad \text{as before}$$

[19] Bureau of the Census, *Sixteenth Census:* 1940. Population and Housing, General Characteristics of Families, Table 31, p. 134. The actual number appearing in the table is 8260, this being just 20 times 413. The sample turned out to be so close to exactly 1/20th of all classes of the population that the inflation factor 20 was used uniformly, there being no necessity for adjusting to marginal totals given by the complete count. See Chapter VII in the author's *Statistical Adjustment of Data* (John Wiley, 1943).

$$\sigma^2 = \Sigma \, (r - np)^2 \binom{n}{r} q^{n-r} p^r$$

$$= \Sigma \, [r(r - 1) + r(1 - 2np) + n^2 p^2] \frac{n!}{(n - r)! \, r!} q^{n-r} p^r$$

$$= n(n - 1)p^2 \Sigma \frac{(n - 2)!}{[(n - 2) - (r - 2)]! \, (r - 2)!} q^{(n-2)-(r-2)} p^{r-2}$$

$$+ \, (1 - 2np) \Sigma \, r \, P(r, n) + n^2 p^2 \Sigma \, P(r, n)$$

$$= n(n - 1)p^2(q + p)^{n-2} + (1 - 2np)np + n^2 p^2$$

$$= n(n - 1)p^2 + (1 - 2np)np + n^2 p^2 = npq \quad \text{as before}$$

Exercise 2. Let $\hat{p} = r/n$ and $\hat{q} = (n - r)/n$ be estimates of p and q. Prove that

$$\frac{\text{C.V. } \hat{p}}{\text{C.V. } \hat{q}} = \frac{q}{p}$$

For measuring small proportions, such as the number of females in some occupation usually followed by males, p will be small, q near unity, q/p large, and the C.V. $\hat{p}$ much greater than the C.V. $\hat{q}$. Thus, while p if small can not be measured precisely without a large sample, its complement q may be measured with great precision.

Exercise 3 (Moment coefficients of the binomial universe). According to Eq. 1a on page 53, the kth moment coefficient of a universe of two cells about its mean will be

$$\mu_k = p_1(a_1 - \mu)^k + p_2(a_2 - \mu)^k$$

Now let

$$a_1 = 0 \quad a_2 = 1 \qquad \text{[These equations define}$$
$$\qquad\qquad\qquad\qquad\qquad \text{a binomial universe}$$
$$p_1 = q \quad p_2 = p \qquad \text{on the 0, 1 basis]}$$

First prove that the mean of this universe is

$$\mu = p$$

and that its variance is

$$\sigma^2 = pq$$

Then prove that

$$\mu_k = q(-p)^k + pq^k$$

These are general results, valid for any value of k. Obviously, if $p = q = \frac{1}{2}$, then $\mu_k = 0$ for all odd values of k. The following special cases should be observed.

$$k = 0, \quad \mu_0 = q \cdot 1 + p \cdot 1 = 1$$

$$k = 1, \quad \mu_1 = -pq + pq = 0 \qquad \text{[Moments being taken}$$
$$\qquad\qquad\qquad\qquad\qquad\qquad \text{about the mean]}$$

$$k = 2, \quad \mu_2 = \sigma^2 = qp^2 + pq^2$$
$$\qquad\qquad = pq(q + p) = pq$$

$$k = 3, \quad \mu_3 = -qp^3 + pq^3$$
$$\qquad\qquad = pq(q - p) \qquad \text{[This is 0 if } p = q]$$

$$k = 4, \quad \mu_4 = qp^4 + pq^4$$
$$\qquad\qquad = pq(q^3 + p^3)$$

Etc. The Pearson measure

$$\beta_2 = \frac{\mu_4}{\sigma^4} = \frac{p^3 + q^3}{pq} = \frac{1}{pq} - 3$$

As p may take any value from 0 to 1, it is obvious that β_2 can take any value from its minimum at $p = q = \frac{1}{2}$ on up to infinity as $p \to 0$ or 1. β_2 is important because it enters the formula that determines the requisite size of sample that is needed for estimating the variance σ^2 of the universe, with a prescribed precision. Further discussion on this point appears later (pp. 340 and 343).

Exercise 4. Find the mean and variance of the hypergeometric series by evaluating the fundamental definitions

$$Er = \sum_0^n r\, P(r, n, N)$$

$$\sigma_r{}^2 = \Sigma\, (r - Er)^2 P(r, n, N)$$

Solution (classical)

By canceling r out of $r!$ and pulling n, Np, and N outside the factorials, also by being careful to make suitable rearrangements, such as $[n - 1] - [r - 1]$ for $n - r$, one finds that

$$Er = \Sigma\, r\, P(r, n) = \sum r\frac{\binom{Np}{r}\binom{Nq}{n - r}}{\binom{N}{n}}$$

$$= n\frac{Np}{N}\sum\frac{\binom{Np - 1}{r - 1}\binom{Nq}{[n - 1] - [r - 1]}}{\binom{N - 1}{n - 1}}\qquad \text{[This summation is unity; why?]}$$

$$= n\frac{Np}{N} = np \qquad\qquad \text{[As before, p. 111]}$$

And likewise,

$$\sigma_r{}^2 = \Sigma\, (r - np)^2 P(r, n)$$

$$= \Sigma\, [r(r - 1) + r(1 - 2np) + n^2p^2]\frac{\binom{Np}{r}\binom{Nq}{n - r}}{\binom{N}{n}} \qquad \text{[Now make cancellations as above]}$$

$$= \frac{n(n - 1)Np(Np - 1)}{N(N - 1)}\sum\frac{\binom{Np - 2}{r - 2}\binom{Nq}{[n - 2] - [r - 2]}}{\binom{N - 2}{n - 2}}$$

$$+ (1 - 2np)\, \Sigma\, r\, P(r, n, N) + n^2p^2\, \Sigma\, P(r, n, N)$$

$$= \frac{n(n - 1)Np(Np - 1)}{N(N - 1)} + (1 - 2np)np + n^2p^2$$

$$= \frac{N - n}{N - 1}\, npq \qquad\qquad \text{[As obtained before, p. 111]}$$

Remark. The student should note that the sums

$$\sum \frac{\binom{Np-1}{r-1}\binom{Nq}{[n-1]-[r-1]}}{\binom{N-1}{n-1}}$$

and

$$\sum \frac{\binom{Np-2}{r-2}\binom{Nq}{[n-2]-[r-2]}}{\binom{N-2}{n-2}}$$

seen in the above development are both unity. The first quantity is the probability of getting $r-1$ red balls in a sample of $n-1$ from a universe of $N-1$, and the second is the probability of getting $r-2$ red balls in a sample of $n-2$ from a universe of $N-2$. The sum of each set of probabilities is unity because some number of red chips from 0 on up must appear in any sample.

Exercise 5 (Comparison of the hypergeometric series with the point binomial). *a.* Take $p = 0.2$, $q = 0.8$, $N = 50$, $n = 10$, and calculate the eleven terms of the hypergeometric series and of the binomial expansion, and compare them after filling out the table below.

	Point binomial $\binom{n}{r}q^{n-r}p^{r}$	*Hypergeometric series* $\dfrac{\binom{Np}{r}\binom{Nq}{n-r}}{\binom{N}{n}}$
0		
1		
2		
3		
4		
5		
6		
7		
8		
9		
10		
Sum	1	1
"Expected" r	$np = 2$	$np = 2$
Variance	$npq = 1.6$	$\dfrac{N-n}{N-1}npq = 1.3$

b. Show that as N increases, the terms of the hypergeometric series approach the corresponding terms of the point binomial.

Solution

$P(r, n, N)$

$$= \frac{\binom{Np}{r}\binom{Nq}{n-r}}{\binom{N}{n}}$$

$$= \frac{Np!}{r!(Np-r)!} \quad \frac{Nq!}{(n-r)!(Nq-n+r)!} \quad \frac{n!(N-n)!}{N!}$$

$$= \binom{n}{r} \frac{\left\{ \begin{matrix} Np \cdot Np-1 \cdot Np-2 \cdots Np-r+1 \\ Nq \cdot Nq-1 \cdot Nq-2 \cdots Nq-n+r+1 \end{matrix} \right\}}{N \cdot N-1 \cdot N-2 \cdots N-n+1}$$

$$= \binom{n}{r} q^{n-r} \cdot p^r$$

$$\times \frac{\left[\begin{matrix} \left(1-\dfrac{0}{Np}\right)\left(1-\dfrac{1}{Np}\right)\left(1-\dfrac{2}{Np}\right)\cdots \\ \left(1-\dfrac{r-1}{Np}\right)\left(1-\dfrac{0}{Nq}\right)\left(1-\dfrac{1}{Nq}\right)\cdots\left(1-\dfrac{n-r-1}{Nq}\right) \end{matrix} \right]}{\left(1-\dfrac{0}{N}\right)\left(1-\dfrac{1}{N}\right)\left(1-\dfrac{2}{N}\right)\cdots\left(1-\dfrac{n-1}{N}\right)}$$

$\ln P(r, n, N)$

$$= \ln \binom{n}{r} q^{n-r}p^r - \frac{1}{Np}(0+1+2+\cdots+\overline{r-1})$$

$$- \frac{1}{Nq}(0+1+2+\cdots+\overline{n-r-1})$$

$$+ \frac{1}{N}(0+1+2+\cdots+\overline{n-1})$$

$$+ \text{terms in } \frac{1}{N^2} \text{ and higher powers} \qquad \left[\begin{matrix} \text{Recall } \ln(1-x) = \\ -x-\dfrac{x^2}{2}-\dfrac{x^3}{3}-\cdots \end{matrix} \right]$$

$P(r, n, N)$

$$= \binom{n}{r} q^{n-r}p^r \left\{ 1 + \frac{1}{2N}\left[n(n-1) - \frac{r(r-1)}{p} - \frac{(n-r)(n-r-1)}{q} \right] \right.$$

$$\left. + \text{terms in } \frac{1}{N^2} \text{ and higher powers} \right\}$$

$$\rightarrow \binom{n}{r} q^{n-r}p^r \text{ as } N \rightarrow \infty$$

Thus, if N is large compared with the sample-size n, the probabilities in sampling without replacement are closely equal individually to what they would be if the sampling were done with replacement. The correction term in the brackets gives an approximate measure of this difference.

It is interesting to note that when summed over all values of r, the first correction term adds up to zero, as can be seen by writing

$$\sum_r \binom{n}{r} q^{n-r} p^r \frac{1}{2N} \left[n(n-1) - \frac{r(r-1)}{p} - \frac{(n-r)(n-r-1)}{q} \right]$$

$$= \sum_r \binom{n}{r} q^{n-r} p^r \frac{1}{2pqN} [np^2 + (q-p)r - (r-np)^2]$$

$$= \frac{1}{2pqN} [np^2 + (q-p)np - \sigma_r^2]$$

$$= \frac{1}{2pqN} [np^2 + npq - np^2 - npq]$$

$$= 0$$

The terms in $1/N^2$ and higher powers must also add up to 0. Why?

Exercise 6. Define the correlation between the x- and y-coordinates of N points x_i, y_i $(i = 1, 2, \cdots, N)$ by the usual formula

$$\rho = \frac{E x_i y_i}{\sigma_x \sigma_y}$$

wherein

$$\Sigma x_i = 0 \quad \sigma_x^2 = \frac{1}{N} \Sigma x_i^2$$

$$\Sigma y_i = 0 \quad \sigma_y^2 = \frac{1}{N} \Sigma y_i^2$$

for convenience each coordinate being measured from the centroid. Now let a series of samples of n points each be drawn without replacement; prove that the "expected" correlation between the sample means is equal to the correlation ρ between the original points. In other words, if

$$\rho_n = \frac{E \bar{x} \bar{y}}{\sigma_{\bar{x}} \sigma_{\bar{y}}}$$

[$\bar{x}$ and $\bar{y}$ are random variables. $\bar{x}\bar{y}$ is also and is to be averaged over the series of samples.]

then

$$E \rho_n = \rho$$

Solution (due to Jerome Cornfield)

For any one sample,

$$\bar{x} = \frac{1}{n} (\alpha_1 x_1 + \alpha_2 x_2 + \cdots + \alpha_N x_N)$$

$$\bar{y} = \frac{1}{n} (\alpha_1 y_1 + \alpha_2 y_2 + \cdots + \alpha_N y_N)$$

wherein α_i is a random variable such that

$$\alpha_i = 1 \quad \text{if Point } i \text{ is in the sample}$$
$$= 0 \quad \text{otherwise}$$

while $x_1, x_2, \cdots, x_N$ are fixed. Then, because α_i is either 0 or 1, it follows that

$$E\alpha_i = \frac{n}{N}$$

$$E\alpha_i^2 = \frac{n}{N}$$

Moreover,

$$E\alpha_i\alpha_j = \frac{n}{N}\frac{n-1}{N-1} \qquad \text{[Because the points are drawn without replacement]}$$

Then

$$E\bar{x}\bar{y} = \frac{1}{n^2}E(\alpha_1 x_1 + \alpha_2 x_2 + \cdots + \alpha_N x_N)(\alpha_1 y_1 + \alpha_2 y_2 + \cdots + \alpha_N y_N)$$

$$= \frac{1}{n^2}\Big[\sum_i E\alpha_i^2 x_i y_i + \sum_i \sum_j E\alpha_i\alpha_j x_i y_j\Big] \qquad \begin{array}{l}[i \text{ runs from 1 to } N; \\ j \text{ also, but } j \neq i]\end{array}$$

$$= \frac{1}{nN}\Big[\Sigma\, x_i y_i + \frac{n-1}{N-1}\Sigma\Sigma\, x_i y_j\Big]$$

$$= \frac{1}{nN}\Big[\quad\text{``}\quad + \frac{n-1}{N-1}\{\Sigma\, x_i\, \Sigma\, y_i - \Sigma\, x_i y_i\}\Big] \qquad \begin{array}{l}[\text{Because } \Sigma\, x_i = 0 \\ \text{and } \Sigma\, y_i = 0]\end{array}$$

$$= \frac{1}{nN}\Big[1 - \frac{n-1}{N-1}\Big]\Sigma\, x_i y_i$$

$$= \frac{1}{nN}\frac{N-n}{N-1}\Sigma\, x_i y_i$$

From Eq. 13' on page 102 it is known that

$$\text{Var } \bar{x} = \frac{N-n}{N-1}\frac{\sigma_x^2}{n} \qquad \text{Var } \bar{y} = \frac{N-n}{N-1}\frac{\sigma_y^2}{n}$$

and therefore

$$E\rho_n = \frac{\dfrac{1}{nN}\dfrac{N-n}{N-1}\Sigma\, x_i y_i}{\dfrac{N-n}{N-1}\dfrac{\sigma_x\sigma_y}{n}} = \rho \qquad\qquad Q.E.D.$$

Exercise 7. There are two areas, 1 and 2, containing N_1 and N_2 households respectively. At first, only Area 1 was to be sampled, and a sampling plan for Area 1 was designed to give a desired coefficient of variation C for a particular characteristic of the population, viz., the proportion of men of age 20–29 engaged in nonagricultural employment. After the plans for the sample of Area 1 had been frozen (or, in the actual incident, after the sample had been taken), a decision was made to cover Area 2 as well, aiming at the same coefficient of variation C for the two areas combined. The problem arose of determining the required size of sample in Area 2. The sampling was to be done by drawing households at random from complete lists of all the households in the two areas.

Solution

For Area 1,

$$\frac{1}{\mu_1{}^2} \frac{N_1 - n_1}{N_1 - 1} \frac{\sigma_1{}^2}{n_1} = C^2$$

Herein μ_1 is the average number of men per family, aged 20–29 and engaged in nonagricultural employment. $N_1\mu_1$ would be the total number of such men. $\sigma_1{}^2$ is the variance between households in Area 1 for this particular population characteristic, and n_1 is the number of households in the sample. The subscript 1 refers to Area 1: the subscript 2 refers to Area 2.

The problem is to determine n_2 so that

$$\frac{N_1{}^2 \dfrac{N_1 - n_1}{N_1 - 1} \dfrac{\sigma_1{}^2}{n_1} + N_2{}^2 \dfrac{N_2 - n_2}{N_2 - 1} \dfrac{\sigma_2{}^2}{n_2}}{(N_1\mu_1 + N_2\mu_2)^2} = C^2$$

The numerator is the variance of the sample in the combined areas, and the denominator is the square of the number of men in both areas having the aforesaid characteristic. By equating the two values of C and performing some algebraic reduction, first simplifying the finite multipliers to $1 - n_1/N_1$ and $1 - n_2/N_2$ respectively, the result is found to be

$$\frac{N_2}{n_2} = 1 + \left(\frac{\sigma_1}{\mu_1} \frac{\mu_2}{\sigma_2}\right)^2 \left(2 \frac{\mu_1}{\mu_2} + \frac{N_2}{N_1}\right) \left(\frac{N_1}{n_1} - 1\right)$$

This is not the value of n_2 explicitly but is the quantity really desired, because it gives the sampling interval to be applied in Area 2.

Exercise 8 (Cornfield's solution for the overexposure of large families [20]).

A sample of workers' families is to be drawn from the card-files of a factory. If two or more people of the same family work in the factory, this family is "overexposed." Cornfield provided a correction in the following form, which the student is to verify. If p is the fraction of all workers to be included in the sample, then the probability of a family with i earners being included in the sample is $1 - q^i$, where $q = 1 - p$. If there are N_i families with i earners in the factory, the "expected" number of these families to be found in the sample is $(1 - q^i)N_i$, whereas if each family had but one worker the "expected" number would be pN_i. If therefore the families of the sample are grouped by the number of earners per family, the weight of the ith group will be $p/(1 - q^i)$.

[20] Jerome Cornfield, "On certain biases in samples of human populations," *J. Amer. Stat. Assoc.*, vol. 37, 1942: pp. 63–8.

Exercise 9. A universe consists of N members, Nq of which lie at or below some value ξ, and Np of which lie above this value. A sample of n is to be drawn in such manner that all members of the universe have equal probabilities.

a. The probability that r members of the sample will exceed ξ is

$$P = \binom{n}{r} q^{n-r} p^r \qquad \text{with replacement}$$

$$= \frac{\dbinom{Nq}{n-r} \dbinom{Np}{r}}{\dbinom{N}{n}} \qquad \text{without replacement}$$

b. Put $r = n$ to find the probability that the smallest member of the sample will exceed ξ.

The N numbers might be the breaking strengths of N test-pieces. The probability just called for is the probability that all n pieces of a random sample therefrom will fall below some critical strength ξ.

c. The probability is only $(\frac{1}{2})^{10}$ that all 10 members of a sample of 10 drawn with replacement will fall below the median of a universe of N numbers.

Some instructors may wish to intercalate Parts A and B of Chapter 13 at this point.

C. THE PROPAGATION OF ERROR

An illustration of simple propagation of error. One could almost make the statement that the mathematical theory of sampling is largely a succession of problems of finding the variance of a function in terms of the variances and correlations of the random variables in the function, and the reader will perhaps become more and more convinced that this statement is not far from the truth. Thus far, for example, we have been concerned with the variances of the mean ($\bar{x}$) and sum ($n\bar{x}$) of n random variables, and of $N\bar{x}$ or the estimated population A of the universe. Further on, many more functions will be encountered, and various methods. At this point we pause *i.* to examine some general principles in the propagation of error, in order to attain a better understanding of the results and developments that will take place later; *ii.* to derive a simple method, called the *differential method*, for calculating the error in a function in terms of the errors in its variables; *iii.* to calculate some results that will be needed later, and in the process to recalculate some that have been derived on previous pages by other methods.

One or more variables are observed, all subject to the errors of sampling, and these errors are propagated into any function of these variables. A diameter is measured, perhaps several times: the measurements vary, and the area calculated from each measured diameter also varies. The measurements on the diameter might be random variables with an "expected" value and a variance. What is the "expected" area and its variance? Under such circumstances let x_i be a measurement on the radius, and let

$$Ex_i = \mu \quad \text{for any } i \tag{40}$$

and

$$E(x_i - \mu)^2 = \sigma^2 \quad \text{for any } i \tag{41}$$

Also let

$$x_i = \mu + (x_i - \mu)$$

$$= \mu + \Delta x_i \qquad [\Delta x_i = x_i - \mu = \text{the error in } x_i]$$

$$S^* = \pi\mu^2 \qquad [S^* = \text{the "true" area}]$$

$$S_i = \pi x_i^2 \qquad [S_i = \text{a calculated area}]$$

$$\Delta S_i = \pi(x_i^2 - \mu^2) \qquad [\text{The error in } S_i]$$

$$= \pi(2\mu + \Delta x_i)\,\Delta x_i \tag{42}$$

It is now to be noted that if the error Δx_i is small compared with μ or Ex_i, then approximately

$$\Delta S \doteq 2\pi\,\mu\,\Delta x \qquad [\text{Subscripts omitted for convenience}]$$

$$= \frac{dS}{dx}\,\Delta x \qquad [\text{Compare with Eq. 45}] \tag{43}$$

the derivative dS/dx or $2\pi x$ being evaluated in the neighborhood of $x = \mu$. Moreover, division of both sides by S or πx^2 shows that approximately

$$\frac{\Delta S}{S} \doteq 2\frac{\Delta x}{x} \tag{44}$$

saying that an error of 1 percent in the measurement of the radius corresponds to about 2 percent error in the area.

In general, if the variables x, y, z are subject to small errors $\Delta x, \Delta y, \Delta z$, a function f will be subject to the error

$$\Delta f = f_x\,\Delta x + f_y\,\Delta y + f_z\,\Delta z + R \tag{45}$$

It will be assumed that the remainder R, arising from the higher powers of the Taylor series, may be neglected.

The derivatives f_x, f_y, f_z are to be evaluated in the neighborhood of $x = Ex$, $y = Ey$, $z = Ez$.

Exercise. The area of an ellipse is $v = \pi xy$. Show that approximately

$$\frac{\Delta v}{v} \doteq \frac{\Delta x}{x} + \frac{\Delta y}{y}$$

The right-triangular relation connecting the mean square error, the variance, and the bias of an estimate. In Chapter 1 the idea of a *preferred procedure* was introduced (p. 15), and this led to the definition of a bias as an "expected" departure from the "expected" result of the preferred procedure. An important relation exists between the "expected" mean squares. Thus, let

$a = $ the "expected" value of a preferred procedure

$Ex = $ the "expected" value of some particular procedure of sampling, interviewing, and estimating, not necessarily preferred

Then, as the student has already proved on page 72,

$$E(x - a)^2 = E[(x - Ex) + (Ex - a)]^2$$
$$= E(x - Ex)^2 + (Ex - a)^2 \qquad (46)$$

That is, term for term,

$$\delta_x^2 = \sigma_x^2 + b_x^2 \qquad (47)$$

wherein δ_x is the *root-mean-square error* of x, σ_x is the *standard error* of x, and b_x is the *bias* of x (more precisely, the root-mean-square error, standard error, and bias of the *procedure* that gives x). The relation

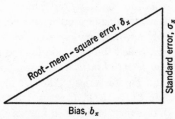

Bias, b_x

FIG. 9. The triangular relation connecting the bias, standard error of sampling, and the resultant root-mean-square error of an estimate x. It is to be noted that the leg representing the standard error may be shortened by increasing the size or complexity of the design of the sample, whereas the leg representing the bias may be shortened only by introducing an improved procedure of survey, such as a revised questionnaire or a new method of asking for the information, or better performance in the field-work, or some other improvement. The resultant or root-mean-square error in x can not be shorter than the leg representing the bias, and in fact for moderately small sampling errors, the resultant will be but little longer than the bias.

connecting these three quantities δ_x, σ_x, and b_x is obviously the Pythagorian relation connecting the legs and hypotenuse of a right triangle, as shown in Fig. 9. This relationship has already been referred to in Chapter 2, where it was stated that the aim in sample design should be to control the hypotenuse and not merely the leg representing the sampling error.

Figure 6 on page 72 may be of further interest here. It may be interpreted to show that, for a fixed sampling procedure and sample-size, and hence for a given value of σ_x, the hypotenuse of the triangle of Fig. 9 is increased but little beyond the standard error σ_x by the presence of small biases. However, large biases may completely eclipse the sampling error.

The propagation of mean square error and variance. Let both members of Eq. 45 be squared and the expected value of every term be taken. Then

$$\sigma_f^2 = (f_x\sigma_x)^2 + (f_y\sigma_y)^2 + (f_z\sigma_z)^2 + 2f_xf_y\rho_{xy}\sigma_x\sigma_y + 2f_xf_z\rho_{xz}\sigma_x\sigma_z$$
$$+ 2f_yf_z\rho_{yz}\sigma_y\sigma_z \quad (48)$$

wherein

$$
\left.
\begin{array}{ll}
\sigma_f^2 = E\,\Delta f^2 = E(f - Ef)^2 & \text{[The variance of } f\text{]} \\[2mm]
Ef = f(Ex, Ey, Ez) & \begin{array}{l}\text{[An approximation to the}\\ \text{``expected'' value of } f\text{]}\end{array} \\[2mm]
\sigma_x^2 = E\,\Delta x^2 & \text{[The variance of } x\text{]} \\[2mm]
\rho_{xy} = \dfrac{E\,\Delta x\,\Delta y}{\sigma_x\sigma_y} & \begin{array}{l}\text{[The correlation between}\\ \Delta x \text{ and } \Delta y\text{]}\end{array}
\end{array}
\right\} \quad (49)
$$

When Δx, Δy, and Δz are all uncorrelated, the terms involving ρ drop out. Equation 48 is the equation for the *propagation of mean square error*. The distinction between mean square error and variance involves only moment coefficients of Δx, Δy, Δz already in the remainder, and Eq. 48 is therefore often called the equation for the *propagation of variance*. It is called the *differential method* because it is based on the use of the differentials in the Taylor series. Another method, already used in Eq. 10, is to calculate $E(f - Ef)^2$ directly, but this can not be done unless the probability of every variable is known. The characteristic function, not to be treated here, provides still another device, but there the problem is to find the distribution of f and not merely its second moment. The differential method has the advantage of giving results even when the distributions of x, y, z and the function f are not known; but because the higher powers of Δx, etc., are neglected, it has the disadvantage for nonlinear functions in giving results whose validity is sometimes difficult to calculate. In spite of the approximate character of the derivation for nonlinear functions, Eq. 48 nevertheless often gives remarkably good results for most of the functions and the sample-sizes met in practice. This happens, for example, when x, y, z are means of samples, and Δx, Δy, Δz are statistical fluctuations above and below the "expected" values of x, y, z. By making the sample-sizes big enough, the remainder, consisting of the terms in the second and higher powers of Δx, Δy, Δz in the Taylor series, can be reduced below any prescribed small numbers, although for any given function it is difficult to calculate in advance just what size of sample is required, as the answer depends on the distributions of Δx, Δy, Δz.

Exercise 1. A universe consists of N numbers $a_i (i = 1, 2, \cdots, N)$, not necessarily all different. A sample of n is to be drawn at random. Let the members of the sample be $x_1, x_2, \cdots, x_n$. As x_i is a random variable, let

$$\left. \begin{aligned} Ex_i &= \mu \\ \sigma_i^2 = E(x_i - \mu)^2 &= \sigma^2 \end{aligned} \right\} \text{for every } i$$

Put

$$f = x_1 + x_2 + \cdots + x_n$$

Find the "expected" value and variance of f. As $Ex_i = \mu$ for all i, it follows that

$$Ef = n\mu$$

whether the drawings be made with or without replacement. For the variance,

$$\sigma_f^2 = n\sigma^2 + \sum_{i \neq j}^n \rho_{ij}\sigma_i\sigma_j$$

wherein ρ_{ij} denotes the correlation between two successive drawings, and $\sigma_i = \sigma_j = \sigma$. By symmetry, ρ_{ij} is constant regardless of ij, wherefore ρ_{ij} may be put equal to ρ and

$$\sigma_f^2 = n\sigma^2 + n(n - 1)\rho\sigma^2$$
$$= n\sigma^2[1 + (n - 1)\rho]$$

  a. With replacement. Here the drawings are independent, and $\rho = 0$ so

$$\sigma_f^2 = n\sigma^2$$

as was seen before (p. 102).

  b. Without replacement. First evaluate ρ by noting that if the universe were exhausted by the sample, f would equal the total population A of the universe, regardless of the order in which the N members of the sample were drawn. Then as

$$x_1 + x_2 + \cdots + x_N = A$$

it follows that because A is constant,

$$\sigma_A{}^2 = 0 = N\sigma^2 + \Sigma\,\rho\sigma^2$$
$$= N\sigma^2 + N(N-1)\rho\sigma^2$$

whence

$$\rho = -\frac{1}{N-1}$$

Then, for a sample of n,

$$\sigma_f{}^2 = n\sigma^2\left[1 - \frac{n-1}{N-1}\right]$$
$$= \frac{N-n}{N-1}\,n\sigma^2$$

agreeing with Eq. 12' on page 102.

Exercise 2. Prove directly that $\rho = -1/(N-1)$, as ρ is defined in the preceding exercise.

Solution

$$\rho = \frac{E(x_i - \mu)(x_j - \mu)}{\sqrt{E(x_i - \mu)^2 E(x_j - \mu)^2}} \qquad \text{[Definition]}$$

$$= \frac{\dfrac{1}{N(N-1)}\displaystyle\sum_{i=1}^{N}\sum_{j=1}^{N}(a_i - \mu)(a_j - \mu)}{\sigma^2} \qquad [j \neq i]$$

$$= \frac{\dfrac{1}{N(N-1)}\left\{[\Sigma(a_i - \mu)]^2 - \Sigma(a_i - \mu)^2\right\}}{\sigma^2}$$

$$= \frac{\dfrac{1}{N(N-1)}\left\{0 - N\sigma^2\right\}}{\sigma^2}$$

$$= -\frac{1}{N-1} \qquad\qquad Q.E.D.$$

Exercise 3. Neglect the bias (if any) in the following functions (which is the same as assuming $\sigma_f = \delta_f$ for any function), and by using Eq. 48 obtain the following relations between variances.

a. Let $r = n\sin^2\theta$.

If

$$Er = np \qquad \text{as for the binomial (p. 111)}$$
$$\sigma_r{}^2 = npq \qquad \text{`` `` `` ``}\ \Big\}$$

then

$$\sigma_\theta{}^2 = \frac{1}{4n} \qquad\qquad [\theta \text{ in radians}]$$

Solution

If $r = n \sin^2 \theta$, then $\sin \theta = \sqrt{r/n}$ and $\cos \theta = \sqrt{1 - r/n}$. Then, by Eq. 48,

$$\sigma_r = \left(\frac{dr}{d\theta}\right)_{r=Er} \sigma_\theta$$

$$= 2n \, \sigma_\theta [\sin \theta \cos \theta]_{r=Er}$$

$$= 2n \sqrt{\frac{Er}{n}} \sqrt{1 - \frac{Er}{n}} \, \sigma_\theta$$

$$= 2n\sqrt{pq} \, \sigma_\theta$$

But $\sigma_r = \sqrt{npq}$, so

$$\sigma_\theta = \frac{1}{2\sqrt{n}}$$

which is equivalent to the result sought. This is the standard error for the arc sine transformation which is to be used later.

b. Linear function

$$f = ax + by + cz$$

$$\sigma_f{}^2 = (a\sigma_x)^2 + (b\sigma_y)^2 + (c\sigma_z)^2 + 2ab\rho_{xy}\sigma_x\sigma_y + 2ac\rho_{xz}\sigma_x\sigma_z$$
$$+ 2bc\rho_{yz}\sigma_y\sigma_z$$

c. Product

$$f = axyz$$

$$c_f{}^2 = c_x{}^2 + c_y{}^2 + c_z{}^2 + 2\rho_{xy}c_xc_y + 2\rho_{xz}c_xc_z + 2\rho_{yz}c_yc_z$$

wherein c denotes coefficient of variation.

d. Quotient

$$f = a\frac{xy}{wz}$$

$$c_f{}^2 = c_x{}^2 + c_y{}^2 + c_w{}^2 + c_z{}^2 + 2\rho_{xy}c_xc_y - 2\rho_{xw}c_xc_w - 2\rho_{xz}c_xc_z$$
$$- 2\rho_{yw}c_yc_w - 2\rho_{yz}c_yc_z + 2\rho_{wz}c_wc_z$$

e. Product of powers

$$f = ax^\alpha y^\beta z^\gamma$$

$$c_f{}^2 = \alpha^2 c_x{}^2 + \beta^2 c_y{}^2 + \gamma^2 c_z{}^2 + 2\alpha\beta\rho_{xy}c_xc_y + 2\alpha\gamma\rho_{xz}c_xc_z + 2\beta\gamma\rho_{yz}c_yc_z$$

f. In particular, if $f = x^{\frac{1}{2}}$, $c_f = \frac{1}{2}c_x$. This theorem will be encountered a number of times in the book. For example, on page 340, C.V. $s^2 \doteq \sqrt{(\beta_2 - 1)/k}$, and it is stated there that C.V. s is closely half as much.

See also pages 526–28, where the series for Var s^2 and Var s may be compared; also Var δ^2 and Var δ.

g. If $f = \ln x/a$

$$\sigma_f = c_x$$

Exercise 4. Verify the fact that 10 percent is the resultant coefficient of variation in the estimated total of all the uses of forest timber listed in the left-hand column. (This table arose in the design of a sample for timber-drainage in the state of Illinois, 1948, and is presented through the kindness of Messrs. Edward C. Crafts and Roy A. Chapman of the Forest Service.)

Commodity	Estimated percent of total	Allowable coefficient of variation
Lumber	45	2
Fuel wood	30	25
Fence posts	10	40
Tight cooperage	3	75
Mining timber	3	75
Veneer	2	100
Pulpwood	2	100
Handle stock	1	100
Poles and piling	1	100
Slack cooperage	1	100
Miscellaneous	2	100
Total	100	10

CHAPTER 5. MULTISTAGE SAMPLING, RATIO-ESTIMATES, AND CHOICE OF SAMPLING UNIT

Research in statistical theory and technique is necessarily mathematical, scholarly, and abstract in character, requiring some degree of leisure and detachment, and access to a good mathematical and statistical library. The importance of continuing such research is very great, although it is not always obvious to those whose interest is entirely in practical applications of already existing theory. Excepting in the presence of active research in a pure science, the applications of the science tend to drop into a deadly rut of unthinking routine, incapable of progress beyond a limited range predetermined by the accomplishments of pure science, and are in constant danger of falling into the hands of people who do not really understand the tools that they are working with and who are out of touch with those that do. . . . . It is in fact rather absurd, though quite in line with the precedents of earlier centuries, that scientific men of the highest talents can live only by doing work that could be done by others of lesser special ability, while the real worth of their most important work receives no official recognition.—Harold Hotelling, *Memorandum to the Government of India*, 24 Feb. 1940 (by permission of the author).

A. MULTISTAGE SAMPLING

Advantages of two or more stages of sampling. Up to this point the sampling under consideration has been single-stage sampling: units were drawn into the sample and the population of each unit was determined by canvassing it completely without recourse to further sampling. But unless the units are small, a prescribed precision may sometimes be met cheaper by drawing more units into the sample and then estimating their populations by *subsampling* them. Such a plan is a *2-stage* plan. The original units into which the universe is divided are *primary* units. Each primary unit that falls into the sample is subdivided into *secondary* units in preparation for the 2d stage of sampling. In 3-stage sampling there will be primary, secondary, and tertiary units. Sometimes four stages are used. Sampling in stages has in fact long been used in one way or another, but it has recently been exploited by Hansen and colleagues with new techniques, one being the device of creating large heterogeneous primary units, with advantages that may become apparent later.

The *frame*, as in Chapters 2 and 4, is a workable description of the sampling units. In multistage sampling a frame must be found or constructed for every sampling unit that is to be sampled. Thus, to start with, there must be a frame that describes all the primary units in the universe. Then, for each primary unit that falls into the sample there must be a frame that describes the secondary units. For each secondary unit that falls into the sample there must be ,a frame that describes the tertiary units. At every successive stage the sampling units become smaller and smaller, and the frames more and more detailed. Finally, at the last stage, the frame may show the ultimate units, which might be small areas, single households, or several successive households.

One of the main advantages of multistage sampling is that preparation of the frames for the next stage is required only in the units that have already fallen into the sample.

The subunits within any larger unit must be exhaustive; together they must account for the whole of the larger unit, so that every square foot of area, every person, every farm, every business establishment, every article, every bale, etc., is in one and only one primary unit, and in one and only one subunit at the 2d, 3d, or any other stage. Otherwise the probability of any one person, d.u., farm, etc., coming into the sample will not correspond with the premises on which the mathematical theory will be built, and the calculations of the errors of sampling will be invalid.

From the standpoint of sheer efficiency of sampling, small units such as households or very small areas, widely dispersed, drawn by a sample of one stage, appear to be excellent. There are three reasons, however, for hesitating to recommend such a plan. First, the cost of travel would be too high in proportion to the amount of interviewing that is done. Second, control of the nonsampling errors would be difficult and costly (Ch. 2): a supervisor should know what is going on in the areas under his charge, and if the areas are small and widely dispersed he may spend nearly all of his time traveling and scarcely any time actually supervising. Third, a probability-sample of small units drawn at one stage, no matter how thin the sample, requires a frame which lists *all* the small units: such a frame may be costly beyond reason. These arguments happily break down when sampling is used as part of a complete census: the complete census then provides the frame and certain basic marginal totals; it provides also field-workers and supervisors, who would be employed anyway, sample or no sample.

It is sometimes supposed that through the use of a mailed questionnaire it is possible to avoid the problems just mentioned. It is to be noted, however, that if a probability-sample by mail is intended, an even more careful basic listing of addresses is required than is needed for interviews. An interviewer may collect information from two or more households that are found listed under one address (one sampling unit), but a mailed questionnaire can not so split and direct itself. Any household not listed has no chance of getting into a mailed sample. Moreover, if the sample is to be a probability-sample, it must be expected that field-work will be required to seek interviews in households that do not respond at all or in full, or from which some of the information is apparently faulty. A well-executed single-stage mailed questionnaire may thus turn out to be an expensive affair unless by good fortune the costs of the basic listing and of the field-work are in some way provided by outside sources.

When the cost of providing the frame is chargeable to the sample, it is necessary to remember that the cost of listing the households over an entire city increases with the size of the city, whereas the sample that is to be drawn and interviewed remains nearly constant, being affected only through the finite multiplier $(N - n)/(N - 1)$ of Chapter 4. For a city of 500,000 inhabitants the cost of listing may far outrun all the other costs; and for a bigger city the discrepancy is even more glaring. A point in population is reached at which it is better to draw out a sample of several hundred areas and perform the listing of households only in these areas [1]—in other words, to introduce two stages of sampling. For reasons mentioned in the preceding paragraph, the argument is the same whether the survey is to be accomplished by interviews or by mail.

Multistage sampling is no less useful in the testing of incoming or outgoing industrial product than it is in surveys of inventories and in social and economic surveys, and the theory is the same. In the sampling of industrial product the primary unit may be bags, bales, or boxes, or groups or layers of bags, bales, or boxes. The secondary sample may be bags, bales, or boxes, or handfuls, small boxes, or other samples such as individual articles drawn according to certain rules from within those primary units which have themselves been drawn into the sample. Once defined, the sampling units possess certain variances (σ_e^2, σ_b^2, σ_w^2, defined in the notation, p. 142) for any particular population characteristic. For any given plan of subsampling the magnitude of the sampling error can be computed in terms of these variances. Moreover, with information on the costs of interviewing and traveling from one secondary unit to another, theory tells how many primary units and how many secondary units to draw into the sample to achieve the greatest possible precision for an allowable cost.

For a national sample of households the country might first be divided into primary areal units, e.g., counties or combinations of dissimilar counties to achieve heterogeneity. Small counties might be tied to larger ones to form a new "county," and were it possible, extra-large counties might be broken into smaller ones to roughly equalize the populations in the primary units. From each stratum a predesignated

[1] *A Chapter in Population Sampling* (Bureau of the Census, 1947), p. 13 and elsewhere.

number of "counties" might be drawn into the sample.[2] A primary unit of this description is not too large to be supervised with care, but it is too large to be completely enumerated. Subsampling provides a way out of the difficulty. Each primary unit (county) is divided into subunits of some kind, and a sample of these subunits is selected, which in turn might be subdivided and sampled again. Eventually, after a nest of sampling operations, the ultimate units are reached in the form of households. The number of ultimate units will be large, while the primary units may be made the right size for economical and careful supervision, yet sufficient in number to constrain the sampling error within the allowable tolerances.

A typical 2-stage plan for a city might be described as follows: *i*. Obtain a map and divide the city into primary units of roughly equal numbers of inhabitants or d.us. (Remark 1, p. 82). The primary units may be numbered in serpentine fashion from 1 to M. *ii*. Draw a specified number m of primary units into the sample—this is usually done by marking the map in a systematic manner, using a constant interval rounded downward from the computed value of M/m. *iii*. Prepare maps or lists of each of the m primary units and subdivide them into secondary units of some kind, such as small areas or single d.us. or possibly uniform "clusters" of d.us. (to be studied later in this chapter): this may be done from detailed commercial maps or from maps made hastily on the spot (see Fig. 10). *iv*. Draw a sample of the secondary units and canvass them to elicit the information desired. (These secondary units will in turn sometimes be further subdivided for another stage of sampling.)

Alternative point of view. The following description has often been helpful to the author in illustrating what takes place when steps are taken to decrease the external variance $\sigma_e{}^2$ by forming large heterogeneous primary units.

Let us return to the ideal-bowl experiment of Chapter 4 and see what would happen if, before any samples are drawn, red and white chips could be paired off in approximately the proportions in which they actually exist initially in the bowl. As a concrete example, suppose that for every red chip there are 3 white chips; then initially $p = \frac{1}{4}$, $q = \frac{3}{4}$. Suppose now that each red chip be consolidated with 3 white ones to make a new sampling unit. The new units will be uniform, in this ideal case, and it will be observed that samples of any number

---

[2] This is a brief description of the sample devised by Morris H. Hansen for the Current Population Survey (giving the Monthly Report on the Labor Force), a monthly miniature sample census of the United States. This monthly survey was initiated in 1939 and planned along somewhat different lines by Lester R. Frankel and J. Stevens Stock.

of them will give the proportion red in the universe without any sampling error whatever. The reason is that the elements are now all alike, having been made so by joining the red and the white in the proper proportions.

Of course, in practice one cannot combine chips of dissimilar colors in precisely the proportions in which they are found in the universe, but

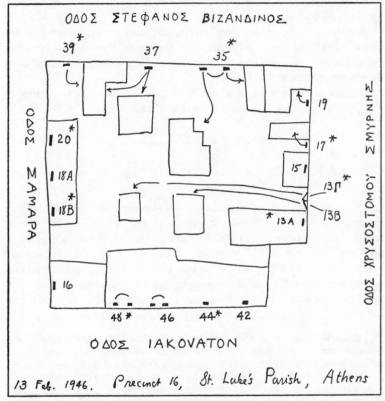

FIG. 10. A "map-list" which was made by the author on a cold, blustery, rainy day in Athens in February 1946. A map-list is usually rough, as it need only definitely identify every d.u. or small group of d.us. within a prescribed area, so that any d.u. or group, if drawn into the sample, may be found unequivocally. Here the d.us. or groups of d.us. were identifiable by legible street numbers on the gates or doors, but sometimes the maker of the map must describe them by location or other physical characteristics, and by assigning numbers arbitrarily. A rough but entirely serviceable map-list like this can be made quickly on the spot, without ringing door bells or asking questions. The preassigned sampling interval is then applied with a random start to select the d.us. for the sample, whereupon the interviewing may be commenced at once. In this particular area the counting interval was 2 and the d.us. in the sample were marked by an asterisk. The author has successfully made similar map-lists elsewhere, as in Bangalore in 1947 for a pilot-study of a survey of the consumption of food in wage-earners' homes.

he can take steps in this direction. Such steps are accomplished by the statistician through his ingenuity at finding and making maps, lists, and data giving population and agricultural characteristics of various areas, and through uses of lists of business establishments, giving not only names and addresses, but characteristics as well, such as type of business, number of employees, total annual sales, inventory, annual pay-roll, capitalization.

Of course, maps, census information, and lists are expensive, but the cost of obtaining them and using them may actually result in large net savings in sampling enterprises, particularly if the costs can be distributed over a number of surveys.

Another way to view this aspect of sample-design is to return to Eq. 16 of Chapter 4, in which it was found that

$$\text{C.V. } \bar{x} = \frac{\text{C.V. universe}}{\sqrt{n}}$$

When sampling without replacement there is of course the finite multiplier to be introduced, but for simplicity it will be ignored here.

It will be observed from this equation that there are two ways of reducing the coefficient of variation of $\bar{x}$; one is to *increase* the sample-size n, and the other is to *decrease* the numerator on the right-hand side—that is, decrease the coefficient of variation in the universe. For many years statisticians have exploited the simple idea of decreasing the variance of $\bar{x}$ by increasing the size of the sample, but it is only recently that effort has been directed toward obtaining more information per unit cost through clever definitions of the sampling units.

Remark. Areas bounded by streets (called blocks in America) will for many surveys be a natural primary unit for urban population enquiries. Since such areas may contain very unequal numbers of d.us., people, and business establishments, equalization may be attempted by subdivision, or the blocks may be stratified, or drawn with probability in proportion to size (Ch. 12). If no census information is at hand by which approximate sizes can be ascertained block by block, a quick cruise by automobile and a very rough count of the d.us. (or business establishments in a business enquiry) is usually sufficient (Ch. 12).

Unpopulated and sparsely populated areas should be tied to populated areas or sampled separately; they must not be omitted, because what appears to be unpopulated on the map may since have become very much populated. In fact, huge housing projects are a perennial source of difficulty as they must be discovered, isolated, and sampled separately.

Subdivision of blocks and the tying of one block to another may be indicated on the map with a heavy pencil in hand during a quick cruise. Fortunately, as already remarked, only approximate uniformity is required even if no stratification is used, and for stratified sampling or drawing with probability in proportion to size only approximate prior measures of size are required. Obsolete census figures, unless great changes have taken place, are good enough. With proper design and tabulation, nonuniformity and inexactness of prior information cancel out and do not cause bias in the final results. Even an old city directory can be used to advantage.[3]

Notation for multistage sampling. Fig. 11 shows a universe which might be a county or city in which M primary areal sampling units have been delineated. Exclusions of territory known to be uninhabited are presumed to have been made (Ch. 2). In the sampling of industrial product,[4] the illustration might refer to a lot or shipment that has been sectioned off into layers or groups of containers, or even single containers numbered from 1 to M. Within each primary unit are a number (N_j) of secondary units, which might be households, groups of households, small areas, or (in industrial sampling) bales, boxes, or individual articles to be tested.

It will be assumed here that in the drawing of the m primary units, any particular unit has the same probability as any other of coming into the sample. Likewise, within any primary unit, any particular secondary unit has the same probability as any other. In other words, all the sampling is to be done with equal probabilities. In good sample-design equal probabilities are assigned only on the premise that the populations of the primary units and of the secondary units within a primary unit are roughly equal, or at least not so dissimilar that the variances σ_e^2 and σ_j^2 will be large, producing large sampling errors in the final results. The delineation and definition of both primary and secondary

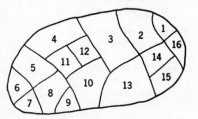

FIG. 11. Delineation of primary sampling units. Here are M = 16 primary sampling units, and they are listed in serpentine fashion so as to gain the benefits of geographic stratification when the sample of primary units is drawn.

sampling units should therefore be carried out with the aim of making them as nearly alike as feasible. If natural boundaries define one or more primary units that are thought to contain roughly twice the population of the others, these units may be given two chances of coming into the sample. This can be done by subdividing a large primary unit into two primary units, preferably each part interpenetrating the other, as by assigning the odd-numbered subunits within it to one primary unit and

---

[3] Some remarks concerning the use of directories were given in Chapter 4, page 82 in particular.

[4] Some of the difficulties of devising suitable units in industrial product are pictured by Shewhart on pages 408–17 of his *Economic Control of Quality of Manufactured Product* (Van Nostrand, 1931).

TABLE OF NOTATION FOR SAMPLING IN TWO STAGES

A universe consists of M primary units. From a random sample of m of them, random subsamples are drawn, consisting of n_j secondary units from the primary unit that was drawn at the jth draw. n_j is a number predetermined by some rule for each of the M primary units.

Name	Universe	Sample
Number of primary units	M	m
Number of secondary units in the jth primary unit	N_j	n_j
Total number of secondary units	$N = \sum_1^M N_j$	$n = \sum_1^m n_j$
Average number of secondary units per primary unit	$\bar{N} = \dfrac{N}{M}$	$\bar{n} = \dfrac{n}{m}$
Population of the x-characteristic		
In the kth secondary unit of the jth primary unit	$a_{jk} \begin{cases} j = 1, 2, \cdots, M \\ k = 1, 2, \cdots, N_j \end{cases}$ (Listed in some specified order)	$x_{jk} \begin{cases} j = 1, 2, \cdots, m \\ k = 1, 2, \cdots, n_j \end{cases}$ (Listed in the order drawn)
In the jth primary unit	$A_j = \sum_{k=1}^{N_j} a_{jk}$	$x_j = \sum_{k=1}^{n_j} x_{jk}$ $X_j = N_j \bar{x}_j = \dfrac{N_j}{n_j} x_j,$ an unbiased estimate of A_j because $EX_j = A_j.$
In all primary units	$A = \sum_1^M A_j$	$x = \sum_1^m x_j$ $X = \dfrac{M}{m} \sum_1^m X_j$ an unbiased estimate of A, because $EX = A.$

Name	Universe	Sample
		$$X' = N\frac{\sum_1^m \cdot X_j}{\sum_1^m N_j}$$ If N_j is variable this is a biased but consistent * estimate of A, as $EX' = A$ if $m = M$.
Average population per primary unit	$\overline{A} = \dfrac{A}{M}$ $= \dfrac{1}{M}\sum_1^M A_j$ $= \dfrac{1}{M}\sum_1^M N_j\mu_j$	$\overline{X} = \dfrac{X}{M}$ $= \dfrac{1}{m}\sum_1^m X_j$ $= \dfrac{1}{m}\sum_1^m N_j\bar{x}_j$ an unbiased estimate of $\overline{A}$, because $E\overline{X} = \overline{A}$. $\overline{X}' = \dfrac{X'}{M}$ a biased but consistent estimate of $\overline{A}$.
Average population per secondary unit in the jth primary unit	$\mu_j = \dfrac{A_j}{N_j}$	$\bar{x}_j = \dfrac{x_j}{n_j}$ an unbiased estimate of μ_j because $E\bar{x}_j = \mu_j$.
Average population per secondary unit, weighted	$\mu = \dfrac{A}{N} = \dfrac{\overline{A}}{\overline{N}}$	$\bar{x} = \dfrac{X}{N} = \dfrac{\overline{X}}{\overline{N}}$ an unbiased estimate of μ because $E\bar{x} = \mu$. $\bar{x}' = \dfrac{\sum_1^m x_j}{\sum_1^m n_j} = \dfrac{x}{n}$ a ratio-estimate of μ, biased but consistent.*

* R. A. Fisher, "On the mathematical foundations of theoretical statistics," *Phil. Trans. Royal Soc.,* vol. A222, 1922: pp. 309–68.

Name	Universe	Sample
Average population per secondary unit, unweighted	$\mu_u = \dfrac{1}{M} \displaystyle\sum_1^M \mu_j$ (If N_j is a constant, μ_u is the same as μ)	$\bar{\bar{x}} = \dfrac{1}{m} \displaystyle\sum_1^m \bar{x}_i,$ an unbiased estimate of μ_u because $E\bar{\bar{x}} = \mu_u.$
For the populations of the y-characteristic, replace	a_{jk} by b_{jk} A_j by B_j A by B $\bar{A}$ by $\bar{B}$	x_{jk} by y_{jk} x_j by y_j X_j by Y_j X by Y $\bar{X}$ by $\bar{Y}$
Ratio of the x-population to the y-population		
In the jth primary unit	$\varphi_j = \dfrac{A_j}{B_j}$	$f_j = \dfrac{X_j}{Y_j},$ a biased but consistent estimate of φ_j.
In all primary units	$\varphi = \dfrac{\displaystyle\sum_1^M A_j}{\displaystyle\sum_1^M B_j}$	$f = \dfrac{\displaystyle\sum_1^m X_j}{\displaystyle\sum_1^m Y_j},$ a biased but consistent estimate of φ.
Variances		
Internal variance of the population per secondary unit within the jth primary unit	$\sigma_j{}^2 = \dfrac{1}{N_j} \displaystyle\sum_{k=1}^{N_j} (a_{jk} - \mu_j)^2$	$s_j{}^2 = \dfrac{1}{n_j} \displaystyle\sum_{k=1}^{n_j} (x_{jk} - \bar{x}_j)^2$
External variances		
Of the population per primary unit	$\sigma_e{}^2 = \dfrac{1}{M} \displaystyle\sum_1^M (A_j - \bar{A})^2$	$s_e{}^2 = \dfrac{1}{m} \displaystyle\sum_1^m (X_j - \bar{X})^2$
Of the population per secondary unit		
Weighted	$\sigma_b{}^2 = \dfrac{1}{M\bar{N}} \displaystyle\sum_1^M N_j(\mu_j - \mu)^2$	$s_b{}^2 = \dfrac{1}{m\bar{n}} \displaystyle\sum_1^m n_j(\bar{x}_j - \bar{\bar{x}})^2$
Unweighted	$\sigma_{bu}{}^2 = \dfrac{1}{M} \displaystyle\sum_1^M (\mu_j - \mu_u)^2$	$s_{bu}{}^2 = \dfrac{1}{m} \displaystyle\sum_1^m (\bar{x}_j - \bar{\bar{x}}_u)^2$

Name	Universe	Sample
Average internal variance		
Weighted	$\sigma_w{}^2 = \dfrac{1}{N} \displaystyle\sum_1^M N_j \sigma_j{}^2$	$s_w{}^2 = \dfrac{1}{n} \displaystyle\sum_1^m n_j s_j{}^2$
Unweighted	$\sigma_{wu}{}^2 = \dfrac{1}{M} \displaystyle\sum_1^M \sigma_j{}^2$	$s_{wu}{}^2 = \dfrac{1}{m} \displaystyle\sum_1^m s_j{}^2$
Total variance of the population per secondary unit		
Weighted	$\sigma^2 = \dfrac{1}{M\overline{N}} \displaystyle\sum_1^M \sum_{k=1}^{N_j} (a_{jk} - \mu)^2$ $= \sigma_b{}^2 + \sigma_w{}^2$	$s^2 = \dfrac{1}{m\overline{n}} \displaystyle\sum_1^m \sum_{k=1}^{n_j} (x_{jk} - \bar{x})^2$ $= s_b{}^2 + s_w{}^2$
Unweighted	$\sigma_{tu}{}^2 = \dfrac{1}{M} \displaystyle\sum_1^M \dfrac{1}{N_j} \sum_{k=1}^{N_j} (a_{jk} - \mu_u)^2$ $= \sigma_{bu}{}^2 + \sigma_{wu}{}^2$	$s_{tu}{}^2 = \dfrac{1}{m} \displaystyle\sum_1^m \dfrac{1}{n_j} \times \sum_{k=1}^{n_j} (x_{jk} - \bar{x})^2$ $= s_{bu}{}^2 + s_{wu}{}^2$

Not all of these variances will be utilized in the text: some are more applicable in the design of experiment. They are introduced here to show some distinction between various types of statistical problems.

The definition of "population" as given on page 84 should be reread carefully. In a population enquiry, a_{jk} is not necessarily the total number of people in a secondary unit; it is rather by definition the number of people in it who possess a particular characteristic, such as male, 20–29, employed. A household might contain 5 people, but no male, 20–29, employed, in which case $a_{jk} = 5$ in a count of the total population and 0 in a count of the number of males 20–29, employed. In other words, a_{jk} depends on the particular characteristic that is being evaluated, and there are as many values of a_{jk} as there are characteristics. For the sake of simplicity, the discussion will proceed as if only one characteristic were of interest. In counting defectives in a container of industrial product, a_{jk} will be the number of defectives in a tray or container of some sort. If the secondary unit is an individual article, $a_{jk} = 1$ if the kth article is defective, 0 if not defective.

the even-numbered ones to another. The new primary areas are thus dispersed over the entire parent area. A similar plan can be carried out by assigning three chances to any primary unit that is thought to contain roughly three times the population of those primary units that are to have one chance of coming into the sample: any such triple-sized primary unit might be subdivided into three primary units, preferably by assigning secondary units 1, 4, 7, 10, $\cdots$ to one unit; 2, 5, 8, 11, $\cdots$ to another; 3, 6, 9, 12, $\cdots$ to another. The actual subdivision need be carried out only for those primary units that actually fall into the sample.

This plan amounts to sampling with probability in proportion to size, a recent invention of my colleague Morris H. Hansen. Sampling with probability in proportion to size largely eliminates the need for delineating primary areas of roughly equal size and permits the use of natural and convenient boundaries, even though the primary units are of widely differing populations. The formulas [5] for sampling with probability in proportion to size will not be developed here although the procedure will be illustrated in Chapter 12.

It will be necessary to have two sets of "expected" values, because there are two stages of sampling. E_j will denote an "expected" value arising from the sampling of primary units. E_k will denote an "expected" value arising from the sampling of secondary units within a particular primary area, and E_{jk} or merely E will denote an "expected" operation of both operations; $E_{jk} = E_j E_k$. Then by reference to the notation wherein X_j and X are defined it should be clear that

$$\underset{k}{E} x_{jk} = \sum_{k=1}^{N_j} \frac{1}{N_j} x_{jk} = \frac{A_j}{N_j} = \mu_j \qquad (1)$$

[All the secondary units in Area j have the same probability, viz., $1/N_j$; hence $\underset{k}{E} x_{jk} = \sum_{k=1}^{N_j} \frac{1}{N_j} x_{jk}$.]

$$\underset{k}{E} X_j = \frac{N_j}{n_j} \underset{k}{E} \sum_{k=1}^{n_j} x_{jk}$$

$$= \frac{N_j}{n_j} \sum_{k=1}^{n_j} \underset{k}{E} x_{jk} = \frac{N_j}{n_j} n_j \underset{k}{E} x_{jk}$$

$$= \frac{N_j}{n_j} n_j \mu_j = N_j \mu_j = A_j \qquad (2)$$

$$EX = \frac{M}{m} E \sum_{1}^{m} X_j = \frac{M}{m} m\, EX_j$$

[All the primary units have the same probability, viz., $1/M$; hence $EX_j = \underset{j\,k}{E E} X_j = \frac{1}{M} \sum_{1}^{M} \underset{k}{E} X_j$ and $\underset{k}{E} X_j = A_j$.]

$$= \frac{M}{m} m \sum_{1}^{M} \frac{1}{M} \underset{k}{E} X_j$$

$$= \frac{M}{m} \frac{m}{M} \sum_{1}^{M} A_j$$

$$= A \qquad (3)$$

Thus the estimates X_j and X are unbiased.

[5] The reference is to the Hansen-Hurwitz paper in *Annals Math. Stat.*, vol. xiv; 1943: pp. 333–62.

In developing the formulas for the variances of $\bar{x}$ and X in Part A of Chapter 4 the variance σ^2 made a very important appearance. Now, in two-stage sampling, *two* variances will appear, viz., the *external* variance σ_e^2 and the *internal* variances $\sigma_j^2(j = 1, 2, \cdots, M)$, as defined in the notation.

Exercise 1. Prove that

$$E\bar{X} = E\frac{X}{M} = \bar{A}$$

$$\underset{k}{E}\bar{x}_j = \mu_j$$

$$E\bar{x} = \mu$$

$$E\bar{x}_u = \mu_u$$

Exercise 2. Prove that

$$\sigma^2 = \sigma_b^2 + \sigma_w^2$$

$$s^2 = s_b^2 + s_w^2$$

$$\sigma_{tu}^2 = \sigma_{bu}^2 + \sigma_{wu}^2$$

$$s_{tu}^2 = s_{bu}^2 + s_{wu}^2$$

Exercise 3. *a.* A universe (e.g., a county or province) is divided into any number M of areal units, of which m units are to be drawn at random for a sample and canvassed completely. Show that, regardless of the relative sizes of these units, even though some units contain 10 times as many people as others, the probability that a particular person will fall into the sample is m/M.

b. Let the jth primary unit in the sample be divided into N_j secondary units of which n_j are to be drawn at random for the sample. These secondary units are to be completely canvassed. Show that, regardless of the relative sizes of the secondary units, the probability that a particular person residing in primary unit j will be in the sample is mn_j/MN_j.

Development of the variance in two-stage sampling.[6] The sampling will be done without replacement, and the theory already developed in Part A of Chapter 4 tells us that if each primary unit in the sample were covered completely then would

$$\text{Var } X = \left(\frac{M}{m}\right)^2 \text{Var} \sum_1^m X_j$$

$$= \left(\frac{M}{m}\right)^2 \frac{M-m}{M-1} m\sigma_e^2 \quad \text{[From Eq. 13', p. 102]} \quad (4)$$

[6] This derivation was first shown to me by my colleague William N. Hurwitz, about 1940.

It will turn out that a second stage of sampling increases this variance by the addition of another term. The variance of X_j arising from the secondary sample taken from within the primary area j will be

$$\operatorname*{Var}_{k} X_j = \left(\frac{N_j}{n_j}\right)^2 \frac{N_j - n_j}{N_j - 1} n_j \sigma_j{}^2 \quad \text{[From Eq. 13', p. 102]} \quad (5)$$

Moreover, by Eq. 30 on page 71,

$$EX_j{}^2 = \operatorname*{Var}_{k} X_j + A_j{}^2 \quad \text{[Note } c, \text{ p. 149]} \quad (6)$$

where $\operatorname*{Var}_{k} X_j$ denotes the variance arising from the secondary sample within the particular primary unit j. Then

$$\begin{aligned}
\operatorname{Var} X &= E(X - EX)^2 \\
&= EX^2 - (EX)^2 \\
&= \left(\frac{M}{m}\right)^2 E\left(\sum_1^m X_j\right)^2 - A^2 \\
&= \left(\frac{M}{m}\right)^2 E\left\{\sum_1^m X_j{}^2 + \sum_{\substack{j=1 \\ (j' \neq j)}}^m \sum_{j'=1}^m X_j X_{j'}\right\} - A^2 \quad \begin{array}{l}\text{[Note } a, \text{ next} \\ \text{page]}\end{array} \\
&= \left(\frac{M}{m}\right)^2 \left\{\sum_1^m EX_j{}^2 + \quad`` \quad EX_j X_{j'}\right\} - A^2 \quad \text{[Note } b\text{]} \\
&= \left(\frac{M}{m}\right)^2 \left\{m \sum_1^M \frac{1}{M} \operatorname*{E}_{k} X_j{}^2 \right. \\
&\quad\quad \left. + m(m-1) \sum_{\substack{j=1 \\ (j' \neq j)}}^M \sum_{j'=1}^M \frac{1}{M(M-1)} A_j A_{j'}\right\} - A^2 \quad \begin{array}{l}\text{[Notes } c \\ \text{and } d\text{]}\end{array} \\
&= \left(\frac{M}{m}\right)^2 \left\{\frac{m}{M} \sum_1^M \left[\left(\frac{N_j}{n_j}\right)^2 \frac{N_j - n_j}{N_j - 1} n_j \sigma_j{}^2 + A_j{}^2\right] \right. \\
&\quad\quad \left. + \frac{m}{M} \frac{m-1}{M-1} \left[\left(\sum_1^M A_j\right)^2 - \sum_1^M A_j{}^2\right]\right\} - A^2 \\
&= \left(\frac{M}{m}\right)^2 \left\{\frac{m}{M} \sum_1^M \left(\frac{N_j}{n_j}\right)^2 \frac{N_j - n_j}{N_j - 1} n_j \sigma_j{}^2 \right. \\
&\quad\quad \left. + m\left[1 - \frac{m-1}{M-1}\right] \frac{1}{M} \sum_1^M A_j{}^2\right\} \\
&\quad\quad + \left(\frac{M}{m}\right)^2 \frac{m}{M} \frac{m-1}{M-1} M^2 \bar{A}^2 - M^2 \bar{A}^2
\end{aligned}$$

$$= \left(\frac{M}{m}\right)^2 \left\{ \frac{m}{M} \sum_1^M \left(\frac{N_j}{n_j}\right)^2 \frac{N_j - n_j}{N_j - 1} n_j \sigma_j^2 \right.$$

$$+ m\left[1 - \frac{m-1}{M-1}\right]\left[\frac{1}{M}\sum_1^M A_j^2 - \bar{A}^2\right]\right\}$$

$$= \left(\frac{M}{m}\right)^2 \left\{ \frac{m}{M} \sum_1^M \left(\frac{N_j}{n_j}\right)^2 \frac{N_j - n_j}{N_j - 1} n_j \sigma_j^2 + \frac{M - m}{M - 1} m \sigma_e^2 \right\} \qquad (7)$$

This is the desired result, but before going on it will be rewritten with the two terms interchanged, as it will be easier to remember the equation and extend it to three or more stages. Thus

$$\mathrm{Var}\, X = \left(\frac{M}{m}\right)^2 \left\{ \frac{M - m}{M - 1} m\sigma_e^2 + \frac{m}{M} \sum_1^M \left(\frac{N_j}{n_j}\right)^2 \frac{N_j - n_j}{N_j - 1} n_j\sigma_j^2 \right\} \qquad (8)$$

A useful approximation obtained by dropping the -1 in the finite multipliers is written in the form

$$\mathrm{Var}\, X \doteq M^2 \left(1 - \frac{m}{M}\right)\frac{\sigma_e^2}{m} + M \sum_1^M N_j^2 \left(1 - \frac{n_j}{N_j}\right)\frac{\sigma_j^2}{mn_j} \qquad (9)$$

Any term in the summation on the right vanishes when the corresponding area is subsampled 100 percent, and the entire sum vanishes when all the M areas are subsampled 100 percent.

NOTES.   *a.* $(a + b + c)^2 = (a^2 + b^2 + c^2)$

$$+ (ab + ba + ac + ca + bc + cb)$$

$$= \text{sum of squares} + \text{sum of cross-products}$$

b. Interchange E and Σ (p. 71).

c. The "expected" value $\underset{j}{EX_j}$ is, as before, the sum of all the A_j, each multiplied by the probability $1/M$ of drawing it. The same statement holds for $\underset{j}{EX_j^2}$. The probability of drawing any particular area is $1/M$ because there are M of them. Thus $\sum_1^m \underset{j}{EX_j^2} = m \sum_1^M \frac{1}{M} A_j^2$. But the "expected" operation $\underset{k}{E}$ must also be taken, as $\underset{jk}{EX_j^2}$ is wanted. The two operators $\underset{j}{E}$ and $\underset{k}{E}$ can be used in either order, so $\sum_1^m \underset{k}{EX_j^2} = \sum_{j=1}^m \underset{k}{E}\underset{j}{E}X_j^2$

$$= m\sum_1^M \frac{1}{M} \underset{k}{EX_j^2}, \text{ and } \underset{k}{EX_j^2} = \underset{k}{\mathrm{Var}}\, X_j + A_j^2.$$

d. There are $M(M - 1)$ permuted cross-products $A_j A_{j'}$, as each of the M primary units can be paired off with $M - 1$ others. Hence the probability of drawing any one of them is $1/M(M-1)$. The secondary samples within the two primary units j and j' are independent, wherefore

$$\underset{k}{E}X_jX_{j'} = \underset{k}{E}X_j\underset{k}{E}X_{j'} = A_jA_{j'} \qquad \text{[P. 72]}$$

So

$$EX_jX_{j'} = \underset{j,j'}{E}A_jA_{j'} = \frac{1}{M(M-1)}(A_1A_2 + A_2A_1 +$$
$$\text{all other cross-products})$$

This is true for any pair of primary units, hence true for all pairs, and as there are $m(m-1)$ possible permuted pairs in the sample,

$$\sum_{j=1}^{m}\sum_{j'=1}^{m} EX_jX_{j'} = \frac{m}{M}\frac{m-1}{M-1}\sum_{j=1}^{M}\sum_{j'=1}^{M} A_jA_{j'} \qquad [j' \neq j]$$

Allocation of the sample by means of simple cost-functions. The main reason for deriving the equation of variance in multistage sampling is to seek guidance on the very important question of how many primary units (m) to take into the sample, and what size of secondary sample (n_j) to take out of primary unit j.

A very interesting approximate result can be seen quickly.[7] Under the assumption that it costs the same to add a new household to the sample regardless of where it is located, and that the total cost is merely proportional to the number of interviews, the total cost of the sample in Area j will be cn_j, where c is the cost of interviewing one household.

The total cost of the entire sample would then be $c\sum_{1}^{m} n_j$. Then if the sample-size n_j were about constant from one area to another (as will be roughly true in applications of this theory), cmn_j would be the cost of the entire sample, and it will be observed that if n_j is cut to half, m may be doubled without changing the total cost. Now the quantity mn_j appears in the summation of Eq. 9; hence, save for the effect of the finite multipliers $1 - n_j/N_j$, which will frequently all be about equal anyhow, the second term remains unchanged as more primary areas are added to the sample, *provided* the sample-size n_j is cut accordingly to keep the total cost constant. The first term, however, does not remain constant; it decreases as m increases. It therefore appears that under the assumptions made above the best design is one in which a great number of primary units are drawn into the sample, and the secondary samples within primary units are light.

The assumption that the cost of interviewing a household in one area is the same as that of interviewing a household in another area, regard-

[7] This result was first demonstrated to me by Morris H. Hansen in 1939. See also P. V. Sukhatme, Report on crop-cutting, Imperial Council of Agricultural Research (Allahabad, 1943–44), p. 11.

less of how many primary units are in the sample is, of course, not exact when the primary units are so large that a salaried supervisor is placed in charge of each. There is then an overhead cost in starting up operations in any area, or (in industrial sampling) in opening a box; and conceivably in a bad design the entire allowable expenditure might be eaten up by overhead, leaving nothing for interviews or testing. Hence another formulation of the problem will be attempted. The new formulation will take account of the overhead costs of operating in the primary areas or, in industrial sampling, of moving and opening the containers to get at the articles to be tested. To avoid some formidable mathematical difficulties it will be assumed that the secondary samples are all equal in size, for which the symbol $\bar{n}$ will be used. So let the total cost of the sample be

$$P = c_1 m + c_2 m \bar{n} = m(c_1 + c_2 \bar{n}) \tag{10}$$

wherein c_1 is the overhead cost of operating one primary unit for the survey, and c_2 is the additional cost of one interview.

With the assumption that n_j is constant and equal to $\bar{n}$ the equation for Var X can be simplified. First, it may as well be assumed that N_j and $\sigma_j{}^2$ are also constant and equal to $\bar{N}$ and $\sigma_w{}^2$ respectively, because in practice n_j would not be approximately uniform unless these other assumptions were also pretty closely satisfied. With $n_j = \bar{n}$, $N_j = \bar{N}$, and $\sigma_j = \sigma_w$, Eq. 8 reduces approximately to

$$\text{Var } X = \left(\frac{M}{m}\right)^2 \left\{ \frac{M-m}{M-1} m\sigma_e{}^2 + m\left(\frac{\bar{N}}{\bar{n}}\right)^2 \frac{\bar{N}-\bar{n}}{\bar{N}-1} \bar{n}\sigma_w{}^2 \right\}$$

$$\doteq M^2 \left\{ \left(1 - \frac{m}{M}\right) \frac{\sigma_e{}^2}{m} + \bar{N}^2 \left(1 - \frac{\bar{n}}{\bar{N}}\right) \frac{\sigma_w{}^2}{m\bar{n}} \right\} \tag{11}$$

What is desired is a minimum in Var X for an allowable cost P; or alternatively, a prescribed Var X at minimum cost. First, let Var X be minimized for a given allowable cost P. From Eq. 10

$$m = \frac{P}{c_1 + c_2 \bar{n}} \tag{10'}$$

whence by substitution

$$\text{Var } X = M^2 \left\{ -\frac{\sigma_e{}^2}{M} + \frac{c_1 + c_2 \bar{n}}{P} \left[(\sigma_e{}^2 - \bar{N}\sigma_w{}^2) + \frac{\bar{N}^2}{\bar{n}} \sigma_w{}^2 \right] \right\}$$

Now replace $\bar{n}$ by a continuous variable y between 1 and $\bar{N}$. The minimizing value of y is found by inspecting the derivative

$$\frac{1}{M^2} \frac{d \text{ Var } X}{dy} = \frac{c_2}{P} (\sigma_e{}^2 - \bar{N}\sigma_w{}^2) - \frac{c_1 \bar{N}^2}{P y^2} \sigma_w{}^2$$

If the quantity $\sigma_e{}^2 - \overline{N}\sigma_w{}^2$ is negative or 0, the derivative is always negative, indicating that Var X reaches no proper minimum but decreases continuously, reaching its lowest point when $y = \bar{n} = \overline{N}$, in which case m, the number of primary units to be drawn into the sample, is $P/(c_1 + c_2\overline{N})$ as is obtained from Eq. 10 by setting $\bar{n} = \overline{N}$. In this case the secondary sample is complete, for $\bar{n} = \overline{N}$.

If $\sigma_e{}^2 - \overline{N}\sigma_w{}^2$ is positive, the derivative vanishes when

$$y = \overline{N}\sqrt{\frac{c_1}{c_2}}\frac{1}{\sqrt{\left(\dfrac{\sigma_e}{\sigma_w}\right)^2 - \overline{N}}} \tag{12}$$

$$= \sqrt{\frac{c_1}{c_2}}\frac{1}{\sqrt{\left(\dfrac{\sigma_b}{\sigma_w}\right)^2 - \dfrac{1}{\overline{N}}}} \tag{12'}$$

indicating that Var X would then have its lowest value when $\bar{n}$ is an integer near to the right-hand member just written. In this case there is a second stage of sampling.[8]

Thus $\bar{n}$ and hence the calculated sampling interval $\overline{N}/\bar{n}$ are determined. The number m of primary units to be in the sample is now to be found from Eq. 10'. The main aspects of the design would be completed by computing Var X in Eq. 11 and noting whether it is small enough. If not, the allowable cost P must be increased, permitting more primary units to be brought into the sample. Or, perhaps some way can be found to reduce costs; if not, the survey must be abandoned.

It is to be noted that $\bar{n}$ does not depend on P, the total cost of the survey; hence if the total cost P is doubled, $\bar{n}$ and $\overline{N}/\bar{n}$ are unaffected but m is doubled. This result for the proper allocation of the sample for greatest efficiency could hardly have been foreseen without mathematics; certainly without mathematics quite different advice has been given in the past.

For the second problem, viz., to minimize the cost P for a fixed value of Var X, we see from Eq. 11 that

$$m = \frac{M^2[\sigma_e{}^2 - \overline{N}\sigma_w{}^2 + \overline{N}^2\sigma_w{}^2/\bar{n}]}{\text{Var } X + M\sigma_e{}^2} \tag{11'}$$

[8] Eq 12' was first published by Hansen, Hurwitz, and Margaret Gurney, "Problems and methods of the sample survey of business," *J. Amer. Stat. Assoc.*, vol. 41, 1946: pp. 173–89. The simpler form shown later on as Eq. 23 seems to have been published almost simultaneously by L. H. C. Tippett, *The Methods of Statistics* (Williams & Norgate, 1931), p. 177, and by Walter A. Shewhart, *The Economic Control of Quality of Manufactured Product* (Van Nostrand, 1931), p. 389.

This value substituted into Eq. 10 gives

$$P = \frac{M^2(c_1 + c_2\bar{n})}{\text{Var } X + M\sigma_e^2}\left[\sigma_e^2 - \overline{N}\sigma_w^2 + \frac{\overline{N}^2\sigma_w^2}{\bar{n}}\right] \tag{13}$$

If $\sigma_e^2 - \overline{N}\sigma_w^2$ is negative or 0, P has its lowest value when $\bar{n} = \overline{N}$, in which case the secondary sample is complete. If $\sigma_e^2 - \overline{N}\sigma_w^2$ is positive, P has its lowest value when $\bar{n}$ is again an integer near to the right-hand member of Eq. 12. As soon as the best value of $\bar{n}$ is calculated, the minimum cost P can be computed by the last equation. If this calculated minimum cost is larger than permissible, it will be necessary to take a smaller value of m and to accept a greater variance in X, or else to abandon the survey, unless some way can be found to reduce costs.

It is to be noted that $\bar{n}$ does not depend on the prescribed value of Var X. However, from Eq. 11' it is seen that the number m of primary units in the sample is to be increased if Var X is decreased. Again, this allocation of the sample ($\bar{n}$ and m) could hardly have been foreseen without mathematics.

It should be noted also that in order to compute the optimum secondary sampling interval $\overline{N}:\bar{n}$, neither c_1, nor c_2, nor σ_e, nor σ_w, need be known in absolute terms, but only in the ratios $\sigma_e:\sigma_w$ and $c_1:c_2$. However, the computation of m and of the total cost and of the expected precision does require absolute numbers for c_1, c_2, σ_e, and σ_w. This is typical of what often happens in sample-design: many questions of procedure, netting huge savings, can be settled with only meager information, provided one knows how to use it. Theory indicates what this information must be, and how far wrong the advance calculations of cost and precision will be when only broad limits can be placed on the advance estimates of the ratios $\sigma_e:\sigma_w$, $c_2:c_1$, or on the absolute values of these four quantities.

As stated earlier in the chapter (p. 136) the overhead cost c_1 for one primary unit should be sufficient to employ a full-time areal supervisor, or at least one on half-time. c_1 would then be his salary, travel allowance, office rent, and other expenses for the survey. If one survey is to be completed per week, on the average, c_1 would be these costs spread over a week. If there were no costs of supervision, c_1 would be 0, and the solution would reduce to the earlier one in which the total cost was equal to cmn_j and in which it was indicated that many primary units should be brought into the sample.

The theory just expounded is oversimplified. In practice the numbers N_j (the number of secondary units in primary unit j) may not all be the same, and certainly σ_j will vary from one primary unit to another. Moreover, the assumptions made in Eq. 10 relating the total cost P to the number of primary and secondary units surveyed can be only approximate. Nevertheless the foregoing theory has taught us some very important lessons regarding the allocation of the sample. For the theory to apply it is only necessary that the numbers N_j and σ_j be not too heterogeneous: $\overline{N}$ is then to be taken as an average value of N_j, and σ_w as an average value of σ_j.

Under conditions where the interviewers or inspectors are to reach m small primary units one after another by some devious route (supposedly so chosen that travel-costs are minimized), the supervision being handled from some central headquarters and not in each primary unit separately, the cost-function will hardly be that given by Eq. 10, but rather of the form shown later on by Eq. 98 (p. 204). This is more difficult to handle mathematically than Eq. 10, and it will usually suffice to treat the problem arithmetically by inserting various likely and unlikely values of the constants A and B in Eq. 98 and of the variances in Eq. 11, and computing the corresponding values of P to find a plan of minimum cost. A few hours devoted to such deliberations and calculations will often result in a clear indication of an optimum plan, or perhaps several good plans between which there is little choice so far as efficiency of sampling is concerned. An example is given further on.

Obviously no cost-function and no proper allocation of the sample are possible without some knowledge of what the costs are. The importance of keeping records on costs, not merely to determine total cost, but to discover the various components of cost (supervision, interviewing, travel) is evident.

A word on tabulation plans. A self-weighting sample is a great convenience in tabulation. For the secondary sample one specifies the same sampling interval in all the m primary units. When the cost-function $P = m(c_1 + c_2\bar{n})$ of Eq. 10 is used, the *calculated sampling interval* will be $\overline{N}:\bar{n}$, wherein $\bar{n}$ comes from Eq. 12; or,

$$\frac{\overline{N}}{\bar{n}} = \sqrt{\frac{c_2}{c_1}} \sqrt{\left(\frac{\sigma_e}{\sigma_w}\right)^2 - \overline{N}} \qquad (14)$$

as may be seen from the preceding pages. The *actual sampling interval* b to be used, however, may be some convenient integer (such as 2, 3, 4, 5, 10, 15, 20, 50, 100) close to the calculated value of $\overline{N}/\bar{n}$.

For drawing the sample of primary units, the *calculated sampling interval* M/m will be found by dividing M by the value of m given by Eq. 10' or 11'. The *actual interval* a may be fractional if desired, or again it may be some integer conveniently near the calculated value of M/m.

A fractional interval presents no difficulty. Thus if M/m is calculated to be 2.47, it may be rounded to 2.5 or 5/2. The rule for the selection of primary units may then be: Take 1, skip 1, take 1, skip 2. Or, out of every 5 successive units, take 2 at random.

The estimated population of the universe will be

$$X = \frac{N}{n} \sum_{j=1}^{m} \sum_{k=1}^{n_j} x_{jk} \quad \left[\frac{N}{n} = \frac{M}{m} \frac{N_j}{n_j}, N_j : n_j \text{ being constant} \right]$$

$$= \frac{N}{n} \times \text{(population in the sample)} \tag{15}$$

N/n is called the *weighting factor*. N and n are the total number of secondary units in the universe and in the sample, respectively. The same weighting factor N/n applies to all the characteristics estimated by the sample, such as total number of inhabitants, total male, total female. In fact, after the returns from the survey are processed, the various populations in the sample are usually found by addition and verified by comparing them with various marginal totals also given by the sample, after which the whole table of sample results is multiplied by the factor N/n to get the final table of estimates for the universe.

As a matter of fact, the weighting factor that is actually used is hardly ever so simple as N/n. Even in the most careful surveys there will be a few nonresponses and other gaps in the information that is brought in. The decision on how to handle these gaps will affect the weighting factor as well as the results. Perhaps the missing information will be filled in by arbitrary rules, such as substituting average values in the gaps, or by making "educated guesses" from ancillary information,[9] or by duplicating the information elicited from "similar" households or individuals. A small amount of this kind of manipulation must be expected, but if too much of it is done, the sample is no longer a probability-sample.

For ease in multiplication the weighting factor may be adjusted to some convenient integer, and the sample likewise raised or lowered in proportion. This is done before the processing of the sample results is begun in earnest. Thus, if the weighting factor turned out to be 20.1 before adjustment, it may be altered into the convenient factor 20 by drawing a sample of 1 in 200 returns, duplicating them, and adding the

[9] In the population census of 1940 unknown and missing ages were supplied in the editing operation by an elaborate device using ancillary information, such as marital status, occupation, relationship to head of family. The savings in tabulation were computed at about $100,000. The procedure is described in a bulletin by W. Edwards Deming, "On the elimination of unknown ages in the 1940 population census," (Bureau of the Census, January 1942; out of print). A similar plan was used by Nathan Keyfitz in the Dominion Bureau of Statistics for use in the Canadian census of 1941. A memorandum on handling other types of small gaps in the information returned by the interviewers is given in Appendix C of *A Chapter in Population Sampling* (Bureau of the Census, 1947).

duplicates to the original sample. If the weighting factor were 19.9 before adjustment, a sample of 1 in 200 returns would be drawn and excluded.[10] The adjusted weighting factor would in both cases be 20.

Extension to three stages and to stratified sampling. Perceiving that a secondary sample adds a summation term to the variance given in Eq. 8, we may conjecture (as is true) that a third stage would merely add another term and give

$$\text{Var } X = \left(\frac{M}{m}\right)^2 \left\{ \frac{M-m}{M-1}\, m\sigma_e^2 + \frac{m}{M} \sum_{j=1}^{M} \left(\frac{N_j}{n_j}\right)^2 \left[\frac{N_j - n_j}{N_j - 1}\, n_j \sigma_j^2 \right. \right.$$
$$\left. \left. + \frac{n_j}{N_j} \sum_{k=1}^{N_j} \left(\frac{N_{jk}}{n_{jk}}\right)^2 \frac{N_{jk} - n_{jk}}{N_{jk} - 1}\, n_{jk}\, \sigma_{jk}^2 \right] \right\} \quad (16)$$

Here there are N_{jk} tertiary units (households or articles) in the kth secondary unit of the jth primary unit, and n_{jk} of them are drawn into the sample. Extension to a fourth stage involves no new difficulty.

In stratified sampling there are a number of strata, say R, in each of which is a number M_i of primary units, supposedly similar in certain characteristics. In two stages of sampling the formula for the variance is

$$\text{Var } X = \sum_{i=1}^{R} \left(\frac{M_i}{m_i}\right)^2 \left\{ \frac{M_i - m_i}{M_i - 1}\, m_i \sigma_{ei}^2 \right.$$
$$\left. + \frac{m_i}{M_i} \sum_{j=1}^{M_i} \left(\frac{N_{ij}}{n_{ij}}\right)^2 \frac{N_{ij} - n_{ij}}{N_{ij} - 1}\, n_{ij}\sigma_{ij}^2 \right\} \quad (17)$$

in which (it is hoped) the notation is an obvious modification of the previous symbols. Stratified sampling is treated elsewhere in Chapters 4, 6, and 12.

Interpretation of theory. The theory that has been worked out in the last few paragraphs, and indeed throughout the chapter, is really only a start in the practice of sampling. However, it offers one way to start. Having learned some theory, the next problem is how to use it: how to attain greater reliability at less cost. Theory indicates what we need to know about the universe (county, city, region) and what we need to do to it in order to design a good sample (low variance, low cost). For example, the symbols σ_e, σ_j, M, N_j have occurred, and still others will make their appearance. These are not absolute properties of the universe, but they depend in part on how we cut it up into sampling units. The greater the supply of prior knowledge in the form of maps and census information, the more information a sample can be made to deliver for a given cost (compare with a later section, p. 183).

---

[10] Cf. page 82 in *A Chapter in Population Sampling* (Bureau of the Census, 1947).

Advance estimates of the cost and expected precision to be obtained from a proposed survey are only as good as the advance estimates of σ_e, σ_w, c_1, c_2. Unfortunately, treatises on the subject-matter of demography, sociology, agriculture, industrial production, etc., seldom provide clues to this very necessary type of quantitative information, and the statistician must seek it out the best he can. After one or more surveys have been taken, excellent estimates can be obtained from the returns, and these estimates serve as advance knowledge for improving and lowering the cost of the next survey.[11]

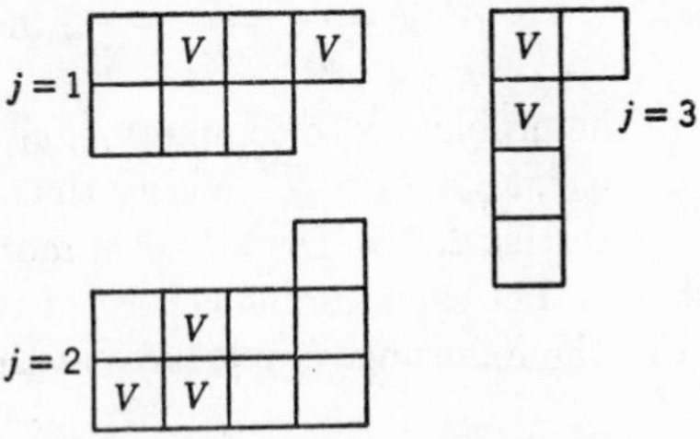

FIG. 12. For an illustrative exercise in the text, a universe consists of three primary units, in each of which are a number of d.us., some of which are vacant (V). Samples are to be drawn, and the total number of vacant d.us. in the universe is to be estimated from the samples.

Theory indicates that if the primary units are to be drawn with equal probabilities, then effort must be made to delineate primary sampling units having roughly equal populations, so as not to run into a high value of σ_e and a high variance of X, or high cost, and theory indicates how the costs will increase with failure to accomplish this aim. Similar remarks hold for the secondary units.

Exercise 1. The universe pictured in Fig. 12 consists of three primary units. Each primary unit consists of d.us. (dwelling units) some of which are vacant (V). Number the d.us. in each primary unit in serpentine fashion. Let k be the running index signifying a particular d.u.; then

k will take the values 1, 2, $\cdots$, 7 in primary unit 1

k " " " " 1, 2, $\cdots$, 9 " " " 2

k " " " " 1, 2, $\cdots$, 5 " " " 3

The double index jk signifies a primary unit and a particular d.u. within it. Let

$$a_{jk} = 1 \quad \text{if the d.u. is vacant } (V)$$
$$= 0 \quad \text{otherwise}$$

By counting it is easily seen that

$$A_1 = 2, \quad A_2 = 3, \quad A_3 = 2, \quad A = \sum_1^M A_j = 7$$

[11] A splendid example of such a series of experiments is Mahalanobis's surveys for the acreage of jute in Bengal, commencing in 1935. The references are given later on.

$$N_1 = 7, \quad N_2 = 9, \quad N_3 = 5, \quad N = \sum_1^M N_j = 21$$

$$M = 3$$

$$\mu = \frac{A}{N} = \frac{7}{21} = \frac{1}{3}, \text{ the overall vacancy ratio}$$

The problem is to compute $\bar{A}$, μ_1, μ_2, μ_3, σ_e^2, σ_b^2, σ_1^2, σ_2^2, σ_3^2, σ_w^2, σ^2, μ_u, σ_{bu}^2, σ_{wu}^2, σ_{tu}^2, $\bar{N}$. Verify that $\sigma^2 = \sigma_b^2 + \sigma_w^2$.

Exercise 2.   a. Draw four or more samples from the universe pictured here.   Let each sample consist of two primary units, drawn at random. Use the following sample-sizes at the second stage.

From primary unit 1, draw $n_1 = 3$ d.us.

"   "   " 2, " $n_2 = 4$ "

"   "   " 3, " $n_3 = 2$ "

[These numbers are approximately proportional to N_1, N_2, N_3, concerning which the student may wish to turn ahead to the subject of allocation in Ch. 6.]

Tippett's numbers or any other set of random numbers may be used for drawing the two primary units and for drawing the d.us. from within whichever primary units fall into the sample.

b. For each sample compute

$$X_j = \frac{N_j}{n_j} \sum_1^{n_j} x_{jk} \quad \text{for each of the two primary units} \\ \text{in the sample}$$

$$X = \frac{M}{m} \sum_1^m X_j = \frac{3}{2}(\alpha_1 X_1 + \alpha_2 X_2 + \alpha_3 X_3)$$

= the estimated total number of vacant d.us. in the universe

Here

$$\alpha_j = 1 \quad \text{if primary unit } j \text{ is in the sample}$$
$$= 0 \quad \text{otherwise}$$

(In every sample, two of the α_j will be 1, and the third one will be 0.)

The instructor may wish to ask each student to draw and compute four samples, then to assemble all the results in the form on the next page before assigning Parts c and d.

c. For the ensemble compute the average value of X and compare it with the known value $A = 7$.

d. Compute the mean square deviation of X from its average.   Compare it with Var X, which is to be computed from Eq. 8 in which $m = 2$ and in which σ_e, σ_1, σ_2, σ_3 have the values computed in the previous exercise.

RECORD OF SAMPLES

Sample number	Primary unit 1 (*if in the sample*)		Primary unit 2 (*if in the sample*)		Primary unit 3 (*if in the sample*)		Estimated total vacancies
	$x_1 = \sum_1^{n_1} x_{jk}$	X_1	$x_2 = \sum_1^{n_2} x_{jk}$	X_2	$x_3 = \sum_1^{n_3} x_{jk}$	X_3	$X = \frac{3}{2} \sum_1^3 \alpha_j X_j$
1 2 . . .							
Average	xxx		xxx		xxx		
Known value		2	xxx	3	xxx	2	7

Estimates of the population per secondary unit. There are many problems in which special interest attaches to the population per secondary unit (μ_j or μ) rather than to the total population (A_j or A). This is so in many social and economic studies, but excellent examples are furnished also by quality control, field trials in agricultural experiments, and in the sampling of bulk materials of industry to estimate certain intensive properties in which there is no definite subunit. Determinations of the percentage of sucrose in raw sugar, the percentage of clean wool in a shipment of grease wool, the percentage of ash in coal, the percentage defective in a shipment of manufactured articles, are examples. The number of secondary units (N_j) per bag or bale or carload of bulk material is indefinite, but fortunately the finite multiplier $(N_j - n_j)/(N_j - 1)$ in Eq. 8 can be replaced by $1 - n_j/N_j$, which is equivalent to $1 - w_j/W_j$, where W_j is the weight of a bag or bale or other unit, and w_j is the weight drawn out as a sample. As a matter of fact, to correspond with most practice, in the theory about to be developed the sample will be assumed so small that w_j/W_j can be neglected and the finite multiplier replaced by unity. Whenever this is not the best assumption, the factor $1 - w_j/W_j$ can easily be restored. The symbol $\bar{x}$ will denote some intensive quality of a lot or shipment which is to be determined by drawing samples and performing tests. The most economical sampling procedure for determining $\bar{x}$ within prescribed limits (such as within 1 percent) is an important problem, to which the foregoing theory will now be applied. The solution will be seen to depend on the variance $\sigma_w{}^2$ within the bale or other unit and on the variance $\sigma_b{}^2$ from bale to bale. Under conditions in which the bales are approximately the same weight and the sample is small, $\sigma_e{}^2$ is replaced by $\bar{N}^2\sigma_b{}^2$, N_j by $\bar{N}$, and the finite multiplier $(N_j - n_j)/(N_j - 1)$ by unity. Then because $N\bar{x} = X$ it is seen from Eq. 8, 9, or 11 that approximately

$$\text{Var } \bar{x} = \frac{M - m}{M - 1} \frac{\sigma_b{}^2}{m} + \frac{\sigma_w{}^2}{m\bar{n}} \tag{18}$$

Here M is the number of bales in the shipment, m the number to be sampled (cored), $\bar{n}$ the number of samples per bale. The cores should be randomly selected from within the bale and of equal size, although a systematic selection in which all parts of the bale have the same chance of coming into the sample is more practicable. For the coefficient of variation of $\bar{x}$ the last equation when divided through by μ^2 gives

$$(\text{C.V. } \bar{x})^2 = \frac{M - m}{M - 1} \frac{\gamma_b{}^2}{m} + \frac{\gamma_w{}^2}{m\bar{n}} \tag{19}$$

wherein $\gamma_b = \sigma_b/\mu$ and $\gamma_w = \sigma_w/\mu$, these being the external and internal coefficients of variation per secondary unit in the universe.

Exercise. By making $\bar{n} = 1$ in all the primary units of the sample, show that the basic formula of Chapter 4 for sampling in a single stage still holds for two stages. Hence if $\bar{n} = 1$ the variance behaves nearly as if the N secondary units were listed and drawn at a single stage. (*Hint:* Place $\bar{n} = 1$, and recall $\sigma_b{}^2 + \sigma_w{}^2 = \sigma^2$; then Eq. 19 for two stages reduces to Eq. 16' in Ch. 4. First indicated to me by Morris H. Hansen, 1949.)

Sampling of bulk products. Wool will serve as an illustration. Wool is imported in highly compressed bales weighing from 200 to 1000 pounds, depending on the country of origin. Duty and the selling price are both based on "grade" and "clean content." The clean content of a sample of wool is defined as the fraction of "clean" wool remaining after certain operations of scouring and chemical reduction have been performed on the sample to bring it to a prescribed standard of moisture and ash content. A routine laboratory procedure for determining the clean content of a sample of wool was developed in the Bureau of Customs by Wollner and Tanner,[12] who also worked on the problem of sampling. The samples tested are cores from $1\frac{1}{8}$ to 2 inches in diameter

[12] H. J. Wollner and Louis Tanner, "Sampling of imported wool for the determination of clean wool content," *Ind. Eng. Chem.*, vol. 13, 1941: pp. 883–87. A standard method of test of wool has been promulgated by the A.S.T.M., publication D584-43, based on the work of Wollner and Tanner.

and from 8 to about 18 inches in length, depending on the density of the bale, and bored in the direction of compression, which is the direction of greatest variability. Cores can be taken from either side of the bale, centered at any point. The assumption here will be that random x, y coordinates will be used for the selection of these points, and that bales for coring can be selected at random out of the shipment.[13] Given certain variances σ_w^2 and σ_b^2 and a desired precision for the determination of the clean content ($\bar{x}$) of the shipment of M bales, the question is how many bales (m) are to be bored, and how many cores ($\bar{n}$) from each bale? Equation 18 applies. At the present writing there is a dearth of data on the necessary variances, but for illustration the following values will serve:

$$\sigma_b = 2 \text{ percent} \qquad \sigma_w = 1 \text{ percent}$$

These are actual figures on some Australian wool kindly furnished by Mr. Louis Tanner of the Customs Laboratory in Boston.

For very small shipments (M small), in which every bale must be in the sample, $m = M$, and the finite multiplier in Eq. 18 will be 0 because a core will be taken from every bale, leaving only the second term involving σ_w, whence

$$M = \frac{1}{\bar{n}} \left(\frac{\sigma_w}{\sigma_{\bar{x}}} \right)^2 \tag{20}$$

For shipments smaller than this value of M, all bales are to be in the sample, and the number $\bar{n}$ of cores per bale must be altered to satisfy the equation

$$\bar{n} = \frac{1}{M} \left(\frac{\sigma_w}{\sigma_{\bar{x}}} \right)^2 \qquad \text{[Small shipment; every bale sampled]} \tag{21}$$

For larger shipments it is convenient to rewrite Eqs. 18 and 19 in the form

$$m = \frac{\dfrac{M}{M-1} \sigma_b^2 + \dfrac{\sigma_w^2}{\bar{n}}}{\sigma_{\bar{x}}^2 + \dfrac{\sigma_b^2}{M-1}} = \frac{\dfrac{M}{M-1} \gamma_b^2 + \dfrac{\gamma_w^2}{\bar{n}}}{C_{\bar{x}}^2 + \dfrac{\gamma_b^2}{M-1}} \tag{22}$$

$\bar{n}$ is first to be determined from the costs of handling a bale and of taking

---

[13] In practice the nearest approach to a random selection of bales is to take cores from every nth bale as the shipment is unloaded: any hope that a shipment of 1000 bales, once in the warehouse, will be sampled at random is a vain one indeed, unless it is all loaded again later for reshipment. Random x, y patterns for the cores could probably be followed by the workmen if a set of cards were made up showing random selections of centers, the cards to be used in turn as bale after bale is cored.

a core out of a bale. Suppose for illustration that:

$c_1 = 50\cent$, the cost of pulling aside an occasional bale for coring during the unloading of the shipment, and later restoring the bale to the remainder of the shipment

$c_2 = 35\cent$, the cost of taking 1 core, including wages and repairs on the machine

The next step is to return to Eq. 12′, neglect the $\overline{N}$ under the radical, and perceive that what is left reduces to

$$\bar{n} = \frac{\sigma_w}{\sigma_b} \sqrt{\frac{c_1}{c_2}} = \frac{\gamma_w}{\gamma_b} \sqrt{\frac{c_1}{c_2}} \tag{23}$$

which in this problem gives

$$\bar{n} = \frac{1}{2} \sqrt{\frac{50}{35}} = 0.6 \tag{24}$$

As $\bar{n}$ can not be less than 1, we are assured that 1 core per sample bale is sufficient.

We might well pause to note that $\bar{n}$, the number of cores per bale, is independent of the size of lot (M) and the number m of bales to be sampled; also independent of the total cost and the desired precision.

The standard error of sampling in the determination of clean content might be prescribed as $\frac{1}{3}$ percent, which allows 1 percent either way as the 3-sigma tolerance. With $\sigma_w = 0.01$, $\sigma_{\bar{x}} = 0.0033$, and $\bar{n} = 1$, Eq. 20 gives $M = 9$, and so for shipments of 9 or fewer bales, every bale would be sampled, and in fact 9 cores will be required from any shipment of fewer than 9 bales. For $M > 9$, Eq. 22 yields values for m, the number of bales to be in the sample. The proper sampling interval will be M/m rounded to some convenient integer.

$$\sigma_{\bar{x}} = 0.0033, \quad \sigma_b = 0.02, \quad \sigma_w = 0.01, \quad \bar{n} = 1 \quad \text{if} \quad M \geqq 9$$

Size of shipment, M bales	9	25	50	100	250	500	1000
Size of sample, m bales	See note 9	18	26	33	40	42	43
Recommended sampling interval	Take all		2	3	5 or 6	12	20

NOTE. For a shipment of fewer than 9 bales, take 9 cores, distributed as uniformly as possible amongst the bales.

To get some idea concerning the samples required in another case, assume that

$$\sigma_b = 3 \text{ percent} \qquad \sigma_w = 3 \text{ percent}$$

For such variability we might relax the prescribed standard error to $\frac{1}{2}$ percent. First of all, Eq. 23 shows that 1 core per bale is again correct, and Eq. 20 indicates that a minimum of 36 cores is required. Hence $M = 36$ is the minimum shipment in which sampling can be used. Eq. 22 gives the proper sampling plan for shipments bigger than 36. The samples required are not much greater than on the previous assumptions, although it must be borne in mind that the prescribed precision $(\sigma_{\bar{x}})$ is slightly looser in the second example.

$$\sigma_{\bar{x}} = 0.005, \quad \sigma_b = 0.03, \quad \sigma_w = 0.03, \quad \bar{n} = 1 \quad \text{if} \quad M \geqq 36$$

Size of shipment, M bales	36	50	100	250	500	1000
Size of sample, m bales	See note 36	42	53	63	67	70
Recommended sampling interval	Take all		2	4	8	15

NOTE. For a shipment of fewer than 36 bales, take 36 cores, distributed as uniformly as possible amongst the bales.

The illustrations for wool apply equally well to a host of other problems in industrial sampling. As with many sampling problems, theory shows what needs to be known in order to design a plan that will deliver the precision required at the lowest cost.

Naturally the costs c_1 and c_2, and the standard deviations σ_b and σ_w, will vary from one type of problem to another. Sometimes their ratios will be such that for the most economical sampling plan, $\bar{n}$ as calculated by Eq. 23 turns out to be 1 (as in the above examples): at other times $\bar{n}$ turns out to be 2, 3, 4, or more.

The first step in introducing a sampling plan is to get some reliable figures on costs and variances. All that is required is proper records, some small statistical computation, and a few control charts to enable assignable causes of variability to be identified. The cost is trivial compared with the savings.

Exercise. A certain laboratory engaged in the testing of wool, when sampling a lot of 500 bales, makes a practice of taking 4 cores from each of 50 bales. (a) Show that if $\sigma_w = 3$ percent and $\sigma_b = 6$ percent, this laboratory is providing a 3-sigma sampling tolerance of 2.4 percent (which is pretty high). (b) Show that if the ratio of costs $c_1:c_2 = 10:7$, this laboratory is paying 89 percent more than necessary for sampling a lot of 500 bales.

Remark. In Chapter 10 methods will be presented for estimating σ_w and σ_b on the assumption that records are available concerning the tests of individual cores. It is possible to obtain estimates of σ_w and σ_b without the use of records of individual cores, and a method for going about this may be discussed at this time. Suppose that (as actually happened) a manufacturer of textiles has records on hand showing $\bar{x}$ for each of numerous lots of various types of wool. In some of the lots $\bar{n}$ (the number of cores per bale) was 2, in others 4, but the tests were not made for the individual cores.

Let one "experiment" be described as the determination of $\bar{x}$ for each of (e.g.) 50 lots of wool of a particular type. In the sampling plan that was used for this "experiment" $\bar{n} = 2$. The variance of the actual 50 numerical values of $\bar{x}$ may be computed for this "experiment": call this variance A.

In another "experiment" with the same type of wool there are 100 lots, and $\bar{n} = 4$. The variance of these 100 numerical values of $\bar{x}$ may likewise be computed: call this variance B. Then for determining σ_w and σ_b there will arise from Eq. 18 the two following simultaneous equations, linear in the two unknowns σ_w and σ_b:

$$A = \frac{M - m}{M - 1} \frac{\sigma_b^2}{m} + \frac{\sigma_w^2}{m\bar{n}} \qquad [m = 50, \bar{n} = 2]$$

$$B = \frac{M - m}{M - 1} \frac{\sigma_b^2}{m} + \frac{\sigma_w^2}{m\bar{n}} \qquad [m = 100, \bar{n} = 4]$$

A difficulty arises in the fact that M (the size of a lot) may differ from 1 lot to another. However, the factor $(M - m)/(M - 1)$ may not differ greatly from 1 lot to another, and it may be entirely practicable to replace it by an average value. When this is not satisfactory for the whole 50 or 100 lots of an "experiment," the lots may be grouped into smaller "experiments." There will then arise more than 2 equations, perhaps 6 or 8 equations, for the 2 unknowns: the method of least squares, or the method of averages, or any other convenient method for solution may be utilized for effecting a solution.

The particular numbers $\bar{n} = 2$ and $\bar{n} = 4$, $m = 50$ and $m = 100$, that were used above happen to be actual in a particular example, but in another example they might be entirely different. It is only necessary that $\bar{n}$ be different in at least 2 experiments.

In the equations written above, A and B are subject to the errors of sampling, as the 50 or 100 lots in an "experiment" constitute but a sample from a whole consignment of wool. From some theory that will be given in Chapter 10, and from the fact that $\bar{x}$ will be about normally distributed, the coefficients of variation of A and B may be taken as $1/\sqrt{2m}$, wherein (in the situation described) m is 50 for A and 100 for B. It follows that the coefficients of variation for A and B will here be about 10 percent and

7 percent, respectively. These errors will be propagated into the determinations of σ_w and σ_b in a complicated way, as the student may wish to investigate.[14]

B. RATIO-ESTIMATES AND CALIBRATING SAMPLES

Ratio-estimates. The total population A possessing a particular characteristic is to be estimated. The particular estimate denoted as X in the notation is unbiased, and its variance has been developed as Eq. 8. Another type of estimate, called a ratio-estimate, will sometimes be more precise, and it is important to know when and when not to use it. To begin our study of a *ratio-estimate* we define

$$f = \frac{\sum_{1}^{m} X_j}{\sum_{1}^{m} Y_j} \tag{25}$$

in which X_j and Y_j stand for *any two numbers* associated with specified characteristics of primary unit j. More general forms of ratio-estimates are possible, but will not be treated here.[15] The distinguishing feature of a ratio-estimate is that both numerator $\sum_{1}^{m} X_j$ and denominator $\sum_{1}^{m} Y_j$ are random variables. Specifically,

X_j = the estimated total x-population in primary unit j

Y_j = the estimated total y-population in primary unit j

The x-characteristic might be the characteristic of being alive, in which case X_j is the number of people in primary unit j. The y-characteristic might be the characteristic of being a d.u. (dwelling unit); then Y_j is

---

[14] This and similar problems are treated in the author's *Statistical Adjustment of Data* (John Wiley, 1943), p. 176 and elsewhere.

[15] A more general form of ratio-estimate is $f = \Sigma\, w_j X_j Y_j / \Sigma\, w_j Y_j^2$, wherein w_j is some function of X_j and Y_j. If $w_j = 1/Y_j$, the fraction just written reduces to the form used in the text. $w_j = 1/Y_j^2$ reduces it to $f = \dfrac{1}{m} \Sigma \dfrac{X_j}{Y_j}$. The value of w_j is determined by special assumptions regarding the universe of points X_j, Y_j: cf. the author's *Statistical Adjustment of Data* (John Wiley, 1943), p. 31. The form treated here seems to be generally suitable and in particular is relatively free from dangers of bias and high errors of sampling arising from freakish values of X_j and Y_j. A full discussion of the errors and biases in a ratio-estimate has never yet been produced.

the number of d.us. in primary unit j (N_j in the notation); and the ratio f would be the estimated average number of people per d.u.

The total population in the universe having the x-characteristic may be estimated in the form

$$X' = fB \tag{26}$$

which is called a *ratio-estimate* because it involves the estimated ratio f. The quantity Y is called the *base* of the ratio-estimate. The sample from which f is derived is frequently called a *calibrating* sample because

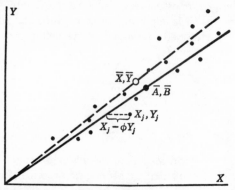

Fig. 13.   Illustrating a ratio-estimate. In the universe there are M points, and the sample picks out m of them at random. The point $\bar{A}$, $\bar{B}$ is the centroid of the universe, and $\bar{X}$, $\bar{Y}$ is the centroid of the sample. The slope of the solid line (measured from the Y-axis) is ϕ, and the slope of the dashed line is f, a random variable. One sample point X_j, Y_j is marked, and the corresponding line-segment $X_j - \phi Y_j$ is shown. The quantity V as defined in Eq. 46 is the root-mean-square value of all M line-segments in the universe, each measured in units of $\bar{A}$.

the m values of X_j are used to calibrate the previous measurements Y_j, or to bring them up to date.

Although X_j and Y_j are both estimated from the present sample for $j = 1, 2, \cdots, m$, it is assumed that the total $B = \sum_1^M B_j$ for the universe is known from other sources. This is the case in estimating proportions of subclasses of the population in the Current Population Survey in the United States (yielding the Monthly Report on the Labor Force). The proportion of employed males, aged 20–29, is estimated as $f = \sum_1^m X_j \Big/ \sum_1^m Y_j$, X_j being the estimated number of employed males 20–29 in primary unit j, and Y_j being the estimated total number of people (of all classes) in the same unit. The total employed males 20–29 is estimated by Eq. 26, in which the base B, the total number of people in the continental United States, does not come from the sample but from other sources: it is calculated month to month very accurately in the Bureau of the Census by taking the total count of the last census and bringing it up to date by adding and subtracting ascertainable figures on births, deaths, emigration, and immigration over the time-interval that has elapsed since the last census. As the base B is not subject to errors of sampling, then from Exercise 1d on page 72 it may be seen that

$$\text{C.V. } X' = \text{C.V. } f \qquad \begin{array}{l}\text{[Only if } Y \text{ is a constant,} \\ \text{not subject to the er-} \quad (27) \\ \text{rors of sampling]}\end{array}$$

Under this circumstance, if f is very precise, as it will be if X_j and Y_j are highly correlated (Eq. 47 *infra* on p. 176, also Fig. 13), then the estimate X' will also be very precise and the ratio-estimate will be advantageous. The gain arising from a ratio-estimate is sometimes remarkable because correlations as high as 0.9 and even higher are frequently encountered. Disappointments are often encountered also, because it is not always easy to prejudge the correlation between X_j and Y_j. It should be understood, however, that after the sample is drawn and tabulated the sampling errors in f and X' are not matters of guess-work and conjecture but can be calculated (*vide infra*) and indeed in a professional job of sampling *are* calculated (Chs. 1, 6, 10, 12).

Remark 1. A particular example of a disappointment comes to the author's mind. It was supposed (1947) in the planning of an important monthly survey on "building starts" (commencement of actual construction on a house or a building intended for apartments or flats) that the ratio between the "starts" in any one county in a particular month of 1947 to the number of d.us. in the county in 1940 (the date of the last census) would be closely the same as in another county, and that an efficient ratio-estimate could be formed, X_j being the number of "starts" in County j in (e.g.) February 1947, Y_j the number of d.us. in this county in 1940, B the total number of d.us. in the whole country 1940 (known from the census), X' the estimated number of "starts" in the whole country during February 1947. It was supposed that counties that had the most d.us. in 1940 would show the greatest number of "starts" in any subsequent month, but the actual correlation between X_j and Y_j was near 0.5, and the ratio-estimate was about as good but no better than the unbiased estimate.

Remark 2. It is important to note that any persistent bias by which the measurements of Y_j are all too low or too high by a fairly constant percentage cancels out and leaves the estimate X' free of this bias. The student should perceive that if for instance B and every Y_j be doubled in Eq. 26, the estimate X' is unaffected.

Sometimes the base B in the ratio-estimate (Eq. 26), instead of being assumed known, is itself obtainable as another ratio-estimate which may be written as

$$Y' = \frac{\displaystyle\sum_1^m Y_j}{\displaystyle\sum_1^m Z_j} \, C = f'C \tag{28}$$

wherein Z_j is a third characteristic which is known primary unit by primary unit and whose total C is known for the universe.   For example, it might be that

$$X_j = \frac{N_j}{n_j} \sum_{k=1}^{n_j} x_{jk} = \text{the estimated total of the } x\text{-character-}$$
$$\text{istic for primary unit } j, \text{ today (e.g.,}$$
$$\text{male, 20–29, employed)} \tag{29}$$

$$Y_j = \frac{N_j}{n_j} \sum_{k=1}^{n_j} y_{jk} = \text{the estimated total of the } y\text{-character-}$$
$$\text{istic for primary unit } j, \text{ today (e.g.,}$$
$$\text{the total number of inhabitants)} \tag{30}$$

$$Z_j = \text{the total of the } z\text{-characteristic for primary unit } j \text{ (e.g.,}$$
$$\text{the number of inhabitants at the last census)} \tag{31}$$

$$C = \sum_1^M Z_j = \text{the total of the } z\text{-characteristic for the entire}$$
$$\text{universe, supposed known (e.g., the total}$$
$$\text{number of inhabitants at the last census)} \tag{32}$$

Then

$$f = \frac{\displaystyle\sum_1^m X_j}{\displaystyle\sum_1^m Y_j}$$

$$= \text{(e.g.) the estimated proportion of the population that}$$
$$\text{today is male, 20–29, employed}$$

$$= \frac{\displaystyle\sum_{j=1}^m \sum_{k=1}^{n_j} x_{jk}}{\displaystyle\sum_{j=1}^m \sum_{k=1}^{n_j} y_{jk}} \qquad \begin{array}{l}\text{[Provided } N_j/n_j \text{ is constant and cancels} \\ \text{out, which is a great advantage in} \\ \text{tabulation because numerator and} \\ \text{denominator are then merely sample-} \\ \text{totals]}\end{array} \tag{33}$$

$$f' = \frac{\sum\limits_{1}^{m} Y_j}{\sum\limits_{1}^{m} Z_j}$$ = the estimated ratio of the total number of inhabitants in the entire country today to the total number at the last census $\qquad$ (34)

The total of the x-characteristic for the universe may then be estimated as

$X' = ff'C$ = the estimated total of the x-characteristic for the whole country, today (e.g. the number of males 20–29, employed) $\qquad$ (35)

By writing the last equation in the form

$$X' = \frac{\sum\limits_{1}^{m} X_j}{\sum\limits_{1}^{m} Z_j} C \qquad (36)$$

it may be seen again from Exercise 1d on page 72 that as Z is a constant not subject to the errors of sampling,

$$\text{C.V. } X' = \text{C.V. } \frac{\sum\limits_{1}^{m} X_j}{\sum\limits_{1}^{m} Z_j} \qquad \begin{array}{l}\text{[Only if } C \text{ is a constant;}\\\text{compare with Eq. 27]}\end{array} \qquad (37)$$

whence (as will be evident later) the estimate X' will be very precise if X_j and Z_j are highly correlated. The ratio $\sum\limits_{1}^{m} X_j \Big/ \sum\limits_{1}^{m} Z_j$ might be of little interest; in fact it is often a hybrid difficult to define. At the same time both the ratios f and f' might be of major importance. Ordinarily there will be only one value of f' but many values of f—one for every population-characteristic. The sample of Greece, described in Chapter 12, provides an example of such a plan. f was calculated for every subclass of the population as a ratio to the total number of inhabitants, and the total number of inhabitants was itself estimated by the ratio-estimate $Y' = f'C$, where f' was the estimated ratio of the total number of inhabitants in 1946 to the total in 1940.

The accompanying table of possible ratio-estimates should be examined at this point.   Actually, a ratio may be formed whenever a pair of numbers X_j and Y_j exists for any primary unit drawn into the sample.

The question of the variance of f is difficult, but a treatment believed to be satisfactory for most purposes will be given presently.   It should be noted at this time, however, that the mean square error of the ratio-estimate in the form of Eq. 26 will be less than the variance of the unbiased estimate if the correlation coefficient between X_j and Y_j is greater than $\frac{1}{2}C_y:C_x$, where the letter C denotes a coefficient of variation. This statement will be apparent from Eq. 52 ahead.

LIST OF SOME POSSIBLE APPLICATIONS OF THE RATIO-ESTIMATE

In each case $f = \dfrac{\sum\limits_1^m X_j}{\sum\limits_1^m Y_j}$, $f' = \dfrac{\sum\limits_1^m Y_j}{\sum\limits_1^m Z_j}$, $B = \sum\limits_1^M Y_j$, and $C = \sum\limits_1^M Z_j$,

$X' = fB$ or $ff'C$ if C is involved.

Characteristic to be estimated	Definitions of X_j, Y_j, Z_j
f = estimated proportion of people possessing the particular characteristic (In this case the ratio-estimate X' would not be formed, as B would not be known. The unbiased estimate X might as well be used.)	Let X_j = estimated number of people having a particular characteristic in primary unit j, from today's sample Y_j = estimated total number of inhabitants in primary unit j, same sample
f = as above f' = estimated ratio of the total number of inhabitants today, to the total number at the time of the last census $X' = ff'C$ = estimated total number of inhabitants today	X_j = as above Y_j = as above Z_j = number of inhabitants in primary unit j at the last census
f = estimated number of inhabitants per d.u. appearing on the map (or directory) $X' = ff'C$ = the estimated number of people in the city	X_j = estimated number of people in primary unit j Y_j = estimated number of d.us. in primary unit j Z_j = number of d.us. appearing on the map or city directory covering primary unit j. (NOTE. As remarked elsewhere, in using a city directory the recommended unit is an *interval* extending from a particular address up to but not including another.

Characteristic to be estimated	*Definitions of X_j, Y_j, Z_j*
	Note also that to obtain the number Z, the total number of d.us. appearing on the map or in the directory must be counted.)
f = estimated number of bushels of wheat per acre of wheat f' = estimated proportion of cropland in wheat $X' = ff'C$ = estimated total number of bushels of wheat	X_j = estimated number of bushels of wheat on the farms of primary unit j Y_j = estimated number of acres of wheat harvested in primary unit j Z_j = number of acres of cropland (all crops) in primary unit j
f = estimated inventory of flour per retail dealer X' = estimated total inventory of flour of all dealers	X_j = estimated inventory of flour for the retail dealers of primary unit j Y_j = number of dealers in primary unit j
f = estimated average annual sales per dealer X' = estimated annual sales for all dealers	X_j = estimated annual sales of the dealers in primary unit j Y_j = number of dealers in primary unit j
f = estimated month-to-month change, 15th February to 15th March	X_j = estimated number of people (or employed, or sales, or net income, or total payroll in dollars) in primary unit j on the 15th February Y_j = same for 15th March
f = estimated change from the base period X' = estimated total for the 15th March	X_j = as above Y_j = same for the base period (possibly a census)
f = calibration factor X' = estimated total forest-stand (or number of hogs or d.us.), corrected	Y_j = forest-stand (or the number of people or hogs or number of d.us.) on primary unit j estimated by some cheap method, perhaps now obsolete X_j = same, but estimated by a more expensive and accurate method
f = calibration factor X' = estimated total, corrected	Y_j = weight or other physical determination on piece j, Y_j now suspected of being obsolete or in the wrong units X_j = recent redetermination, considered accurate

Remark 3. It is often asserted by the laity that the total number of inhabitants can not be estimated by a sample, that this number must be obtained by a complete count, that sampling is useful only for estimating the proportions of subclasses within the population (proportion male, female, 20–29, employed, etc.). Such statements probably stem from two sources: *i.* instinctive trust in ratio-estimates which are assumed to give proportions with great precision no matter how crude or biased the sampling procedure may be; *ii.* failure to use proper design and procedures of tabulation by which the minimum cost and precision of obtaining an estimate of the total number of inhabitants or of any other figure can be calculated.

The real points are that: *i.* the number of inhabitants *can* be estimated by sampling; but *ii.* the cost of obtaining a desired precision (e.g., a coefficient of variation of 1 percent) for a small area (possibly defined as one having 75,000 or fewer inhabitants) may be as costly as a complete count; *iii.* a sample that gives too little precision for the number of inhabitants (e.g., a coefficient of variation of 3 percent) may nevertheless through ratio-estimates give excellent precision for proportions of subclasses of the population, such as the proportion of males, 20–29, employed; *iv.* for larger areas, sampling will be cheaper than a complete count. It is difficult to cite actual figures, because so much depends on the physical facilities and census information available, but experience shows that an area having 250,000 inhabitants can be sampled so as to estimate the number of inhabitants with a coefficient of variation of 1 percent (a really very precise result, comparable with the accuracy of a complete count) at perhaps a fifth of the cost of a complete count. But of course the complete count gives detail by small areas, which a sample cannot do.

Remark 4 (Possible checks and controls on the tabulation). In the process of tabulating the returns of a survey, the numbers (populations) in any exhaustive set of subclasses of the sample in primary unit j, when summed, will equal the total population (of all classes) counted in the sample for that primary unit. Multiplication of *all* these numbers by N_j/n_j or any other weighting factor will not destroy this relationship. It is obvious that an elaborate system of checks and controls on the tabulations can be formulated by which arithmetic errors in any stage of the work can be caught early and corrected before they have gone far enough to do a lot of damage. Of course, in a large enterprise, it is to be expected that a few errors will escape undetected through a maze of controls, only to be picked up later after considerable damage has been done.

Remark 5. It should be mentioned that a ratio-estimate is slightly biased, owing to the fact that the denominator in Eq. 25 is a random variable. The ratio written in Eq. 25 is nevertheless consistent, as it gives the true ratio φ for the universe when the sample is total. A simple illustration of the bias and variance in this estimate is given in a later section. By contrast, the estimate $X = (M/m) \sum_1^m X_j$ is unbiased.

The variance of a ratio-estimate. The formulas

$$\operatorname{Var} f \doteq \frac{M - m}{M} \frac{1}{m\bar{B}^2} \frac{1}{m - 1} \sum_1^m (X_j - fY_j)^2 \quad [\bar{B} = EY_j] \quad (38)$$

and

$$(\text{C.V.} f)^2 \doteq \frac{M - m}{M} \frac{1}{m(m - 1)} \sum_1^m \left[\frac{X_j - f Y_j}{\overline{X}} \right]^2 \qquad (38a)$$

or the equivalent form

$$(\text{C.V.} f)^2 \doteq \frac{M - m}{M} \frac{1}{m(m - 1)} \sum_1^m \left[\frac{X_j}{\overline{X}} - \frac{Y_j}{\overline{Y}} \right]^2 \qquad (38b)$$

for estimates of the variance of the ratio-estimate f as defined in Eq. 25 will be found satisfactory under most conditions met in practice. For a proof let [16,17]

$$x = \sum_1^m X_j, \quad y = \sum_1^m Y_j, \quad F(x, y) = \frac{x}{y} = f$$

The problem is to find Var x/y, i.e., Var f. The proof depends on the validity of the linear terms of a Taylor expansion in powers of $x - Ex$ and $y - Ey$, which is

$$F(x, y) = F(Ex, Ey) + (x - Ex)F_x + (y - Ey)F_y + R \qquad (39)$$

The derivatives F_x and F_y are to be evaluated at $x = Ex$ and $y = Ey$; whence $F_x = 1/y$ has the value $1/Ey$, and $F_y = -x/y^2$ has the value $-Ex/(Ey)^2$. R denotes the remainder, which will now be neglected, and the effect of the neglect discussed later. Then

$$E[F(x, y) - F(Ex, Ey)]^2$$

$$\doteq E[(x - Ex)F_x + (y - Ey)F_y]^2 \qquad (40)$$

$$= E \left[\frac{x - Ex}{Ey} - \frac{Ex}{(Ey)^2} (y - Ey) \right]^2$$

$$= E \frac{y^2}{(Ey)^2} \left[\frac{x^2}{y^2} - 2 \frac{x}{y} \frac{Ex}{Ey} + \left(\frac{Ex}{Ey} \right)^2 \right] \quad \begin{array}{l} \text{[After squaring and} \\ \text{reducing]} \end{array} \quad (41)$$

That is,

$$E \left[\frac{x}{y} - \frac{Ex}{Ey} \right]^2 \doteq E \frac{y^2}{(Ey)^2} \left[\frac{x}{y} - \frac{Ex}{Ey} \right]^2$$

$$= \frac{1}{(Ey)^2} E \left[x - y \frac{Ex}{Ey} \right]^2 \qquad (42)$$

---

[16] I owe this proof to my colleague William N. Hurwitz.

[17] The symbols x and y as used here are out of line with the notation, but it is hoped that no confusion will arise.

It is to be noted that $Ex/Ey = \varphi$, while $x/y = f$. The left-hand side is thus the mean square error in f and in fact is $\mathrm{Var}\, f + (\mathrm{Bias}\, f)^2$, where $\mathrm{Bias}\, f = Ef - \varphi$ (Eq. 47, p. 129). But as $(\mathrm{Bias}\, f)^2 = (Ef - \varphi)^2$ involves only the "expected" values of terms in the fourth and higher degrees of $x - Ex$ and $y - Ey$ combined, which have already been absorbed in the remainder R, it follows that the left-hand member of the last equation may be replaced by $\mathrm{Var}\, f$, whence it appears that

$$\mathrm{Var}\, f \doteq \frac{1}{\left(E\sum_1^m Y_j\right)^2} E\left[\sum_1^m X_j - \varphi \sum_1^m Y_j\right]^2$$

$$= \frac{1}{(mEY_j)^2} E\left[\sum_1^m Y_j(\varphi_j - \varphi)\right]^2 \tag{43}$$

in which $X_j = \varphi_j Y_j$.

It will now be assumed that there is no subsampling: [18] the mathematical derivation is beyond control otherwise, although in the author's opinion the final form of the estimate of $\mathrm{Var}\, f$ as already given in Eq. 38 may be used to estimate $\mathrm{Var}\, f$ with and without subsampling. Going on,

$$\mathrm{Var}\, f = \frac{1}{(m\, EY_j)^2}\left\{E\sum_1^m Y_j^2(\varphi_j - \varphi)^2\right.$$

$$+ E\sum_{j=1}^m \sum_{\substack{j'=1 \\ (j' \neq j)}}^m Y_j(\varphi_j - \varphi)Y_{j'}(\varphi_{j'} - \varphi)\right\}$$

$$= \frac{1}{(m\, EY_j)^2}\left\{\frac{m}{M}\sum_1^M Y_j^2(\varphi_j - \varphi)^2\right.$$

$$+ \frac{m}{M}\frac{m-1}{M-1}\sum_{j=1}^M \sum_{\substack{j'=1 \\ (j' \neq j)}}^M Y_j Y_{j'}(\varphi_j - \varphi)(\varphi_{j'} - \varphi)\right\}$$

$$= \frac{1}{(m\, EY_j)^2}\left\{\frac{m}{M}\sum_1^M Y_j^2(\varphi_j - \varphi)^2\right.$$

$$+ \frac{m}{M}\frac{m-1}{M-1}\left[\left\{\sum_{j=1}^M Y_j(\varphi_j - \varphi)\right\}^2 - \sum_1^M Y_j^2(\varphi_j - \varphi)^2\right]\right\}$$

[18] The symbols X_j and Y_j will be used even though only one stage of sampling is considered. There are two reasons: first, convenience; second, the fact that the final product (Eq. 38) probably holds pretty well even when there are higher stages of sampling.

$$= \frac{1}{(m\,EY_j)^2} \left\{ \frac{m}{M} \sum_1^M Y_j{}^2(\varphi_j - \varphi)^2 \right.$$

$$\left. + \frac{m}{M} \frac{m-1}{M-1} \left[0 - \sum_1^M Y_j{}^2(\varphi_j - \varphi)^2 \right] \right\}$$

$$= \frac{1}{(m\,EY_j)^2} \frac{M-m}{M-1} \frac{m}{M} \sum_1^M Y_j{}^2(\varphi_j - \varphi)^2$$

$$= \frac{M-m}{M-1} \frac{1}{m(EY_j)^2} \frac{1}{M} \sum_1^M (X_j - \varphi Y_j)^2 \tag{44}$$

In practice $\sum_1^M (X_j - \varphi Y_j)^2$ must be estimated, and to accomplish

this estimate it is satisfactory to replace $\dfrac{1}{M-1} \sum_1^M (X_j - \varphi Y_j)^2$ by

$\dfrac{1}{m-1} \sum_1^m (X_j - fY_j)^2$. When this is done, the result is Eq. 38,
already written. If the number m of primary units is large, and the
burden too great to use all of them, particularly if there is need for
speed, it will almost always be satisfactory to select a secondary sample
of 50 or 75 primary units, replacing $\dfrac{1}{m-1} \sum_1^m (X_j - fY_j)^2$ by

$\dfrac{1}{m'-1} \sum_1^{m'} (X_j - f'Y_j)^2$ in Eq. 38, where now m' is the 50 or 75
primary units selected at random from the m primary units of the

sample, and $f' = \sum_1^{m'} X_j \Big/ \sum_1^{m'} Y_j$. When necessary, EY_j may be esti-

mated from the sample; but usually, when the ratio-estimate is advantageous, this quantity is known beforehand from the map or census
data. It may be recalled from the table of notation that EY_j is denoted
by B/M or $\bar{B}$, as seen in Eq. 38.

Remark. A word should be said regarding the remainder R which was
dropped. It contains $1/m^2$ and higher powers of $1/m$; hence, as $x - Ex$
and $y - Ey$ are bounded, there is some sample-size m beyond which the
estimate of Var f as given in Eq. 38 will differ from the actual value of
Var f by any desired quantity however small, but it is impossible from the
above development to say in general what this size is. The answer to this
question depends on the distributions of X_j and Y_j. For the universes and
sample-sizes met in the author's practice so far, Eq. 38 gives good results.

It is interesting to note that the last equation may be written

$$(\text{C.V.} f)^2 \doteq \frac{M - m}{M - 1} \frac{V^2}{m} \tag{45}$$

which has the familiar appearance of Eq. 13′ on page 102 of Chapter 4. Here

$$V^2 = \frac{1}{M} \sum_1^M \left(\frac{X_j - \varphi Y_j}{\bar{A}}\right)^2 \qquad [\bar{A} = EX_j] \tag{46}$$

and V^2 may be interpreted as a coefficient of variation, viz., the mean square deviation in the x-direction of the M points X_j, Y_j from the line of slope φ drawn through the origin, each deviation being measured in units of $\bar{A} = EX_j$. In Fig. 13 $\bar{A}$ is the universe-average of the x-characteristic per primary unit, as defined in the notation, and $\bar{B} = EY_j$ connotes the same for the y-characteristic. $\bar{X} = \frac{1}{m} \sum_1^m X_j$ and $\bar{Y} = \frac{1}{m} \sum_1^m Y_j$ are corresponding averages obtained from the sample. It is obvious from the figure and from the use of V in Eq. 45 that high correlation between X_j and Y_j will give precise estimates of f, in which connexion the reader should consult the text following Eq. 53.

Another formula for the sampling error in f which is fairly well known and identical with Eq. 38 but perhaps less readily adaptable to computation is this:

$$(\text{C.V.} f)^2 \doteq \frac{M - m}{M} \frac{1}{m - 1} (C_x'^2 + C_y'^2 - 2r\,C_x'C_y') \tag{47}$$

wherein C_x' and C_y' denote the coefficients of variation of the m sample-values of X_j and Y_j respectively, and r is the computed correlation coefficient between these m pairs. Specifically,

$$C_x' = \frac{s_x}{\bar{X}} \qquad \text{[The coefficient of variation of the } m \text{ sample-values of } X_j]$$

$$\bar{X} = \frac{1}{m} \sum_1^m X_j \qquad \text{[The mean of the } m \text{ sample-values of } X_j]$$

$$s_x^2 = \frac{1}{m} \sum_1^m (X_j - \bar{X})^2 \qquad \text{[The variance of the } m \text{ sample-values of } X_j]$$

$$r = \frac{\frac{1}{m} \sum_1^m (X_j - \bar{X})(Y_j - \bar{Y})}{s_x s_y} \qquad \text{[The correlation coefficient computed for the } m \text{ pairs of } X_j \text{ and } Y_j]$$

To derive the above formula for Var f, return to Eq. 40, still neglecting the remainder R, square the right-hand side, and get

$$E[F(x, y) - F(Ex, Ey)]^2$$

$$= E[(x - Ex)^2 F_x{}^2 + (y - Ey)^2 F_y{}^2 - 2(x - Ex)(y - Ey)F_x F_y]$$

$$= \left(\frac{Ex}{Ey}\right)^2 E\left[\left(\frac{x - Ex}{Ex}\right)^2 + \left(\frac{y - Ey}{Ey}\right)^2 - 2\frac{(x - Ex)(y - Ey)}{ExEy}\right]$$

$$= \varphi^2(c_x{}^2 + c_y{}^2 - 2\rho\, c_x c_y) \tag{48}$$

wherein c_x and c_y denote the coefficients of variation of the probability-distributions of x and y, ρ is the correlation between x and y, and $\varphi = Ex/Ey = EX_j/EY_j$ as before. Now, as x and y are both summations taken over the m primary units of the sample, which is assumed to have been drawn without replacement, $c_x{}^2$ and $c_y{}^2$ are both smaller by the factor $(M - m)/(M - 1)m$ than the corresponding coefficients of variation C_x and C_y of the individual values X_j and Y_j of the universe. Specifically, the symbols C_x and C_y are defined by the equations

$$C_x{}^2 = \frac{E(X_j - EX_j)^2}{(EX_j)^2} \quad \text{and} \quad C_y{}^2 = \frac{E(Y_j - EY_j)^2}{(EY_j)^2} \tag{49}$$

In regard to the correlation ρ it should be recalled from Exercise 6 on page 124 that the correlation between means of samples and hence the correlation between sums of samples of points drawn without replacement are the same as the correlation between the individual points X_j and Y_j of the universe. Therefore, if the left-hand side of Eq. 48 be replaced by Var f (to which it is very nearly equal) and if the other suggested replacements be made and both members be divided through by φ^2, the result is

$$(\text{C.V.}\,f)^2 \doteq \frac{M - m}{M - 1}\frac{1}{m}\,(C_x{}^2 + C_y{}^2 - 2\rho C_x C_y) \tag{50}$$

This is algebraically identical with Eq. 45 because it was derived from the same starting-point. In practice, C_x, C_y, and ρ must be estimated from the sample of m points, and Eq. 47 follows immediately by writing r as an estimate of ρ and by substituting $mC_x'^2/(m - 1)$ as an estimate of $MC_x{}^2/(M - 1)$ with a similar substitution involving $C_y'^2$ (a partial justification for which may be apparent in the next chapter).

In view of the fact that Eqs. 45 and 50 are identical with each other, it must be that V, as defined in Eq. 46, may also be written as

$$V^2 = C_x{}^2 + C_y{}^2 - 2\rho C_x C_y \tag{51}$$

The numerical illustration in the next section may help toward an understanding of this equation.

From Eqs. 50 and 37 it is clear that

$$(\text{C.V.}\,X')^2 = (\text{C.V.}\,f)^2 \underset{>}{\overset{\leq}{=}} \frac{M-m}{M-1}\frac{C_x{}^2}{m} = (\text{C.V.}\,X)^2$$

according as

$$\rho \underset{<}{\overset{>}{=}} \frac{C_y}{2C_x} \tag{52}$$

This important inequality shows under what conditions the ratio-estimate X' is more precise than the unbiased estimate X.

From the foregoing formulas for Var f or C.V. f it is simple to compute 2- or 3-sigma limits for f. Another possibility is to compute Var f and to use tests of significance with probability levels, or to compute confidence limits for φ (Chs. 9 and 17), for which Fieller's device should be mentioned.[19]   For computing upper and lower 0.02 confidence limits of φ, for example, he puts

$$t_1 = \frac{f-\varphi}{f(\text{Est'd C.V.}\,f)} \tag{53}$$

The required estimate of the C.V. f in the denominator may be made by Eq. 45 or 50, and t_1 is taken from Fisher's table of t or from the nomograph of Nekrassoff (p. 555), for $m-1$ degrees of freedom. This equation gives a quadratic in f, whose two roots are the desired upper and lower confidence limits for φ.

$$\frac{f-\varphi}{\sigma_f}$$

has the t-distribution which will be studied later.

Numerical illustration of the bias and variance of a ratio-estimate. The calculations should be undertaken as an exercise.

First example

The problem here will be to estimate φ, the number of people per d.u. in the universe. The universe will consist of but 3 primary units, each having subunits, which are merely d.us., and in these d.us. are people. Samples of $m = 2$ primary units will be taken; no subsampling.

---

[19] E. C. Fieller, *J. Royal Stat. Soc.*, vol. vii, 1940–41: pp. 1–64, p. 19 in particular. R. C. Geary proved that if X_j and Y_j are normally distributed random variables, then the fraction on the right of Eq. 53 is normally distributed for large samples. The reference is to his article "On the frequency distribution of the quotient of two normal variates," *J. Royal Stat. Soc.*, vol. xciii, 1930: pp. 442–6.

$$\sum_1^m X_j \text{ will be the number of people in the sample}$$

$$\sum_1^m Y_j \quad `` \quad `` \quad `` \quad `` \quad `` \text{ d.us.} \quad `` \quad `` \quad ``$$

$$f = \frac{\displaystyle\sum_1^m X_j}{\displaystyle\sum_1^m Y_j} \quad `` \quad `` \quad `` \text{ estimated number of people per d.u.}$$

There are only 3 possible samples, and it will be easy to compute their variance and to compare it with the result given by Eq. 38. The universe is described in Table 1, and the samples in Table 2. The 3 estimates of Var f are shown in Table 3.

TABLE 1. THE UNIVERSE

Primary unit j	a_{jk}, population in subunit k			X_j, population in primary unit j	Y_j, number of secondary units in primary unit j
	1	2	3		
1	1	xxx	xxx	1	1
2	1	2	xxx	3	2
3	1	2	3	6	3
Totals				$A = 10$	$B = 6$

$$\varphi = \frac{A}{B} = \frac{10}{6} = \frac{5}{3} \text{ people per d.u.}$$

$$EY_j = \frac{6}{3} = 2 \ (= \bar{B} \text{ in Eq. 38})$$

TABLE 2. SAMPLES OF TWO PRIMARY UNITS

Sample consisting of primary units	X_j		Y_j		$\sum_1^m X_j$	$\sum_1^m Y_j$	$f = \dfrac{\displaystyle\sum_1^m X_j}{\displaystyle\sum_1^m Y_j}$
	$j = 1$	$j = 2$	$j = 1$	$j = 2$			
1 and 2	1	3	1	2	4	3	1.333
1 and 3	1	6	1	3	7	4	1.75
2 and 3	3	6	2	3	9	5	1.8
Averages					$6\frac{2}{3}$	4	1.628

TABLE 3.   CALCULATION OF VAR f BY EQ. 38

fY_j		$X_j - fY_j$		$\sum_1^m (X_j - fY_j)^2$	Var f calculated by Eq. 38
$j = 1$	$j = 2$	$j = 1$	$j = 2$		
1.333	2.667	−0.333	0.33	0.222	0.009
1.75	5.25	−.75	.75	1.125	.047
3.6	5.4	−.6	.6	0.72	.032
Average					0.029

It should first of all be noted from Table 2 that the Av $(M/m) \sum_1^m X_j$ $= \frac{3}{2} \text{Av} \sum_1^m X_j = 1.5 \times 6\frac{2}{3} = 10$, which is exactly the total number of people in the universe. Likewise Av $(M/m) \sum_1^m Y_j = \frac{3}{2} \text{Av} \sum_1^m Y_j = 1.5 \times 4 = 6$, which is exactly the total number of d.us. in the universe. These statements illustrate the fact that

$$X = \frac{M}{m} \sum_1^m X_j$$

is an unbiased estimate of the total population of the universe, as has been declared a number of times. Next it should be noted also from Table 2 that Av $f = 1.628$, while the true value of φ is $\frac{5}{3}$ or 1.67. Thus, the ratio-estimate is slightly biased, as was stated earlier. This bias, however, is smaller than the standard error of

$$\frac{1}{B} \frac{M}{m} \sum_1^m X_j \qquad [B = 6, M = 3, m = 2]$$

which is the *unbiased* estimate of φ. $\left(\sum_1^m X_j \right.$ is shown in Table 2, and the student may easily observe the truth of this statement.)

Eq. 38 here takes the form

$$\text{Var} f = \frac{1}{3} \frac{1}{2 \times 2^2} \frac{1}{2 - 1} \sum_1^m (X_j - fY_j)^2$$

$$= \frac{1}{24} \sum_1^m (X_j - fY_j)^2$$

which is used in Table 3.   The actual Var f is

$$\tfrac{1}{3}[(1.333 - 1.628)^2 + (1.75 - 1.628)^2 + (1.8 - 1.628)^2] = 0.044$$

Of course Eq. 38 can not be expected to provide a good estimate of Var f when there are only two primary units in the sample.

As an exercise the student should calculate $\mu_u = 1.5$ and the variances σ^2, $\sigma_j{}^2$, $\sigma_w{}^2$, $\sigma_e{}^2$, $\sigma_b{}^2$, $\sigma_{tu}{}^2$, $\sigma_{wu}{}^2$, $\sigma_{bu}{}^2$ for the universe described in Table 1, page 179.   (Note that $\sigma_1{}^2$ does not exist, as there is only 1 secondary unit in the primary unit $j = 1$.)

Second example

Let a universe consist of two primary units as shown in Fig. 14. To bring out the distinction between the unbiased estimate X and the ratio-estimate X', the two primary units are purposely made very unequal.   Each primary unit consists of d.us., some vacant (V) and some not vacant.   The characteristic to be measured is vacancy: a d.u. counts 1 if vacant, 0 otherwise, as on page 157.   Here, B_j will be used in place of N_j, in conformity with the notation that has been used for the ratio-estimate.

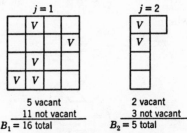

Fig. 14.   For an illustrative exercise on the ratio-estimate, a universe consists of two primary units, in each of which are a number of d.us., some of which are vacant (V). Samples of 1 primary unit are to be drawn, and from each sample the proportion vacant and the total number of vacant d.us. in the universe are to be estimated.

$B_1 = 16$, the number of d.us. in the 1st primary unit

$B_2 = 5$, " " " " " " 2d " "

$a_{jk} = 1$ if the jkth d.u. is vacant

$\quad\ = 0$ otherwise

$A_1 = \sum_1^{16} a_{1k} = 5$, total vacancies in the 1st primary unit

$A_2 = \sum_1^{5} a_{2k} = 2$, " " " " 2d " "

$A = A_1 + A_2 = 5 + 2 = 7$, the total number of vacancies in the universe

$\varphi = \dfrac{A}{B} = \dfrac{5 + 2}{16 + 5} = \dfrac{7}{21} = \dfrac{1}{3}$, the overall proportion vacant

Imagine that a sample of one primary unit is drawn and canvassed completely; no secondary sampling. (Multistage sampling may of course be used, but it only confuses the discussion at the moment.) For $m = 1$ there are but two possible samples, corresponding to the two primary units, either of which may be in the sample, and it will be supposed that each has the same probability.

Two estimates of A and φ will be used, and these estimates compared.

$$\bar{x} = \frac{X}{B}, \qquad \text{the unbiased estimate of } \varphi$$

where

$$X = \frac{M}{m} \sum_1^m X_j, \quad \text{``} \qquad \text{``} \qquad \text{``} \qquad \text{`` } A$$

$$f = \frac{\displaystyle\sum_1^m X_j}{\displaystyle\sum_1^m Y_j}, \text{ the ratio-estimate of } \varphi$$

$$X' = fY, \qquad \text{``} \quad \text{``} \qquad \text{``} \qquad \text{`` } A$$

(Of course, as there is but one primary unit in the sample, the summations contain but a single term.) The results are shown in Table 4.

TABLE 4. RESULTS IN THE SECOND EXAMPLE

$$\frac{M}{m} = 2, \quad B = 16 + 5 = 21$$

Sample	$\sum_1^m X_j = X_j$	$\sum_1^m Y_j = Y_j$	$\bar{x}$	f	X	X'
$j = 1$	5	16	$\frac{10}{21}$	$\frac{5}{16}$	$2 \times 5 = 10$	$\frac{5}{16} \times 21 = 6.6$
$j = 2$	2	5	$\frac{4}{21}$	$\frac{2}{5}$	$2 \times 2 = 4$	$\frac{2}{5} \times 21 = 8.4$
Average	xxx	xxx	$\frac{1}{3}$	$\frac{171}{480}$	7	7.4
Bias	xxx	xxx	0	$\frac{11}{480}$	0	0.4
Standard deviation	xxx	xxx	$\frac{1}{7}$	$\frac{1}{23}$	3	.9

The student should note the small bias (0.4) of the ratio-estimate X' and should also compare its standard deviation with the standard deviation of the unbiased estimate X, calculated with the formula

$$\text{Var } X = \left(\frac{M}{m}\right)^2 \left\{ \frac{M - m}{M - 1} m\sigma_e{}^2 + 0 \right\} \qquad \begin{array}{l}\text{[Eq. 8 with no} \\ \text{subsampling]}\end{array}$$

The second term is 0 because there is no subsampling. In this case $m = 1$ and $M = 2$.

Importance of censuses for subsequent samples. A ratio-estimate provides one of the most effective ways of making use of prior information (census, maps, parish records, previous weighings, etc.). The sample is then used *only to bring the prior information up to date*. The sample provides a set of correction-factors (f), one for each population characteristic.

Besides ratio-estimates, there are still other ways of using prior information, as for stratifying the sampling units by various predominant characteristics and by size-classes as of some previous census.

It may now be obvious why sampling enthusiasts are always strong supporters of complete counts. Census information is extremely useful in reducing the costs and increasing the precision of subsequent samples. Complete counts give data by small areas, and such data are useful not only for their intended prime purpose of providing statistics for social, economic, business, marketing, and other studies, but for sampling purposes as well. A complete count provides

1. Lists (of people, d.us., business establishments, farms) for subsequent sample studies and for controls of both samples and later complete counts. Any such lists rapidly become obsolete, but retain their utility when supplemented by a small sample of areas to correct for biases in the list.

2. Characteristics of small areas, showing the number of inhabitants, d.us., households, business establishments, farms, etc., by important characteristics. This information cuts the cost of future sample-studies because it permits sample-designs embodying:

Stratification of the areas.

Ratio-estimates.

Measures of size (population, number of d.us., sales and employment in business establishments, etc.) for constructing areal units of roughly equal size, or for drawing areal units with probability in proportion to size (theory not given here).

Measures of size for defining areal units of roughly equal numbers of inhabitants, d.us., business establishments, etc.

The fact is that, so far as subsequent samples are concerned, the information furnished by a complete count *need not be accurate*. Even a wretched or obsolete complete count is usually extremely useful in the planning of samples. Any census no matter how good becomes out of date anyhow, later, but by proper sample-design (as already explained) the inaccuracies of prior census information *do not cause bias in the estimates prepared from the sample*.[20]

Application of the ratio-estimate to industrial weighing and other quality determinations. The foregoing theory applies to many types of

[20] The bias of a ratio-estimate is ignored here because it is usually so extremely small.

THIRTY-SIX BALES OF GREASY WOOL FROM ARGENTINA LANDED IN BOSTON

Bale number	Y_j "Marked" weight, Buenos Aires (kgs.)	X_j Weight as determined by the Bureau of Customs, Boston (lbs.)
1366	447	980
1367	445	978
1368	446	975
1369	446	982
1370	449	986
1371	447	978
1372	447	981
1373	449	982
1374	445	978
1375	448	980
1376	449	988
1377	448	981
1378	449	983
1379	446	977
1380	447	979
1381	449	987
1382	447	979
1383	448	979
1384	447	979
1385	449	983
1386	447	983
1387	448	987
1388	448	987
1389	445	979
1390	448	980
1391	447	980
1392	447	979
1393	448	985
1394	447	980
1395	450	987
1396	449	981
1397	449	984
1398	449	987
1399	446	977
1400	441	970
1401	435	959
Totals	16,092	35,300
Samples cored before weighing		17
		35,317

industrial sampling. In general terms, we shall suppose that for each article, bale, bag, or other unit in a shipment or purchase there has been a previous determination Y_j of quality, perhaps marked on the article or container or otherwise recorded on proper papers. Except for changes in moisture content, or possible persistent bias in the quality determinations, the quality should still be the same. It must be remembered, however, that immediate redeterminations would not give exact agreement, either, particularly if the scale or instruments are sensitive and are properly handled. Often it is not necessary to make complete new determinations of quality at the later date when received, but only on a limited amount of material properly selected. Breakage and pilferage are supposed to be obvious, and such containers thrown out for separate treatment. Statistically speaking, the problem is to determine what size of sample m will suffice. The supposition is that the M numbers Y_1, Y_2, $\cdots$, Y_M are free or very cheap, whereas the m numbers X_1, X_2, $\cdots$, X_m, representing new determinations, are costly. When this is not so there is no problem: the sensible solution would then be to reweigh the whole shipment. In terms of the foregoing theory the solution lies in using the ratio-estimate

$$X' = fY \qquad \text{[Eq. 26]}$$

What should the size of sample (m) be in order that C.V. X' shall be within prescribed tolerances, e.g., 0.0017? (This value 0.0017, incidentally, corresponds to a 3-sigma interval of $\frac{1}{2}$ percent, an extremely narrow tolerance for many quality-determinations.) This problem presents a case where Y is not subject to sampling error, as Y is merely the sum of all the weights as measured previously.

For an illustration I shall turn to the Bureau of Customs. Wool, tobacco, duck feathers, goose feathers, cheese, olive oil, and many other commodities are imported in bales or bags or boxes, on which it is customary for the seller abroad to mark the weight as he determines it at the time of shipment. These form the series Y_j, which are free. The Bureau of Customs and the buyer reweigh the units or as many as are necessary as they are landed in New York, Newport News, or elsewhere, thus establishing a sample of X_j values. A set of data is shown in the accompanying table. Therein it is seen that

$$\Sigma X_j = 35{,}300 \text{ lbs.} \qquad\qquad \Sigma Y_j = 16{,}092 \text{ kgs.}$$

$$\bar{A} = \frac{35{,}300}{36} = 980.56 \text{ lbs.} \qquad\qquad \bar{B} = \frac{16{,}092}{36} = 447 \text{ kgs.}$$
$$\text{per bale} \qquad\qquad\qquad\qquad\qquad \text{per bale}$$

$$\varphi = \frac{\bar{A}}{\bar{B}} = 2.1936$$

Note that if the 17 lbs. of wool that were taken out in cores are replaced, the ratio of the total number of pounds as determined in Boston to the total number of kilograms as determined in Buenos Aires is $35{,}317 \div 16{,}092 = 2.1947$, which is close to the usual conversion factor 2.2046 for converting kilograms to English pounds. However, the two numbers differ by barely 3 sigma as we shall see presently, and the difference, small as it is, is probably significant of slight evaporation of moisture or difference in calibration or use of scales at either end (Rio de Janeiro or Boston).

The student is asked to verify the following calculations (slide rule used).

Variances $\qquad\qquad \sigma_x^2 = 28.2 \quad \sigma_y^2 = 6.89$

Coefficients of variation $\qquad C_x = \dfrac{\sqrt{28.2}}{980.56} = 0.00541 \quad C_y = \dfrac{\sqrt{6.89}}{447.00} = 0.00587$

Correlation between X_j and Y_j

$$\rho = \frac{\dfrac{1}{M}\sum_1^M (X_j - \overline{X})(Y_j - \overline{Y})}{\sigma_x \sigma_y}$$

$$= \frac{\dfrac{1}{M}\sum_1^M X_j Y_j - \overline{X}\,\overline{Y}}{\sigma_x \sigma_y} = 0.903$$

It will be noted that C_y for weights in kilograms is slightly larger than C_x for weights in pounds, whereas it should be surmised that a coefficient of variation should be independent of the units used, or any persistent biases. As each bale was weighed only to a whole kilogram or a whole pound, the difference may arise from errors in rounding. As the kilogram is the larger unit, these errors would be more pronounced for kilograms, giving a higher value to C_y, as observed.

The quantity V will now be computed.

$$V^2 = C_x^2 + C_y^2 - 2\rho C_x C_y \qquad\qquad \text{[Eq. 51]}$$
$$= 10^{-8}\{54^2 + 59^2 - 2 \times 0.903 \times 54 \times 59\}$$
$$= 10^{-8}\{(54{-}59)^2 + 2 \times 0.097 \times 54 \times 59\}$$
$$= 643 \times 10^{-8} \qquad\qquad \text{[By slide rule]}$$
$$V = 0.0025 \text{ or } 0.25 \text{ percent} \qquad\qquad (54)$$

In exercises appearing shortly the student is asked to compute C.V. X_j/Y_j and V by using Eq. 46.

In this illustration, V being so small, it is clear from Eq. 47 that the weighing of but 1 bale drawn at random from this shipment would have provided an estimate of X' with a standard error of 0.25 percent. Two random bales would certainly have given all the precision needed. However, before one may safely specify in advance the number of weighings, he must be able to predict the value of V on the basis of past experience and knowledge of the source and type of the material. Control charts on $X_j : Y_j$ would need to be kept for each type of wool over a period of months in order to learn what numerical value is suitable for any particular type of wool, or indeed, whether sampling is practicable at all for some particular types of wool. Control as exhibited by a control chart would be interpreted to mean that $X_j : Y_j$ is *stable*—that is, steady enough to be predictable for sampling a *future* shipment. Lack of control would force settlement on a value of $X_j : Y_j$ by engineering judgment rather than by the objective estimates provided by the control chart. Smaller samples are possible with safety when $X_j : Y_j$ exhibits control. This problem has now been extensively and successfully studied by Louis Tanner of the Customs Laboratory, Boston, and Churchill Eisenhart of the National Bureau of Standards. Applications to many other raw and manufactured products are also being developed in industry.

The average error in weighing a sample of m bales instead of the entire M bales of a lot can be computed by Eq. 46 or Eq. 51, either of which gives

where

$$\left.\begin{aligned} m &= \frac{M}{M + m_\infty - 1}\, m_\infty \\[2mm] m_\infty &= \left(\frac{V}{\text{C.V. } f}\right)^2 \end{aligned}\right\} \tag{55}$$

wherein C.V. f represents an allowable sampling error, and m_∞ denotes the sample-size when the lot-size is infinite.

A reasonable sampling tolerance might be $\frac{1}{2}$ percent, or 1 lb. in 200, which fixes the C.V. f at 0.005/3 or 0.0017. We may now calculate a table of required sample-sizes under various assumptions for V (which denotes the coefficient of variation of the universe of values of $X_j : Y_j$: Fig. 13, p. 166). The student should verify the calculations in the accompanying table. Interpolation or recalculation for any particular type of wool not covered by the table can easily be made. In application to testing and purchasing in industry, the sample of articles or other units might be chosen systematically off the production line or as the material is received. In this particular example (weighing wool) the same plan would be satisfactory, but a more economical plan, if satisfactory, would be to weigh the first m bales as they are unloaded. The advantage gained is that the weighing crew may thereupon proceed to other work, such as weighing another sample from another shipment on another wharf. Fewer weighing crews would thus be required.

ILLUSTRATIVE TABLE OF REQUIRED SAMPLE-SIZES FOR WEIGHING BALES OF WOOL

3 C.V. $f = 0.005 = \frac{1}{2}$ percent

V assumed for the coefficient of variation of X_j/Y_j in the universe	Required sample, m, for lot-size M			
	$m_\infty = \left(\dfrac{V}{\text{C.V. } f}\right)^2$	$m = \dfrac{M}{M + m_\infty - 1}\, m_\infty$		
	$M = \infty$	$M = 500$	$M = 100$	$M = 50$
0.0025	2	2	2	2
Suggested sampling interval		1 in 250	1 in 50	1 in 25
0.0050	9	9	9	8
Suggested sampling interval		1 in 50	1 in 10	1 in 5
0.0075	20	19	16	14
Suggested sampling interval		1 in 25	1 in 6	1 in 4
0.0100	36	34	26	21
Suggested sampling interval		1 in 15	1 in 4	1 in 2
0.0125	56	50	36	26
Suggested sampling interval		1 in 10	1 in 3	1 in 2

Applications of these principles abound in industry, and the possible savings are enormous. While it is true that "spot checking" is in common use, very little of it as practiced could be called sampling. *Sampling*, in modern techniques, *means a specified degree of protection* to both buyer and seller, at minimum cost (Ch. 1).

It should be noted that the ratio-estimate is really a calibration of one scale in terms of another, and that the use of the method is not in the slightest degree impaired if the "marked weights" are in kilograms, Spanish pounds, Mexican pounds, Turkish pounds, okes, gallons, litres, or any other unit.

Exercise 1. By direct computation find the average value of the 36 values of $X_j : Y_j$ in the table of weights, and the standard deviation and coefficient of variation of $X_j : Y_j$. Explain the difference between

$$\sum_1^M X_j \Big/ \sum_1^M Y_j \text{ and the average of } X_j : Y_j.$$

Exercise 2. Use Eq. 46 to compute V.

Outline

$$\varphi = \frac{35,300}{16,092} = 2.1936 \text{ lbs. per kg.} \quad \text{(as before)}$$

$$\bar{A} = \frac{35,300}{36} = 980.56 \text{ lbs. per bale} \quad \text{(as before)}$$

Now compute $X_j - \varphi Y_j$ and its square for every one of the 36 points $(j = 1, 2, \cdots, 36)$. Then compute

$$\sum_1^{36} (X_j - \varphi Y_j)^2 = 220.84$$

whence by Eq. 46

$$V^2 = \frac{220.84}{36 \times 980.56^2}$$

$$= 638 \times 10^{-8}$$

and

$$V = 0.0025$$

as before.

C. SOME THEORY IN THE CHOICE OF SMALL UNITS

Some problems in the choice of sampling unit. From the theory of Part A and from the first term of Eq. 8 in particular, it is obvious that if the number of units in the sample is severely limited, they should be as large and heterogeneous as possible in order to reduce the external variance σ_e^2 and thus to reduce the sampling variance of the estimated population. The larger the unit, the easier it is to get a good map by which to delineate and describe the area. But in the various stages of subsampling, the units become smaller and smaller and more homogeneous; also more and more difficult and expensive to delineate and describe. Some further theory is needed for getting a view of the effect of homogeneity of small units, and a start will be made here.

In the sampling of human populations, which includes farms and commercial establishments that serve human populations, the ultimate unit may be one of the following examples:

A d.u. or all the d.us. lying within a certain small area, or the inhabitants thereof, or some particular members of the families in these d.us.

A farm or group of farms, or the inhabitants thereof, or some particular members of the families living on these farms.

A business establishment of a particular kind (retail hardware), or a group of business establishments (perhaps a "shopping center," bazaar, or all the business of a particular town).

In industrial sampling the ultimate unit may be:

An article, a yard of cloth or film, a test panel or piece.
All the articles in a container.

The terms "primary unit," "secondary unit," etc., are relative. In this part of the chapter the term "primary unit" will refer to the penultimate unit, the one containing the ultimate units described above, although there may have been several previous stages of sampling. The theory given here is intended to throw some light on the choice of subunits.

Because of the small additional cost of interviewing the family next door, it is a temptation to concentrate a sample into *clusters* of 2, 3, 6, or 10 successive d.us. On the other hand, from a practical angle of getting information, *the number of d.us. canvassed is not the real issue; the standard error of the results is.*

Dissimilarities or heterogeneity within a sampling unit leads to gains in efficiency: similarities lead to losses. The question is whether the losses arising from the similarities within a cluster (as of 3 or 5 consecutive households) are offset by the decreases in the cost of collecting the information, as opposed to the higher cost of collection from households more widely dispersed. It may pay to use clusters, taking enough of them to attain the precision desired.

One of the main practical problems in cluster-sampling is to delineate and define the clusters. This is usually done with the aid of detailed maps. A city directory can also be used for the area that it covers. Unfortunately directories and maps often stop at some civil boundary even though the area beyond is heavily populated. Some other type of plan beyond the coverage of the directory or map is then required. The fact is that *any* sampling plan is usually a patch-quilt of plans.

In many physical situations, in the sampling of both human populations and industrial product as well, once the equations are derived, sensible choices of sampling unit will be more or less obvious from the theory that follows and from specialized knowledge which the statistician must possess concerning how the subunits were built, and why, and how much time will be involved in interviewing and traveling. For example, levels of rent and income do not usually vary sharply from one door to the next, and there are good psychological and other basic reasons why they do not. In line with this, small clusters of 1, 2, or 3 d.us. in urban areas will usually be more efficient than bigger clusters for studies of rent and income. Industrial product may be and usually is piled in bales, bins, bags, or layers in order of the date of manufacture, source of raw material, etc.

In summary, in arriving at a decision whether to use a cluster of 2, 3, 5, or 10 d.us. as a subunit in preference to listing all the d.us. and taking every nth (a systematic or patterned sample, p. 83) or to choose the simpler alternative of using some larger and perhaps more variable area containing maybe 10–50 or more d.us. as a subunit because it does not require a special job of mapping, some of the problems that the statistician faces are these:

> How much will the standard error of some important character-istic (estimated population or particular proportion) be increased by using clusters of size $\bar{N} = 2, 3, 6, 10$ in place of the individual d.u.? What will be the increase with a still larger and more vari-able unit? On what plan will the total cost of attaining the pre-scribed precision be least?

> How can clusters or larger areas be delineated and defined un-equivocally; and how much will it cost to use them (special maps, lists, extra supervision)?

Theory will be presented here to show what needs to be known in order to decide what size of cluster will give the most information for the money. Mahalanobis gave a complete solution for the use of grids in agricultural sampling. Quotations from his work are given toward the end of this chapter.

Remark. The decision regarding whether to use a single dwelling unit or a cluster of dwelling units often resolves to the decision of what clusters can be formed without too great an expenditure of funds, and within the limitations of time. Maps and lists cost money, whether purchased or created on the spot. This expense is part of the cost of cluster sampling; and, if it is too great, some naturally existing unit must be used, which may unfortunately be too large or too small or too homogeneous for great efficiency.

The ideal arrangement exists when the first cost of maps or lists can be distributed amongst a number of surveys, and not charged to any one in particular.

Notation for cluster-sampling. A system of notation is needed. Here, the primary unit will play the role of the universe which is to be sampled, and an attempt will be made to maintain the functional characteristics of the symbols that were used in Part A.

In practice, the primary unit is a city or metropolitan district, or as much of it as is covered by detailed maps or directories. The subunits are to be comparatively small groups (*clusters*) of d.us., to be canvassed completely (no subsampling).

TABLE OF NOTATION FOR THE THEORY OF CLUSTER-SAMPLING

μ = mean population per element of the universe (see text)
σ = standard deviation of the population per element of the universe

Name	Primary unit	Sample
Number of elements (d.us. or articles)	$N = M\overline{N}$	$n = m\overline{N}$
Number of clusters	M	m
Average number of elements per cluster	$\overline{N} = N/M$	$\bar{n} = n/m$
Number of elements in Cluster j	N_j	n_j
	(Actually, it will be assumed in the equations that $N_j = \overline{N}$ for all clusters, and that $n_j = \bar{n} = \overline{N}$, as there will be no subsampling within a cluster.)	
Population of the kth element of the jth cluster	$a_{jk}\begin{cases} j = 1, 2, \cdots, M \\ k = 1, 2, \cdots, \overline{N} \end{cases}$ (Listed in some specified order.)	$x_{jk}\begin{cases} j = 1, 2, \cdots, m \\ k = 1, 2, \cdots, \overline{N} \end{cases}$ (Listed in the order drawn into the sample.)
Population of the jth cluster	$A_j = \displaystyle\sum_{k=1}^{\overline{N}} a_{jk}$	$X_j = \displaystyle\sum_{k=1}^{\overline{N}} x_{jk}$
Total population	$A = \displaystyle\sum_{1}^{M} A_j$	$x = \displaystyle\sum_{1}^{m} X_j$ $X = \dfrac{M}{m} x = N\bar{x}$ an unbiased estimate of A, as $EX = A$
Average population per cluster	$\overline{A} = \dfrac{A}{M}$	$\overline{X} = \dfrac{x}{m} = \dfrac{X}{M}$ an unbiased estimate of $\overline{A}$, as $E\overline{X} = \overline{A}$.
Average population per element in the jth cluster	$\mu_j = \dfrac{A_j}{\overline{N}}$	$\bar{x}_j = \dfrac{X_j}{\overline{N}}$
Average population per element	$\mu = \dfrac{A}{N}$	$\bar{x} = \dfrac{x}{m\overline{N}} = \dfrac{X}{N}$ an unbiased estimate of μ, as $E\bar{x} = \mu$.

Name	Primary unit	Sample
Total variance of the secondary units	$\sigma^2 = \dfrac{1}{M\bar{N}} \displaystyle\sum_{j=1}^{M} \sum_{k=1}^{\bar{N}} (a_{jk} - \mu)^2$ $= \sigma_b{}^2 + \sigma_w{}^2$	xxx
Internal variance in Cluster j	$\sigma_j{}^2 = \dfrac{1}{\bar{N}} \displaystyle\sum_{k=1}^{\bar{N}} (a_{jk} - \mu_j)^2$	xxx
Average internal variance	$\sigma_w{}^2 = \dfrac{1}{M} \displaystyle\sum_{1}^{M} \sigma_j{}^2$	xxx
External variance per secondary unit	$\sigma_b{}^2 = \dfrac{1}{M} \displaystyle\sum_{1}^{M} (\mu_j - \mu)^2$	xxx
Variance between the populations of clusters	$\sigma_e{}^2 = \dfrac{1}{M} \displaystyle\sum_{1}^{M} (A_j - \bar{A})^2$	xxx
Intraclass correlation coefficient	$\rho = \dfrac{1}{\sigma^2} \, \mathrm{Av} \displaystyle\sum_{j=1}^{M} \sum_{k \neq k'}^{\bar{N}} (a_{jk} - \mu)(a_{jk'} - \mu)$ $= \dfrac{1}{\sigma^2} \left\{ \sigma_b{}^2 - \dfrac{\sigma_w{}^2}{\bar{N} - 1} \right\}$	xxx

From the elementary theory of Chapter 4 it is known that, if m clusters are drawn at random without replacement and canvassed completely and $\bar{x}$ is computed, the variance of the probability-distribution of $\bar{x}$ will be

$$\mathrm{Var}\ \bar{x} = \frac{M - m}{M - 1} \frac{\sigma_b{}^2}{m} \qquad \text{[M clusters; } m\bar{N} \text{ d.us.]} \qquad (56)$$

whereas, if the sampling had been done by drawing the same number ($m\bar{N}$) of d.us. individually at random and canvassing them, the sampling variance of the new estimate $\bar{x}$ would have been

$$(\mathrm{Var}\ \bar{x})' = \frac{M\bar{N} - m\bar{N}}{M\bar{N} - 1} \frac{\sigma^2}{m\bar{N}}$$

$$= \frac{M - m}{M - 1} \frac{N - \bar{N}}{N - 1} \frac{\sigma^2}{m\bar{N}} \qquad \begin{array}{l} \text{[} m\bar{N} \text{ d.us. drawn} \\ \text{at random with-} \\ \text{out replacement} \\ \text{from a universe} \\ \text{of } M\bar{N} \text{ d.us.]} \end{array} \qquad (57)$$

Before proceeding into the main purpose of this section it may be noted that for the unbiased estimate $X = N\bar{x}$ of the total population of the primary unit,

$$\text{Var } X = N^2 \text{ Var } \bar{x}$$

and that the

$$\text{C.V. } X = \text{C.V. } \bar{x}$$

(58)

Now let y be the relative excess variance arising from sampling by clusters; then

$$1 + y = \frac{\text{Var } \bar{x}}{(\text{Var } \bar{x})'} \qquad \text{[Definition of } y] \quad (59)$$

$$= \frac{M - m}{M - 1} \frac{\sigma_b^2}{m} \div \frac{M - m}{M - 1} \frac{N - \bar{N}}{N - 1} \frac{\sigma^2}{m\bar{N}} \qquad (60)$$

$$= \frac{N - 1}{N - \bar{N}} \frac{\bar{N} \sigma_b^2}{\sigma^2}$$

$$= 1 + \frac{\bar{N} - 1}{\sigma^2} \left\{ \frac{M}{M - 1} \sigma_b^2 - \frac{1}{\bar{N} - 1} \sigma_w^2 \right\} \qquad (61)$$

When M is large enough, as it usually is in practice, the fraction $M/(M - 1)$ may be placed equal to unity, and what is left may be written as

$$1 + y = 1 + (\bar{N} - 1)\rho \qquad (62)$$

wherein

$$\rho = \frac{1}{\sigma^2} \left\{ \sigma_b^2 - \frac{\sigma_w^2}{\bar{N} - 1} \right\} \qquad \text{[As given in the notation]} \quad (63)$$

is the *intraclass correlation coefficient* for the clusters within the primary unit. Turned around, the last relation gives

$$\sigma_b^2 = \frac{\sigma^2}{\bar{N}} [1 + (\bar{N} - 1)\rho] \qquad (64)$$

whereupon by substitution into Eq. 56 the variance of cluster-sampling may be expressed as

$$\text{Var } \bar{x} = \frac{M - m}{M - 1} \frac{\sigma^2}{m\bar{N}} \{1 + (\bar{N} - 1)\rho\} \qquad (65)$$

The following conclusions may be drawn:

i. The term $(\bar{N} - 1)\rho$ in the last equation very closely measures the relative excess variance arising from the use of m clusters of size $\bar{N}$ instead of the same number $(m\bar{N})$ of d.us. drawn individually.

ii. If $\rho = -1/(N - 1)$ the bracket $1 + (\bar{N} - 1)\rho$ has the value

$(N - \bar{N})/(N - 1)$, and the sampling variance of $\bar{x}$ is precisely equal to the variance shown in Eq. 57 for a sample of $m\bar{N}$ d.us. drawn individually.

iii. If $\rho < -1/(N - 1)$ there will be a gain in precision by sampling in clusters, and if $\rho > -1/(N - 1)$ there will be a loss.

In the next section we shall see under what conditions ρ may be "expected" to have a value equal to, above, or below $-1/(N - 1)$.

Remark 1. Great gains in efficiency of the unbiased estimate of the number of inhabitants can be achieved if the clusters are fairly uniform in size: $\sigma_b{}^2$ in the equations is then small, and the sampling variance of the population-count is small.

No map can be expected to distinguish unequivocally between 1-, 2-, and 3-family houses, or to show definitely whether any particular building contains dwellings where population may be found. Likewise, no existing map shows the number of d.us. in multistory apartment buildings; the number must be estimated and the clusters numbered by some rule and code. Maps on which most of the dwellings can be identified and the number of flats can be estimated in multiunit buildings are also good enough to permit fair uniformity in the number of d.us. per cluster. Changes are continually taking place, and no map can be entirely up to date. The map may show 6 d.us. where actually at present there are more or less— sometimes none and sometimes many. Moreover, there are of course sometimes d.us. in buildings shown as offices or factories, and such areas must be included in one cluster or another. The actual labor and details are tedious yet extremely interesting, but they cannot be described here for lack of space.

Hansen and Hurwitz in the Census aimed originally at an average cluster of 6 d.us. for most population sampling [21] in urban areas. Experience shows that with care and use of the excellent Sanborn maps [22] the coefficient of variation for this size of cluster as measured in actual d.us. found is about $\frac{1}{2}$. The coefficient of variation (σ_b/μ) in the number of inhabitants will of course be slightly greater than that for the number of d.us. per cluster because of variability from one d.u. to another within the cluster. The figure $\frac{1}{2}$ is only a rough way of conveying the idea of what sort of coefficient of variation is attainable. Actually, it varies from one city to another, and as it is usually less than $\frac{1}{2}$, the figure $\frac{1}{2}$ may be used for computing the sampling error of a population count as well as a count of the number of d.us. Thus a sample of 600 clusters (about 3600 d.us.) would give a coefficient of variation of $0.5/\sqrt{600}$ or 2 percent.[23] The same sample would usually give much better results for the proportion of the population in various age, sex, employment, and other classes, provided the classes are not too small, because for proportions it is usually possible to use a ratio-estimate.

[21] At the present writing, now that better figures on the variances and costs of interviewing in various sizes of cluster are available, there is a trend toward smaller clusters, particularly for economic studies.

[22] Published by the Sanborn Map Company, 60 Cedar Street, New York 6.

[23] By comparison, "complete" censuses in many parts of the world are no more reliable than this.

Remark 2. It is important to note that variability in size of cluster *does not cause bias* in the final results. It does of course increase the sampling error, particularly if the unbiased estimate is used.

Remark 3. Distinct gains in precision arise through the use of small clusters; also some disadvantages in the office and field. The smaller the clusters, the harder it is to delineate them unequivocally on the map, and the more areas there will be on which the final delineation of the clusters must be deferred until the interviewing takes place, at which time the area must first be mapped and listed on the spot before the clusters can be defined.

When sufficiently good detailed maps can be found or made, and when the similarity of neighboring d.us. is cause for uneasiness, as it will be in studies of rent, income, type of housing, and the like, the smallest possible cluster, viz., clusters of one d.u., may be used. It is to be noted that a cluster of one d.u. really refers to the *area* that *apparently* contains one d.u. No matter how big or how small a cluster is intended to be, the interviewer should be warned that when he arrives on the scene he may find 0, 1, 2, 3, or any other number of d.us. As stated before, he is to canvass *all that he finds* in the area or interval assigned to him (p. 82). It is strict application of this rule that gives the sample its unbiased character, by which the sample gives a population count independent of a census.

Remark 4. It is to be remembered that a cluster that produces a small correlation ρ for one characteristic may give a larger ρ for another. Thus, for a sample that is intended to measure the sex-ratio (males : females) in "normal" families, the dwelling unit is better than an individual, because for this characteristic, $\rho < 0$ (see the exercises further on). But for estimating the number of people (male plus female) in an age-group like 20–29, a sample of dwelling units, which will include a certain number of individual people, will not be as efficient as the same number of individuals drawn at random. This is because $\rho > 0$ when the characteristic under consideration is the total number of people in an age-group.

Remark 5. $E\bar{x} = \mu$ under the assumption that the cluster-size N_j is constant, but if the cluster-size varies widely then the ratio-estimate

$$\bar{x}' = \frac{\sum\limits_{j=1}^{m} X_j}{\sum\limits_{j=1}^{m} N_j} \quad \text{[Compare with Eq. 25]} \quad (66)$$

would be used, and the total population of the primary unit would be estimated as

$$X = N\bar{x}' \quad \text{[Compare with Eq. 26]} \quad (67)$$

As N is a constant, determined not by the sample but by counting d.us. on the map, the C.V. X = C.V. $\bar{x}'$ (see page 72).

Exercise. Given the definition

$$\rho = \frac{1}{\sigma^2} \operatorname{Av} \sum_{j=1}^{m} \sum_{k \neq k'}^{\overline{N}} (a_{jk} - \mu)(a_{jk'} - \mu) \qquad [N_j = \overline{N}]$$

show that

$$\rho = \frac{1}{\sigma^2} \left\{ \sigma_b{}^2 - \frac{\sigma_w{}^2}{\overline{N} - 1} \right\}$$

as written in the notation, page 193.

Solution

With no element repeated ($k \neq k'$) in the numerator of ρ there will be $\overline{N}(\overline{N} - 1)$ pairs of terms like $(x_{jk} - \mu)(x_{jk'} - \mu)$ in Cluster j. Therefore

$$\operatorname{Av} \sum_{j=1}^{M} \sum_{k \neq k'}^{\overline{N}} (x_{jk} - \mu)(x_{jk'} - \mu)$$

$$= \frac{1}{M} \sum_{j=1}^{M} \frac{1}{\overline{N}(\overline{N} - 1)} \left\{ \left[\sum_{k=1}^{\overline{N}} (x_{jk} - \mu) \right]^2 - \sum_{k=1}^{\overline{N}} (x_{jk} - \mu)^2 \right\}$$

$$= \frac{1}{M} \sum_{j=1}^{M} \frac{1}{\overline{N}(\overline{N} - 1)} \left\{ \sum_{k=1}^{\overline{N}} [(x_{jk} - \bar{x}_j) - (\mu - \bar{x}_j)] \right\}^2 - \frac{M\overline{N}\sigma^2}{M\overline{N}(\overline{N} - 1)}$$

$$= \frac{1}{M} \sum_{j=1}^{M} \frac{1}{\overline{N}(\overline{N} - 1)} \left\{ 0 - \sum_{k=1}^{\overline{N}} (\mu - \bar{x}_j) \right\}^2 - \frac{\sigma^2}{\overline{N} - 1}$$

$$= \frac{1}{M} \sum_{j=1}^{M} \frac{1}{\overline{N}(\overline{N} - 1)} \left\{ \overline{N}(\bar{x}_j - \mu) \right\}^2 - \frac{\sigma^2}{\overline{N} - 1}$$

$$= \frac{\overline{N}}{\overline{N} - 1} \sigma_b{}^2 - \frac{\sigma^2}{\overline{N} - 1}$$

$$= \sigma_b{}^2 - \frac{\sigma^2 - \sigma_b{}^2}{\overline{N} - 1} = \sigma_b{}^2 - \frac{1}{\overline{N} - 1} \sigma_w{}^2$$

Therefore

$$\rho = \frac{1}{\sigma^2} \left\{ \sigma_b{}^2 - \frac{1}{\overline{N} - 1} \sigma_w{}^2 \right\} \qquad Q.E.D.$$

A hypothetical mechanism for the formation of clusters. The following hypothesis for the formation of clusters will be studied. It is not intended as a description of how clusters are really formed, but rather to provide a basis for predicting something about the variance of cluster-sampling in particular situations. Known departures from the mechanism to be described will provide some guidance in the choice of sampling unit.

Let the universe consist of $N = M\bar{N}$ elements (d.us. or individual articles), having a mean population μ per element and variance σ^2. M clusters are to be formed by drawing M random samples of $\bar{N}$ elements each, without replacement. These M samples exhaust the universe. A sample of m clusters is now to be drawn and canvassed with the aim of estimating μ. As the M clusters are only a rearrangement of the original N d.us., there has been no change in the mean μ or standard deviation σ; and if the M clusters are formed over and over again in the manner assumed above, the statistical measures σ_j^2, σ_w^2, σ_b^2, ρ will take on the character of random variables. The "expected" values of σ_b^2, σ_w^2, and ρ can be found. Thus, by definition,

$$E\sigma_b^2 = E\frac{1}{M}\sum_1^M (\bar{x}_j - \mu)^2$$

$$= \text{Var } \bar{x}_j = \frac{N - \bar{N}}{N - 1}\frac{\sigma^2}{\bar{N}} \qquad \text{[Ch. 4]} \quad (68)$$

Next,

$$E\sigma_w^2 = E\sigma^2 - E\sigma_b^2 \qquad \begin{array}{l}\text{[Because } \sigma_w^2 + \sigma_b^2 = \sigma^2\text{, and } \sigma^2 \text{ is} \\ \text{constant]}\end{array}$$

$$= \sigma^2 - \frac{N - \bar{N}}{N - 1}\frac{\sigma^2}{\bar{N}} = \frac{N}{N - 1}\frac{\bar{N} - 1}{\bar{N}}\sigma^2 \qquad (69)$$

In this model the random variables σ_w^2 and σ_b^2 thus furnish unbiased estimates of the original σ^2, and these estimates may be written in the form

$$\sigma'^2 = \frac{N - 1}{N - \bar{N}}\bar{N}\sigma_b^2 \qquad \begin{array}{l}\text{[Here } \sigma'^2 \text{ denotes the } ex\text{-} \\ ternal \text{ estimate of } \sigma^2\text{]}\end{array} \quad (70)$$

$$\sigma''^2 = \frac{\bar{N}}{\bar{N} - 1}\frac{N - 1}{N}\sigma_w^2 \qquad \begin{array}{l}\text{[Here } \sigma''^2 \text{ denotes the} \\ internal \text{ estimate of } \sigma^2\text{]}\end{array} \quad (71)$$

It follows from Eq. 63 that

$$\rho = \frac{1}{\sigma^2}\left\{\frac{N - \bar{N}}{N - 1}\frac{\sigma'^2}{\bar{N}} - \frac{N}{N - 1}\frac{\sigma''^2}{\bar{N}}\right\}$$

$$= \frac{1}{(N - 1)\sigma^2}\{(M - 1)\sigma'^2 - M\sigma''^2\} \qquad (72)$$

Then, as $E\sigma'^2 = E\sigma''^2 = \sigma^2$, it follows directly that

$$E\rho = -\frac{1}{N - 1} \qquad (73)$$

Also, by taking the "expected" value of each member of Eq. 43, it is seen that

$$E \operatorname{Var} \bar{x} = \frac{M - m}{M - 1} \frac{E\sigma_b{}^2}{m} = \frac{M - m}{M - 1} \frac{N - \bar{N}}{N - 1} \frac{\sigma^2}{m\bar{N}}$$

$$= \frac{M\bar{N} - m\bar{N}}{M\bar{N} - 1} \frac{\sigma^2}{m\bar{N}} \qquad (74)$$

The same equation is derived from Eq. 65 by replacing ρ by $E\rho = -1/(N - 1)$ as in Eq. 73.

This result, as it is identical with Eq. 57 for the variance of a sample of individual d.us. drawn at random, shows that if the clusters are formed *as if* by chance, there is no "expected" loss or gain in the variance of cluster-sampling over the sampling of individual d.us. In other words, when the N elements are thoroughly mixed, the sample may be drawn in chunks or a little at a time; it makes no difference. As there is certainly a saving in the cost of interviewing in clusters, preference would definitely go to cluster-sampling under such circumstances.

By recalling that y or $(\bar{N} - 1)\rho$ measures the excess variance of cluster-sampling over random sampling, it is now clear from the last equation that if the mechanism for the formation of clusters is such that the two estimates σ' and σ'' should agree, on the average, then the "expected" excess is zero. If, on the other hand, there are forces that attract people of dissimilar characteristics into proximity and repel people of similar characteristics, variability within the cluster is relatively great, and variability from cluster to cluster is reduced (as the sum $\sigma_w{}^2 + \sigma_b{}^2 = \sigma^2$ is constant). These forces may be sufficient to bring about the superiority of sampling by clusters, which actually then gives smaller variance than the random sampling of the same number $(m\bar{N})$ of individual elements.[24] An example is afforded by the measurement of the sex-ratio (male : female) in an area, because the normal family is founded upon the dissimilarity of sex (cf. the exercises further on). Another example is the age-distribution of the population: for either example the d.u. or a cluster of d.us. is a better sampling unit than the same number of individual people drawn at random. Usually, however, the reverse is true: people of similar incomes, color, and tastes prefer to flock together, reducing the internal variance for these characteristics so that cluster-sampling has the higher variance. Yet clusters averaging (e.g.) six d.us. will for many characteristics provide the best sample by giving the precision desired at less cost than if a larger or smaller cluster were used.

Remark 1. In the last chapter the distribution of the ratio $\sigma':\sigma''$ will be studied for samples from a normal universe. The reader may wish to

[24] The exercises on pages 209-12 show some details of the calculations.

turn also to another discussion of this ratio on page 575, wherein a very simple criterion is laid down for judging whether σ' is "significantly" greater than σ''.

Remark 2. A considerable amount of research in the past has been devoted to the "significance" of the ratio $\sigma':\sigma''$ and its variability for different sizes of cluster. Significant departures from unity are interpreted as indicating the presence of outside or assignable forces and not mere chance in the formation of the clusters. In sample-design, however, the problem is not whether σ' and σ'' are "significantly" different, but how the ratio $\sigma':\sigma''$ varies with the size $(\bar{N})$ of cluster, and what possible savings are to be gained by using clusters.

Remark 3. The number $\bar{N}$ in the equations refers to the number of d.us. per cluster *at the date of the sample*. This figure can not be controlled absolutely, and its control is not worth more than a reasonable effort. As has been mentioned, the number of d.us. appearing on the map or directory is one figure, but the number actually encountered is another. Not only will the actual number (N_j) vary slightly from one cluster to another, but the actual average $(\bar{N})$ will not be the average apparent from the map or directory. Even though the map be perfect, there will be two difficulties: (a) no absolutely constant size of cluster, except clusters of size 1, can be forced into an inextensible area or other unit (such as a city "block" or apartment-house); there will usually be a remainder which must be a cluster all by itself or be tied in with an adjacent cluster, either of which violates the condition of constancy. (Sometimes, but only sometimes, the remainder can be tied in with another remainder to form a new cluster of apparent size $\bar{N}$ or close to it.) Moreover, (b) changes take place—new buildings, demolitions, alterations.

The theory for a variable N_j in place of $\bar{N}$ is not difficult. For calculating the excess, previously defined as $y = (\bar{N} - 1)\rho$, it is only necessary to redefine ρ so that

$$\rho = \frac{1}{\sigma^2} \left\{ \frac{1}{M\bar{N}} \sum_1^M N_j(\bar{x}_j - \mu)^2 \right.$$

$$\left. - \frac{1}{M\bar{N}(\bar{N} - 1)} \sum_{j=1}^M \sum_{k=1}^{N_j} (a_{jk} - \mu_j)^2 \right\}$$

$$= \frac{1}{\sigma^2} \left\{ \sigma_b{}^2 - \frac{1}{\bar{N} - 1} \sigma_w{}^2 \right\} \tag{75}$$

wherein

$$\sigma_b{}^2 = \frac{1}{M\bar{N}} \sum_1^M N_j(\bar{x}_j - \mu)^2 \tag{76}$$

$$\sigma_w{}^2 = \frac{1}{M\bar{N}} \sum_1^M \sum_{k=1}^{N_j} (a_{jk} - \mu_j)^2 \tag{77}$$

$$\sigma^2 = \frac{1}{M\bar{N}} \sum_1^M \sum_{k=1}^{N_j} (a_{jk} - \mu)^2 \tag{78}$$

It can then be shown that Eqs. 70–74 still hold, as the student may undertake to show as an exercise.

Exercise 1. Show that as

$$\left(\frac{\sigma_b}{\sigma_w}\right)^2 \overset{<}{\underset{>}{=}} \frac{N - \overline{N}}{N(\overline{N} - 1)} = \frac{M - 1}{N - M} \doteq \frac{1}{\overline{N}} \quad \begin{matrix} \text{[Approximately, if} \\ M \text{ and } N \text{ are} \\ \text{large]} \end{matrix} \quad (79)$$

cluster-sampling is $\begin{bmatrix} \text{better than} \\ \text{equal to} \\ \text{worse than} \end{bmatrix}$ random samples of the same number of ultimate units (e.g., d.us.) drawn individually.

Exercise 2. By making use of Eq. 72 show that Eq. 65 reduces to

$$\text{Var } \bar{x} = \frac{M - m}{M - 1} \frac{\sigma^2}{m\overline{N}} \left\{ 1 + \frac{\overline{N} - 1}{N - 1} \frac{(M - 1)\sigma'^2 - M\sigma''^2}{\sigma^2} \right\} \quad (80)$$

$$\doteq \quad `` \quad `` \quad \left\{ 1 + \frac{\overline{N} - 1}{\overline{N}} \frac{\sigma'^2 - \sigma''^2}{\sigma^2} \right\} \quad \begin{matrix} \text{[Approximately, if} \\ M \text{ is large]} \end{matrix} \quad (81)$$

wherein σ' and σ'' represent the external and internal estimates of σ as indicated in Eqs. 70 and 71.

Remark. In the author's experience these two alternative forms of Eq. 65 are easier to use in the planning stages than Eq. 65 itself. The reason lies in the fact that here the two estimates of σ^2 are laid out where they can be seen and their difference compared with σ^2. Thus in a particular problem there might be evidence that the means of clusters of size $\overline{N} = 4$ will be about 1.5 times as variable as the clusters themselves: that is, $\sigma'^2 = 1.5\sigma''^2$. Then approximately

$$\sigma_b{}^2 = \frac{1}{\overline{N}} \sigma'^2 \quad \begin{bmatrix} \text{From Eq. 70, } \dfrac{N - \overline{N}}{N - 1} \text{ being} \\ \text{replaced by unity]} \end{bmatrix} \quad (82)$$

$$\sigma_w{}^2 = \frac{\overline{N} - 1}{\overline{N}} \sigma''^2 \quad \begin{bmatrix} \text{From Eq. 71, } \dfrac{N - 1}{N} \text{ being} \\ \text{replaced by unity]} \end{bmatrix} \quad (83)$$

Addition gives

$$\sigma^2 = \left[\frac{1}{\overline{N}} \left(\frac{\sigma'}{\sigma''} \right)^2 + \frac{\overline{N} - 1}{\overline{N}} \right] \sigma''^2$$

$$= [\tfrac{1}{4} \times 1.5 + \tfrac{3}{4}]\sigma''^2 = \tfrac{9}{8}\sigma''^2 \quad (84)$$

whence from Eq. 81

$$\text{Var } \bar{x} = \frac{M - m}{M - 1} \frac{\sigma^2}{m\overline{N}} \left\{ 1 + \frac{\overline{N} - 1}{\overline{N}} \frac{\sigma'^2 - \sigma''^2}{\sigma^2} \right\}$$

$$= \quad `` \quad `` \quad \left\{ 1 + \frac{3}{4} \frac{1.5 - 1}{\frac{9}{8}} \right\}$$

$$= \quad `` \quad `` \quad \{ 1 + \tfrac{1}{3} \} \quad (85)$$

The term $\frac{1}{3}$ in the braces corresponds to the increment y or $(\overline{N} - 1)\rho$ in Eqs. 62 and 65. Interpreted, this means that under the conditions stipulated, the sample drawn in clusters of size $\overline{N} = 4$ would require a third more households than if it were drawn in the form of single households. On the other hand, the savings in the lower cost of the frame that is required for defining clusters of approximately 4 households instead of single households, and the decreased costs of travel, might more than offset the extra households.

The student may wish to calculate the value of ρ in this example and show that $\rho = \frac{1}{9}$.

Illustration with a binomial universe. A useful illustration is furnished by 0, 1 variates, whereby x_{jk} is to be either 0 or 1, as when a house is either occupied or vacant, or a man is either employed or not. Each cluster will then have a simple distribution like the one shown in Fig. 15 for Cluster j. The horizontal lines represent stacks of poker chips, each chip being an element of the cluster. Those marked 0 are for elements (d.us.) of population 0 (as for occupied), and those marked 1 are for elements of population 1 (as for vacant). The mean of the universe is at p, and its variance is pq, for which σ^2 will occasionally be substituted.

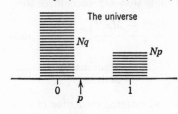

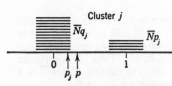

Fig. 15. A primary unit contains elements labeled 0 and 1 in the ratio $q:p$. Cluster j is formed by drawing $\overline{N}$ elements at random without replacement. The arrows show the position of the means. p_j is a random variable.

Let Cluster j be formed by drawing $\overline{N}$ chips without replacement. Denote by q_j and p_j the proportions of 0 and 1 elements in it. Then the mean of Cluster j is at p_j, and its variance is $p_j q_j$. Fig. 15 may assist. It follows from Eq. 63 that

$$\rho = \frac{\dfrac{1}{M} \sum_{1}^{M} (p_j - p)^2 - \dfrac{1}{M(\overline{N} - 1)} \sum_{1}^{M} p_j q_j}{\sigma^2} \tag{86}$$

Now p_j and q_j are random variables concerning which it can be said that

$$E p_j = \mu = p \tag{87}$$

For variety, $E\sigma_w^2$ will be derived before $E\sigma_b^2$ here. By definition, $\sigma_j^2 = p_j q_j$, so

$$E\sigma_w{}^2 = E\sigma_j{}^2 = Ep_jq_j = Ep_j(1 - p_j)$$

$$= p - Ep_j{}^2$$

$$= p - [\text{Var } p_j + (Ep_j)^2]$$

$$= p - \left[\frac{N - \overline{N}}{N - 1}\frac{pq}{\overline{N}} + p^2\right]$$

$$= \frac{N}{N - 1}\frac{\overline{N} - 1}{\overline{N}}\,pq = \frac{N - M}{N - 1}\,pq \qquad (88)$$

That is,

$$\frac{\overline{N}}{\overline{N} - 1}E\sigma_w{}^2 = \frac{\overline{N}}{\overline{N} - 1}Ep_jq_j = \frac{N}{N - 1}pq \qquad (89)$$

Also,

$$E\sigma_b{}^2 = E\frac{1}{M}\sum_1^M (p_j - p)^2 = \text{Var } p_j$$

$$= \frac{N - \overline{N}}{N - 1}\frac{pq}{\overline{N}} = \frac{M - 1}{N - 1}pq \qquad [\text{Ch. 4}] \quad (90)$$

Thus, internal and external estimates of $\sigma^2 = pq$ may be written as follows:

$$\sigma''^2 = \frac{N - 1}{N}\frac{\overline{N}}{\overline{N} - 1}\frac{1}{M}\sum_1^M p_jq_j = \frac{N - 1}{M - 1}\sigma_w{}^2 \qquad (91)$$

$$\sigma'^2 = \frac{N - 1}{N - \overline{N}}\overline{N}\frac{1}{M}\sum_1^M (p_j - p)^2 = \frac{N - 1}{M - 1}\sigma_b{}^2 \qquad (92)$$

as before. Then, following Eq. 86, it is seen that

$$\rho = \frac{1}{pq}\left\{\frac{1}{M}\sum_1^M (p_j - p)^2 - \frac{1}{\overline{N} - 1}\frac{1}{M}\sum_1^M p_jq_j\right\} \qquad (93)$$

which is Eq. 63 for the binomial case. The "expected" value of ρ is then seen to be

$$E\rho = \frac{1}{N - 1}\frac{1}{\overline{N}pq}\{(N - \overline{N})pq - Npq\}$$

$$= -\frac{1}{N - 1} \qquad (94)$$

as in Eq. 73. Now let $\hat{p}$ be the sample-estimate of p defined by the equation

$$\hat{p} = \frac{1}{m\overline{N}}\sum_{j=1}^m\sum_{k=1}^{\overline{N}} x_{jk} \qquad (95)$$

From Eq. 56, page 193, it is seen that

$$\text{Var } \hat{p} = \frac{M - m}{M - 1} \frac{1}{mM} \sum_{j=1}^{M} (p_j - p)^2 \tag{96}$$

If the clusters are formed by chance,

$$E \text{ Var } \hat{p} = \frac{M - m}{M - 1} \frac{1}{m} E \frac{1}{M} \sum_{j=1}^{M} (p_j - p)^2$$

$$= \frac{M - m}{M - 1} \frac{N - \bar{N}}{N - 1} \frac{pq}{m\bar{N}} \qquad \text{[By Eq. 90]} \quad (97)$$

which is the variance of $\hat{p}$ as it would be if the $m\bar{N}$ d.us. had been drawn individually at random.

The examples that follow show some calculations of variances for several plans of cluster-sampling.

Illustration of the use of a particular cost-function in the choice of size of cluster.   Eq. 65 or its equivalent, Eqs. 80 and 81, relate Var $\bar{x}$ to the variances within and between clusters.   Often (as already stated) it is the limitations of possible frames that fix the size of the cluster, but there have been instances in the author's experience when splendid detailed frames were already on hand permitting the use of any desired size of cluster, or indeed of a completely random selection of the ultimate units in a single stage of sampling.   In one such experience the problem was to determine the average condition of 150,000 telephone poles and aerial equipment belonging to a public utility company and scattered all over a state.   There had been no previous experience in the problem, and (as so often happens) there was no time for experimental work; hence it was necessary to rely on conjectures regarding the variances and correlations that were needed to determine the most economical plan of sampling.   After careful but rapid deliberation with officials and workmen in the company who had knowledge of costs and the history of the purchasing and repairing of the articles to be inspected, and with the generous aid of my colleague Morris H. Hansen, the following equation for cost was arrived at:

$$P \text{ (dollars)} = \text{Total cost} = A \sqrt{m} + B \, m\bar{N} \tag{98}$$

Herein $m\bar{N}$ is the total number of articles to be inspected, and B is the cost of inspecting a single article once the inspecting crew has arrived on the scene.   The first term on the right represents travel-costs.   This form of cost-function is useful when the primary units are to be reached by some devious route one after another, with supervision from a central headquarters, and where there are no fixed salaries involved at each primary unit.[25] These are different stipulations from those that led to Eq. 10 (p. 151) in which there was a fixed charge mc_1 directly proportional to the number of primary units. Here it was conjectured that B must be close to \$1.50 and that A must be somewhere between \$20 and \$40. (It turns out that although these limits are broad, they are close enough to point definitely to the most economical plan.) Various guesses were made at likely and unlikely values of ρ for use in Eq. 65, or equivalent relations between σ' and σ'' in Eqs. 80 and 81. It turned out that even with wide latitudes of choice the most economical sampling job would be done with clusters not bigger than 2 articles. As this size of cluster would bring in nearly all the primary units (tax districts) of the state, and thus involve considerable travel anyhow, the plan actually determined upon was in fact the very safe one of sampling single poles in one stage of sampling, which was possible because of the perfect inventory in the form of maps.

More precisely, the plan was finalized in the form of 10 interpenetrating samples each of 1 article in 1000. Tippett's numbers (p. 87) gave 10 random starts between 1 and 999: these starts augmented successively by 1000 gave the 10 samples. These 10 samples could be compared with each other, permitting estimation not only of the sampling error but of the differences between crews of observers.

Previous to this formulation, clusters of 10 had been intended—a procedure that would have cost three times as much as the plan adopted. This is an excellent example of how the use of theory, aided by quick and educated guessing, can point to an economical plan and to a proper size of sample.

Empirical studies of the variances of sampling by clusters. It is possible from the returns of a sample to calculate the two estimates σ' and σ'', and also Var $\bar{x}$, for the characteristics of chief interest in a particular survey. Such calculations are of great help in designing another survey involving the same population characteristics, the same type of universe, and clusters of approximately the same size. But the changes in variance to be expected from using some other size of cluster can not be calculated closely without some further theory. What is needed is a formula connecting σ^2 with $\bar{N}$. One might surmise, from theories of condensation in physics, that such a relation could be devised. A masterful piece of

---

[25] The use of the term in $\sqrt{m}$ is justified in a note by my colleague Eli S. Marks "A lower bound for the expected travel among m random points," *Annals Math. Stat.*, vol. xix, 1948: pp. 419–22.

work along this line was performed by Mahalanobis,[26] who found that

$$\text{Var } \bar{x} = \frac{b}{a^g} \qquad (99)$$

wherein a is the number of acres per grid, and b and g are constants. In Mahalanobis's surveys a square grid of area a acres is placed at random on the map, and the areas under jute enclosed within the grid are measured by the map. If a farm is only partially included within the grid, its contribution to jute is recorded as the product of the actual area of that farm included by the grid, multiplied by the fractional part of the whole farm that is under jute. On the basis of surveys covering several districts in Bengal over the years 1935–40 Mahalanobis found that

$$g = 0.4748 - 0.6686 \, p \qquad (100)$$

wherein p is the estimated proportion of the whole area under jute. With a parabolic term added, the equation found was

$$g = 0.4954 - 1.0387 \, p + 0.9642 \, p^2 \qquad (101)$$

The value of b turned out to be from 0.034 to 0.057 acre. Mahalanobis surmised that this figure represents the smallest practicable area that can be put into jute.

It should be noted that if the area under jute were distributed at random in small cells over a district, g in Eq. 99 would be unity. The fact

[26] P. C. Mahalanobis, *Report on the Sample Census of Jute in 1939* (Indian Central Jute Committee, Calcutta, 1940; confidential). Several earlier reports were issued but need not be cited, because Mahalanobis greatly extended this work in an article of fundamental importance, "On large-scale sample surveys," *Phil. Trans. Royal Soc.*, vol. B231, 1944: pp. 329–451. The equations and other data quoted here are from this article. A summary of his large-scale sample surveys in Bengal, together with some important new developments in theoretical statistics, was delivered before the Royal Statistical Society in London, 16 July 1946, and published in the *J. Royal Stat. Soc.*, vol. cix, Part iv, 1946: pp. 325–78. Independently and at about the same time, H. Fairfield Smith was concerned with the problem of yield in his studies of the design of field experiments in agriculture. His publication was entitled "An empirical law describing heterogeneity in the yields of agricultural crops," *J. Agric. Soc.*, vol. 28, 1938: pp. 1–23, which may also be consulted with profit. Smith independently arrived at the same relation as Mahalanobis's for Var $\bar{x}$, and, like Mahalanobis, he also introduced a cost-function, for which Smith used the form

$$K_1 + K_2 x + K_g (A + Bx)$$

as the cost per plot, finding that the value of N (in my notation) which gives the most information for the money is

$$N = \frac{GK_1}{(1 - b)K_2}$$

where b, K_1, and K_2 are constants.

that g is around 0.5 and less is significant of the correlation between neighboring fields, and the fact that g decreases still further as the proportion p increases is in accordance with expectations.

In regard to costs, Mahalanobis found that the time required for enumeration (e) plus the time for journey (j) between grids is well enough represented by the function

$$t(e + j) = c_0 + wc_1 + 0.0793\ wa - 0.5906\ w^2 \qquad (102)$$

Year	c_0	c_1
1939	0.4437	2.5740
1940	0.2857	2.7946
1941	0.4531	2.6319

wherein w is the average density of grids per square mile, and a is the number of acres per grid. In spite of the fact that workers vary amongst themselves and from year to year, and that the character of the fields differs materially from year to year, the general features of the cost-function seem to be fairly well maintained.

Mahalanobis then proceeded as follows to find the optimum size and density of grid, and I quote some of the most brilliant passages ever recorded in theoretical statistics.

267. The fact that the numerical value of g is much less than unity shows that the physical field in the case of the jute crop in Bengal is definitely of a nonrandom type. From an abstract point of view this immediately justifies the use of the theory of grid sampling. In order to appreciate the magnitude of the difference in costs made in practice it is, however, necessary to consider numerical examples. Suppose that it is desired to attain the final estimate with a percentage error of 1% over a particular region of 5507 sq. miles. Adopting the 1941 cost figures and the actual distribution of size of grid and density used in 1941 on a proportionate basis, it is found that the total cost would be 90,652 investigator-hours for the field work. If, instead of using the above size-density distribution, one uniform size of 20-acre grids had been adopted, then the cost for attaining the same precision would have been 239,926 investigator-hours. If 40-acre grids had been used, the cost would have risen still higher to 281,168 investigator-hours. This is a typical example. It is clear that under actual conditions of work in Bengal there is no doubt that the approach adopted in the present work has been definitely more efficient.

268. I may mention here in passing that in 1938 at a meeting of the Jute Census Committee an eminent agricultural expert gave it as his considered opinion that grids of size less than 36 acres must not be used, as otherwise results would be unreliable. The above numerical example shows that had his advice been accepted then the cost for the field portion of the work alone would have been easily three times as great. To obtain a final estimate with the same precision in 1941 the actual cost in working with grids of 36 acres would have been something like Rs. 2,47,400 (£18,555) against Rs. 79,800 (£5925) in the method actually adopted, which means a saving

of Rs. 1,67,600 (£12,630) in field-work alone in one single year. Further remarks are scarcely necessary.

269. Explanation will now be given of a graphical-numerical method of obtaining in particular cases the optimum size and density of grids at a given cost level, using the special forms which were adopted for the area survey work for acreage under jute in Bengal in 1940. The following graduation equations were used for the total cost T and the variance V of the estimated total area under jute:

$$V = \sum_{k=1}^{5} \frac{A_k b_k}{w_k a_k{}^g} \tag{269.1}$$

$$T = \sum_{k=1}^{5} [A_k(c_0 + c_1 w_k + c_2 w_k a_k + c_3 w)] \tag{269.2}$$

where in this particular case the summation k was over five zones depending on five different ranges into which p (the proportion under jute) was divided; c_0, c_1, c_2, and c_3 are cost parameters supposed to be constant over the different zones; A_k is the area of the kth zone; a_k and w_k are the size and density of the grids in the kth zone in any conventional units, say acres and number per square mile respectively; b_k is a zonal constant; and g is a pooled value over all zones and is obtained in the following way.

270. The variance function in the kth zone is graduated by the formula

$$V_k = \frac{A_k b_k}{w_k a_k{}^{g_k}} \quad (k = 1, 2, 3, 4, 5) \tag{270.1}$$

(In any zone, for purposes of graduation a number of size-density combinations in the exploratory state were used, though in the final design only one combination was used, namely, the optimum.) Having obtained $b_k(k = 1, 2, \cdots, 5)$, the whole variance material is graduated by the equation

$$\frac{V_k}{b_k} = \frac{A_k}{w_k a_k{}^g} \tag{270.2}$$

It is to be noted here that for any size and density there may be a number of values of observed variances; each such value is to be divided by the b_k of the zone from which the variance has been observed; having got the values of V_k/b_k from the whole area, the value of g was fitted by the formula (270.2). This is, as already mentioned, a pooled g, the pooling being as explained above. For optimum then

$$\delta V + \frac{1}{\lambda} \delta T = 0 \tag{270.3}$$

which in the present case leads to the equations

$$(c_1 + c_2 a_k + 2c_3 w_k)a_k{}^g w_k{}^2/b_k = \lambda, \quad c_2 w_k{}^2 a_k{}^{g+1}/b_k g = \lambda \tag{270.4}$$

It follows that

$$a_k = (c_1 + 2c_3 w_k)g/(1 - g)c_2 \quad (k = 1, 2, 3, 4, 5) \tag{270.5}$$

Substituting this in any of the equations (270.4), it is found that

$$c_2 w_k{}^2 \left\{ \frac{(c_1 + 2c_3 w_k)g}{(1 - g)c_2} \right\}^{g+1} \frac{1}{g} = b_k \lambda \equiv \mu_k \quad (k = 1, 2, 3, 4, 5) \tag{270.6}$$

271. There are eleven unknowns, λ, a_k, w_k ($k = 1, 2, 3, 4, 5$). There are also ten equations (270.4), while the total cost T given by (269.2) must be kept fixed at a given value, which furnishes a further equation. It can easily be seen that these equations are compatible and independent. The equations (269.2) and (270.4) cannot, however, be algebraically solved. Recourse is had to a graphical-numerical procedure for solving these equations.

Research and empirical observations now in progress in Calcutta, Washington, Ames, Raleigh, Rothamsted, and doubtless other places on the variances and costs association with different sizes of clusters will make it possible to specify the optimum design of sample for many other characteristics of human populations, housing, farms, commerce, etc.

Exercise 1. Suppose that a sample is to be taken in a certain city to determine the proportion of males. A question arises regarding the ultimate sampling unit: which is better, the household or the individual person?

Solution

First make these simplifying assumptions:[27] (a) The city consists entirely of 4-person families—husband, wife, and 2 children. (b) The sexes of the children are binomially distributed with equal probabilities. Thus a fourth of the families have 2 boys, half of them have a boy and a girl, and the remaining fourth have 2 girls.

Here, the cluster will be the family, and $\overline{N} = 4$. A female is to count 0, and a male 1. The mean values and variances of a group of the 3 kinds of families are shown in the accompanying table. For ease in calculation the city will be assumed to consist of some integral multiple of 4 families, say 100,000 families. The variances will then be those computed in the table.

CALCULATIONS FOR 1 GROUP OF 4 FAMILIES

1 family = parents + 2 children

i		p_i	$(p_i - p)^2$	$p_i q_i$
Serial number of family	Number of male children	Mean number of males in the family	Contribution to external variance	Internal variance
1	0	$\frac{1}{4}$	$\frac{1}{16}$	$\frac{3}{16}$
2	1	$\frac{1}{2}$	0	$\frac{4}{16}$
3	1	$\frac{1}{2}$	0	$\frac{4}{16}$
4	2	$\frac{3}{4}$	$\frac{1}{16}$	$\frac{3}{16}$
Average	1	$p = \frac{1}{2}$	$\sigma_b^2 = \frac{1}{32}$	$\sigma_w^2 = \frac{7}{32}$

$$\sigma^2 = \sigma_b^2 + \sigma_w^2 = \frac{8}{32} = \frac{1}{4}$$

[27] These are the assumptions made in a calculation carried out by Hansen and Hurwitz, "Relative efficiencies of various sampling units in population inquiries," *J. Amer. Stat. Assoc.*, vol. 37, 1942: pp. 89–94. This assumption gives results that are in good agreement with experience.

It follows from the calculations in the table that

$$\rho = \frac{\sigma_b{}^2 - \dfrac{1}{\overline{N} - 1} \sigma_w{}^2}{\sigma^2} \qquad \text{[Table of notation, p. 193]}$$

$$= \frac{\dfrac{1}{32} - \dfrac{1}{4 - 1}\dfrac{7}{32}}{\frac{8}{32}} = -\frac{1}{6}$$

Thus the brace in Eq. 65 has the value

$$1 + (\overline{N} - 1)\rho = 1 - (4 - 1)\tfrac{1}{6} = 1 - \tfrac{1}{2}$$
$$= \tfrac{1}{2}$$

Therefore the variance of the proportion male in samples consisting of whole families will be only half what it would be in samples of individuals, even though the same number of individuals is included in each type of sample.

With the family as the sampling unit the variance of the estimated proportion male will be

$$\mathrm{Var}\,\hat{p} = \frac{M - m}{M - 1}\frac{1}{m}\frac{1}{32} \qquad \text{[From Ch. 4]}$$

The same result comes from Eq. 65, because there

$$\mathrm{Var}\,\hat{p} = \frac{M - m}{M - 1}\frac{\sigma^2}{m\overline{N}}\left\{1 + (\overline{N} - 1)\rho\right\}$$

$$= \frac{M - m}{M - 1}\frac{\frac{1}{4}}{m4}\left\{1 - \frac{1}{2}\right\}$$

$$= \frac{M - m}{M - 1}\frac{1}{m}\frac{1}{32}$$

Exercise 2. Prove that if the problem is to determine the proportion of male children in the city, the individual child is negligibly better as a sampling unit than the 2 children of a family.

Solution

Here $\overline{N} = 2$, $p = \tfrac{1}{2}$, and the calculations may be made again for a group of 4 families.

i		p_i	$(p_i - p)^2$	$p_i \varsigma_i$
Serial number of family	Number of male children	Mean number of male children in the family	Contribution to external variance	Internal variance
1	0	0	$\tfrac{1}{4}$	0
2	1	$\tfrac{1}{2}$	0	$\tfrac{1}{4}$
3	1	$\tfrac{1}{2}$	0	$\tfrac{1}{4}$
4	2	1	$\tfrac{1}{4}$	0
Average	1	$\tfrac{1}{2}$	$\sigma_b{}^2 = \tfrac{1}{8}$	$\sigma_w{}^2 = \tfrac{1}{8}$

$$\sigma^2 = \sigma_b{}^2 + \sigma_w{}^2 = \tfrac{1}{4}$$

From the values seen in the table,

$$\rho = \frac{\sigma_b{}^2 - \dfrac{1}{\overline{N} - 1}\sigma_w{}^2}{\sigma^2}$$

$$= \frac{\dfrac{1}{8} - \dfrac{1}{2 - 1}\dfrac{1}{8}}{\frac{1}{4}} = 0$$

Thus the brace of Eq. 65 is unity, and the sampling of clusters will be practically identical with sampling individuals. Strictly, the two would be equal only if ρ were $-1/(N - 1)$, so there is a small but entirely negligible balance in favor of the individual child.

Remark. These two exercises perhaps overstress the possibility of negative or zero values of ρ. While such values are not impossible, and are actually met in experience, it is much more usual to encounter positive values of ρ, as has been explained.

More on the use of enlarged primary units. Reduced or even negative intraclass correlation can sometimes be induced in the formation of primary units by enlarging the sampling unit, as by combining urban and rural areas or any other two dissimilar areas. The areas that are combined need not be adjacent, but if not adjacent, an eye must be kept on travel-costs. To see the effect of enlarged sampling units, let $\sigma_{1b}{}^2$ be the external variance before the sampling units (districts, blocks, townships, counties, etc.) are enlarged by combining them in pairs. Let $\sigma_{2b}{}^2$ be the external variance between pairs after combination. Then

$$\sigma_{2b}{}^2 = \tfrac{1}{2}\sigma_{1b}{}^2(1 + \rho) \tag{103}$$

where ρ is the correlation between the areas that are paired.[28]

Proof of Eq. 103

Let μ be the mean of all $2N$ areas, all composed of an equal number of d.us. Then by definition

$$\sigma_{1b}{}^2 = \frac{1}{2N}\sum_1^{2N}(x_i - \mu)^2$$

Now combine the $2N$ areas into N pairs, and number them 1, 2 for the first pair; 3, 4 for the second; etc.; up to $2N - 1$, $2N$.

[28] Due to Hansen and Hurwitz, "On the theory of sampling from finite populations," *Annals Math. Stat.*, vol. xiv, 1943: pp. 333–62, p. 337 in particular.

Then

$$\sigma_{2b}^2 = \frac{1}{N} \sum_i \left[\frac{1}{2}(x_i + x_{i+1}) - \mu \right]^2 \qquad [i = 1, 3, 5, \cdots, 2N - 1]$$

$$= \frac{1}{N} \Sigma \left[\tfrac{1}{2}(x_i - \mu) + \tfrac{1}{2}(x_{i+1} - \mu) \right]^2$$

$$= \frac{1}{4N} \Sigma \left[(x_i - \mu)^2 + (x_{i+1} - \mu)^2 + 2(x_i - \mu)(x_{i+1} - \mu) \right]$$

$$= \tfrac{1}{2}\sigma_{1b}^2 (1 + \rho)$$

wherein

$$\rho = \frac{\dfrac{1}{N} \Sigma (x_i - \mu)(x_{i+1} - \mu)}{\sigma_{1b}^2} \qquad\qquad Q.E.D.$$

If the pairing is done at random, the "expected" value of ρ is $-1/(2N - 1)$; the proof follows the same argument by which Eq. 73 was deduced. Usually ρ will be positive if the paired areas are required to be contiguous. But if the pairing is done deliberately with the purpose of combining unlike areas, ρ will be small; it will even be negative if there is not too much worry about travel-costs, as when dissimilar noncontiguous areas may be paired if desired. Thus the variance between the pairs of combined areas may be made less than the original external variance, or perhaps only half as great or less. And although the internal variance within the paired areas is almost certainly increased by the consolidations, the increase is in practice slight, and the total variance is almost certainly decreased.

Exercise 3. Let $(\sigma_{\bar{N}b})^2$ be the external variance of the population per d.u. when the clusters are of size $\bar{N}$. Prove that

$$(\sigma_{\bar{N}b})^2 = \frac{1}{\bar{N}} \sigma_{1b}^2 [1 + (\bar{N} - 1)\rho]$$

where ρ is the average intraclass correlation between the pairs of elements within the clusters.

CHAPTER 6. ALLOCATION IN STRATIFIED SAMPLING

It is better not to know so much than to know so many things that ain't so.—Josh Billings.

How the problem arises. Thus far we have learned how to compute the variance of a sum or mean under certain sampling conditions, and emphasis has been placed on the importance of attaining maximum precision per unit cost. Stratified sampling is sometimes useful in achieving these aims. Stratified sampling is a plan by which the universe is divided into M bowls (strata or classes), and a sample is taken from each class as if it were a separate universe. Strata may be created on the basis of geographic location, rent, percent nonwhite, size of city, type of farming, etc. For information on commerce, the commercial concerns might be divided up by area, type of business, and size. In industrial sampling, the contractor, department, order of production, and source of raw material, are most important. The material to be sampled often presents natural breaking points, either by size or type. The aim in stratification is to break up the universe into classes that are fundamentally different in respect to the average or level of some quality-characteristic. A volume could be written on this subject, but the first step toward an understanding is the theory which tells what properties of the various classes govern the variance of the estimate of the mean (μ) or the total (ξ) of the entire universe. This book will go no further. As a matter of fact, this is by far the most interesting part of the story, because as will be seen by theory, under many circumstances it is not very important to stratify, yet under other conditions it is very important to do so. Theory will show that many common "intuitive" ideas regarding the desirability and supposed benefits of stratification are untenable.

The problem to be attempted here is the most efficient distribution of a sample amongst the several classes (strata) into which a universe may have been divided, in order to achieve maximum precision per unit cost for estimated characteristics of the universe. It would be possible by cut-and-try methods to increase the size of sample in one class and decrease it in another, each time recomputing the total cost and expected variance, and thus eventually to arrive at a satisfactory allocation of the sample amongst the classes, but this is tedious and it is much better to have a direct indication beforehand of just what to do. The solution of this problem, the proper allocation of the sample among the several classes of the universe, is the aim of this chapter. There can be no talk of gains arising from stratification without a statement regarding the intended allocations, and the procedures of estimation.

Formulation of the problem. The characteristic to be measured might be the inventory of nails or flour, the total annual sales of nails or flour or of all commodities combined; or it might be the total unemployed, or employed at so many hours per week, or the number of males or females or both under 18 and in school. The universe of elements to be sampled might consist of a list of all the dealers of some specific or general type (all hardware dealers, all establishments selling flour at retail) within a given region. Or it might consist of a list of all the dwelling units, or of all the small areas into which the region could be divided for purposes of sampling. The dealers, dwelling units, or areas, are supposed to have been classified or stratified, possibly on the basis of location and likely also on some other significant characteristic such as inventory, sales, number of employees, or number of occupied dwelling units reported at the latest census. In industrial sampling the universe might be a shipment of material, in bulk or in containers, which could be classed by lot, layer, producer, date of production, size of bale, color, or any other characteristic that might be thought important. The characteristic being measured might be any extensive property, such as weight or clean wool-content.

In each class at the time the sample is to be taken there will be an (unknown) average inventory or sales per dealer or per area; or an average number of people employed, unemployed, or in school, either per household or per area. Imagine that the universe consists of N elements (chips) divided into M bowls (strata or classes), and that a sample is taken from each. The notation is explained in the accompanying table. By definition

$$\mu = \frac{\Sigma N_i \mu_i}{\Sigma N_i} \tag{1}$$

TABLE OF NOTATION

	Universe				Sample		
Bowl	Size	Mean	Standard deviation		Size	Mean	Standard deviation
1	N_1	μ_1	σ_1		n_1	$\bar{x}_1$	s_1
2	N_2	μ_2	σ_2		n_2	$\bar{x}_2$	s_2
.					.		
.					.		
.					.		
M	N_M	μ_M	σ_M		n_M	$\bar{x}_M$	s_M
Totals	N	xxx	xxx		n	xxx	xxx

It will be supposed that the sampling units in any bowl have equal probabilities, whence

$$E\bar{x}_i = \mu_i \tag{2}$$

and that they are drawn independently at random so that by Eqs. 13 and 13' on page 102,

$$
\begin{aligned}
\text{Var } \bar{x}_i &= \frac{N_i - n_i}{N_i - 1} \frac{\sigma_i{}^2}{n_i} \quad \text{without replacement} \\
&= \frac{\sigma_i{}^2}{n_i} \quad \text{with replacement}
\end{aligned}
\Biggr\} \tag{3}
$$

The Greek letters $\mu_1, \mu_2, \cdots, \sigma_1, \sigma_2, \cdots$ denote actual averages today in the various classes, supposedly obtainable if a complete count were taken. The total population for the entire universe will be

$$\xi = N_1\mu_1 + N_2\mu_2 + \cdots \text{ through all classes } = N\mu \tag{4}$$

wherein N_i is the (supposed known) number of members in Class i. For an estimate of ξ we shall first use the unbiased estimate [1]

$$X = N_1\bar{x}_1 + N_2\bar{x}_2 + \cdots \tag{5}$$

There will be a contribution of variance from every class, and the samples from the various classes are independent, wherefore from Eq. 48 on page 130

$$
\begin{aligned}
\text{Var } X &= N_1{}^2 \text{ Var } \bar{x}_1 + N_2{}^2 \text{ Var } \bar{x}_2 + \cdots \\
&= N_1{}^2 \frac{N_1 - n_1}{N_1 - 1} \frac{\sigma_1{}^2}{n_1} + N_2{}^2 \frac{N_2 - n_2}{N_2 - 1} \frac{\sigma_2{}^2}{n_2} + \cdots
\end{aligned} \tag{6}
$$

It should be noted that the problem to be solved here is associated with a particular kind of sampling and estimation. A random sample of n_1 dealers, households, or areas is to be taken from Class 1, n_2 from Class 2, etc. Moreover, and very important, the estimate X is to be formed first by straight averaging to get $\bar{x}_1, \bar{x}_2, \cdots$, followed then by multiplication by $N_1, N_2, \cdots$ as Eq. 5 indicates. In symbols, if x_{ij} denotes the inventory of Dealer j in Class i, then

$$\bar{x}_i = \frac{1}{n_i} \sum_{j=1}^{n_i} x_{ij}$$

[The n_i dealers or dwelling units are supposedly numbered $j = 1, 2, \ldots, n_i$ after the sample is drawn]

In practice, instead of forming these averages one by one, it is usually more expeditious to form the terms $N_i\bar{x}_i$ directly by noting that

$$N_i\bar{x}_i = \frac{N_i}{n_i} \sum_{j=1}^{n_i} x_{ij}$$

[Note that $\sum_{j=1}^{n_i} x_{ij}$ is merely the sum of the inventories of the sample-dealers in Class i]

[1] The use of a biased estimate will be treated toward the end of the chapter.

Here it is the sampling interval N_i/n_i, or the ratio of universe to sample-size in Class i, that is used as a multiplier of the sample inventory $\sum_{j=1}^{n_i} x_{ij}$. To find the best values of N_i/n_i is the aim of this chapter.

If some other procedure of estimation were introduced, for example, the ratio of the inventories of the sample-dealers of Class i at two different dates (the census-date and the sample-date), all that follows would be affected. That is to say, the best allocation under one procedure of sampling and estimating may not be the best under another procedure. Thus, although in this chapter we shall solve a very important problem in optimum allocation, the solution should be regarded as applicable only in the specific procedure described.[2] It should be said, though, that most of the remarks made here will hold good in other procedures as well.

The cost of conducting the survey will be

$$C = n_1 c_1 + n_2 c_2 + \cdots \tag{7}$$

wherein c_i is the cost of a questionnaire in Class i (including collection, follow-up, and tabulation, with a prorated share of overhead). The problems to be solved may be stated as follows:

Problem I. Find what sampling intervals N_1/n_1, N_2/n_2, etc., will produce minimum Var X for a fixed allowable budget C.

Problem II. Find what sampling intervals N_1/n_1, N_2/n_2, etc., will produce a prescribed Var X at minimum cost.

Fortunately, the two problems almost completely overlap.

Solution of the first problem: cost prescribed. The solution, when finally arrived at, will determine a set of values of n_1, n_2, $\cdots$ which constitute the best sizes of sample in the several strata (i.e., *best* in the sense of giving minimum Var X for a given budget C). These will be termed the "equilibrium" values from the analogy with problems of equilibrium in mechanics. Let δ denote a variation from an equilibrium value. Then, if Var X is at a proper minimum, its variation will be zero to within higher powers of δn_i; i.e.,

$$\delta \operatorname{Var} X = \frac{\partial \operatorname{Var} X}{\partial n_1} \delta n_1 + \frac{\partial \operatorname{Var} X}{\partial n_2} \delta n_2 + \cdots = 0 \tag{8}$$

It will be observed that the procedure is to perform the differentiations exactly as if n_1, n_2, $\cdots$ were continuous variables, when in fact they can proceed only by steps of unity. This procedure can nevertheless be relied upon to give results for n_1, n_2, $\cdots$ correct within a unit; in fact, the results (Eq. 13 ahead) are probably exact.

[2] A more complicated problem, first sampling areal units and then subsampling dwelling units from them, is treated in *A Chapter in Population Sampling* (Bureau of the Census, 1947).

Now differentiation of the first term on the right-hand side of Eq. 6 gives

$$\frac{\partial}{\partial n_1} N_1{}^2 \frac{N_1 - n_1}{N_1 - 1} \frac{\sigma_1{}^2}{n_1} = \frac{\partial}{\partial n_1} \frac{N_1{}^3 \sigma_1{}^2}{(N_1 - 1)n_1} - \frac{\partial}{\partial n_1} \frac{N_1{}^2 \sigma_1{}^2}{N_1 - 1}$$

$$= -\frac{N_1{}^3 \sigma_1{}^2}{(N_1 - 1)n_1{}^2} - 0 \qquad (9)$$

The second term on the right is 0 because n_1 is not contained in the term directly above it.

Every term on the right of Eq. 6 will contribute a derivative like the one just written; hence from Eq. 8 it follows that

$$\delta \operatorname{Var} X = \frac{N_1{}^3 \sigma_1{}^2}{(N_1 - 1)n_1{}^2} \delta n_1 + \frac{N_2{}^3 \sigma_2{}^2}{(N_2 - 1)n_2{}^2} \delta n_2 + \cdots = 0 \quad (10)$$

Now differentiate Eq. 7 to get

$$\delta C = c_1 \delta n_1 + c_2 \delta n_2 + \cdots = 0 \qquad (11)$$

The 0 on the right expresses the fact that, however the sample-sizes are permitted to vary, the total cost C is to be kept fixed. This equation thus imposes a condition on the variations δn_1, δn_2, $\cdots$: they can not all be independent; one of them is fixed by Eq. 11.

Next, multiply Eq. 11 through by $-\lambda^2$ and add it to Eq. 10, collecting the coefficients of δn_1, δn_2, $\cdots$. The result is

$$\left[\frac{N_1{}^3 \sigma_1{}^2}{(N_1 - 1)n_1{}^2} - \lambda^2 c_1 \right] \delta n_1 + \left[\frac{N_2{}^3 \sigma_2{}^2}{(N_2 - 1)n_2{}^2} - \lambda^2 c_2 \right] \delta n_2 + \cdots = 0 \quad (12)$$

Now suppose I let δn_2, δn_3, $\cdots$ take on any values whatever, forcing Eq. 11 to hold by a proper choice of δn_1, at the same time choosing λ so as to make the coefficient of δn_1 vanish in Eq. 12; the first term is then 0 no matter what δn_1 may be. But this requires that the sum of all the remaining terms be 0 for any values whatever of δn_2, δn_3, $\cdots$, which can be so only if each coefficient is separately equal to 0. And thus every coefficient vanishes, giving

$$\lambda^2 c_i = \frac{N_i{}^2}{n_i{}^2} \sigma_i{}^2 \frac{N_i}{N_i - 1} \quad \text{for all } i \qquad (13)$$

$$\doteq \frac{N_i{}^2}{n_i{}^2} \sigma_i{}^2 \qquad (14)$$

The last form arises by neglecting the -1 compared with N_i, an approximation that will be assumed in what follows and for which no apology need be made.

Interpretation of the result. The equations just written are extremely important; they express the allocation of the sample for minimum variance. For turned around Eq. 14 appears as

$$\frac{n_i}{N_i} = \frac{\sigma_i}{\lambda \sqrt{c_i}} \tag{15}$$

This equation is a special case of much more general results obtained by Mahalanobis.[3] It tells us that if Var X is to be a minimum, the sample-size in any class should not only be in proportion to the size of that class, but also in proportion to the standard deviation of the inventories in that class. Moreover, if the unit cost c_i of collecting information within a particular class is relatively very great, the sample in that class should be cut down because of $\sqrt{c_i}$ in the denominator.

In practice one must not forget that differential sampling ratios may introduce complications into the field- and office-procedures. There is simplicity in strict proportionality, wherein n_i/N_i is a constant for all classes, and the statistician should therefore not hastily prescribe differential sampling ratios in accordance with Eq. 15 without some preliminary calculations on gains and costs. It is usually possible to determine in advance what the relative costs and sampling errors will be under various plans and to make a rational choice of sampling ratios (*vide infra*).

By evaluating λ it is possible to assign an absolute numerical magnitude to n_i. To do this, we go back to Eq. 14 and multiply each side by n_i, then add for all classes, using the summation sign Σ to indicate the operation of summing. The result is

$$\lambda^2 \Sigma n_i c_i = \sum \frac{N_i{}^2}{n_i} \sigma_i{}^2$$

But by Eqs. 7 and 15 this gives

$$\lambda^2 C = \lambda \Sigma N_i \sigma_i \sqrt{c_i}$$

whence

$$\lambda = \frac{\Sigma N_i \sigma_i \sqrt{c_i}}{C} \tag{16}$$

Everything on the right is supposed known or approximately determinable beforehand, wherefore a value for λ can be calculated and used in Eq. 15 to settle in advance the sample-sizes n_i to be drawn from the various classes.

It is possible to make a direct calculation of the minimized value of the Var X. This is done by going back to Eq. 6 and rewriting it in the

---

[3] See the quotation on pages 206–9.

form

$$\text{Var } X \doteq N_1 \left(\frac{N_1}{n_1} - 1\right)\sigma_1{}^2 + N_2 \left(\frac{N_2}{n_2} - 1\right)\sigma_2{}^2 + \cdots \qquad (17)$$

Going on, the equilibrium value of N_i/n_i is introduced from Eq. 15 with the result that the *minimized*

$$\text{Var } X = \left(\frac{\lambda\sqrt{c_1}}{\sigma_1} - 1\right) N_1\sigma_1{}^2 + \left(\frac{\lambda\sqrt{c_2}}{\sigma_2} - 1\right) N_2\sigma_2{}^2 + \cdots$$

$$= \lambda \Sigma N_i \sigma_i \sqrt{c_i} - \Sigma N_i \sigma_i{}^2 \qquad (18)$$

Again, λ is to be determined from Eq. 16, then used here. And thus the Var X to be expected from the wisest expenditure of the budget C can be calculated pretty closely beforehand, the only limitation being the prior determinations of N_1, N_2, $\cdots$, σ_1, σ_2, $\cdots$. The second term on the right arises from the finite multipliers $(N_i - n_i)/(N_i - 1)$ and may be neglected if the sample-sizes n_i/N_i are all small (0.1 or less).

Exercise. The property of a large public utility company is to be appraised to determine its over-all physical condition. The property consists of 14 main accounts—poles and aerial equipment maintained thereon, manholes and underground cable and conduit, meters, terminals, switches, etc. Samples of items in each account are to be drawn at random and examined by skilled inspectors, and each item is to be rated as 100, 90, 80, 70, etc. Let w_i be the proportional book value of the ith account when new; then $\Sigma w_i = 1$. If $\bar{x}_i$ is the average grade of the items inspected in the ith account, then

$$\bar{x} = \Sigma w_i x_i$$

will be an unbiased estimate of the average weighted condition of the entire 14 accounts.

Let n_i be the number of items to be inspected in the ith account, and let c_i be the cost of inspecting one item therein, including travel and tabulation. Let σ_i be the standard deviation of the numerical grades in this account. Then as the samples in the different accounts are entirely independent, there is no correlation between them, wherefore from Eq. 48 on page 130 it follows that

$$\sigma_{\bar{x}}{}^2 = \Sigma (w_i \sigma_i)^2$$

The finite multipliers are neglected here: they may be restored if circumstances call for them.

Assuming that the binomial sampling theory applies, show that the most economical plan would be to adjust the sample-sizes in accordance with the rule by which

$$n_i = \frac{w_i \sigma_i}{\lambda \sqrt{c_i}}$$

The symbol w_i may be translated into dollars, and one sample item allowed, on the average, for (e.g.) every \$25,000 on the books. The factor $\sigma_i/\sqrt{c_i}$ is usually highly variable from one account to another, and the sample should be adjusted up or down in each account in accordance with the values of $\sigma_i/\sqrt{c_i}$ that are likely to be encountered.

After the observations are completed, the results will provide an estimate of the variance $\sigma_i{}^2$ in each of the 14 accounts. These estimates may then be pooled in the above equation $\sigma_{\bar{x}}{}^2 = \Sigma\,(w_i\sigma_i)^2$ to calculate an estimate of $\sigma_{\bar{x}}$. Although only 9 degrees of freedom will be provided in each account when each one is sampled by the Tukey plan with 10 independent subsamples (p. 96), nevertheless the estimate of $\sigma_{\bar{x}}$ for the grand mean of the 14 accounts will possess a pleasingly small coefficient of variation.

> **Remark.** For any account so small or so uniform or so costly to inspect that n_i turns out to be less than 25 items, in my own practice I either put a floor at 25 or omit such an account altogether, giving it a value of $\bar{x}_i$ so low that there can be no controversy. If public education were not involved, there would be no problem: n_i as computed is most economical and provides an unbiased procedure. However, much time and effort may be lost trying to explain to the court why a sample of (e.g.) only 8 items is sufficient in a particular account, and it may be cheaper to increase the sample to 25 or even higher, or to omit the sample altogether and to fix this particular value of $\bar{x}_i$ arbitrarily and adversely to the company. This latter procedure saves the trouble of selecting the sample, preparing the forms, and processing them, and may be the cheapest way out. This is an example of the use of a *controlled bias*, to be discussed further in this chapter.

A simplifying assumption—equal unit costs. The variation in unit costs may be considerable, as when one class of household in the sample is urban and the cost of traveling from one household to another is small while another class is rural in a sparsely populated area and the cost of traveling from one household to another is considerable. Again, one class might be handled by mail at low cost, while another class requires personal interviews at high cost. However, in many enquiries, such as mail reports from business firms, the unit costs of collection and tabulation are nearly equal in all classes; and as cost enters only through the factor $\sqrt{c_i}$ in the denominator of Eq. 15 anyhow, it is then convenient and satisfactory under such circumstances to put

$$c_1 = c_2 = \cdots = c \qquad (19)$$

and

$$\lambda\sqrt{c} = k \qquad (20)$$

whereupon from Eq. 15

$$\frac{n_i}{N_i} = \frac{\sigma_i}{k} \qquad \text{[Neyman sampling]} \quad (21)$$

so that

$$\frac{n_1}{N_1} : \frac{n_2}{N_2} = \sigma_1 : \sigma_2 \tag{22}$$

Thus, if $\sigma_1 = 2\sigma_2$, then the sampling interval [4] N_1/n_1 in Class 1 is only half that of N_2/n_2 in Class 2.

Historical note. The advantages of allocating the sample according to Eq. 21 were discovered by Neyman. The reference is to page 580 of his article entitled, "On the two different aspects of the representative method," *J. Roy. Stat. Soc.*, vol. xcvii, 1934: pp. 558–606.

If the unit costs are so nearly equal that Eq. 19 may be used, then by solving Eq. 21 for n_i and summing over all classes it follows that

$$k = \frac{\Sigma N_i \sigma_i}{\Sigma n_i} = \frac{\Sigma N_i \sigma_i}{n} \tag{23}$$

whence upon returning to Eq. 21 we see that it may be written

$$\frac{n_i}{N_i} = \frac{n\sigma_i}{\Sigma N_i \sigma_i} \tag{24}$$

wherein n, the total sample, must satisfy the relations

$$n = n_1 + n_2 + \cdots = \frac{C}{c} \tag{25}$$

Under the assumption of equal unit costs (Eq. 19), Eq. 18 for the minimized Var X reduces to

$$\begin{aligned}
\text{Var } X &= \left(\frac{k}{\sigma_1} - 1\right) N_1 \sigma_1{}^2 + \left(\frac{k}{\sigma_2} - 1\right) N_2 \sigma_2{}^2 + \cdots \\
&= k \,\Sigma\, N_i \sigma_i - \Sigma\, N_i \sigma_i{}^2 \\
&= \frac{(\Sigma N_i \sigma_i)^2}{n} - \Sigma\, N_i \sigma_i{}^2 \qquad \text{[Neyman sampling assumed]}
\end{aligned} \tag{26}$$

Along with Eqs. 15 or 21, this equation is one of the most important ones in the chapter. It is very handy for calculation, for it evaluates at once the minimized Var X without the intermediate step of first finding n_1, n_2, $\cdots$ and substituting them into Eq. 6. As a matter of fact, if it is known that the sampling fractions n_i/N_i are all going to be

---

[4] This interpretation is introduced here for the very practical reason that a systematic selection employing a constant sampling interval with a random start is permissible and possibly even advantageous in place of a random selection in many types of sampling problems: cf. Chapter 4.

small, the second term in the last equation may be neglected, at least for a rough evaluation of the expected precision, for which the form

$$\text{C.V. } X = \frac{1}{\sqrt{n}} \frac{\Sigma N_i \sigma_i}{\xi} \tag{27}$$

may be preferred. This form requires an advance estimate of the total inventory ξ.

The right-hand fraction in the last equation can be written

$$\frac{\Sigma N_i \sigma_i}{\xi} = \frac{N_1 \sigma_1 + N_2 \sigma_2 + \cdots}{N_1 \mu_1 + N_2 \mu_2 + \cdots} \tag{28}$$

which may be regarded as a generalized coefficient of variation of the stratified universe. Eq. 27 then appears as

$$\text{C.V. } X = \frac{\text{C.V. universe}}{\sqrt{n}}$$

exactly as was encountered in Chapter 4 in consideration of sampling from an unstratified universe.

If there is only one class in the universe, the right-hand side of Eq. 28 reduces to $\sigma : \mu$ (dropping the subscript), and this is the ordinary definition of the coefficient of variation of a distribution.

Remark 1. In sample-design one is often confronted with a decision between two plans (equal unit costs assumed):

Plan B: Sample-sizes proportional to N_i.
Plan C: Sample-sizes proportional to $N_i \sigma_i$ (Neyman sampling; Eq. 15 or 21).

One should not too hastily specify Plan C. It will always show a gain (*vide infra*) but it is essential to bear in mind that the variable sampling ratios of Plan C also require variable weighting factors, and that the increased costs and complications in the field-work and tabulation may wipe out the apparent savings. The expected savings are not difficult to compute and do not require accurate advance estimates of the standard deviations $\sigma_1, \sigma_2, \cdots$ within classes. Preliminary calculations should always be carried out before making a decision. Simple examples are given further on.

It is often possible to classify the elements of the universe into strata in such manner that the standard deviations σ_i are almost sure to be approximately equal. Plans B and C are then identical, and maximum efficiency and simplicity are attained. This is often accomplished well enough by grouping the elements on equal class-intervals as of a previous census.

Remark 2. In practice almost every survey is multipurpose, and some times two or more characteristics are competitive, in which case the sample-design that is optimum for attaining the desired reliability in one characteristic may not be optimum for another (Ch. 4, p. 82). Proportionate sampling (Plan B) then attains maximum simplicity and, almost always, approximately maximum efficiency (Ch. 12).

Remark 3. Eqs. 21–27 correspond to minimum variance with a fixed number of questionnaires, whether the unit costs are equal or not.

A further simplifying assumption, often useful. In a wide variety of sampling problems it may be assumed that when the elements (dealers, areas) of the universe are grouped according to some measure of size (inventory, sales, number of employees, number of dwelling units) as of a past census, the standard deviations in the various classes at a later date will be nearly proportional to their means. In symbols,

$$\frac{\sigma_i}{\mu_i} = h, \quad \text{a constant} \qquad \substack{\text{[First exploited by Han-}\\ \text{sen and Hurwitz]}} \qquad (29)$$

This assumption is often admittedly rough, yet extremely useful. Substitution into Eq. 26 gives

$$\text{Var } X = h^2 \left(\frac{\xi^2}{n} - \Sigma N_i \mu_i^2 \right) \qquad (30)$$

in which approximations to h and ξ give a ready value of $\text{Var } X$. If the finite multipliers in Eq. 6 are all nearly unity, the second term in parentheses will be small compared with the first and will serve its purpose even though approximated crudely. In fact, if the second term is neglected altogether, there is left the very simple relation

$$\text{C.V. } X = \frac{h}{\sqrt{n}} \qquad (31)$$

wherein n, the total sample, comes from Eq. 25.

By introducing Eq. 29 when applicable, the Neyman sampling fractions n_i/N_i may be found quickly by noting that

$$\frac{n_i}{N_i} = \frac{n\sigma_i}{\Sigma N_i \sigma_i} \qquad \text{[Eq. 24]}$$

$$= \frac{n}{\xi} \mu_i \qquad (32)$$

h has disappeared, and it is necessary only to assume a mean μ_i for each class in order to calculate the Neyman allocations.

These means (μ_i) should preferably be advance estimates referring to today, rather than to the last census; then today's estimate of ξ is simply $\Sigma N_i \mu_i$ (see the numerical illustrations further on), and $\sigma_i : \mu_i$ in Eq. 29 represents today's coefficient of variation estimated in advance for Class i.

Actually, in Eqs. 30 and 31 it is only necessary that h be an average value of $\sigma_i : \mu_i$. In using Eq. 32 it is possible to make allowances for departures from the average (see Step iv on the next page).

The assumption contained in Eq. 29 is the constancy of the coefficient of variation of the distribution of inventories, sales, or employed, from one class to another. It is a reasonable one in saying that the absolute variability in time amongst big dealers is bigger than the absolute variability amongst small ones. The assumption of strict proportionality, however, can not be expected to hold very closely, but fortunately the usefulness of the assumption is not seriously impaired if some classes depart moderately[*] from strict proportionality, provided we have some knowledge of these departures and are able to make allowances for them (again see Step iv below). For example, it is usual to find considerably greater values of $\sigma_i:\mu_i$ in the smaller classes, a good example being the "blocks" [5] occupied by none or only 1 or 2 families at the time of a census: a few years later there may be many new houses on these blocks where were only vacant lots at the time of the census. Such areas should be "oversampled"—i.e., the sample assigned to such areas should be double or treble the sample-size calculated from Eq. 32, and none should be omitted from the sample. If there is any danger that a huge housing project may have been built meanwhile, any such areas should be thrown out for a separate sampling procedure [cf. (*i*) *A Chapter in Population Sampling* (Bureau of the Census, 1947), pp. 9 and 10; (*ii*) Ch. 11 in this book, p. 371.]

Steps in the use of the above results in sample-design. The assumption will be made here that the unit costs are the same in all classes, as specified by Eq. 19. Then, if the sample is to be allocated according to Eq. 21, the procedure is as follows:

Step i. Prescribe an allowable cost C and calculate the total allowable sample-size $n = C/c$. c is assumed known.

Step ii. Arrive at suitable advance approximations for ξ, h, and μ_i if Eq. 29 is usable; otherwise arrive at suitable approximations for N_i and σ_i. This step requires considerable knowledge of the universe, as might have been acquired in previous surveys or perhaps in a pilot study.

Step iii. Calculate the minimized expected Var X or C.V. X from Eq. 26, 27, 30, or 31 (whichever is applicable).

Step iv. If the expected precision, just calculated, appears to be good enough, proceed to calculate the sampling fractions n_i/N_i by Eq. 15, 21, or 32, the last if applicable, in which case raise or lower the sample-sizes by appropriate amounts in those classes wherein the value h assumed for $\sigma_i:\mu_i$ is known to be too low or too high.

If, on the other hand, the expected Var X as computed in Step iii is not considered to be small enough, then it will be necessary to increase the allotted budget from C to some new value C' and recompute. If additional funds are not to be had, it will be necessary either to accept a lower standard of precision (greater Var X) or to abandon the survey as not being worth while for the allowable funds.

If the assumption of equal unit costs is not justifiable, these steps require modifications.

---

[5] "Block" is used in America to designate the small urban areas bounded by streets

Solution of the second problem: variance prescribed. Here C is to be made a minimum, the n_i to be of such magnitude that $\text{Var } X$ has a prescribed value (such as C.V. $X = 2$ percent, 5 percent, 25 percent). The mathematical solution is identical with the first problem down to the point where λ is determined. It is now to be found from Eq. 18, which gives

$$\lambda = \frac{\text{Var } X + \Sigma N_i \sigma_i^2}{\Sigma N_i \sigma_i \sqrt{c_i}} \tag{33}$$

Everything on the right is supposed to be known, so λ can be calculated. Then from Eq. 16 the cost is seen to be

$$C = \frac{1}{\lambda} \Sigma N_i \sigma_i \sqrt{c_i} \tag{34}$$

The required sample-sizes n_i would be computed from Eq. 15 at the same time. The total sample-size will of course be

$$n = \Sigma n_i \qquad \text{[As in Eq. 25]}$$

Simplifications may be introduced under the same conditions as those mentioned in earlier paragraphs. In the first place, if the finite multipliers in Eq. 6 can all be replaced by unity, the second term in the numerator of Eq. 33 can be thrown away in the calculation of λ. Second, if the unit costs are the same in all classes, so that Eq. 19 can be used, then in place of Eqs. 33 and 34 it would be simpler to write

$$k = \lambda\sqrt{c} = \frac{\text{Var } X + \Sigma N_i \sigma_i^2}{\Sigma N_i \sigma_i} \qquad \begin{array}{l}\text{[The second term, arising from} \\ \text{the finite multipliers, can} \\ \text{sometimes be neglected, as} \\ \text{mentioned]}\end{array} \tag{35}$$

and

$$C = \frac{c}{k} \Sigma N_i \sigma_i \tag{36}$$

The sample-sizes n_i would be computed from Eq. 21 at the same time. The total sample will be

$$n = \Sigma n_i = \frac{C}{c} = \frac{\Sigma N_i \sigma_i}{k} \tag{37}$$

If, in addition, the constancy of the ratio $\sigma_i : \mu_i$ may be usefully assumed, as expressed by Eq. 29, then Eq. 35 gives

$$k = \lambda\sqrt{c} = \frac{\text{Var } X + h^2 \Sigma N_i \mu_i^2}{h\xi} \qquad \begin{array}{l}\text{[The second term, arising} \\ \text{from the finite multi-} \\ \text{pliers, can sometimes be} \\ \text{neglected, as mentioned]}\end{array} \tag{38}$$

whereupon the cost

$$C = \frac{c}{k} \Sigma N_i \sigma_i \qquad \text{[Eq. 36]}$$

simplifies to

$$C = \frac{ch\xi}{k}$$

$$\doteq \frac{ch^2\xi^2}{\text{Var } X} = \frac{ch^2}{(\text{C.V. } X)^2} \qquad \begin{array}{l}\text{[Neglecting the second}\\ \text{term in Eq. 38]}\end{array} \qquad (39)$$

The sample-sizes would then be found from Eq. 32.

When the variance is prescribed, and the unit costs c_i are all assumed to be equal, the steps to be applied in allocating the sample according to Eq. 21 may be outlined as follows:

Step i. Decide on the prescribed value of Var X, or perhaps, if easier, think in terms of C.V. X (2 percent, 5 percent, 25 percent, etc.).

Step ii. Arrive at suitable advance approximations for ξ, h, and μ_i if Eq. 29 is usable; otherwise arrive at suitable approximations for N_i and σ_i. This step requires considerable knowledge of the universe, as might have been acquired in previous surveys or perhaps in a pilot study.

Step iii. Calculate λ from Eq. 33 and then the minimum cost C from Eq. 34, or preferably from Eq. 36, or from Eq. 39 if applicable.

Step iv. If it is apparent that the cost is not going to be too heavy, proceed to allocate the sample by Eq. 15, 21, or 32, the last if applicable, in which case raise or lower the sample-sizes by appropriate amounts in those classes wherein the value h assumed for $\sigma_i : \mu_i$ is known to be too low or too high.

If, on the other hand, the expected minimum cost C as computed in Step iii is too high, it will be necessary to accept a lower level of precision than initially prescribed and to recompute the cost. If this plan is not acceptable, the survey must be abandoned, as being too costly.

If the assumption of equal unit costs is not justifiable, these steps require modifications.

Gains compared for different allocations. Two simple illustrations will help. Let the problem be the determination of the total inventories of 3000 dealers, which have been divided (for simplicity) into just two classes, as of some previous census. σ_i and μ_i represent advance estimates, as of today. One universe is shown in Table 1. It will be assumed that the unit costs are the same in both classes, wherefore the total costs of both plans will be equal because the sample-sizes are equal (120).

TABLE 1. A UNIVERSE TO BE SAMPLED BY TWO PLANS

Class	N_i	μ_i	$N_i\mu_i$ (All in tons)	σ_i	Sample-sizes, n_i	
					Plan B n_i proportional to N_i (proportionate sampling)	Plan C n_i proportional to $N_i\sigma_i$ (Eq. 32) (Neyman sampling)
1	1000	100	100,000	90	40	60
2	2000	50	100,000	45	80	60
Total	3000	xxx	$\xi = 200{,}000$	xxx	120	120

Equation 6 will give the variance of the total estimated inventory. For simplicity the finite multipliers will be neglected; then

$$\text{Var } X = N_1{}^2\frac{\sigma_1{}^2}{n_1} + N_2{}^2\frac{\sigma_2{}^2}{n_2} \qquad \begin{array}{l}\text{[Approximation to}\\ \text{Eq. 6]}\end{array} \qquad (40)$$

Under Plan B this gives

$$\text{Var } X = 1000^2 \times \frac{90^2}{40} + 2000^2 \times \frac{45^2}{80}$$

$$= \frac{1000^2 \times 45^2}{40}(4+2) = \frac{1000^2 \times 45^2}{120} \times 18$$

Under Plan C the same formula gives

$$\text{Var } X = 1000^2 \times \frac{90^2}{60} + 2000^2 \times \frac{45^2}{60}$$

$$= \frac{1000^2 \times 45^2}{60}(4+4) = \frac{1000^2 \times 45^2}{120} \times 16$$

The loss in using Plan B is thus $(18 - 16)/16$ or $\frac{1}{8}$. That is, Plan B will produce an eighth more variance at the same cost or will cost an eighth more for the same variance. Under other conditions (different N_i and σ_i) the gains arising from Plan C may be much greater, but here the gain is not much more than enough to offset the additional costs and complications in the field and tabulations arising from the introduction of different sampling ratios in the two classes.

This illustration is introduced to show the danger of jumping too rapidly into one plan or the other without some preliminary computations. Even rough guesses at the standard deviations σ_1 and σ_2 may be sufficient to indicate which type of allocation will best serve the purpose, or whether stratification should be used at all.

The student should satisfy himself that Eq. 26 with the second term neglected gives the same result for Plan C. Thus,

$$\text{Var } X = \frac{(\Sigma N_i \sigma_i)^2}{n} \qquad \begin{array}{l}\text{[Eq. 26, second term}\\\text{neglected]}\end{array} \qquad (41)$$

$$= \frac{(1000 \times 90 + 2000 \times 45)^2}{120}$$

$$= \frac{1000^2 \times 45^2}{120} \times 16 \quad \text{as before}$$

Comparison with the result for Var X using no stratification at all is interesting. The total variance of the universe is

$$\sigma^2 = P\sigma_1{}^2 + Q\sigma_2{}^2 + PQ(\mu_2 - \mu_1)^2 \qquad \text{[P. 58]} \quad (42)$$

$$= \tfrac{1}{3} \times 90^2 + \tfrac{2}{3} \times 45^2 + \tfrac{1}{3} \times \tfrac{2}{3}(100 - 50)^2$$

$$= \frac{41,450}{9}$$

P and Q here denote the proportions in the two classes; $P = N_1/N$ and $Q = N_2/N$, where $N = N_1 + N_2$. An unstratified random sample of 120 will give

$$\text{Var } X = N^2 \frac{\sigma^2}{n} \qquad \begin{array}{l}\text{[Plan A, the finite multiplier}\\\text{neglected]}\end{array} \qquad (43)$$

$$= 3000^2 \times \frac{41,450}{120 \times 9} = \frac{1000^2 \times 45^2}{120} \times 20.5$$

It is now possible to compare the variances arising from all three sampling procedures. Table 2 illustrates a very important point, viz., that under certain conditions the gains arising from stratified sampling will be disappointing.

TABLE 2. RESULTS OF SAMPLING THE UNIVERSE IN TABLE 1

Sampling procedure	Var X		
Plan A: unstratified random sampling	$(1000^2 \times 45^2/120)$	20.5	by Eq. 43
Plan B: sample-sizes proportional to N_i	"	18	by Eq. 40
Plan C: sample-sizes proportional to $N_i \sigma_i$	"	16	by Eq. 40 or 41

Another simple illustration may be helpful. Let a new universe be defined and sampled as described in Table 3. The variances arising from Plans A, B, and C are shown in Table 4, which the student should calculate as an exercise. It will be observed that in this case proportionate sampling shows a heavy gain over no stratification, while Neyman's allocation shows little gain over proportionate sampling. These results arise from the fact that the means of the two strata are wide apart, while the standard deviations are not so far apart.

TABLE 3. ANOTHER UNIVERSE TO BE SAMPLED

Class	N_i	μ_i	$N_i \mu_i$ (All in tons)	σ_i	Plan B n_i proportional to N_i (proportionate sampling)	Plan C n_i proportional to $N_i \sigma_i$ (Eq. 32) (Neyman sampling)
					Sample-sizes, n_i	
1	1000	200	200,000	100	40	60
2	2000	50	100,000	50	80	60
Total	3000	xxx	$\xi = 300,000$	xxx	120	120

The student should now give attention to Eqs. 47 and 51 and observe that they predict the results shown in Tables 2 and 4.

TABLE 4. RESULTS OF SAMPLING THE UNIVERSE IN TABLE 3

Sampling procedure	Var X
Plan A: unstratified random sampling	$(1000^2 \times 50^2/60)18$ by Eq. 43
Plan B: sample-sizes proportional to N_i	" 9 by Eq. 40
Plan C: sample-sizes proportional to $N_i \sigma_i$	" 8 by Eq. 40 or 41

Note in regard to advance estimates of the variances. It has been explained that the variances $\sigma_1{}^2$, $\sigma_2{}^2$, $\cdots$ within the several classes at the time the sample is taken are assumed to be known in advance, approximately at least, as has been explained. Oftentimes considerable ingenuity is required to evaluate these variances beforehand. One should make use of census information and other previous studies, consultation with experts in the subject-matter, and sometimes a pilot study for estimating some of the variances as well as for testing the questionnaire and instructions. There always seems to be a way out if an honest effort is made, particularly in view of certain favorable features of the problem.

For instance, it is important to note that successful allocation does not require accurate knowledge of the variances involved. This is indeed fortunate. It is so because Var X does not increase rapidly from its minimum value as n_1, n_2, $\cdots$ are moved only small distances δn_1, δn_2, $\cdots$ away from their equilibrium values: this can be seen from Eq. 8, in which the increase in Var X arising from the *first* powers of δn_1, δn_2, $\cdots$ has been placed equal to zero (the condition for a proper minimum), wherefore Var X is affected only through the squares and higher powers of δn_1, δn_2, $\cdots$. Now, if δn_1 is small, its square is still smaller, and the effect on Var X may be negligible. Hence it is not necessary to come out with exactly the equilibrium values of n_1, n_2, $\cdots$: approximations will still serve the purpose tolerably well, which means that only reasonably good values of σ_1^2, σ_2^2, $\cdots$ are required.

This point is brought out in the first numerical illustration, wherein two different allocations of the sample did not produce greatly different values of Var X. In the second illustration the two allocations of the sample were sufficiently different to produce distinctly different values of Var X.

Another noteworthy feature is that Neyman allocation is obtained if only the advance estimates of σ_1, σ_2, $\cdots$ are in the *right proportions*. If all are estimated too low or too high by, for example, 20 percent, as when h in Eq. 29 is set too high, the sampling fractions n_i/N_i are unaffected, as is easy to see from Eq. 24 or 32. Damage is done, however, to the advance estimate of the Var X or the estimated cost. For instance, if the variances were all initially overestimated by 20 percent, then for a prescribed cost C (Problem 1) the advance calculation of the C.V. X will also be too high by about 10 percent, as will be discovered when the variances are evaluated from the returns (cf. the section, "A word on the final determination of the precision reached," *infra*). In other words, the error band will turn out to be 10 percent smaller than expected. The same thing will happen when a particular C.V. X has been prescribed; the cost will then be about 20 percent greater than was necessary to reach the prescribed precision. The additional expenditure goes for unneeded precision, purchased interestingly enough at the cheapest possible rate.

Exact formulation of the gains arising from stratification. Here we shall compare three plans of allocation, the same three in fact that were just illustrated with numerical examples. Now, however, we shall seek an exact formulation of the comparisons. The variances associated with the three different plans are those shown below.

Plan B: proportionate sampling in which n_i is proportional to N_i. Here

$$\text{Var } X \doteq \Sigma\, N_i(b-1)\sigma_i^2 \qquad \text{[From Eq. 17]} \quad (44)$$

wherein for all classes

$$\frac{N_i}{n_i} = b \qquad \begin{array}{l}\text{[The sampling inter-}\\\text{val for all classes]}\end{array} \quad (45)$$

Plan C: Neyman sampling in which n_i is proportional to $N_i \sigma_i$. Here

$$\text{Var } X = \frac{(\Sigma N_i \sigma_i)^2}{n} - \Sigma N_i \sigma_i^2 \qquad \text{[Eq. 26]}$$

Plan A: in which there is no stratification at all. Here

$$\text{Var } X = N^2 \frac{N-n}{N-1} \frac{\sigma^2}{n} = N(b-1)\sigma^2 \quad \begin{matrix} [N-1 \text{ replaced} \\ \text{by } N] \end{matrix} \quad (46)$$

wherein, as in Eq. 45,

$$b = \frac{N}{n}$$

Let these three variances be designated by B, C, A. Then

$$A - B = (b-1)(N\sigma^2 - \Sigma N_i \sigma_i^2)$$
$$= N(b-1)(\sigma^2 - \Sigma P_i \sigma_i^2)$$
$$= N(b-1) \Sigma P_i(\mu_i - \mu)^2 \qquad \begin{matrix} \text{[Note that } \sigma_i \\ \text{is absent]} \end{matrix} \quad (47)$$

Herein

$$P_i = \frac{N_i}{N}, \quad \begin{matrix} \text{the proportion of the} \\ \text{universe in Class } i \end{matrix} \qquad (48)$$

and

$$\sigma^2 = \Sigma P_i \sigma_i^2 + \Sigma P_i(\mu_i - \mu)^2 \quad \text{[Ex. 2, p. 59]} \quad (49)$$

The relative gain of Plan B over Plan A will be

$$\frac{A-B}{A} = \frac{\Sigma P_i(\mu_i - \mu)^2}{\sigma^2} \qquad (50)$$

The result shows that stratification with proportionate sampling (Plan B) removes the variance between classes. Obviously, this plan will be appreciably better than straight random unstratified sampling only if one or more of the large strata have means differing widely from the mean of the universe, so that at least one term $N_i(\mu_i - \mu)^2$ will contribute sufficiently to the difference $A - B$. Thus Plans B and A are compared. Next, to compare Plans B and C we note that

$$B - C = \frac{N}{n} \Sigma N_i \sigma_i^2 - \frac{(\Sigma N_i \sigma_i)^2}{n} \qquad \text{[Finite multipliers all neglected]}$$

$$= \frac{1}{n} [\Sigma N N_i \sigma_i^2 - (\Sigma N_i \sigma_i)^2]$$

$$= \frac{1}{n} [\Sigma N_i^2 \sigma_i^2 + \Sigma \Sigma N_i N_j \sigma_i^2 - \Sigma N_i^2 \sigma_i^2 - \Sigma \Sigma N_i N_j \sigma_i \sigma_j]$$

$$\qquad\qquad\qquad\qquad\qquad [N = \Sigma N_i; \text{ also } i \neq j]$$

$$= \frac{1}{n} \Sigma \Sigma N_i N_j(\sigma_i - \sigma_j)^2 \qquad\qquad [i < j]$$

$$= bN \Sigma \Sigma P_i P_j(\sigma_i - \sigma_j)^2 \qquad \text{[Note that } \mu_i \text{ is absent]} \quad (51)$$

The relative gain $(B - C)/C$ will be worth while only if two or more of the σ_i are greatly different, and only then provided neither class is very small. For although $(\sigma_i - \sigma_j)^2$ be large for one pair of classes, the factor $P_i P_j$ may reduce the term to some negligible amount if either Class i or Class j is relatively very small, and the extra cost of using sample-sizes proportional to $N_i \sigma_i$ rather than the uniform density of Plan B might not be worth while.

In particular, if two classes have equal variances they contribute nothing at all to the advantage of Plan C over Plan B. To go further, if all classes have the same variance, Plans B and C are identical.

The student should return to the numerical illustrations of the preceding section and test out Eqs. 50 and 51 for the differences between $A - B$ and $B - C$.

For the solutions of certain problems in allocating a sample to strata that are to be sampled in 2-stages, see pages 238–41; also pages 26 ff. in *A Chapter in Population Sampling* (Bureau of the Census, 1947).

Remark on the simplicity of proportionate sampling in industry. In manufacturing, and in general in the sampling of industrial product as it is loaded or unloaded or as it moves by conveyor from one department to another, the usual practice of testing one or more articles from every truck-load, or from every (e.g.) half-hour's production, is proportional sampling,[6] either systematic or randomized. In drawing samples from production lines it is wise to avoid being too systematic. Every operation and every item should be subject to test, but the whole purpose of testing is defeated if anyone is able to predict which items are to be tested. In the testing of imported raw sugar, the buyer, the seller, and the Bureau of Customs all settle upon the worth of the shipment by means of polarization-tests performed upon a sample which is compounded from many small equally sized specimens of sugar drawn by means of a pointed instrument which is plunged into every 7th bag as the sugar is unloaded. This is one example of proportionate sampling in randomized blocks, as each sling contains 7 bags, all alike in appearance. The possibility that bias might arise from the particular interests of the buyer or the seller is supposedly removed by exchanging "samplers" ever hour.

A word on the final determination of the precision reached. At the risk of repetition, a word is offered here on the importance of evaluating the precision actually reached, for at least some of the characteristics computed from the sample. What has gone before in this chapter is advice on sample-design, which in the modern sense implies *advance* calculation of the expected variance for some critical characteristic to be obtained from the sample (total inventory, sales, or employment). This advance calculation can only be made on the basis of incomplete knowledge—the best obtainable at the time, but incomplete nevertheless. For example, one never knows what snags the field-work will

---

[6] It is assumed that the truckloads and rates of movement are equal.

run into, and these snags affect the variances $\sigma_1{}^2$, $\sigma_2{}^2$, $\cdots$. A subsample taken from the returns will give estimates of these variances as they actually occurred in the sampling. Substitution into Eq. 6 then gives a good estimate of Var X. The proper sample to take for estimating the variances from the returns will be discussed in Chapter 10.

This last stage—the final evaluation of Var X—be it noted, is independent of the assumptions that went into the design, however crude, and hence independent of whether the design was really a good one. The cost is trifling.

A further argument lies in the experience to be gained by computing some actual variances under the conditions met, in order to have them available for future sample-designs.

Discussion of possible aims in an actual problem (contributed by Richard H. Blythe, Jr., Federal Trade Commission). The discussion so far in this chapter has centered around the aim of designing the best sample for estimating a total or average figure of some kind (employment, sales, inventories) over all classes combined. Other aims, however, are sometimes encountered, or at least suggested, in practice. An actual example, often encountered, is a proposal to obtain figures of equal absolute accuracy in all classes, or equal percentage errors in all classes. Of course, samples can be designed to meet such requirements and many others, but it is interesting to see how much these whims actually cost.

The problem is exemplified by a program intended to collect financial data from American manufacturing corporations. A number of financial items, such as total sales, cost of goods sold, and income, were to be covered by the survey. In designing the sample, figures on net income before deducting federal taxes were taken as the controlling characteristic, because net income was deemed most important, and also because it was the most variable of any of the important items. Thus if a sample could be designed to meet the requirements of precision (standard error) for net income, the estimates of other characteristics would almost certainly be sufficiently precise.

Prior data were available which made possible the stratification of manufacturing corporations based upon their size as measured by total assets. As often happens, the potential users of the results were vitally interested not only in the total overall size-classes, but also in the estimates for each of a number of size-classes separately.

The first step in designing the sample was to assemble the necessary preliminary estimates of the variances of net income for each of the asset size-classes. This information is shown in Table 5. Only a glance is needed to see the pattern of concentration of a large proportion of the total net income amongst a small number of large companies. This pattern of concentration occurs frequently in populations of economic and social variables.

Using the data in Table 5 and the methods described earlier in this chapter, a sample based on Plan C was calculated (n_i proportional to

TABLE 5. DESCRIPTION OF THE UNIVERSE

(1)	(2)	(3)	(4)
		Estimated average	Standard deviation
Stratum limits in terms of total assets (thousands of dollars)	Number of corporations	net income (thousands	of net income of dollars)
	N_i	μ_i	σ_i
Unknown	5,600	1	5
Under 50	28,700	1	5
50–99	11,100	5	8
100–249	13,000	15	20
250–499	7,500	50	65
500–999	5,100	100	130
1000–4999	5,800	300	390

$N_i \sigma_i$). The total size of sample was specified, and the aim was to achieve the greatest accuracy from the estimated total net income. The number of samples to be drawn from each stratum under a plan of this kind is shown in the second column of Table 6. The results to be expected

TABLE 6. ALLOCATION OF SAMPLES AND RESULTING COEFFICIENTS OF VARIATION

Plan C means Neyman sampling (n_i proportional to $N_i \sigma_i$)

(1)	(2)	(3)	(4)	(5)
Stratum limits in terms of total assets (thousands of dollars)	Number of sample corporations		Predicted coefficients of variation	
	Plan C	Equal C.V.	Plan C	Equal C.V.
Unknown	54	2390	67.6%	7.7%
Under 50	277	3650	29.9	7.7
50–99	172	416	12.1	7.7
100–249	502	291	5.8	7.7
250–499	942	287	4.0	7.7
500–999	1281	282	3.1	7.7
1000–4999	4372	284	1.0	7.7
All classes	7600	7600	1.1	4.9

from such an allocation are listed in column 4. The expected values of the coefficient of variation of the estimated total of each stratum are seen to vary from 1.0 percent for the largest class to 29.9 percent for the smallest. (The value of 67.6 percent for the class of unknown size is not considered because no interest attaches to this class. It includes companies, mostly small, for which size-data were not available.)

When the meaning of this result was explained to those who were responsible for the program on a policy level, there was general agreement that the sampling errors of the smaller classes were undesirably large, while the error for the total taken over all classes was unnecessarily small. Consequently a compromise was demanded: Why not decrease the sample from the large companies and increase it in the smaller companies? As there was no good reason to favor any one class over another, it was argued that a reasonable aim was to allocate the sample so as to yield the same coefficient of variation for all classes.

The mathematical problem of equalizing the coefficients of variation in all classes requires that the sample-sizes n_i be obtained from the equations

$$\frac{N_1 - n_1}{N_1 - 1} \frac{C_1{}^2}{n_1} = \frac{N_2 - n_2}{N_2 - 1} \frac{C_2{}^2}{n_2} = \cdots \quad \begin{array}{c}\text{[This comes from} \\ \text{Eq. 6]}\end{array} \quad (52)$$

wherein

$$n_1 + n_2 + \cdots = n \qquad \text{[As in Eq. 25]}$$

and C_i stands for the coefficient of variation of net income in Class i. If the finite multipliers could be neglected, the problem would be very simple, because then

$$n_i = \frac{n}{\Sigma\, C_i{}^2} C_i{}^2 \qquad (53)$$

The n_i so obtained may indeed be useful approximations. A systematic solution by which these or any other reasonable approximations can be refined was given by Stock and Frankel.[7] Another method of solution, less systematic, is to examine the finite multipliers and adjust them by hand to achieve the desired equalities in Eq. 52. The strata having the largest and smallest finite multipliers are studied, and the allocation between them is adjusted so as to equalize their coefficients of variation. Usually a small amount of adjustment of the remaining strata will bring all of them into line. Results so reached are shown in Col. 3 of Table 6, the equal coefficients of variation being 7.7 percent, as given in the right-hand column.

Next it is interesting to see what has been gained by this second method of allocation. In 3 of the 7 classes the equalized coefficient of variation (7.7 percent) is smaller than would have been obtained under Plan C (n_i proportional to $N_i \sigma_i$), shown alongside. This is the gain. But what are the losses? In 4 classes the coefficient of variation has

---

[7] J. Stevens Stock and Lester R. Frankel, "The allocation of samplings among several strata," *Annals Math. Stat.*, vol. x, 1939: pp. 288–93, Sec. 4 in particular.

been increased and markedly so. Also the sampling error of the total has jumped from 1.1 percent to 4.9 percent, a serious loss. Considering the situation as a whole, the net "improvement" is a disappointment; the 7.7 percent coefficient of variation is so large as to make all the estimates by classes too poor for many uses.

To reduce the uniform coefficient of variation from 7.7 percent to some usable figure such as 2 or 3 percent, an enormously increased total sample would be required. But the size of the sample could not be increased for reasons of economy and burden of response, and some other

TABLE 7. ALLOCATION OF SAMPLES AND RESULTING COEFFICIENTS OF VARIATION
FOR COMBINED CLASSES

Plan C means Neyman sampling (n_i proportional to $N_i \sigma_i$)

(1)	(2)	(3)	(4)	(5)
Class limits in terms of total assets (*thousands of dollars*)	Number of sample corporations		Predicted coefficients of variation	
	Plan C	Equal C.V.	Plan C	Equal C.V.
Under 250 and unknown	1005	4000	5.7%	2.7%
250–999	2223	2000	2.5	2.7
1000–4999	4372	1600	1.0	2.7
All classes	7600	7600	1.1	1.8

compromise seemed advisable. Another plan suggested was to combine certain classes and make no attempt to publish the figures for all 7 classes. To this end, 3 size-groups were created in place of 7, in the hope of publishing data by each of these classes. The allocation and coefficients of variation resulting from another application of Plan C are shown in Cols. 2 and 4 of Table 7. Once again the error in the group of small corporations was deemed too large, so an allocation yielding equal coefficients of variation for each of the 3 classes was worked out, the results being displayed in Cols. 3 and 5 of Table 7. Considerable gain in the smallest class is in evidence, and the error of the total (1.8 percent) is obviously much reduced below 4.9 percent, the error that was obtained previously by this type of allocation to 7 classes (Table 6). The improvement arises because it was possible to use Plan C within classes. For example, the 4000 corporations in the top group (assets under $250,000, plus assets unknown) were allocated by Plan C to the 4 subclasses comprising this group.

Before acclaiming this system as the final answer, we must note that actually we have achieved an improvement over the application of Plan C in only the top group. In the others and in the total we take a loss by equalizing the coefficients of variation. If the error 1.8 percent of the total is satisfactory (Table 7), let us ask how large a sample would be required to yield this same error by Plan C. The results were achieved by applying Eqs. 33 and 15 and are shown in Table 8, wherein it is seen that only 4024 sample corporations would be required to yield a coefficient of variation of 1.8 percent in the total, whereas a sample of 7600 allocated by the compromise gave this same precision in the total. Thus if we shift from Plan C to the compromise we nearly double the size of the sample needed to achieve equal accuracy in the totals. If 1.8 percent

TABLE 8. ALLOCATION OF SAMPLE AND RESULTING COEFFICIENTS OF VARIATION FOR A SMALLER SAMPLE, WITH ALLOCATION ACCORDING TO PLAN C

(Neyman sampling; n_i proportional to $N_i \sigma_i$)

Class limits in terms of total assets (thousands of dollars)	Number of sample corporations (By Eqs. 33 and 15)	Predicted coefficient of variation
Under 250 and unknown	533	7.9%
250–999	1177	3.6
1000–4999	2314	2.1
All classes	4024	1.8

is a satisfactory level of precision for the total, we must ask ourselves whether it is worth while to double the cost merely to achieve some gains in the accuracy of the estimates of the 3 classes separately.

The cost and performance of a sample intended for any other special need could be computed, but the 4 already worked out are sufficient to choose from:

i. A sample of 7600 allocated by Plan C amongst 7 classes, giving a coefficient of variation of 1.1 percent in the total, and widely differing coefficients of variation for the 7 classes (Table 6, Cols. 2 and 4).

ii. A sample of 7600 allocated amongst the 7 classes so as to yield a uniform coefficient of variation of 7.7 percent in each class, giving 4.9 percent in the total (Table 6, Cols. 3 and 5).

iii. A sample of 7600 allocated amongst 3 classes so as to yield a uniform coefficient of variation of 2.7 percent in each class, giving 1.8 percent in the total (Table 7, Cols. 3 and 5).

Allocation by Plan C amongst and within these 3 classes decreases the coefficient of variation to 1.1 percent, as was obtained earlier (compare Cols. 2 and 4 of Table 6 with Cols. 2 and 4 of Table 7).

iv. A sample of 4024 allocated by Plan C amongst 3 classes, giving a coefficient of variation of 1.8 percent in the total and coefficients of variation not differing widely in the 3 classes.

The second plan has already been discarded. The advantages of the third choice over the fourth probably do not warrant doubling the cost, wherefore either the first or fourth plan would probably be chosen.

This example illustrates the tremendously high price of catering to special aims and special interests in order to obtain figures for particular classes or areas.

It is difficult to resist mentioning specifically a further example, for which no calculations will be required. It is sometimes a temptation to enlarge a regional or a national sample into one that will provide data not only for the nation as a whole, but separately also for some few large cities as well, say for New York, Chicago, Philadelphia, Detroit, Los Angeles, and perhaps one or two others. This can be done, of course, if funds can be found, but every city added to such a list requires about as big a sample as was initially allocated to the entire country.

Optimum allocation in two stages of stratified sampling. In this chapter consideration has been given thus far only to sampling in one stage. For sampling in two stages the basic mathematical principles are similar, but the actual equations are more complex. Various types of situations arise in sampling in two stages. For instance, one may or may not restrict himself to constant overall weighting factors. Each case must be formulated on its own merits and peculiarities. An example in two stages is given on pages 27–31 in *A Chapter in Population Sampling*. There the sampling ratio $M_i:m_i$ for primary units was to be constant for all strata, and the sampling ratio $N_i:n_i$ within primary units was also to be constant.

A very simple and useful rule was expounded by Hansen [8] in 1947. It does not give exactly the theoretical optimum for a uniform overall weighting factor, but in consideration of the savings arising from simplicity, it is probably better than the theoretical optimum.

The problem dealt with by Hansen is a sample of d.us. to be drawn for a survey of an urban population. At the first stage, a sample of "blocks" [9] is to be drawn, and at the second stage, a sample of d.us. is to be drawn from these blocks.

---

[8] Morris H. Hansen, *The Sampling of Human Populations*, read at the International Statistical Conferences, Washington, 6–18 September 1947.

[9] For the benefit of readers not familiar with the American use of the word "block," it should be explained that a block is merely an area bounded by streets or other recognizable boundaries. It may be rectangular or irregular in any fashion. It may contain hundreds of d.us. and business establishments, as in some areas of New York; or it may contain very few d.us. or none at all. It may be a park or vacant property, public or private.

It is costly to list d.us. in an area containing a large number of them. The rule to be stated decreases the total cost of listing by drawing fewer large blocks into the sample than would be drawn with probabilities in proportion to their sizes. Let the blocks of the city be divided into two strata, large and small (subscripts L and S). The three conditions specified for the sample-design are these:

i. A uniform overall weighting factor in both strata, large and small—

$$a_L b_L = a_S b_S \tag{54}$$

ii. A uniform sampling ratio $b_L = N_i : n_i$ is to be applied in all the large blocks, and some other constant ratio b_S is to be applied in all the small blocks.

iii. The rule for drawing primary units is to be expressed by the formula

$$\frac{a_S}{a_L} = \sqrt{\frac{N_L}{N_S}} \tag{55}$$

The symbols have the following meanings:

$a_L = \dfrac{M_L}{m_L}$,   the counting interval for drawing blocks from the L-stratum

a_S,   similarly defined for the S-stratum

$b_L = \dfrac{N_L}{\bar{n}_L}$,   the counting interval for drawing d.us. from the sample-blocks of the L-stratum

b_S,   similarly defined for the S-stratum

Now suppose that it has been decided that a 5 percent sample shall be drawn. Then

$$a_L b_L = a_S b_S = 20 \tag{56}$$

Suppose further that it is known that approximately $\bar{N}_L = 100$ d.us. per block and that $\bar{N}_S = 25$ d.us. per block. It has been decided that a proper work-load will be obtained with $n_S = 5$; i.e., an average of 5 d.us. are to be canvassed in the S-stratum. Then

$$b_S = \frac{\bar{N}_S}{\bar{n}_S} = \frac{25}{5} = 5 \tag{57}$$

$$a_S = \frac{20}{b_S} = \frac{20}{5} = 4 \tag{58}$$

$$a_L = a_S \sqrt{\frac{\bar{N}_S}{\bar{N}_L}} = 4 \sqrt{\frac{25}{100}} = 2 \tag{59}$$

$$b_L = \frac{20}{a_L} = \frac{20}{2} = 10 \tag{60}$$

That is, every 5th d.u. will be drawn from the sample-blocks of the S-stratum, but every 10th d.u. will be drawn from the sample-blocks of the L-stratum. Every 4th block will be drawn from the S-stratum; every other block will be drawn from the L-stratum.

Extensions to three size-strata might be desirable in a city containing some huge blocks of 300 or 400 d.us. The additional equations would not be difficult. An example of allocation of sample in multi-stage sampling occurs on page 379.

Exercise 1. In sampling the dwelling units of a city for estimating the proportion colored, it is possible to classify the dwelling units into two groups, in one of which the proportion colored is 0.1 and in the other of which it is 0.9. Prove that sample-sizes proportional to N_i are identical with sample-sizes proportional to $N_i \sigma_i$; in other words, that Plans B and C are identical. (HINT. $\sigma_1{}^2 = p_1 q_1 = p_2 q_2 = \sigma_2{}^2$)

Exercise 2. If the two groups of dwelling units of Exercise 1 are of equal size, then stratified sampling with a uniform sampling ratio will show a gain of 64 percent over no stratification at all. That is, only 36 percent as many dwelling units would be required in Plan B or C as in Plan A, to reach a prescribed precision.

Exercise 3. (a) If the proportions colored in the two classes are 0.25 and 0.75, then the gains of Plans B and C drop to 25 percent. (b) If the proportions are 0.6 and 0.4, the gains in efficiency are only 4 percent; that is, the gains accomplished by stratification are now almost negligible.

Exercise 4. Two strata have means μ_1 and μ_2, standard deviations σ_1 and σ_2, numbers N_1 and N_2. A total sample of n is to be drawn. How should the sample be allocated between the two strata so as to minimize the variance of $\bar{x}_2 - \bar{x}_1$, which is an estimate of $\mu_2 - \mu_1$? Assume unit costs c_1 and c_2 as in the text.

Answer:

$$n_1{:}n_2 = \frac{\sigma_1}{\sqrt{c_1}} \sqrt{\frac{N_1}{N_1 - 1}} : \frac{\sigma_2}{\sqrt{c_2}} \sqrt{\frac{N_2}{N_2 - 1}}$$

$$= \frac{\sigma_1}{\sigma_2} \sqrt{\frac{c_2}{c_1}} \left\{ 1 + \frac{1}{2} \left(\frac{1}{N_1} - \frac{1}{N_2} \right) + \cdots \right\}$$

The student should note that when $\sigma_1 = \sigma_2$ and $c_1 = c_2$, this result does *not* reduce to proportionate sampling, but to *equal sample-sizes* from the two strata except for the small disparity introduced by the finite multipliers.

Exercise 5. A large corporation decided to ascertain the value of its motor vehicles by examining a sample of 100. The company, of course, keeps a list of all of its property. In the design of the sample the universe was divided into two classes—Class 1, vehicles of $1\frac{1}{2}$ tons capacity and over; Class 2, vehicles of lesser capacity, including passenger cars. After some deliberation it was predicted that the mean value in Class 1 would be about $5000, and that the mean value in Class 2 would be about $2000. Eq. 29 was assumed to hold, viz., standard deviation proportional to the mean. The costs of inspecting the different vehicles were assumed to be equal. Show that under these assumptions the sampling intervals should satisfy the relation

$$\frac{N_1}{n_1} : \frac{N_2}{n_2} = 2:5$$

wherein

$$n_1 + n_2 = 100$$

whence if $N_1 = 93$ and $N_2 = 796$, the sample-sizes will be about $n_1 = 23$ and $n_2 = 77$. In other words, the sampling intervals would be about 4 and 10 with random starts in the two classes.

Exercise 6. Derive the formulas for the optimum allocation of sample shown on pages 27–31 of *A Chapter in Population Sampling* (Bureau of the Census, 1947).

This exercise will provide an extension of the foregoing theory into a case of stratified sampling in two stages in the central city, combined with a single stage of sampling in the surrounding zone, with and without a constant overall weighting factor.

Some common misconceptions regarding stratification. *a.* It is occasionally alleged by the laity that a particular sample is vitiated because some of its elements were "misclassified." A real universe is dynamic, not static, and the information that is used for classification is always to some extent out of date. Moreover, in any real survey a few blunders may occur; they ought not to, but they do, and a unit will occasionally be dropped into a stratum where no one intended it to go. Misclassification—or what appears to be misclassification—is thus to be expected as the natural course of events. The point to be addressed toward allegations of misclassification is that misclassification is irrelevant except to the actual sampling errors, which, after they are evaluated empirically (Ch. 10), are known and are not debatable, regardless of the reasons for classification or misclassification.

Sometimes the universe is so fluid that even the most recent data are quite useless for purposes of stratification; the strata may as well be formed by chance, which is to say that there is no use forming any strata at all. If one nevertheless hopefully goes through the motions of stratification, the final appraisal of precision after the survey is completed will show that under such circumstances nothing was gained: the universe might as well have been left unstratified, and the sample drawn at random. Eq. 50 shows that unless the various means (μ_i) are sufficiently different, stratified proportionate sampling will give the same standard error as unstratified sampling. And Eq. 51 shows that unless the standard deviations (σ_i) are sufficiently different, even Neyman sampling will be disappointing. The above examples illustrated these points.

b. It is sometimes supposed that stratification must always bring gains in precision, but for the reasons just mentioned, this is not so; the results of stratification are often disappointing. On the other hand, under other circumstances, the gains of stratification are great.

c. It is sometimes supposed that if the universe is divided into many strata, and a big enough sample allocated to each stratum, then any kind of sampling within strata is good enough. As one nameless text puts it, "to the extent that we fail in obtaining randomness, we make up for lost accuracy by stratifying." This is, of course, but a fond fancy, apparently invented as a refuge for bad sampling. Actually, in the above theory, the sample of n_i in the ith stratum is to be drawn with the same care as if the stratum were a little universe, as actually it is: otherwise bias is introduced, and the variance is not as calculated. One may *if he knows what he is doing* take a cheap biased sample in some insignificant stratum (small P_i in Eqs. 50 and 51), or leave it out entirely if upper limits to $|\mu_i - \mu|$ can be fixed, showing that the resulting bias is negligible. (This trick is explained to some extent in the following sections.)

Use of a biased formula of estimation. An estimate $\hat{\varphi}$ is the product of a series of operations, ending with a calculation by the use of a formula in which the results of the survey are combined in a prescribed manner with the probabilities of selection. Some very useful formulas of estimation are *biased;* i.e., the "expected" results that they give differ from the "expected" result of an unbiased formula. We have already encountered examples in the ratio-estimate of Chapter 5 and in the exercises beginning on page 87. Again, the bias of the formula by which $\hat{\varphi}$ is calculated (for short, the bias of the estimate $\hat{\varphi}$) is defined as

$$b_{\hat{\varphi}} = E\hat{\varphi} - \varphi \qquad (61)$$

it being assumed that $\hat{\varphi}$ is a random variable, and that φ is the "expected" result arising from the use of an unbiased formula. Then by the same algebraic identity as seen on page 129,

$$\delta_{\hat{\varphi}}{}^2 = \sigma_{\hat{\varphi}}{}^2 + b_{\hat{\varphi}}{}^2 \qquad (62)$$

wherein $\delta_{\hat{\varphi}}$ is the root-mean-square error in $\hat{\varphi}$. *A biased formula is never used unless an upper limit to the bias can be computed from known properties of the universe, nor unless this upper limit is negligibly small.* The actual bias can not in practice be computed exactly; if it could be, it would be subtracted out, and a new and unbiased estimate

$$\varphi' = \hat{\varphi} - (E\hat{\varphi} - \varphi) \qquad (63)$$

formed, which would have the same variance as $\hat{\varphi}$ but no bias.

Remark. It is important to distinguish a *biased formula* from a *biased procedure of selection*. It is the former that we are discussing now: the latter was discussed in Chapters 1 and 2. The bias in a biased formula arises from the manner in which an estimate is prepared.

It should be noted that the bias in any useful biased estimate (the topic of these pages) is exceedingly small compared with other biases, as the student may recall from one illustration on page 180 in Chapter 5.

Several formulas of estimation are shown in the accompanying table.[10] The properties of these estimates, left as exercises, are computed on the assumption that all the sampling units have equal probabilities. It will be observed that Eq. 5 is but a special case of the third estimate (m_3).

Estimate of μ   *Properties*

1. $m_1 = \dfrac{\Sigma \, n_i \bar{x}_i}{\Sigma \, n_i}$

   a. Biased except when $n_1/N_1 = n_2/N_2 = \cdots = n_M/N_M = n/N$, which corresponds to proportional sampling throughout

   b. Consistent *

   c. $\operatorname{Var} m_1 = \dfrac{1}{(\Sigma \, n_i)^2} \sum \dfrac{N_i - n_i}{N_i - 1} n_i \sigma_i^2$

2. $m_2 = \dfrac{1}{M} \Sigma \, \bar{x}_i$

   a. Biased except when $N_1 = N_2 = \cdots = N_M$

   b. Inconsistent * except when $N_1 = N_2 = \cdots = N_M = N/M$

   c. $\operatorname{Var} m_2 = \dfrac{1}{m^2} \sum \dfrac{N_i - n_i}{N_i - 1} \dfrac{\sigma_i^2}{n_i}$

3. $m_3 = \Sigma \, a_i \bar{x}_i$

   where

   $\Sigma \, a_i = 1$

   a. Biased except when $a_i = N_i/N$ (as assumed in Eq. 5)

   b. Inconsistent * except when $a_i = N_i/N$

   c. $\operatorname{Var} m_3 = \Sigma \, a_i^2 \dfrac{N_i - n_i}{N_i - 1} \dfrac{\sigma_i^2}{n_i}$

* See footnote on page 43 of Chapter 5.

Exercise 1. The bias of the 1st estimate in the accompanying table is

$$Em_1 - \mu = E \frac{\Sigma \, n_i \bar{x}_i}{\Sigma \, n_i} - \mu$$

$$= \frac{\Sigma \, n_i \mu_i}{\Sigma \, n_i} - \mu = \frac{\Sigma \, n_i (\mu_i - \mu)}{\Sigma \, n_i}$$

which is zero if the sampling is proportionate, i.e., if $n_i/N_i = n/N$ for all strata.

[10] This table was suggested by Section 15 in P. Thionet's *Méthodes Statistiques Modernes* (Hermann, Paris, 1946).

Exercise 2. The bias of the 2d estimate is

$$Em_2 - \mu = E \frac{1}{M} \Sigma \bar{x}_i - \mu$$

$$= \frac{1}{M} \Sigma \mu_i - \mu = \frac{1}{M} \Sigma (\mu_i - \mu)$$

$$= \sum \left[\frac{1}{M} \mu_i - \frac{N_i}{N} \mu \right]$$

which happens to be independent of n_i. The bias is zero if $N_1 = N_2 = \cdots = N_M$ in which case $\dfrac{N_i}{N} = \dfrac{1}{M}$.

Exercise 3. The inconsistency of the 2d estimate is

$$\frac{1}{M} \Sigma \mu_i - \mu = \frac{1}{M} \Sigma (\mu_i - \mu)$$

which is identical with the bias, because the bias in this case is independent of n_i.

Exercise 4. The bias of the 3d estimate is
$$Em_3 - \mu = E \Sigma a_i \bar{x}_i - \mu = \Sigma a_i(\mu_i - \mu)$$
$$= 0 \quad \text{if } a_i = N_i/N$$

Exercise 5. Prove the statements regarding the variances shown in the table.

Exercise 6. If the sampling is proportional, and if in the 3d estimate $a_i = N_i/N$, then the 1st and 3d estimates are identical and unbiased.

Exercise 7. If the sampling is proportional and if in the 3d estimate $a_i = N_i/N$, then (a) the difference between the variances of the 2d and 3d estimates is

$$\text{Var } m_2 - \text{Var } m_3 = \sum \frac{N_i - n_i}{N_i - 1} \left(\frac{1}{M^2} - \frac{N_i^2}{N^2} \right) \frac{\sigma_i^2}{n_i}$$

$$\doteq \left(1 - \frac{n}{N} \right) \sum \left(\frac{1}{M^2} - \frac{N_i^2}{N^2} \right) \frac{\sigma_i^2}{n_i}$$

wherein $1 - n/N$ is written as an approximation to the finite multipliers, which are practically all equal in proportional sampling. (b) Show also that if the strata are equal in size (N_i all equal) this difference is zero. (c) If the strata are not equal, this difference may be positive or negative, depending on the various N_i/N and σ_i. Thus, under the condition of proportional sampling, and with $a_i = N_i/N$, it is possible for Var m_2 to be equal to, less than, or greater than Var m_3.

This exercise shows that an estimate that seems preferable under one set of conditions may not appear so advantageous under another set of conditions.

Exercise 8. (a) The bias of the 3d estimate is $\Sigma\, a_i\mu_i - \mu$ or $\Sigma\, a_i(\mu_i - \mu)$ or $\displaystyle\sum \left(a_i - \frac{N_i}{N}\right)\mu_i$. (b) The m.s.e. $m_3 = \Sigma\, a_i{}^2\sigma_{\bar{x}_i}{}^2 +$ $[\Sigma\, a_i\mu_i - \mu]^2$. (c) Show that the Var m_3 is a minimum when

$$a_i = \frac{\dfrac{1}{\sigma_{\bar{x}_i}{}^2}}{\displaystyle\sum \dfrac{1}{\sigma_{\bar{x}_i}{}^2}}$$

This value of a_i minimizes the variance of the linear function $\Sigma\, a_i\bar{x}_i$, but not its mean square error. This linear function will be a biased estimate of μ except when under fortuitous circumstances the above value of a_i is equivalent to N_i/N.

(d) Show that the m.s.e. m_3 is a minimum when the a_i satisfy the following symmetric equations, in which the unknowns appear in the heading and the coefficients in the table. For convenience, the equations are written only for a_1, a_2, a_3, and λ, the Lagrange multiplier which enters the solution. The bottom row is the equation $a_1 + a_2 + a_3 = 1$.

a_1	a_2	a_3	$-\lambda$	$=$	1
$\sigma_{\bar{x}_1}{}^2 + \mu_1{}^2$	$\mu_1\mu_2$	$\mu_1\mu_3$	1	$=$	$\mu_1\mu$
$\mu_1\mu_2$	$\sigma_{\bar{x}_2}{}^2 + \mu_2{}^2$	$\mu_2\mu_3$	1	$=$	$\mu_2\mu$
$\mu_1\mu_3$	$\mu_2\mu_3$	$\sigma_{\bar{x}_3}{}^2 + \mu_3{}^2$	1	$=$	$\mu_3\mu$
1	1	1	0	$=$	1

Solution

Let y denote the m.s.e. m_3. Then

$$\frac{\partial y}{\partial a_i} = 2a_i\sigma_{\bar{x}_i}{}^2 + 2\mu_i(\Sigma\, a_i\mu_i - \mu); \qquad \frac{\partial}{\partial a_i}\Sigma\, a_i = 1$$

The solution follows the same argument as found on page 217.

$$\delta y = 0$$

$$\delta\,\Sigma\, a_i = 0$$

Use the multiplier $-\lambda$ on the last equation and add.

$$\tfrac{1}{2}\delta y - \lambda\delta\,\Sigma\, a_i = [a_i\sigma_{\bar{x}_i}{}^2 + \mu_i(\Sigma\, a_i\mu_i - \mu) - \lambda]\,\delta a_i = 0$$

Every one of the M brackets must be 0, giving M equations as written above, to which is appended the bottom row $\Sigma\, a_i = 1$. Altogether there are $M + 1$ equations and $M + 1$ unknowns, a_1, a_2, $\cdots$, a_M, and λ.

Exercise 9. Prove that in the preceding exercise

$$\lambda = \text{Min. m.s.e. } m_3 + \mu \cdot \text{Bias } m_3$$

Solution

Multiply the top row through by a_1, the second row by a_2, $\cdots$, the last by $-\lambda$. Add, remembering that $\Sigma\, a_i = 1$. The result is a quadratic form which reduces to

$$\Sigma\, a_i^2 \sigma_{\bar{x}_i}^2 + (\Sigma\, a_i\mu_i)^2 - \lambda = \mu\, \Sigma\, a_i\mu_i$$

The solution is completed by transposition and a bit of algebraic alteration, and by recalling that Bias $m_3 = \Sigma\, a_i\mu_i - \mu$.

Exercise 10. Find what values of n_i and a_i (i.e., what allocation of sample and what weights) will minimize the m.s.e. m_3.

Answer: The a_i will satisfy the equations of the preceding exercise, while

$$n_i = \frac{a_i\sigma_i}{\sqrt{c_i}} \frac{C}{\Sigma\, a_i\sigma_i\sqrt{c_i}} \sqrt{\frac{N_i}{N_i - 1}}$$

wherein C and c_i are the total and unit costs of questionnaires, as in the text.

CHAPTER 7. DISTINCTION BETWEEN ENUMERATIVE
AND ANALYTIC STUDIES

Any experiment may be regarded as forming an individual of a "population" of experiments which might be performed under the same conditions. A series of experiments is a sample drawn from this population.

Now any series of experiments is only of value in so far as it enables us to form a judgment as to the statistical constants of the population to which the experiments belong. In a greater number of cases the question finally turns on the value of a mean, either directly, or as the mean difference between the two quantities.—Student, "The probable error of a mean," *Biometrika*, vol. vi, 1908: pp. 1–15.

To sum up, the experiments must be conducted in such a way that their results may be capable of being considered to be a random sample of the population to which the conclusions are to apply.—Student, "Mathematics and agronomy," *J. Amer. Soc. Agron.*, vol. xviii, 1926: pp. 703–19.

Enumerative and analytic studies. The distinction between *enumerative* and *analytic* studies is extremely important in the design and analysis of either complete counts or samples. In both types of study the ultimate aim is to provide a rational basis for *action*. A problem exists, and something is to be done about it. In the enumerative problem something is to be done to some portion of the contents of the bowl regardless of the reasons why that portion is so large or so small. In the analytic problem, on the other hand, something is to be done to regulate and predict the results of the cause system that has produced the universe (city, market, lot of industrial product, crop of wheat) in the past and will continue to produce it in the future.

An example of an enumerative problem is the estimation of the number of inhabitants in a certain area and the inventory or the yield of wheat, rice, or other grain, in order to determine the amount of food that must be shipped in to feed the people in that area, or the amount that can be shipped out without detriment to health. Such a problem is enumerative because it depends on determining the number of people in the area, and the inventory and production of grain, and does not involve the (analytic) question of why all these people are there or why the crop is what it is. A familiar example in the United States is Congressional apportionment, which depends on the number of people in a Congressional District, and not at all on why they are there or how many are expected to be there some years hence.

Further illustrations of the enumerative use of data could be given without end. In some states (for example California) a city treasury receives from the state a specified amount of money per head per annum. If the Bureau of the Census charges 10¢ per head for the cost of a census, and if the city receives 75¢ annually per head from the state, then it is obvious that if the population of the city has increased as much as

15 percent, the extra receipts from the state treasurer will in one year pay the cost of a special census.

When certain cities in America were swelled with in-migrants because of war production in the spring of 1944, special censuses [1] were taken with the aim of arriving at equitable allocations of food, gasoline, repair parts for buses and trolley cars, and other necessaries of living. These cities had grown enormously because of war production, and equitable distribution to them was impossible because no one knew just how many people were in them: assertions of editors and chambers of commerce did not provide verifiable figures on growth. The problem was *enumerative* because the action (viz., allocation of food and materials) depended on *how many* people were there and *not why* they were there.

From the standpoint of the consumer, acceptance-sampling (next chapter) of industrial product may be taken as another enumerative use of data; but from the standpoint of the producer, acceptance-sampling may be taken as analytic, as it helps him to control his process. By law the Social Security program is partly an enumerative problem because federal reimbursement to a state depends on the number of inhabitants 65 and over within the state. Public health programs, agricultural adjustments, and other allotments depend on population and acreage and are examples of enumerative uses of data. Administrative problems concerned with the guidance of these programs, however, are analytic.

The enumerative problem calls for a survey to determine how many of certain classes of chips are in the bowl—*this* bowl, just as it is and not as it might have been. The theory that is applicable in the enumerative problem is the theory of sampling without replacement, wherein the finite multiplier $(N - n)/(N - 1)$ on page 103 reduces the sampling variance to 0 as the size of sample is increased to 100 percent. A 100 percent sample from the bowl, drawn without replacement, thus represents the totality of information on this problem. A limiting case of the enumerative use of data is seen in a circumstance wherein each individual report determines the action to be taken in regard to the respondent. An example is an income-tax form. Another example is an application for a loan; another is a social-security record. Assessments and allotments are computed for the individual on the basis of the information supplied for him alone. Samples are not used for such computations. On the other hand, studies of methods of taxation or social security, leading to intelligent legislation affecting tax rates and benefits, are analytic, and sample studies of income-tax forms or of other characteristics of various segments of the population provide the only practicable solution.

[1] The populations of nine cities were estimated by samples; see *A Chapter in Population Sampling* (Bureau of the Census, 1947).

In analytic problems the action is to be directed at the underlying causes that have made the frequencies of various classes of the population what they are, and will govern the frequencies of these classes in time to come. Familiar examples of analytic studies can be found in intelligent city-planning and in the measurement of differential effects of treatments in agriculture and entomology. The control-chart is a splendid example, the purpose being to control the production process. Other examples are medical and social studies wherein interest centers in the causes that produce differences in health, fertility, or death-rate in different segments of a population of people. The Current Population Survey in the United States, in reality a miniature monthly census, aids studies of employment, unemployment, school attendance, etc. A quarterly survey in Canada provides the same information. The monthly sample of deaths by causes, published by the National Office of Vital Statistics in the United States, aids in the control of epidemics and spread of disease.

In analytic problems interest centers in the causes of patterns and variations that take place from year to year, or from area to area, or from class to class, or from one treatment or variety to another. The ultimate reason for studying the causes is to learn to control them and hence to control health, yields of crops, diseases of crops, insects, performance of workers, and quality of industrial product. For such studies even a complete count is still only a sample of the product (people, crops, insects, or industrial product) of the underlying cause system: this will be evident from theory soon to be studied. This fact is extremely important in sample-design and tabulation plans. For analytic uses a small sample taken at frequent time-intervals, by showing trends and changes while they are taking place, will furnish much more information than would be furnished at about the same cost by a much larger sample or even a complete count taken at wider intervals of time. Most of the examples already mentioned are in fact studied by means of small frequent samples (e.g., the control-chart; the Current Population Survey).

The routine collection of enumerative or accounting data often furnishes an extremely important source of data for analytic studies. An example is the records of sales and complaints in a department store, used as a basis for improving service and increasing sales.

In analytic problems the concern is not this one bowl alone but the sequence of bowls that will be produced in the future (or allowably, bowls already produced but not yet measured). A particular bowl is studied only so far as it helps in understanding the causes that gave rise to it. The amount of information that one bowl can supply is limited. A 100 percent sample from any one bowl is all the information that that bowl can supply, and if this is not enough, further information must be supplied by samples from other bowls or from related knowledge.

This means that in using a census-table for analytic purposes, even though the figures come from a complete count, it is absolutely necessary to bear in mind that small numbers (frequencies) in a cell are unreliable in the sense of having a standard error just as if they had arisen in sampling, as indeed they did. This fact is of extreme importance in the planning of a research program and in planning the tabulations of a survey, as well as in interpreting data in tables already made.

Quantitative illustration.[2] Exercise 1 at the close of this chapter provides quantitative answers to the analytic uses of data under four specific assumptions. The results of this exercise, combined with theory already gained in Chapter 4, will provide a foundation for a quantitative distinction between the analytic and enumerative uses of data, so that we may perceive what effect this distinction has on the design and tabulations of a proposed survey. The student should first turn ahead and work through the exercise, then return to this section. The "supply" in Fig. 16 represents the cause system or production process. Repeated random samples drawn therefrom (called *lots*) will show varying proportions of red chips, whose distribution will follow the binomial or hypergeometric series, depending on whether the chips are drawn with or without replacement. A lot drawn from the supply and placed in the lot-container corresponds in practice to an industrial lot, or a crop, or the people in a city, or a segment of the population that has access to certain medical facilities. In the analytic problem what is wanted is an estimate of p, the proportion of red chips in the supply, in order that probability-limits may be placed on the proportion of red chips that will be found in future lots drawn from the same supply. ("Same supply" means a constant value of p.) In the exercise the lot-container is to be filled and sampled in four different ways (Cases A, B, C, D). In actual experience, Case B is probably most important: Case D is next in importance. The next chapter makes extensive use of the results for Case B, and so does all work in quality control and all experimental work in agricultural and medical science.

In the enumerative problem what is wanted is an estimate of x/N, the proportion of red chips in the lot, regardless of what is in the supply or what future lots may contain. By referring ahead to the results of Case B in Exercise 1, and back to results already known from Chapter 4,

---

[2] The author's indebtedness to Frederick F. Stephan for assistance in developing this section extends over several years, as is obvious from a joint publication "On the interpretation of censuses as samples," *J. Amer. Stat. Assoc.*, vol. 36, 1941: pp. 45–9. Further indebtedness to him is expressed for providing solutions to the problem given further on at the close of this section.

several parallel conclusions can be set down side by side, as shown in the accompanying table.

COMPARISON BETWEEN THE PURPOSES, FORMULAS, AND PRECISIONS OF SAMPLES USED IN ANALYTIC AND ENUMERATIVE PROBLEMS

Analytic	*Enumerative*
Purpose. From a sample drawn from the lot-container, to estimate the proportion p of red chips in the *supply* (cause system, production process).	*Purpose.* From a sample drawn from the lot-container, to estimate the initial proportion x/N of red chips in the *lot-container*.
Let a succession of lots be drawn from the supply with replacement, and from each lot let a sample of n be drawn without replacement. If r denotes the number of red chips in a sample, then it is to be proved in the exercise that	Let a succession of samples be drawn without replacement * from the given lot. In this lot, x chips are red and $N - x$ are white. If r denotes the number of red chips in a sample of n, then it is known from Chapter 4 that

$$E \frac{r}{n} = p \qquad (1)$$

$$E \frac{r}{n} = \frac{x}{N} \qquad (1')$$

$$\text{C.V.} \frac{r}{n} = \sqrt{\frac{q}{np}} \qquad (2)$$

$$\text{C.V.} \frac{r}{n} = \sqrt{\frac{N - n}{N - 1} \frac{N - x}{nx}} \qquad (2')$$

* Each sample is to be replaced after it is drawn, and before the next sample is drawn. As heretofore, the words "without replacement" refer to the mode of drawing a particular sample.

Clearly, if on the enumerative side of the table we take a complete count of the lot (i.e., if we make n equal to N), the sample drawn from the lot-container will then determine its contents absolutely; this follows because the finite multiplier $(N - n)/(N - 1)$ on the right reduces to 0. In contrast, note what happens on the left when the aim is analytic—i.e., when the aim is to estimate p and q in the supply. The coefficient of variation of this estimate is $\sqrt{q/np}$ for a sample of size n, and a complete count of the lot-container ($n = N$) still has a coefficient of variation of $\sqrt{q/Np}$, which is different from 0. A complete count gives all the information that the lot is capable of giving concerning p, and if this is not enough information, then other lots must be studied.[3] The estimate of p which is made by combining two complete counts from each of two lots has a coefficient of variation equal to $\sqrt{q/2Np}$, which is smaller than $\sqrt{q/Np}$ but is still not 0. Thus, if a complete count of the

[3] The reader is referred to further discussion in a paper by Yates and Zacopanay, "The estimation of the efficiency of sampling," *J. Agric. Sci.*, vol. xxv, 1935: pp. 547–77.

lot does not furnish enough information concerning p, it is necessary to wait until another lot is produced and classified and counted. If the problem is one dealing with crops or populations of experimental animals or people, the opportunity to study another lot may not arise for some time.

> **Remark.** Actually, the most common circumstance met in practice is not covered here at all. I refer to the case in which the supply is not constant from one lot to another; in effect, p varies indiscriminantly. In the language of quality control, the production process is then unstable or out of control. No prediction can be made about future lots under such circumstances.
>
> In the use of a control chart one is continually watching for changes in the production process, which signify instability. So long as the production process remains constant, successive lots are in effect drawn from the same supply.

Importance of the distinction between analytic and enumerative aims in the planning and tabulation.[4] It may be seen now why the distinction between the enumerative and analytic aims is so important. For the enumerative purpose (i.e., to determine the contents of a lot), any tabulation program however extensive is allowable if it provides information that someone wishes to have and can pay for. As already noted, a complete count, *for enumerative purposes*, possesses no error of sampling. On the other hand, *for analytic uses* the complete count still has a sampling error with coefficient of variation equal to $\sqrt{q/Np}$. A 25 percent sample from the lot has a coefficient of variation only twice the minimum, and a sample of 1 in 16 has a coefficient of variation only 4 times the minimum. It is extremely important to keep these figures in mind, both in planning studies and in interpreting the data therefrom.

The foregoing discussion may be applied now to a comparison of two treatments, such as two medical programs. Thus, suppose that the federal government instituted a program offering free pre- and postnatal medical and hospital care to a certain class of women and to their infants through one year.[5] Certain questions like the following ones arise in regard to the administration and effectiveness of the program and in consideration of its extension or modification.

   a. Does access to free pre- and postnatal medical and hospital care have an appreciable effect on the maternal death-rate?

   b. Is the infant death-rate appreciably lower or higher for the class that had the benefit of prenatal care under the program?

   c. Is the infant morbidity-rate appreciably lower or higher for those who had been under the program's prenatal care?

---

[4] Miss Katherine Simmons assisted materially in the writing of this section.

[5] The argument is general and is not intended to give the results of any particular program.

Answers are desired not only because of the importance of this particular program but also because they provide knowledge on which to predict the effectiveness of any modifications or similar programs to be proposed for the future.

Case histories are on file in the state offices and are sufficiently uniform so that a considerable number of items of information can be tabulated from them for the entire national experience. Now let us see how far the available quantitative data may be expected to go in providing answers to questions of the kind propounded above. It will be observed, in the first place, that all three of the questions are *analytic;* they deal with the cause systems that somehow regulate maternal and infant deaths and morbidity. For the sake of brevity, the discussion will be restricted to consideration of maternal deaths. It will be necessary to find groups of cases to compare with the cases of the medical program under consideration. In this instance, probably the only comparable group would be the remainder of the population by area, color, and age-groups; this information is contained on the transcripts of death-certificates sent into the National Office of Vital Statistics. In comparing (e.g.) the maternal death-rate for a certain color and age-group, there are three "supplies" (or cause systems, or treatments). One is the medical program; the other two are the absence of the medical program with and without hospitalization of confinement cases. Each gives rise in one year to a "lot" of confinement cases, for any color and age-group.

Clearly, even on a national basis, the number of cases in the "lot" arising from the first supply is so limited that even a complete count for any one year cannot possibly provide the information that is needed to answer questions like the ones above. For suppose there were 50,000 confinement cases under the medical program. With an expected death-rate of 2.5 percent there would be 1250 maternal deaths. Even if a complete count is made of all the 50,000 or so confinement cases on record in the state offices for any one year, this "lot" is still to be treated as a sample. The observed 1250 maternal deaths are subject to statistical variability; the coefficient of variation is $\sqrt{q/Np}$, which on the assumption of a death-rate of 2.5 percent and 1250 total deaths will be $\sqrt{0.975/1250}$ or about 3 percent, which corresponds to a 3-sigma error band of about 8 percent. A significant decrease in the maternal death-rate might barely be evident were it not that classification by color would be necessary. For the nonwhite the coefficient of variation would be about 25 percent, which would likely obscure any difference to be met in practice. Tabulations in still finer groups, such as age-groups, would be even less meaningful. In fact, if n is the expected number of cases to fall in a certain cell, the coefficient of variation of this cell will be $\sqrt{q/np}$, and if np is 100 or lower, the coefficient of variation of the observed frequency will be 10 percent or higher. Cell frequencies of less than 100 thus have a pretty high relative error, yet they may nevertheless be useful, as when the magnitude of the absolute error, rather than the relative error, is important.[6] Also, small frequencies are sometimes usefully tabulated so that meaningful combinations can be formed with other cells. Thus, unless the difference between two treatments (such as the presence and absence of a medical program) is striking, the information available in one year is hardly sufficient to point definitely one way or another, and *no tabulation program can overcome this fundamental difficulty.*

It might be well to add a remark to the effect that, in accordance with the foregoing conclusions, there is some justification for tabulating in classes 2 or 3 times as fine as can be used in any one year by itself. The reason is that data may be accumulated over a period of several years, and if the medical program remains fairly rigid, the data should be additive, so that cells that are too thin to be meaningful in one year may become meaningful over the course of several years.

The following exercise is fundamental in a study of the preceding paragraphs.

Exercise 1. Into a "lot-container" (Fig. 16) is placed a sample of N balls which are drawn at random from a supply of Mp red and Mq white balls. In turn, into a "sample-container" is placed a sample of n balls which are drawn at random from the lot-container. Of these n balls, r are found to be red. There are four possible cases, A, B, C, D, as shown in the table. Prove that $Er/n = p$ in all cases, and that the variances of r are those shown in the table.

The N balls are put into the lot-container	The sample of n is drawn from the lot-container	
	With replacement	Without replacement
With replacement	A $\mathrm{Var}\ r = npq\left\{1 + \dfrac{n-1}{N}\right\}$	B $\mathrm{Var}\ r = npq$
Without replacement	C $\mathrm{Var}\ r = npq\left\{1 + \dfrac{n-1}{N}\dfrac{M-N}{M-1}\right\}$	D $\mathrm{Var}\ r = npq\dfrac{M-n}{M-1}$

NOTE. Procedures for drawing with replacement can be worked out with no difficulty. The drawings might be performed by a blindfolded man

[6] This question will be treated in Chapter 10, p. 344.

with proper shuffling between draws. An assistant might watch to see what color of ball is drawn and then immediately place in the supply or in the lot-container (whichever one the drawings are being taken from) a ball of the same color, afterward stirring in preparation for the next draw. (See also the next exercise, p. 259.) More convenient procedures can be devised with random numbers (p. 87).

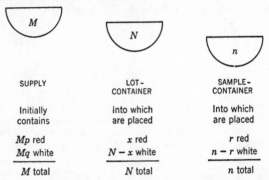

SUPPLY	LOT-CONTAINER	SAMPLE-CONTAINER
Initially contains	Into which are placed	Into which are placed
Mp red	x red	r red
Mq white	$N - x$ white	$n - r$ white
M total	N total	n total

FIG. 16. Showing the arrangement for successive formations of lots and for drawing a sample from each lot. The numbers x and r are random variables.

Solution (*William N. Hurwitz*)

Let the index i refer to a particular lot; then as lot after lot is drawn, i takes the values 1, 2, 3, $\cdots$. Let the index j refer to a particular drawing from a lot; then as a sample of n chips is drawn from a lot, j takes the values 1, 2, $\cdots$, n. Put

$$x_{ij} = 1 \quad \text{if the } j\text{th drawing from the } i\text{th lot is red}$$
$$= 0 \quad \text{otherwise}$$

Also denote by p_i the mean in the ith lot. x_{ij} and p_i are random variables. As draw after draw is made from a particular lot, x_{ij} has the "expected" value $Ex_{ij} = p_i$. Moreover, as lot after lot is drawn from the supply, p_i has the "expected" value $Ep_i = p$. First it is to be noted that

$$\underset{i}{E}\underset{j}{E}x_{ij} = \underset{i}{E}\underset{j}{E}x_{ij} = \underset{i}{E}p_i = p$$

[E means "expected" value with j variable, i constant, as in Ch. 5]

Hence

$$Er = E\sum_{j=1}^{n} x_{ij} = \sum_{j=1}^{n} Ex_{ij} = np$$

and this is true for all four cases. For the variances, the solution may be derived by writing

$$x_{ij} - p = (x_{ij} - p_i) + (p_i - p) \qquad [j = 1, 2, \cdots, n]$$

Now square each member, sum over all n values for the sample of n chips drawn from the ith lot to find that

$$\frac{1}{n} \sum_{j=1}^{n} (x_{ij} - p)^2 = \frac{1}{n} \sum_{j=1}^{n} (x_{ij} - p_i)^2 + (p_i - p)^2 + \frac{2}{n} (p_i - p) \sum_{j=1}^{n} (x_{ij} - p_i)$$

Now with p_i constant (i.e., using the *same* lot over and over) take the "expected" value of each term, assuming that the chips of a sample are not replaced until the sample is completed. The result is

$$E_j \frac{1}{n} \sum_{j=1}^{n} (x_{ij} - p)^2 = E_j \frac{1}{n} \sum_{j=1}^{n} (x_{ij} - p_i)^2 + (p_i - p)^2 + 0$$

$$= \frac{N - n}{N - 1} \frac{p_i q_i}{n} + (p_i - p)^2$$

Now let new lots be formed, over and over, and the above calculations repeated for each lot. Every term just written is a random variable, and if the lots are drawn from the supply without replacement, the "expected" values of these terms, obtained by applying E, will give

$$E_i E_j \frac{1}{n} \sum_{j=1}^{n} (x_{ij} - p)^2 = \frac{N - n}{(N - 1)n} E_i p_i q_i + \operatorname{Var} p_i$$

$$= \frac{N - n}{(N - 1)n} \left(E_i p_i - E_i p_i^2 \right) + \operatorname{Var} p_i$$

$$= \frac{N - n}{(N - 1)n} \left(p - \operatorname{Var} p_i - p^2 \right) + \operatorname{Var} p_i$$

$$= \frac{N - n}{(N - 1)n} \left(pq - \frac{M - N}{M - 1} \frac{pq}{N} \right) + \frac{M - N}{M - 1} \frac{pq}{N}$$

The left-hand side is none other than $E(x_{ij} - Ex_{ij})^2$, which is by definition the variance of $\frac{1}{n} \sum_{j=1}^{n} x_{ij}$. The variance of the number of red chips r will be n^2 times as great, as $r = \sum_{j=1}^{n} x_{ij}$. As both sample and lot were drawn without replacement, the result just written must be Case D, and algebraic reduction will show that the right-hand side if multiplied by n^2 does indeed agree with the answer recorded for Case D in the table on page 254.

For Case B, simply let M approach infinity. The factor $(M - N)/(M - 1)$ approaches 1, and in the limit the right-hand side of the last equation (after it is multiplied by n^2) gives

$$\operatorname{Var} r = npq \left\{ \frac{N - n}{N - 1} \left(1 - \frac{1}{N} \right) + \frac{n}{N} \right\}$$

$$= npq$$

For Case C, M remains finite, but the factor $(N - n)/(N - 1)$ is to be replaced by unity, and the variance-equation gives

$$\operatorname{Var} r = npq \left\{ 1 + \frac{M - N}{M - 1} \frac{n - 1}{N} \right\}$$

again in agreement with the table. For Case A merely replace

$(M - N)/(M - 1)$ by unity in the result just written, and the solution is completed.

It should be noted that:
In Case A,

$$\text{C.V.} \, r = \text{C.V.} \frac{r}{n} = \sqrt{\frac{q}{np} \left\{ 1 + \frac{n - 1}{N} \right\}} \rightarrow \sqrt{\frac{q}{Np}} \quad \text{as } n \rightarrow \infty.$$

In Case B,

$$\text{C.V.} \, r = \text{C.V.} \frac{r}{n} = \sqrt{\frac{q}{np}}$$

into which the lot-size N does not enter.

In Case C,

$$\text{C.V.} \, r = \text{C.V.} \frac{r}{n} = \sqrt{\frac{q}{np} \left\{ 1 + \frac{n - 1}{N} \frac{M - N}{M - 1} \right\}} \rightarrow \sqrt{\frac{M - N}{M - 1} \frac{q}{Np}}$$

as $n \rightarrow \infty$.

Another Solution (Frederick F. Stephan)

Let $P(x)$ denote the probability that there will be x red chips in a lot of size N drawn from the supply, and let $P(r|x)$ be the probability that there will be r red chips in a sample of n drawn from the lot containing x red chips. Show that the probabilities are correct as listed in the table.

Proba-bility	Case A	Case B	Case C	Case D
$P(x)$	$\binom{N}{x} q^{N-x} p^x$	As in Case A	$\dfrac{\binom{Mp}{x} \binom{Mq}{N - x}}{\binom{M}{N}}$	As in Case C
$P(r\|x)$	$\binom{n}{r} \left(1 - \dfrac{x}{N}\right)^{n-r} \left(\dfrac{x}{N}\right)^r$	$\dfrac{\binom{x}{r} \binom{N - x}{n - r}}{\binom{N}{n}}$	As in Case A	As in Case B

Then proceed to find

$$Er = \sum_{r=0}^{n} \sum_{x=0}^{N} r P(r|x) P(x) = np \quad \text{for all cases}$$

and

$$\text{Var} \, r = Er^2 - (Er)^2$$

$$= \sum_{r=0}^{n} \sum_{x=0}^{N} r^2 P(x) P(r|x) - (np)^2$$

Before proceeding on this solution, it will save time to borrow several summations from Chapters 4 and 13. In these sums r is to run from 0 to n, although it may also run over all possible integral values (Ch. 14).

$$i. \quad \sum \binom{n}{r} q^{n-r} p^r = 1 \qquad \text{[Sum of all the terms of the binomial]}$$

$$ii. \quad \sum \frac{\binom{Nq}{n-r}\binom{Np}{r}}{\binom{N}{n}} = 1 \qquad \begin{array}{l}\text{[Sum of all the terms of the} \\ \text{hypergeometric series]}\end{array}$$

$$iii. \quad \sum r \binom{n}{r} q^{n-r} p^r = np \qquad \text{[Mean of the point binomial]}$$

$$iv. \quad \sum r^2 \binom{n}{r} q^{n-r} p^r = npq + n^2p^2 \qquad \begin{array}{l}\text{[2d moment of the point} \\ \text{binomial about } r = 0]\end{array}$$

$$v. \quad \sum r \frac{\binom{Nq}{n-r}\binom{Np}{r}}{\binom{N}{n}} = np \qquad \text{[Mean of the hypergeometric series]}$$

$$vi. \quad \sum r^2 \frac{\binom{Nq}{n-r}\binom{Np}{r}}{\binom{N}{n}} = \frac{N-n}{N-1} npq + n^2p^2 \qquad \begin{array}{l}\text{[2d moment of} \\ \text{the hypergeo-} \\ \text{metric series} \\ \text{about } r = 0]\end{array}$$

The student will be interested to note that for Case B

$$P(x)P(r|x) = \binom{N}{x} q^{N-x} p^x \frac{\binom{x}{r}\binom{N-x}{n-r}}{\binom{N}{n}}$$

$$= \left\{ \binom{n}{r} q^{n-r} p^r \right\} \left\{ \binom{N-n}{x-r} q^{\overline{N-n}-\overline{x-r}} p^{x-r} \right\}$$

whence

$$P(r) = \sum_{x=0}^{N} P(x)P(r|x) = \binom{n}{r} q^{n-r} p^r$$

which is merely the rth term of the binomial $(q + p)^n$, just as if the sample of n balls had been drawn *with replacement* directly from the supply without the introduction of the lot-container between the supply and the sample.

In Case D

$$P(r) = \sum_{x=0}^{N} P(x)P(r|x)$$

$$= \sum_{x=0}^{N} \frac{\binom{Mp}{x}\binom{Mq}{N-x}}{\binom{M}{N}} \frac{\binom{x}{r}\binom{N-x}{n-r}}{\binom{N}{n}}$$

$$= \frac{\binom{n}{r}}{\binom{M}{Mp}} \sum_{x=0}^{N} \binom{M-N}{Mp-x}\binom{N-n}{x-r} \quad \text{[By rearrangement]}$$

$$= \frac{\binom{Mp}{r}\binom{Mq}{n-r}}{\binom{M}{n}}$$

which is the rth term of the hypergeometric series, just as if the sample of n had been drawn directly from the supply without replacement, without the imposition of the lot-container.

The required variances follow easily.

Exercise 2 (Treating a succession of samples). Assume that an infinite sequence of receptacles has been arranged and that to the ith receptacle a finite number N_i has been assigned. Then, starting with the first receptacle, which contains N_1 balls, some white and the rest black, a sampling process is begun as follows:

1. A ball is withdrawn at random.

2. If the ball is white, a white ball from a separate supply is placed in the next receptacle; if it is black, a black ball is placed in the next receptacle.

3. The ball drawn from the first receptacle is replaced in it.

4. The drawing at random with replacement from the first receptacle is repeated until N_2 balls have been placed in the second receptacle.

5. Then the first receptacle is abandoned, and the process is repeated with the second receptacle as the source from which N_3 balls are drawn. N_3 balls are placed in the third container with colors to match the N_3 drawings from the second container.

6. The process continues indefinitely. As soon as N_i balls have been placed in the ith receptacle, it becomes the source from which balls are drawn, and for each ball drawn the $(i+1)$th receptacle receives a ball of like color from the separate supply.

The student will observe that this procedure provides a way of filling each container *with replacement* from the preceding one. Since the drawing is with replacement, N_{i+1} can be greater than N_i; i.e., a container may have more balls placed in it than there were in the preceding container.

Problem.   Prove that as $i \rightarrow \infty$ the probability approaches 1 that all the balls in the ith receptacle are of the same color.

Solution

Let N be the greatest of the N_i.

Q_i be the probability that the balls in the ith receptacle are neither all white nor all black.

$P_i(x)$ be the probability that exactly x white balls will be placed in the ith receptacle.

$P_i(x|y)$ be the probability that exactly x white balls will be placed in the ith receptacle if there are y white balls in the preceding one.

Obviously

$$P_i(x) = \sum_y P_i(x|y)P_{i-1}(y)$$

$$P_i(0|y) = 1 \quad \text{if} \quad y = 0 \left.\right\}$$
$$ = 0 \quad \text{if} \quad y = N_{i-1}$$

$$P_i(N_i|y) = 1 \quad \text{if} \quad y = N_{i-1}\left.\right\}$$
$$ = 0 \quad \text{if} \quad y = 0$$

Also

$$P_i(x|y) = \binom{N_i}{x}\left(\frac{y}{N_{i-1}}\right)^x\left(\frac{N_{i-1} - y}{N_{i-1}}\right)^{N_i - x}$$

$$\geq \left(\frac{1}{N_{i-1}}\right)^{N_i} \quad \text{if} \quad 0 < y < N_{i-1}$$

$$\geq N^{-N}$$

Then

$$P_i(0) = \sum_y P_i(0|y)P_{i-1}(y) \geq P_{i-1}(0) + Q_{i-1}N^{-N}$$

and

$$P_i(N_i) = \sum_y P_i(N_i|y)P_{i-1}(y) \geq P_{i-1}(N_{i-1}) + Q_{i-1}N^{-N}$$

$$Q_i = 1 - P_i(0) - P_i(N_i)$$

$$\leq 1 - P_{i-1}(0) - P_{i-1}(N_{i-1}) - 2Q_{i-1}N^{-N}$$

$$= Q_{i-1}(1 - 2N^{-N})$$

Further, if

$$r = 1 - 2N^{-N}$$

then

$$-1 < r < 1 \quad \text{for} \quad N > 1$$

and

$$\frac{Q_i}{Q_{i-1}} \leq r$$

or

$$Q_i \leq r^{i-1}$$

whence

$$Q_i \to 0 \quad \text{as} \quad i \to \infty$$

This completes the solution of the problem.

Example. Let $N_i = 2$ for $i = 2, 3, \cdots$, and let the first receptacle contain equal numbers of white and black balls. Then derive the results shown in the table below. The last line is the probability that the balls

$i =$	2	3	4	5	$\cdots$	i
$P_i(0) =$	$\frac{1}{4}$	$\frac{3}{8}$	$\frac{7}{16}$	$\frac{15}{32}$	$\cdots$	$\frac{1}{2} - (\frac{1}{2})^i$
$P_i(1) =$	$\frac{1}{2}$	$\frac{1}{4}$	$\frac{1}{8}$	$\frac{1}{16}$	$\cdots$	$(\frac{1}{2})^{i-1}$
$P_i(2) =$	$\frac{1}{4}$	$\frac{3}{8}$	$\frac{7}{16}$	$\frac{15}{32}$	$\cdots$	$\frac{1}{2} - (\frac{1}{2})^i$
$P_i(0 \text{ or } 2) =$	$\frac{1}{2}$	$\frac{3}{4}$	$\frac{7}{8}$	$\frac{15}{16}$	$\cdots$	$1 - (\frac{1}{2})^{i-1}$

in the ith receptacle are all white or all black, and it is to be noted that this probability approaches unity as $i \to \infty$, i.e., as the number of receptacles increases.

Remark. The above proof applies only when the sequence of numbers $N_1, N_2, N_3, \cdots$ is bounded (G. R. Seth).

CHAPTER 8. CONTROL OF THE RISKS IN ACCEPTANCE SAMPLING

The originality of writers is often remarked upon; that of readers seldom; yet the second is often more remarkable than the first.—Wm. Follett, *The Atlantic*, April 1943, p. 114.

I may express the opinion that some of the published explanations are more remarkable than the phenomenon itself.—Hugh M. Smith, *Science*, vol. 82, 16 Aug. 1935: p. 151.

The producer's and the consumer's risks. This chapter will introduce some applications of the theory of probability to the control of the risks that are involved in formulating rules of acceptance based on samples, with particular reference to the risks encountered in the inspection of industrial product.[1] The problem of sample-inspection is to be treated from the standpoint of mass production. Each lot is to be sampled, the sample partially or wholly inspected, and the lot accepted or rejected on the basis of tests made on the sample and in accordance with certain definite rules regarding the allowable number of defectives in the sample. Any particular acceptance-plan is characterized by the rules for sampling, acceptance, and rejection.

Under any plan of acceptance there are four important kinds of risk that should be defined: [2]

The consumer's A-risk: the probability that a lot of quality p_1 will be accepted (p_1 being an unwanted fraction defective).

The consumer's B-risk: the probability that a lot drawn at random from a supply of quality p_1 will be accepted (a supply being a manufacturing process or unlimited supply of lots).

The producer's A-risk: the probability that a lot of quality p_0 will be rejected (p_0 being a desirably low fraction defective).

The producer's B-risk: the probability that a lot drawn at random from a supply of quality p_0 will be rejected.

The theory in this chapter will deal with the **consumer's A-risk** and the **producer's B-risk,** following Dodge and Romig.[3] The producer can

---

[1] The author records here with pleasure the generous assistance of Harold F. Dodge and his co-workers H. G. Romig and Miss Mary N. Torrey, who took the trouble to criticize earlier drafts of this chapter. These friends are of course not to be held responsible for the choice of content, the terminology, or any errors.

[2] These concepts of risk, even to the notation, have been used for many years by the engineers of the Bell System (Dodge, Romig, Shewhart, and others), although in publications the particular risk involved has not always been stated explicitly in these symbols. In some of their publications, p_0 corresponds to the process-average value of fraction defective.

[3] Harold F. Dodge and Harry G. Romig, *Sampling Inspection Tables* (John Wiley,

only control his *process*, not the individual lots;[4] hence it seems that the more practical risk to him is the B-risk. In the words of the preceding chapter, his problem is *analytic*, whereas the consumer's is *enumerative*. Some other writers have chosen to work with the producer's A-risk, which accounts for the appearance of hypergeometric terms in some books, in place of binomial terms in this book, Eq. 5 for example.

Acceptance-plans, the subject of this chapter, deal with inspection of the finished item. But long before the item arrives at this stage, its quality is pretty well determined. If in various important stages of a production process, whether machine or human, a state of statistical control has been achieved (Ch. 14), then there is no evidence of avoidable variability, although a different procedure for selecting the "control-sample"[5] might bring to light new evidence disclosing lack of control in some particular operation. Binomially controlled quality (mentioned earlier), where it exists, is attained with the aid of control-charts.

It is often difficult to draw up meaningful specifications by which an article can be classed as wanted or not wanted. For instance, what properties are there in a clock, in a bottle of perfume, in a rug, or in a dining-room light-fixture, by which one can test it and determine whether it will sell or give the kind of service that the customer deems satisfactory? It is thus often difficult to determine just what procedures in the manufacture need to be kept under statistical control in order to produce a "good" article. Moreover, it is often possible to keep a production process in excellent control at producing the wrong thing or performing the wrong operation, just as one may develop a reliable and reproducible but faulty technique of playing golf or the piano.

It is nevertheless imperative for many items of manufacture to lay down specifications and tests for weight, thickness, diameter, breaking strength, finish, etc., by which an article can definitely be classed as conforming or nonconforming, even if in many cases the specifications and tests must be partly arbitrary. It is presumed in this chapter that

1944). This book consolidates under one cover several articles and tables originally published by Dodge and Romig and by Keeling and Cisne in the *Bell System Technical Journal* for 1929, 1941, 1942.

[4] It is assumed that the producer does not sort or screen his lots before letting them go out. In other words, sorting or screening, with the aim of controlling the quality of lots, is *not* part of the production process. The variation in quality from lot to lot is supposed to arise from the nature of the process and not from artificial control, as by sorting, after the product is made.

[5] A "control-sample" is a small sample tested at intervals in order to furnish data for the control-chart. Data from the control-sample are not to be confused with data from a sample from a lot to determine acceptance of the lot. See *Control Chart Method of Controlling Quality during Production*, Z1.3–1942 (American Standards Association, 70 East 45th St., New York 17), p. 27.

the inspectors make no mistakes (which is of course a myth), so that each article tested can be classified unequivocally conforming or nonconforming. If the test is not destructive, the foregoing assumption requires that every item of any sample will always be classified the same way no matter how many times retested. Statistical control as manifest by control-charts is the best guarantee that quality is being built into the article; nevertheless a final verification of quality through an acceptance-procedure may be desirable, particularly if doubt exists concerning the maintenance of statistical control. The acceptance-procedures about to be described will afford the calculated protection to producer and consumer independently of the state of control; but certainly, if control exists, as evidenced by the manufacturer's control-charts, the amount of acceptance sampling to be required may be reduced or even eliminated.

Several sampling problems will now be stated to be followed later by partial solutions and remarks. They will not be solved either separately or in detail, but enough theory will be given to illustrate how the risks α and β and the gains and losses associated with them can be controlled.

Problem 1. Bowls containing chips, some red and some white, are presented for classification. From the way in which the chips are manufactured, it is known that any bowl must contain either the proportion p_0 or p_1, and the problem is to label each bowl p_0 or p_1 by sampling. Repeated samples from any one bowl will vary; hence a bowl that really contains the proportion p_0 will occasionally be wrongly labeled p_1, and conversely. How big a sample ought to be taken in order to control to prescribed levels (α and β)—

The producer's risk that a bowl containing the proportion p_0 will be labeled p_1 and rejected?

The consumer's risk that a bowl containing the proportion p_1 will be labeled p_0 and accepted?

Note that here there are but two monochromatic proportions, p_0 and p_1, whereas in the next problem and in fact all that follow, the producer puts forth binomially controlled lots,[6] varying from 0 defectives to N on the distribution $(q + p)^N$. The distinction between Problems 1 and 2 should be noted carefully.

Problem 2. A manufacturer delivers electrical fuses in lots, binomially controlled at the quality-level p_0. His purchaser tests samples from every

---

[6] In a binomially controlled product of quality p (called the *process-average*, often abbreviated p.a.), the quality will vary randomly from lot to lot, and the distribution of defectives will follow the binomial series $(q + p)^N$. This is what Dodge and Romig call "uniform" quality, as is defined in the footnote on page 31 and at the top of page 44 of their book, *Sampling Inspection Tables*. If the standard deviation of quality amongst lots is unstable (not in control), the producer's risk is increased even though he maintains the desired *average* quality. Tighter control than binomial is hardly realistic. Binomial control represents a limiting but attainable state and is fortunately well suited to the calculations that need to be made. An unvarying proportion of defectives will be referred to as *monochromatic quality*.

lot, and according to the results of these tests, some lots are accepted and some are rejected. Because of sampling errors there are two kinds of risk:

The producer's risk that some of his lots will be rejected.
The consumer's risk that some bad lots will be accepted.

How can the losses resulting from these risks be controlled at the least expense?

Problem 3. The head of a large corporation (or government department) must examine a sample of every division's personnel actions (hiring, firing, ratings, promotions, demotions) in order to be able to report whether each division is conforming to this or that rule. It is assumed that personnel activities are decentralized, only general regulatory powers being retained at headquarters. Because of sampling errors there are two kinds of risk—

The division's risk that a larger percentage of error will be reported than actually exists, and the division forced to undergo an elaborate, expensive, and unnecessary review of its personnel actions over the past year.
The corporation's risk that a high rate of error existing in some division will not be discovered till later, when certain cases may come to light and cause trouble (e.g., discrimination, disregard of seniority, hiring people at higher grades than they are qualified for, etc.)

How can these risks be controlled satisfactorily and at the least expense? (NOTE: By substituting pay-roll slips in place of personnel actions, another common and important type of problem is described.)

Problem 4. In a government statistical agency (or in the record department of a life-insurance company), questionnaires and other records are processed and transformed into statistical information. One operation is called *coding*, wherein a number or other symbol is assigned to a particular characteristic (0 for single, 1 for married, 3 for widow, 4 for widower, etc.) in preparation for punching a card for each individual person. The coding must be done by hand by people thoroughly trained to recognize some of the nice but important distinctions between various characteristics. Verifiers (inspectors) review a sample of the work of a coder. The coders occasionally make mistakes, although sometimes a "mistake" is no more than an honest difference of opinion between the coder and the verifier (inspector), and the verifier could be wrong, although his word must be accepted. Because of sampling errors there are two kinds of risk [7]—

The coder's risk that some work will be sent back, even though his performance is under control at an acceptable quality-level p_0. The coder is then wrongly demerited, and the cost of verifying his work is unnecessarily increased.
The agency's risk that some work which is actually worse than the

[7] The practical aspects of this problem were described in two papers: W. Edwards Deming and Leon Geoffrey, "On sample inspection in the processing of census returns," *J. Amer. Stat. Assoc.*, vol. 36, 1941: pp. 351–60; and W. Edwards Deming, Benjamin J. Tepping, and Leon Geoffrey, "Errors in card punching," *J. Amer. Stat. Assoc.*, vol. 37, 1942: pp. 525–36.

specified lowest acceptable quality-level [8] will not be caught, to cause embarrassment in the published tabulations.

How can these risks be controlled satisfactorily and at the lowest expense?

The principle of calculated risk. The statistical element in all the above problems is the control of the two types of sampling error in the classification of lots. Control of these errors automatically implies control of the financial risks associated with them. By control the statistician does not mean elimination or even virtual elimination of errors in sampling, but *economic control* by which he achieves minimum economic losses, all costs considered, including the costs of sampling. This is the *principle of calculated risk*.[9] It should be noted in the first place that someone must decide what probability-levels α and β are most economical in view of the costs of sampling and testing and the consequences of the two types of risk. As a result of broad practice, Dodge and Romig place $\beta = 0.1$ and settle upon practicable sample-sizes and values of AOQL (*vide infra*) that will be suitable under various conditions, thus settling upon α indirectly. In contrast, in Peach's tables, shown later on, α and β are both placed equal to 0.05.

Decisions on the sampling plan are partly administrative, as company profits and public relations are at stake, but the practicing statistician finds that he must take an active part in these decisions. He is expected to make advance estimates of the costs of sampling and testing, and it is up to him also to estimate the consequences of the two types of risk, and on the basis of such studies and other considerations to recommend suitable plans.

Remark 1. There is no use drawing up any kind of sample-test unless action will be taken as a result of the test.

Remark 2. It should be noted that the specifications and tests by which an article is to be classed conforming or nonconforming are to be kept distinct from the plan by which lots are to be sampled and accepted or rejected (the acceptance-plan). The latter depends on the purchaser's own manufacturing and marketing practices, and these views may change independently of the tests by which an article is classed conforming or nonconforming. The purchaser may be able to tolerate some items that are nonconforming, and it may pay him to do so, as he may strike a good bargain by purchasing imperfect lots at reduced rates. On the other hand, it may pay him to demand better quality and pay the price for it. Such decisions are reflected in the plan of acceptance, and do not affect the tests by which an article is classed conforming or nonconforming. The contract between the vendor and purchaser should specify the plan of acceptance but should stipulate that any imperfect items found are to be replaced by good ones.

---

[8] For example, the lowest acceptable quality-level p_1 might be specified as 4 wrong symbols in 800 codes.

[9] This term first came to my ears from my friend J. M. Juran.

Choice of theory to be used. From the standpoint of control of the process and calculation of the producer's risk, it seems preferable to use a mathematical theory that corresponds to samples drawn from lots of variable quality. For ease in calculation, binomially controlled variability will be assumed in what follows. In this state of control, lots of size N are randomly distributed about the average quality p on the binomial series $(q + p)^N$. As is known from the last chapter, samples of size n drawn without replacement from such lots are distributed on the binomial series $(q + p)^n$, regardless of the lot-size N. On the other hand, the consumer wishes to have lot-by-lot protection against accepting quality p_1, and the appropriate theory is the hypergeometric series, which of course involves N. Thus, in the language of the preceding chapter, the producer's B-risk is to be treated here as a problem in expectancies over a series of lots, hence as an aid in his *analysis of causes* of variability; whereas the consumer's A-risk is to be treated as a problem in *enumeration or estimation of a finite* lot of size N.

Remark. The reader who has studied the control of quality by the use of the Shewhart control charts will perceive now that the positions of the control limits are ordinarily calculated in accordance with the theory for regulating the producer's B-risk for a continuous process. Thus, on the chart for fraction defective, the average fraction defective in a series of samples of size n, each drawn from its respective lot without replacement, is computed and called $\hat{p}$ for the "central line." The 3-sigma control limits are then calculated by the binomial theory as

$$\hat{p} \pm 3\sigma_{\hat{p}} = \hat{p} \pm 3\sqrt{\frac{\hat{p}\hat{q}}{n}} \qquad (1)$$

This calculation corresponds to Case B on page 254, for which $P(r)$ was seen to be the binomial term $\binom{n}{r} q^{n-r} p^r$, into which the lot-size N did not enter. If the samples are drawn with replacement, or if the lots are formed at random from a finite supply of M articles already manufactured, $P(r)$ for one of the other cases developed in the preceding chapter should be used.

The sample-point and sample-line.[10] Let a sample of n be drawn one at a time from a bowl containing initially N chips, some red and some white, in the ratio $p:q$. The red chips might be the defective articles, and the white chips the good articles in a lot of N presented for inspection. Suppose that a pawn starts from 0 in Fig. 17 and moves one unit to the right when a white chip is drawn into the sample, and one unit upward when a red one is drawn. As the n articles of the sample are

---

[10] For many of the ideas in this section I am indebted to an article by G. A. Barnard entitled, "Sequential test in industrial statistics," *Suppl. J. Royal Stat. Soc.*, vol. viii, no. 1, 1946: pp. 1–26.

drawn and tested one at a time, the pawn moves on a random walk [11] from 0 to the "sample-point" $x = n - r$, $y = r$, r being the number of red chips found in the entire sample and $n - r$ the number of white ones.

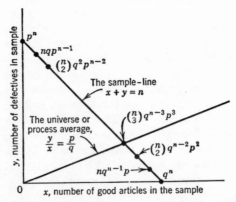

FIG. 17.   Showing the sample-line and the process-average on a sample-space wherein x denotes the number of white chips (conforming items) and y denotes the number of red chips (nonconforming items) in a sample of n chips drawn at random and with replacement from a bowl containing Np red and Nq white chips.

The *sample-point* obviously lies somewhere on the 45-degree *sample-line*

$$x + y = n \tag{2}$$

There are $n + 1$ possible sample-points on this line, corresponding to $y = 0, 1, 2, \cdots, n$. Also there are $\binom{n}{y}$ different paths by which the pawn can move from 0 to the sample-point x, y, and altogether there are 2^n different paths from 0 to the $n + 1$ points. The line

$$\frac{y}{x} = \frac{p}{q} \tag{3}$$

has for its slope the initial ratio of red to white in the bowl. If the sample-point falls at the intersection of the two lines (2) and (3), then the ratio of red to white in the sample agrees precisely with the initial ratio of red to white in the bowl. Intersection is of course impossible if np is not an integer.

For calculating the producer's B-risk of rejection, the bowl whence the sample was drawn will be considered as a lot of N articles (red for defectives, white for good) which are formed on the series $(q + p)^N$ in a binomially controlled process of quality p. The probability·

[11] Other exercises in random walks are given at the end of Chapter 13.

distribution of the number of red chips in the sample is the binomial series $(q + p)^n = \sum \binom{n}{y} q^x p^y$ (cf. the preceding section), in which summation $q^x p^y$ is the probability that the pawn will take any one of the $\binom{n}{y}$ paths from 0 to the sample-point x, y. For calculating the consumer's A-risk of acceptance, p is simply the proportion of defectives in the lot, and the hypergeometric term $\dfrac{\binom{Nq}{x}\binom{Np}{y}}{\binom{N}{n}}$ gives the probability that the pawn will arrive at the point x, y.

As an exercise the student should satisfy himself that the hypergeometric term will give 0 for the probability that $y > Np$ or $x > Nq$.

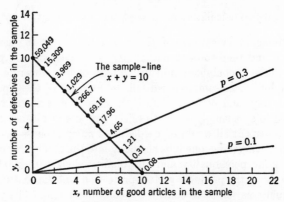

Fig. 18. Showing the sample-line $x + y = n$ for $n = 10$. The numbers alongside the 11 sample-points show the *probability ratios* $q_1{}^x p_1{}^y : q_0{}^x p_0{}^y$, based on the binomial series for $p_1 = 0.3$ and $p_0 = 0.1$. These values of p determine the sloping lines $y/x = 3/7$ and $y/x = 1/9$.

The probability of the pawn taking any particular path from 0 to a particular sample-point depends on p, as just noted; hence the probabilities for two different values of p, such as p_0 and p_1, will be different. As an exercise the student should calculate the ratio of $q^x p^y$ for $p = 0.3$ to its value for $p = 0.1$, for each of the 11 sample-points on the sample-line for $n = 10$ in Fig. 18, thus verifying the numbers appearing there. This ratio can be written

$$(\tfrac{7}{9})^x (\tfrac{3}{1})^y$$

and it is to be calculated for $x = 0, 1, 2, \cdots, 10$.

Controlling the producer's and consumer's risks by single sampling.
It is desired to obtain protection against the two kinds of error, in
other words, to regulate both the producer's risk and the consumer's
risk. Quantitatively, a plan might be desired whereby, for example,
only 1 lot in 20 from a process binomially controlled at the average
quality p_0 (which might be 1 percent defective material) will be rejected,
while 1 in 10 having quality as bad as p_1 (which might be 3 percent
defective material) will be accepted. Here $\alpha = 0.05$ and $\beta = 0.1$. It
is hoped to regulate these risks by finding a sample-size n that is not
too big (hence not too costly) and by designating certain points on the
sample-line $x + y = n$ as rejection points and others as acceptance
points, with the rule that when for any lot the sample-pawn falls on a
rejection point the lot is to be rejected, and when the pawn falls on an
acceptance point the lot is to be accepted.

It is easy enough to lay out a plan for controlling the producer's risk
so that not more than (e.g.) $\frac{1}{20}$th of his lots will be rejected. It is
only necessary to calculate with $p = p_0$ the probability $\binom{n}{y} q^x p^y$ of the
pawn arriving at every one of the $n + 1$ points on the sample-line
$x + y = n$, and then to designate as points of rejection any points on
this line whose probabilities add up to 0.05 or less, the rule being to
reject any lot whose sample-pawn falls on any one of these points. A
still better way, and perfectly safe if the only aim is not to reject too
many good lots, is simply never to reject any lots at all; accept everything. No tests at all would be needed, and much expense saved. But
in accepting everything, one falls headlong into the other error of accepting lots having poor quality and any quality along with good ones, if
any, thus giving the producer perfect protection and the consumer none.

It is equally as simple to control the consumer's risk. The procedure
is the same except that the calculations of probabilities are to be made
with $p = p_1$. Again, one perfectly safe way to guard against accepting
any lots at all of quality as bad as p_1 is simply to reject all lots; accept
nothing. Again, no tests would be needed, and the expense would be
saved. But in rejecting everything, and giving the consumer perfect
protection (and also no material), the producer's risk is maximized;
everything that he produces is rejected.

Clearly, an impasse has been reached: it is impossible to minimize
one risk without maximizing the other. It is possible, however, with
samples, to risk a *little* of one and a *little* of the other, and *to know how
much of both we are risking.*

The next step, as there must be some testing, and some of both risks
accepted, is to try to decide whether it makes any difference which
points on the sample-line are designated points of rejection and which

are designated points of acceptance. A theorem first stated by J. Neyman and E. S. Pearson [12] comes into play here. If in our desire to control
the consumer's risk of accepting a lot of quality p_1 we were permitted
to pick out *but one point* on the line $x + y = n$ as a point of rejection,
we should select that point for which the probability-ratio

$$q_1{}^x p_1{}^y : q_0{}^x p_0{}^y, \quad \text{or} \quad (q_1/q_0)^x (p_1/p_0)^y$$

[Appropriate changes should be
made if hypergeometric probabilities are used]

is a maximum. Such a selection will have the advantage of maximizing
the chance of rejecting lots of quality p_1. And if we were permitted to
select two points, we should hold on to the one already selected and take
in addition that point at which the probability-ratio has its next highest
value. A third point would be added similarly; etc. In this way we
build up a *region of rejection* which will have some prescribed value of
probability. The remaining portion of the sample-line is the *region of
acceptance*.

If β is the probability of accepting lots of quality p_1, then $1 - \beta$
is the probability of rejecting them. β might be 0.1, for instance, as in
the Dodge-Romig tables (*loc. cit.*); then $1 - \beta$ would be 0.9. To determine where on the sample-line $x + y = n$ the region of acceptance leaves
off and the region of rejection begins, probabilities based on p_1 are
summed successively, beginning at the bottom point ($x = n$, $y = 0$),
until the addition of one more point would overshoot the desired
probability-level β. In symbols,

$$\beta \geq \sum_{y=0}^{c} \frac{\dbinom{Nq_1}{x}\dbinom{Np_1}{y}}{\dbinom{N}{n}}$$

[The consumer's A-risk]   (4)

or

$$\beta \geq \sum_{y=0}^{c} \binom{n}{y} q_1{}^x p_1{}^y$$

[The consumer's B-risk]   (4')

The number of points c required to build up the probability β is called
the *acceptance number*. If a sample contains c or fewer defectives, the
sample-pawn stops in the region of acceptance, and the lot is accepted.
If the sample contains $c + 1$ or more defectives, the sample-pawn stops
in the region of rejection, and the lot is rejected. The number c depends
of course on n, p_1, p_0, and β, and if hypergeometric terms are used, on

[12] See, for example, J. Neyman and E. S. Pearson, *Statistical Research Memoirs*
(University College, London, 1936), vol. I: pp. 1–37.

N as well. In some plans c turns out to be 0, in which case the summation written above contains only one term, viz., q^n, and n must be big enough to reduce q^n below β.

Remark 1. By this same plan of acceptance, the probability that a lot having monochromatic quality p will be accepted is simply the summation in Eq. 4 rewritten with p in place of p_1 and summed over the region of acceptance (i.e., from $y = 0$ to $y = c$). The probability so computed is called the *power* of this particular region of acceptance for the quality p. And because this particular region of acceptance maintains β at the desired level for p_1, yet maximizes the probability of acceptance for any $p < p_1$, it is *most powerful*. Being most powerful for *every* quality-level $p < p_1$, it is *uniformly most powerful*. These terms originated with Neyman and E. S. Pearson.

To calculate the producer's risk α of having nis binomially controlled lots of average quality p_0 rejected by this plan, Eq. 4' is rewritten with p_0 in place of p_1 and summed over the region of rejection on the sample-line $x + y = n$, which gives

$$\alpha = \sum_{y=c+1}^{n} \binom{n}{y} q_0{}^x p_0{}^y$$

$$= 1 - \sum_{y=0}^{c} \binom{n}{y} q_0{}^x p_0{}^y \qquad \begin{array}{l}\text{[The producer's B-risk,} \\ \text{regardless of the lot-} \quad (5) \\ \text{size, } N\text{]}\end{array}$$

In a single-sampling scheme with a fixed sample-size n, either β for a given p_1 can be fixed, or α for a given p_0, but not both. However, a proper size n and acceptance number c can be found to meet the requirements that α and β have particular values for given quality-levels p_0 and p_1.

The student should turn to Fig. 17 or 18 and note that the probability β determines an ordinate $y = c$ which divides the sample-line into two portions, a *lower portion for acceptance*, extending from $y = 0$ to and including $y = c$, and an *upper portion for rejection*, extending from $y = c + 1$ to and including $y = n$.

An exceedingly clever observation by Paul Peach [13] enables him to condense a great many single-sampling schemes into one convenient table, which is reproduced here on page 283 with his kind permission. α and β are both 0.05 in this table. After p_0 and p_1 are selected, entry is made with the ratio $R = p_1/p_0$ to find c and np_0. Then, because p_0 is decided upon in advance, n can be found immediately by dividing np_0 by p_0. It should be noted that the Peach tables do not give minimum total inspection: the Dodge-Romig tables do.

---

[13] It is to be noted that α in Peach's tables refers to the producer's A-risk, although there is little distinction numerically between A- and B-risks if $n = 0.1N$ or less.

Remark 2. It is important to note that the consumer's A-risk β can be reduced to 0 by setting $c = 0$ and $n = N$ in the hypergeometric sum Eq. 4, which is to say that the consumer can be completely protected by 100 percent inspection regardless of p_1. In contrast, the producer's B-risk α, although it decreases with increasing sample-size n, can *not* so be reduced to 0. This is because binomial terms do not reduce to 0 when n is placed equal to N.

Remark 3. The simplest device for computing the acceptance number for a single-sampling plan is through the use of the Mosteller-Tukey double square-root paper. A radial line is to be drawn in Fig. 29 (p. 310) at angle $\tan^{-1}\sqrt{p_0/q_0}$; also a parallel line AA at a distance of 1.64σ above it. Any point on the line AA has coordinates $\sqrt{r}$ and $\sqrt{n-r}$ that correspond to a sampling plan of sample-size n and acceptance number $c = r$, having $\alpha = 0.05$ approximately. Next, a radial line is drawn at angle $\tan^{-1}\sqrt{p_1/q_1}$; also a parallel line BB at a distance of 1.28σ below it. Any point on this line gives a sampling plan having $\beta = 0.1$ approximately, provided n is not greater than $0.1N$. The intersection of AA and BB determines a sampling plan having the required risks α and β. The risks can be increased or decreased by raising or lowering the lines AA and BB and reading off the new intersection.

Incomplete sums of the binomial can be made with the help of tables of the incomplete beta function (p. 480). Fortunately also, the Poisson exponential series (p. 418) provides an excellent approximation to both the binomial and hypergeometric series when p and n/N are small. The construction of a set of sampling tables is by no means a small computing project, particularly if the sample-sizes are intended to provide minimum total inspection, including the screening of "rejected" lots (*vide infra*) as is the aim in the Dodge-Romig tables (*loc. cit.*).

It should be remembered that exact calculations of risks or sample-sizes or acceptance-numbers are extravagant, as any plan is to some extent arbitrary, and (as in Ch. 2) there are many errors of inspection besides sampling errors. It is also of some importance to note that the consumer's A-risk correctly calculated with hypergeometric terms is going to be a bit smaller than appears by the use of binomial terms.

Remark 4. A record of the lot-quality is indispensable for control of the process. Now samples that are curtailed when $c + 1$ defectives are found are biased; the "expected" average fraction defective in such samples is higher than the fraction defective in the lots. Likewise the remaining samples that are not curtailed are also biased, their "expected" average fraction defective being too low. Hence for a quality-record or for data with which to plot a control-chart, every 10th sample may be tested through to n tests whether the acceptance number c is exceeded or not, and the number of defectives recorded. In double sampling (*vide infra*), every 10th first sample may be tested through for control purposes.

Minimum total inspection. In practice a rejected lot may not really be rejected. The word "rejected" merely means that the lot is to follow a particular routine, to be screened, regraded, repaired, corrected, or even rejected, or indeed perhaps accepted after argument or waiver of specifications. In screening, the rest of the lot is given 100 percent inspection. Defective items are supposedly removed, and what is left of the lot is supposedly perfect. This brings up the question of the minimum amount of total inspection when rejected lots are screened.

No matter whether the producer or the consumer does the inspecting, the consumer pays the cost. As in all economic problems, however, the producer and the consumer should work together to devise the cheapest possible plans of sampling and inspection that will deliver the required number of good articles at the lowest reasonable price.

If binomially controlled lots of quality p are presented for inspection, and if "rejected" lots are screened, then in a single-sampling plan the average total number of items inspected per lot of size N will be

$$I = n + (N - n) \left[1 - \sum_{y=0}^{c} \binom{n}{y} q^x p^y \right] \quad \begin{array}{l} \text{[Rejected lots are as-} \\ \text{sumed to be screened]} \end{array} \quad (6)$$

When the production-process is not controlled, the average amount of inspection per lot will differ somewhat, but usually not seriously, from the quantity I calculated here. The Dodge-Romig tables not only control β for various quality-levels but also minimize the total average amount of inspection required per lot, for various sizes of lot.

If the tests are destructive, screening is of course not done, in which circumstance the question of the minimum amount of total inspection including screening does not arise.

Double sampling. By introducing a procedure whereby a second sample may be tested, it is possible to cut down the average amount of inspection below that required in a single-sampling plan giving the same protection. Such a plan is called a *double-sampling plan*. The routine of double sampling is not too cumbersome in most circumstances where mass inspection is carried out. The sample-sizes are denoted by n_1 and n_2, and the acceptance numbers by c_1 and c_2, and it must be remembered that c_2 includes c_1. The second sample is tested only if the number of defectives in the first sample exceeds c_1. The lot is accepted if the number of defectives in the first sample is c_1 or fewer, or if the total number of defectives in both samples is c_2 or fewer.

An ingenious table for double sampling, arranged by Paul Peach, is shown on page 284. In this table, α and β are both 0.05. As with single-sampling plans, the Dodge-Romig tables are more extensive because their aim is not only to control β but also to achieve a minimum amount of sampling, including screening.

Remark 1. Besides cutting inspection costs, there is a very interesting psychological factor concerned with double sampling, particularly when extremely high quality is required and the allowable number of defects in a sample (or two samples) is very small. Double sampling always allows one defective before a lot is rejected, whereas single sampling may allow none (as where $c = 0$). It is a human tendency to avoid taking any important action on the evidence of a single isolated defect. Like a first offense, the single defect may be looked upon lightly. But two defectives are another story, and double sampling never calls for rejection of a lot on less than two defectives. The actual story is of course not the story of a single lot, but a whole succession of lots, perhaps hundreds or thousands, and although single sampling is just as sound statistically as double sampling or any other kind of sampling, although not always as efficient, the psychological factor from the nonstatistical producer's angle should not be overlooked.[14]

Remark 2. The first samples in a double-sampling plan can be used for control purposes (cf. Remark 4 *supra*). The second samples can not be used for control purposes even if they are tested to completion; hence the second samples may as well be curtailed (stopped when and if c_2 defectives altogether have been counted).

The OC-curves. Much of what has been said in the preceding sections regarding the control of the producer's and consumer's risks can be expressed graphically through the curve of *operating characteristics*, usually abbreviated *OC-curve*. The probability of acceptance is shown on the vertical scale, and the fraction defective on the horizontal. There are in current use two OC-curves for every sampling plan—one curve calculated with binomial probabilities relating to the probability of acceptance for lots drawn randomly from a *supply* having quality p_0, the other calculated with hypergeometric probabilities relating to the probability of acceptance for an individual *lot* having quality p_1. A schematic exhibit [15] of the OC-curves corresponding to a given plan of acceptance and a particular lot-size is shown in Fig. 19. p_1 is quality that is to be discouraged, and it is to be accepted only with probability β: quality worse than p_1 is to have still less chance of acceptance. For a single-sampling plan an OC-curve showing the producer's B-risk can be passed through any one point, say p_0, $1 - \alpha$, by adjusting n and c; and likewise a curve showing the consumer's A-risk can be passed very nearly through any other, such as p_1, β. Every OC-curve for a single-sampling plan must be labeled with the two numbers for n and c, and the curve for the consumer's risk should also show the lot-size N if the hypergeometric series is required in calculating the probabilities, although for samples not too large (n not greater than $0.1N$), the two OC-curves practically coincide if the probabilities are the same.

For a double-sampling plan it is also possible to choose two points through which an OC-curve is to pass, closely at least (although the determination of n_1, n_2, c_1, c_2 is sometimes very difficult, as there is no direct method of fitting the curve). Every OC-curve for a double-

---

[14] This paragraph will be found in the introduction to the Dodge-Romig book in very nearly the same words.

[15] I do not have on hand OC-curves showing both the producer's B-risk and the consumer's A-risk as functions of p for some particular plan of acceptance, hence a..1 unable to show curves that have been plotted to numerical values. Fig. 19 is pictorial only.

sampling plan must be labeled with four numbers for n_1, n_2, c_1, c_2, and again, the OC-curve for the consumer's risk might show a fifth number for the lot-size N. A rejection number k for the first sample may be part of a double-sampling plan. For a sequential plan, also, it is possible to choose two points through which an OC-curve must pass.

Given any OC-curve, say for double sampling, it is possible to find a single-sampling plan and a sequential plan that will give very closely

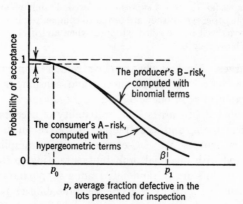

FIG. 19. OC curves, schematic only, corresponding to some particular plan of acceptance. For some binomially controlled quality p_0 (process-average) the producer's B-risk is α, and is computed by Eq. 5. For some particular monochromatic quality p_1, the consumer's A-risk is β, and is computed by the hypergeometric terms in Eq. 4.

the same curve. Hence even an expert can not be expected to glance at an OC-curve and perceive what sort of plan was used to derive it, although he may succeed with more or less effort in reproducing it pretty closely with several plans.

The AOQL. A lot that is perfect will always be passed, because no defectives will be found in the sample. On the other hand, a lot that is of very bad quality will almost certainly be stopped, and if the remainder of the lot is screened, only good items are finally allowed to go out. The inspection department is then turned into a factory or a house of correction, and the cost of 100 good items is exorbitant because the vendor is being paid to make mostly bad ones. Of the lots of intermediate quality, some will be passed, and some stopped for screening. The material that goes out is therefore worse, on the average, from lots initially of intermediate quality than from lots that were initially very bad. Thus, if the average outgoing quality (AOQ) is plotted on the vertical axis against the average fraction defective (p) of binomially controlled lots that are presented for inspection, the curve will start off from 0, go up to a maximum, and come back down to 0 at $p = 1$.

Fig. 20 shows an AOQ curve, reproduced from the Dodge-Romig book by the kind permission of the authors. The ordinate at the maximum is called the AOQL—the average outgoing quality limit. Some particular lot that is passed may be of very poor quality, but the AOQ averaged over all lots can not be worse than the AOQL, no matter how bad the quality presented.

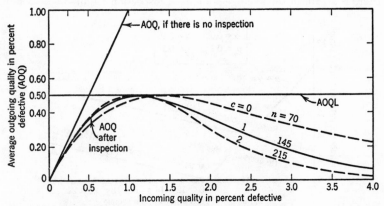

FIG. 20. Showing the average outgoing quality (AOQ) for three plans having the same average outgoing quality limit (AOQL). The horizontal scale shows the quality of binomially controlled lots before inspection, and the vertical scale shows the corresponding average quality after inspection. The rejected lots, assumed to be of size $N = 1000$, are screened. Although all three plans of acceptance give the same AOQL, they will require differing amounts of inspection for any given incoming average quality. The solid line is drawn for $c = 1$ and $n = 145$, which is the plan that gives minimum inspection for incoming binomial quality $p = 0.25\%$. (Reproduced by permission of the Bell Telephone Laboratories, Inc., from *Sampling Inspection Tables* by H. F. Dodge and H. G. Romig, published by John Wiley & Sons, Inc., 1944.)

Every sampling plan has two OC-curves, an AOQ-curve, and an AOQL; this is true for single-sampling, double-sampling, and sequential plans. As with the OC-curve, the shape of the AOQ-curve and hence the magnitude of the AOQL depend on the particular sampling plan. A specified AOQL may be attained with various sampling plans, as is manifest in Fig. 20.

In the Dodge-Romig tables, the AOQL is given for every plan. In fact, some of their plans, for both single and double sampling, are drawn especially to provide for a specified AOQL (.0010, .0025, .0050, .0075, .01, .015, .02, .025, .03, .04, .05, .07, .10), with a minimum total amount of inspection, including the screening.

Sequential sampling. The fundamental idea of double sampling by which control of the risks is attained with a smaller average amount of inspection (by starting off with a small sample and taking a second only if a decision is not reached on the first one) can be extended into multiple schemes.[16] In a *sequential* plan, the testing is to be continued only until the sample-pawn crosses one or the other of two decision boundaries. These boundaries may be infinitely long, but in practice they are usually curtailed to force a decision one way or the other after so many tests. Sometimes the pawn will cross a boundary after only a few tests, and there is a gain in the average amount of inspection required.

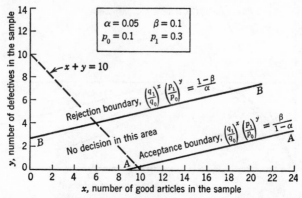

FIG. 21. Showing a sequential plan. The line AA is the boundary of acceptance, and the line BB is the boundary of rejection. So long as the pawn remains between them, there is no decision, except that in practice it is desirable to provide a definite upper limit to the sample-size, with an arbitrary rule for rejection and acceptance. The intercepts and slopes of the lines AA and BB depend upon α and β for particular quality levels p_0 and p_1. In other words, the two lines definitely determine an OC-curve. The sample-line $x + y = 10$ is drawn for reference.

The sequential theory to be described here is based on the probability-ratio test for binomial theory. This theory was developed by A. Wald [17] acting upon a suggestion from Allen Wallis and Milton Friedman.

The regions of rejection and acceptance that were constructed earlier upon the Neyman-Pearson theory with binomial probabilities depended on finding where the probability-ratio $(q_1/q_0)^x(p_1/p_0)^y$ is a maximum and where it is a minimum along the sample-line $x + y = n$. It was found that this ratio increases along the sample-line from bottom to top for all values of $p_1 > p_0$, and this will be true for every n.

The characteristic feature of a double-sampling plan is that the sample-line $x + y = n_1$ for the first sample contains a region of no decision between $y = c_1$ and c_2, a second sample n_2 being required if the pawn arrives in the region of no decision at the end of the first sample. In contrast, the sample-line $x + y = n_1 + n_2$ for both samples is definitely

[16] A brief history is given by Dodge and Romig (*loc. cit.*).

[17] A. Wald, *Sequential Analysis* (John Wiley, 1947); Statistical Research Group, *Sequential Analysis of Statistical Data: Applications* (Columbia, 1946).

divided into regions of acceptance and rejection, their boundary being between $y = c_2$ and $y = c_2 + 1$. It is easy to think of extending the idea of double sampling into multiple tests.

Now suppose that the boundaries of the regions of acceptance for a multiple-sampling plan were joined by a curve, to be called the *acceptance boundary*. The procedure would be to test n_1 items, then n_2, then n_3, etc., declaring the lot accepted when and if the sample-pawn ever strikes the acceptance boundary. The same could be done by bounding the regions of rejection with a *rejection boundary*.

For any acceptance boundary, the probability β of a lot of any quality p_1 being accepted can be calculated, although of course the labor might be prodigious unless considerable judgment were exercised in laying out the boundary. The Wald theory depends on the use of the straight line

$$\left(\frac{q_1}{q_0}\right)^x \left(\frac{p_1}{p_0}\right)^y = a, \quad \text{a constant} \tag{7}$$

for a boundary. In logarithmic form this equation appears as

$$x \log \frac{q_1}{q_0} + y \log \frac{p_1}{p_0} = \log a \tag{8}$$

which, being linear in x and y with constant coefficients, represents a straight line in the x, y plane.

Let there be one such line for acceptance and another for rejection, both infinitely long. Specifically, in Fig. 21 let

$$\left(\frac{q_1}{q_0}\right)^x \left(\frac{p_1}{p_0}\right)^y = b \quad \text{along } BB \text{ for rejection} \tag{9}$$

and

$$\left(\frac{q_1}{q_0}\right)^x \left(\frac{p_1}{p_0}\right)^y = a \quad \text{along } AA \text{ for acceptance} \tag{10}$$

These lines will be parallel, as only the constant (right-hand) terms are different. Next let

$$\left.\begin{aligned}
T_1 &= \sum_{BB} \binom{n}{y} q_1{}^x p_1{}^y \\[2mm]
T_0 &= \sum_{BB} \binom{n}{y} q_0{}^x p_0{}^y \\[2mm]
S_1 &= \sum_{AA} \binom{n}{y} q_1{}^x p_1{}^y \\[2mm]
S_0 &= \sum_{AA} \binom{n}{y} q_0{}^x p_0{}^y
\end{aligned}\right\} \tag{11}$$

T_1 is the total probability of the pawn ever reaching the line BB as the sampling is continued indefinitely, the lot-quality presumed to be p_1. T_0 is similarly defined for quality p_0. S_1 and S_0 are similarly defined for the line AA and for qualities p_1 and p_0. Then *without performing any probability calculations at all* [10] it can be seen from Eqs. 9 and 10 that whatever be the values of T_1, T_0, S_1, and S_0, it is a fact that

$$T_1 : T_0 = b \qquad \text{[By Eq. 9]} \quad (12)$$

$$S_1 : S_0 = a \qquad \text{[By Eq. 10]} \quad (13)$$

Now the probability that the pawn shall arrive in the region of no decision between the parallel lines (9) and (10), after n tests, can be made as small as desired by making n big enough (p. 417). It follows that

$$T_1 + S_1 = 1 - \epsilon_1 \qquad\qquad (14)$$

$$T_0 + S_0 = 1 - \epsilon_2 \qquad\qquad (15)$$

where ϵ_1 and ϵ_2 are probabilities, functions of n, which approach 0 as $n \to \infty$. ϵ_1 and ϵ_2 will now be neglected, and the last four equations solved for the four unknowns, T_1, S_1, T_0, and S_0. Owing to the neglect of ϵ_1 and ϵ_2, none of the following results will be exact, but the inexactness is of no practical consequence.

$$\left.\begin{aligned}
T_1 &= \frac{b(a-1)}{a-b} \\[2mm]
T_0 &= \frac{a-1}{a-b} \\[2mm]
S_1 &= a\frac{1-b}{a-b} \\[2mm]
S_0 &= \frac{1-b}{a-b}
\end{aligned}\right\} \qquad (16)$$

Let

$$\left.\begin{aligned}
S_0 &= 1 - \alpha \\
S_1 &= \beta
\end{aligned}\right\} \qquad (17)$$

to conform with the notation already agreed upon. Then from Eqs. 16 it will be found that

$$\left.\begin{aligned}
b &= \frac{1-\beta}{\alpha} \\[2mm]
a &= \frac{\beta}{1-\alpha}
\end{aligned}\right\} \qquad (18)$$

Hence from Eq. 9 the boundary for rejection, in logarithmic form, is

$$x \log \frac{q_1}{q_0} + y \log \frac{p_1}{p_0} = \log \frac{1-\beta}{\alpha} \qquad (19)$$

and from Eq. 10 the boundary for acceptance is

$$x \log \frac{q_1}{q_0} + y \log \frac{p_1}{p_0} = \log \frac{\beta}{1 - \alpha} \qquad (20)$$

It is obviously very easy to draw up a sequential plan for any desired pair of coordinates p_0, $1 - \alpha$ and p_1, β on the OC curve. Fig. 21 shows boundaries drawn for

$$p_0 = 0.1 \qquad\qquad p_1 = 0.3$$

$$\alpha = 0.05 \qquad\qquad \beta = 0.1$$

Table 3 on page 285 shows a sequential plan. It is taken from Paul Peach's book *Industrial Statistics and Quality Control* (Edwards & Broughton Co., Raleigh, 1947), with the kind permission of the author.

Remark 1. It is to be noted that in the equations of the two decision-lines (19) and (20), p_0 and p_1 appear only as their ratio, and once the ratio $p_1 : p_0$ is given, $q_1 : q_0$ is also given. Hence to draw the two lines it is only necessary to specify

$$R = \frac{p_1}{p_0} \qquad (21)$$

and the desired values of α and β for control of the two types of error. $1 - \alpha$ is the probability of accepting a lot of quality p_0, and β is the probability of accepting a lot of quality p_1.

Remark 2. It should not be assumed without critical and mature consideration that the sequential idea is invariably the best plan in every type of sampling problem. Conditions vary with different kinds of product. Where it is relatively easy to vary the number of tests from one lot to another or when testing is expensive, the sequential plan should have consideration; but where the routine must be kept fairly uniform or when testing is inexpensive, the sequential idea may cause more confusion than it is worth. With the sequential plan the inspectors' workload varies as the quality of the lots presented for inspection varies. When the quality is extremely bad or extremely good, the inspectors' workload is light, because decisions are reached quickly with a small number of tests on each lot. But when the quality is somewhere between p_0 and p_1, the number of tests per lot may be double the number ordinarily required when the quality is staying fairly near normal. Thus as the quality fluctuates the inspectors will alternately be overworked and underworked, with consequent demoralization or need for transfer of personnel from one function to another, which is difficult and sometimes unworkable. With double sampling, the workload will vary, too, but not as much.

Another important consideration to keep in mind is the need for a continuous record of the lot-quality, either for the management or for use in the control-chart (which the student may wish to pursue in some of the references mentioned in the bibliography further on). This is somewhat difficult with sequential sampling, because samples that are curtailed at a decision-boundary are biased; this bias can be overcome, however, by more complicated calculations.[18]

Remark 3. The reader may have noted already that the talk has been in terms of decision boundaries as curves or lines, whereas the actual boundaries must be a set of discrete points in the x, y plane. The distinction is of little practical consequence, although it is carefully made in published sampling tables. If the actual points of decision are on or just interior to the lines AA and BB, the actual protection is slightly greater, or never less, than the α and β used in Eqs. 19 and 20.

Remark 4. The theory of sequential analysis given here is applicable only where binomial sampling theory is applicable. For other sampling theories, the sequential theory must be modified. For adequate treatments the reader is referred to Wald's book.[17]

Definition of a lot. The student will observe from the foregoing theory and from Exercise 4 ahead that for the greatest economy in attaining any desired levels of protection to producer and consumer, the lot-size N should be large. The practical application of this principle, however, requires some skill. With care in the planning, the requirement that lots be large may work a hardship on the small producer, and on the purchaser as well. Moreover, one of the prime purposes of inspection should be directed toward control of the production process—not mere attainment of a desired A.O.Q. One or two untrained or careless operators in a section may be the source of most of the errors or defects that are encountered: the removal or retraining of these operators may raise the quality of the whole section significantly, and it is desirable that the inspection records should disclose such sources of trouble. Yet in many cases the work of one operator must be accumulated over a period of two or three weeks before there is enough of it to inspect economically: meanwhile, his work and the work of others must move on through inspection to the next operation. In such conditions it may be possible to build up large lots from sublots that originate with the various workers. Separate records on each sublot may be kept, as inspection proceeds, and these records may be consolidated and summarized at the end of two or three weeks for use with a control chart for appropriate action on the production process. Meanwhile, no material is detained, and the desired A.O.Q. is attained with the economy of large lots. Material from the small producer may be handled in the same manner.[19]

[18] Girshick, Mosteller, and Savage, "Unbiased estimates for certain binomial sampling problems with applications," *Annals Math. Stat.*, vol. xvii, 1946: pp. 13–23; Blackwell, "Conditional expectation and unbiased sequential estimation," *ibid.*, vol. xvii, 1947: pp. 105–10; Blackwell and Girshick, "A lower bound for the variance of some sequential estimates," *ibid.*, vol. xviii, 1947: pp. 277–80.

[19] Leslie E. Simon, "The industrial lot and its sampling implications," *J. Franklin Inst.*, vol. 237, 1944: pp. 359–70. See also his book in the bibliography.

TABLE 1.   FOR SINGLE-SAMPLING PLANS

$$\alpha = \beta = 0.05$$

(Courtesy of Paul Peach)

R	c	np_0
58.0	0	0.051
13.0	1	0.355
7.5	2	0.818
5.7	3	1.366
4.6	4	1.970
4.0	5	2.61
3.6	6	3.29
3.3	7	3.98
3.1	8	4.70
2.9	9	5.43
2.7	10	6.17
2.5	12	7.69
2.37	14	9.25
2.03	21	14.89
1.81	30	22.44
1.61	47	37.20
1.51	63	51.43
1.335	129	111.83
1.251	215	192.41

DIRECTIONS FOR USE OF TABLE 1

1. Calculate $R = p_1/p_0$.
2. Find R in the table.   If it does not appear, use the next larger value shown.
3. Read directly the acceptance number c.
4. Divide np_0 by p_0 to get n, the sample-size.

BIBLIOGRAPHY

K. A. BROWNLEE, *Industrial Experimentation*, His Majesty's Stationery Office, 1948.

COMMITTEE, *Control Chart Method of Controlling Quality*, Z1.3–1942, American Standards Association, 70 East 45th St., New York 17.

OWEN L. DAVIES, *Statistical Methods in Research and Production*, Oliver and Boyd, 1947.

HAROLD F. DODGE and H. G. ROMIG, *Sampling Inspection Tables*, John Wiley, 1944.

B. P. DUDDING and W. J. JENNETT, *Quality Control Charts* (Standard 600R), British Standards Institute, 1942.

H. A. FREEMAN, *Industrial Statistics*, Wiley, 1942.

EUGENE GRANT, *Statistical Quality Control*, McGraw-Hill, 1946.

J. M. JURAN, *Management of Inspection and Quality Control*, Harper & Bros., 1945.

PAUL PEACH, *Industrial Statistics and Quality Control*, Edwards & Broughton Co., Raleigh, 1947.

E. S. PEARSON, *Industrial Standardization and Quality Control* (Standard 600), British Standards Institute, 1935.

TABLE 2.   FOR DOUBLE-SAMPLING PLANS

$$\alpha = \beta = 0.05$$

(Courtesy of Paul Peach)

R	c_1	k_1	c_2	$n_1 p_1$
15.1	0	1	1	0.207
8.3	0	2	2	0.427
5.1	1	3	4	1.00
4.1	2	4	6	1.63
3.5	2	5	7	1.99
3.0	3	7	9	2.77
2.6	5	11	13	4.34
2.3	6	13	16	5.51
2.02	9	17	23	8.38
1.82	13	23	32	12.19
1.61	21	34	50	20.04
1.505	30	45	69	28.53
1.336	63	83	138	60.31

DIRECTIONS FOR USE OF TABLE 2

1. Calculate $R = p_1/p_0$.
2. Find R in the table. If it does not appear, use the next larger value shown.
3. Read off c_1, k_1, c_2 directly.
4. Divide np_0 by p_0 to get n_1, the first sample-size. Discard any fraction that may result from the division.
5. Second sample-size n_2 is the same as n_1.

WILLIAM B. RICE, *Control Charts in Factory Management*, Wiley, 1947.

WALTER A. SHEWHART, *The Economic Control of Quality of Manufactured Product*, Van Nostrand, 1931.

LESLIE E. SIMON, *An Engineer's Manual of Statistical Methods*, Wiley, 1941.

STATISTICAL RESEARCH GROUP (edited by Eisenhart, Hastay, and Wallis), *Selected Techniques of Statistical Analysis*, McGraw-Hill, 1947.

STATISTICAL RESEARCH GROUP (edited by Freeman, Girshick, Wallis, and others), *Sequential Analysis of Statistical Data: Applications*, Columbia, 1945.

STATISTICAL RESEARCH GROUP (edited by Freeman, Friedman, Mosteller, and Wallis), *Sampling Inspection*, McGraw-Hill, 1948.

A. WALD, *Sequential Analysis*, Wiley, 1947.

EXERCISES

Exercise 1. *a.* If the quality in the lots submitted for inspection were constant (as is impossible), and if rejected lots were really rejected and not screened and resubmitted, the AOQ would be identical with the quality submitted. This is true no matter what plan is used. Thus, a sampling plan all by itself is no guarantee that the quality accepted will be an improvement, or that it will measure up to any desired level.

TABLE 3. FOR SEQUENTIAL-SAMPLING PLANS

$$\alpha = \beta = 0.05$$

(Courtesy of Paul Peach)

R	A	B	C	np_0
15.0	1.1196	6.576	0.1876	0.60
10.0	1.3168	5.741	.2516	1.05
7.5	1.5048	5.265	.3067	1.70
6.0	1.6922	5.133	.3557	2.50
5.0	1.8839	5.040	.4002	3.40
4.5	2.0159	4.972	.4279	4.20
4.0	2.1871	4.949	.4604	5.30
3.5	2.4203	4.991	.4998	7.00
3.0	2.7599	5.138	.5483	10.00
2.6	3.1732	5.395	.5964	14.30
2.3	3.6403	5.743	.6401	20.10
2.0	4.3743	6.353	.6927	31.40
1.8	5.1584	7.053	.7344	46.50
1.6	6.4511	8.258	.7832	76.00
1.5	7.4779	9.239	.8108	106.00
1.33	10.5395	12.223	.8630	224.00

DIRECTIONS FOR USE OF TABLE 3

1. Calculate $R = p_1/p_0$.
2. Find R in the table, interpolating if necessary.
3. Obtain h and S from the formulas

$$h = A - p_0B$$
$$S = C/p_0$$

4. Obtain the maximum sample-size n by dividing np_0 by p_0. Truncate at this number.
5. Continue as directed in the text.

NOTE. The above table was calculated for $\alpha = \beta = 0.046$; the effect of truncation is to increase the risks to about 0.05.

b. Prove that if the quality in the submitted lots varies, any sampling plan really effects an improvement (but see Ex. 4).

c. If the "rejected" lots are screened and then resubmitted, the average outgoing quality is improved whether the lots are of uniform or variable quality.

Exercise 2. Explain the fallacy in the following plan (once widely used). The inspector forms inspection lots of various sizes, and from each inspection lot he draws a sample consisting of 10 percent of the items in the inspection lot. Each item in the sample is subjected to a quality test. If 2 percent or fewer of the items in the sample fail, the inspection lot is accepted; otherwise, the inspection lot is rejected. (See Ex. 4.)

Exercise 3. Discuss the following statements, which were extracted from an inspection manual privately communicated to the author.

a. Whenever possible, it is desirable that the entire inspection lot be formed before sampling inspection is begun. This is not always done; for example, the inspector may select his sample from an assembly line as the items produced reach him. If inspection is done on such a moving inspection lot, it will sometimes happen that a rejection number is reached before the entire inspection lot has been completely produced. The rejection of the entire inspection lot would then involve the rejection of items not yet completely produced. This appears unreasonable, particularly when the source of the defectives found by the inspector is located and removed before completion of the remaining items in the inspection lot. One alternative to rejection of the entire inspection lot, including the unfinished items, is to reject only the part of the inspection lot that was available to the inspector for sampling. This alternative is also undesirable. First, it in effect changes the protection given by the sampling plan since the uncompleted parts of rejected inspection lots have more than one chance of being accepted. Second, rejection of part of inspection lots gives the supplier less incentive to improve the quality of his product than rejection of entire inspection lots.

b. When inspecting material from a moving inspection lot, if the rejection number has been reached prior to the completion of the production of the inspection lot, the inspector should report to the supplier for corrective action the defect or defects which caused rejection. When the inspector is satisfied that corrective action has been taken, a new inspection lot shall be started at the point in the line at which the defect or defects occurred. Material which has been produced prior to the corrective action should be included in the previous inspection lot and rejected.

Exercise 4 (The inefficiency of small lot-sizes [20]). Trays containing 50 articles each come through a particular operation and at a certain point are inspected. If no defective is found in the sample, the tray is accepted; if one or more defectives are found, the tray is sent back (rejected). This is a "single-sampling" plan with $c = 0$ for acceptance

TABLE 4. THE DISTRIBUTION OF DEFECTIVE ARTICLES IN THE TRAYS PRESENTED FOR INSPECTION. THE PROCESS AVERAGE IS 4% DEFECTIVE.

Number of defective articles per tray of 50 articles	Number of trays having quality p	Percent defective (p)
0	100	0
1	200	2
2	400	4
3	200	6
4	100	8
5 or more	0	...

[20] This demonstration, or one equivalent thereto, was first shown to me by G. Rupert Gause, now of the Bell Telephone Laboratories.

TABLE 5.   SAMPLES OF 3: DISPOSITION OF 1000 TRAYS PRESENTED FOR INSPECTION

Percent defective (p)	Number of trays having quality p	Number of trays rejected	Number of defective articles in the trays that are	
			Rejected	Accepted
0	100	0	0	0
2	200	12	12	188
4	400	46	92	708
6	200	34	102	498
8	100	22	88	312
10 or more	0	0	0	0
Totals	1000	114	294	1706
Percent defective, average			5.16	3.85

TABLE 6.   SAMPLES OF 5: DISPOSITION OF 1000 TRAYS PRESENTED FOR INSPECTION

Percent defective (p)	Number of trays having quality p	Number of trays rejected	Number of defective articles in the trays that are	
			Rejected	Accepted
0	100	0	0	0
2	200	19	19	181
4	400	74	148	652
6	200	53	159	441
8	100	34	136	264
10 or more	0	0	0	0
Totals	1000	180	462	1538
Percent defective, average			5.13	3.75

TABLE 7.   SAMPLES OF 10: DISPOSITION OF 1000 TRAYS PRESENTED FOR INSPECTION

Percent defective (p)	Number of trays having quality p	Number of trays rejected	Number of defective articles in the trays that are	
			Rejected	Accepted
0	100	0	0	0
2	200	40	40	160
4	400	145	290	510
6	200	99	297	303
8	100	60	240	160
10 or more	0	0	0	0
Totals	1000	344	867	1133
Percent defective, average			5.04	3.45

number. The distribution of defectives [21] in the incoming product is exhibited in Table 4. Compute the results shown in Tables 5, 6, 7. Explain the failure of the samples to discriminate between trays containing various numbers of defective articles even when as many as 10 or 20 percent are tested. Note that the percentage defective in the rejected trays is scarcely greater than in the accepted trays, and that improvement in quality in the accepted trays is barely perceptible as the sampling is increased from 4 to 10 and 20 percent. What is the remedy? (Bigger samples are required for better discrimination. The lot-size should also be greatly increased, so that the sampling fraction $n:N$ will be small and economical.)

Exercise 5. Prove the following statement, which is taken from Gen. Simon's book, *An Engineer's Manual of Statistical Methods* (Wiley, 1941), "Small samples from a moderately defective lot are better than the lot more frequently than they are poorer than the lot."

Exercise 6. Criticize the following instructions for inspection, which are copied from a set of instructions that was issued by a large corporation.

> The crates of articles shall be similar in all respects. During the process of manufacture the inspector representing the purchaser shall make whatever inspection and tests as are necessary to determine that all parts are being manufactured in accordance with the requirements of the contract. The inspector shall also ascertain that the conditions of interchangeability of all parts not permanently joined together are being maintained. The duties of the inspector described herein shall not serve to relieve the manufacturer of his responsibility in furnishing products that are in complete accordance with the requirements of the contract, order or requisition, drawings, and specifications.

[21] Through inadvertence this distribution is a bit tighter than binomial, but I shall not bother to loosen it and recompute.

CHAPTER 9. THE SAMPLE AS A BASIS FOR ACTION

At Hanoi the night passed off quietly except for some Viet-Namese artillery fire.—News item in *The Statesman*, Delhi, 3 Jan. 1947.

During the previous night R.A.F. Bisley bombers started large fires at Gafsa and Sbeitla. One Bisley gunner reported seeing a large explosion fifty miles from the target.—*The Times* (London), quoted in the *Math. Gazette*, vol. xxix, 1945: p. 108, from Inst. Lt. F. J. North, R.N.

A. ESTIMATES AND ERROR BANDS

The purpose in making surveys. In reiteration, the prime purpose in carrying out a survey or experiment is to find answers to certain questions in order to provide a basis for rational *prediction*, usually in order that *action* may be taken on the sources of the data—i.e., action on the causes that make the universe what it is. The results of the sample, once they have been brought in, must be translated and interpreted in such manner that they will be maximally useful. Estimates of various statistical characteristics of the universe must be prepared; the reliability of these estimates must be measured (Ch. 10), and the estimates interpreted. For instance, in population sampling the action that is to be taken often depends on how many men and women in the universe in particular age-classes are unemployed, retired, disabled, engaged in agricultural labor for wages, or interested in purchasing a particular commodity or service, or how many acres are under cultivation of a particular crop.

One of the most important problems of sample-design is to settle on the formula or formulas for preparing estimates from the data of the sample. The formula chosen must preserve all the prior information concerning the universe and must also make the most use possible of the data of the survey. The purpose in this chapter is to present some elementary practical theory and views on estimation.

An estimate and the measure of its reliability lead to prediction and action. It may turn out that a prediction is wrong, and a decision is wrong. A wrong decision may result in total or partial loss of crop; maldistribution of grain, either in time or space; loss of profits, as by locating a plant in the wrong area through misjudging a market; wrong prediction of an election; purchase of materials that are unsuited to a particular use; loss of a political campaign; wrong result in an experiment in pure science wherein the aim is to increase knowledge; or in fact any wrong answer to the problem in whose solution the survey or experiment was designed to give guidance. A wrong decision not only brings losses of this kind, or worse, but also discredits the evidence on which the decision was based. A wrong recommendation, based on a survey, thus discredits the survey and all who had a hand in it. An estimate, to be useful, *must not be too far wrong;* hence an estimate is no good without some measure of its reliability. One important piece of statistical evidence of reliability (or the lack of it) is the standard error of an estimate. The calculation of a standard error enables the risks and costs of various decisions to be calculated and balanced. A standard error is of course only one link in the evidence for reliability, as was explained in Chapters 1 and 2, and the reader is reminded that a standard error can be calculated from the sample-design only if the sample is a probability-sample (Ch. 1).

It is of course often necessary to make estimates from bad data. The statistician's mathematical training is then no help to him, and except for his broad experience in many branches of subject-matter into which his versatility has led him, he is no abler than anyone else to make estimates or to test hypotheses. It should be said, though, that he is quick to discover the defects and limitations of numerical data, and he will not make the mistake of misusing probability-theory to calculate misleading limits on bad data. One of his important contributions is the development of theory (control chart, run chart, and other techniques) by which bad data can be detected and defects in methods of testing and surveying discovered and corrected.

Another statistical procedure of analysis that leads to action is the testing of hypotheses (see Exercise 1, p. 312).

The problem of estimation. Next we shall think in a little more detail on the problem of estimation, which will be described as the question of how best to perform calculations on the data of the sample, perhaps in combination with prior data,[1] to determine *within calculable limits* what answer would have been obtained from a complete coverage had one been taken with the same definitions, the same procedures of selecting and training the interviewers or assistants, the same questionnaire, and the same thoroughness as was possible with the sample. A statistical estimate is expected to be objective, scientific, and demonstrable. It is not obtained by crystal-gazing and is not somebody's expert opinion. It is not merely a figure that someone *hopes* is not too far wrong.

This is not to gainsay the virtue of expert opinion in its proper place, nor to imply that a statistical estimate is always good, nor that one must consult a statistician at every turn, but rather to make clear that

---

[1] An example is the use of a ratio-estimate, treated in Chapter 5.

the statistician, when he makes an estimate, must know and declare how good his estimate is and be ready to exhibit the evidence for his calculations.

The method by which estimates will be calculated is part of the sampling plan. No sampling plan is complete without a statement of the procedure of estimation. Usually there are several methods of estimation that can be used. They will vary in cost and precision. The statistician will often find ways of cutting costs by decreasing the size of his sample but regaining the lost precision by introducing a more elaborate system of estimation (normally carried out in the tabulations). Sometimes it pays to do the opposite, viz., to take a bigger sample and use "inefficient" but less costly methods of estimation.[2] The decision on the plan of estimation should be viewed from the standpoint of obtaining the greatest amount of information per unit cost; but the total cost, it must be remembered, includes both the collection of the information and the estimating. The sample-design will depend greatly on the plan of estimation, and conversely. Thus, if a ratio-estimate (Ch. 5) is to be used, it is not necessary to go to extra heavy expense to delineate areas of equal populations; but, on the other hand, as will be seen, a ratio-estimate is not always a possibility. Again, if the tabulation procedure is to be kept as simple as possible, as when an unbiased estimate with uniform weight is to be used, the preparation and selection of the sample must be more elaborate.

The method of maximum likelihood, illustrated with binomial sampling. Suppose that n drawings have been made with replacement from a bowl containing red and white chips, and that the stirring and the drawing are carried out in such manner that it is useful to assume that $Er/n = p$ and that $\text{Var } r = npq$. A sample when examined shows r red chips, and r may be 0 or 1 or 2 or any number to n. Unless $r = 0$ or n, it is obvious that the bowl must surely have contained both red and white chips. But what else can be inferred? Suppose that the proportion red in the bowl is needed as a basis for action. It might be agreed to accept the proportion r/n as the proportion in the bowl. Various other agreements would be possible. For example, it might be agreed to accept the average of the first and last drawings, or of the first two drawings, etc. One might conceivably form a biased estimate by adding $\frac{1}{4}$ to r/n if $r/n < \frac{1}{4}$ and subtracting $\frac{1}{4}$ from r/n if $r/n > \frac{3}{4}$. Some methods of estimation will be biased and some unbiased, and some will have higher variances than others.

One method of estimation that can often be used and which possesses important properties, such as being asymptotically normally distrib-

[2] Frederick Mosteller, "On some useful 'inefficient' statistics," *Annals Math. Stat.*, vol. xvii, 1946: pp. 377–408. See also page 564 in this book.

uted, asymptotically unbiased, and of minimum variance, and various optimum properties in small samples is Fisher's *method of maximum likelihood*.[3] In the binomial problem for a particular value of p, the probability of r successes occurring in n trials in Bernoulli sampling is $\binom{n}{r} q^{n-r} p^r$ (p. 113). This expression, considered as a function of p for an observed r, is identical with what Fisher calls the *likelihood* of the

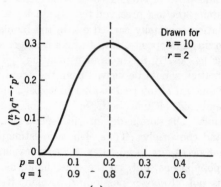

FIG. 22.   Showing how the rth term $\binom{n}{r} q^{n-r} p^r$ of the point binomial varies with p. This term takes its maximum value where p has the value of r/n.

given value of p. On this basis p_2 is "twice as likely" as p_1 for an observed r if $\binom{n}{r} (1 - p_2)^{n-r} p_2^r$ is double $\binom{n}{r} (1 - p_1)^{n-r} p_1^r$. The *maximum likelihood* or *optimum* value of p is that which gives $\binom{n}{r} q^{n-r} p^r$ its maximum value; it is easily found by setting the derivative $\frac{d}{dp} \binom{n}{r} q^{n-r} p^r$ equal to zero and solving for p—a simple problem in maxima and minima. As $\binom{n}{r}$ is constant, independent of p, the derivative of the rth term of the point binomial with respect to p is

$$\frac{d}{dp} \binom{n}{r} q^{n-r} p^r = \binom{n}{r} \frac{d}{dp} (1 - p)^{n-r} p^r$$

$$= \binom{n}{r} q^{n-r} p^r \left[-\frac{n - r}{1 - p} + \frac{r}{p} \right] \tag{1}$$

[3] R. A. Fisher, *Messenger of Math.*, vol. 41, 1912: pp. 155–60; *Phil. Trans. Roy. Soc.*, vol. A222, 1922: pp. 309–68, page 326 in particular. The student may wish to pursue Cramér's presentation in his *Mathematical Methods of Statistics* (Princeton, 1946), Chapter 32.

This vanishes if the part in the bracket vanishes, i.e., if

$$\frac{n - r}{1 - p} = \frac{r}{p} \tag{2}$$

which is fulfilled if $p = r/n$. This is the *most likely* or *optimum* value of p for the one experiment yielding r red chips. It is often written

$$\hat{p} = \frac{r}{n} \tag{3}$$

This estimate of p is characterized as *optimum* or *most likely* because it is obtained by maximizing the probability of what was observed in the

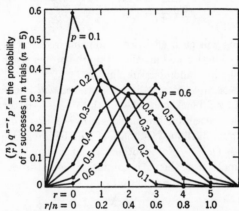

FIG. 23.   Illustrating the meaning of likelihood in the binomial. The expansion of $(q + p)^5$ for various values of p. The ordinates show the probabilities of getting 0, 1, 2, 3, 4, or 5 red chips in five trials, for various values of p. These polygons show that the value of p having maximum likelihood for a sample showing r red chips is simply the fraction r/n. For example, at the abscissa $r = 2$, for which $r/n = 0.4$, the polygon having the highest ordinate is that for which $p = 0.4$; at the abscissa $r = 1$, for which $r/n = 0.2$, the polygon having the highest ordinate is that for which $p = 0.2$; etc. The polygons for $p = 0.7$, 0.8, and 0.9 are omitted to avoid confusion; they would be the same as those for $p = 0.3$, 0.2, and 0.1 with the ordinates taken in reverse order from left to right. The polygons for $p = 0$ and 1 would consist of single points of unit height at the abscissas $r = 0$ and 1 respectively, all the other ordinates being zero.

sample. Evidently for what was actually observed, the bowl that gives a greater probability than any other bowl could give is that in which the proportion p of red chips is precisely the fraction r/n observed in the sample.

_Figs. 22 and 23 may help to clarify the idea.   In Fig. 22 $\binom{n}{r} q^{n-r} p^r$ is plotted as a function of p, and the maximum is seen to occur when $p = r/n$. Fig. 23 shows the six terms of the expansion of $(q + p)^5$, one polygon for each value of p. As the legend explains, the maximum ordinate at any abscissa r belongs to the polygon for which $p = r/n$.

From the theory of Chapter 4 it is known that the distribution of $\hat{p}$ or of r/n will be binomial between $\hat{p} = 0$ and $\hat{p} = 1$ with mean at

$$E\hat{p} = E\frac{r}{n} = \frac{1}{n} Er = \frac{1}{n} np = p \tag{4}$$

and with

$$\text{Var } \hat{p} = \frac{1}{n^2} \text{Var } r = \frac{1}{n^2} npq = \frac{pq}{n} \tag{5}$$

The

$$\text{C.V. } \hat{p} = \sqrt{\frac{\text{Var } \hat{p}}{(E\hat{p})^2}} = \sqrt{\frac{q}{np}} \tag{6}$$

These equations will be used in the following sections in dealing with fixed-interval prediction and in calculating the proper size of sample for a desired error band in binomial sampling.

Exercise. A sample of n balls is drawn *without replacement* from a container holding x black and $N - x$ white balls. It turns out that r of the n balls in the sample are black, and $n - r$ are white. Show then that the maximum likelihood estimate of x is the integer equal to or just less than $(N + 1)r/n$. Stated otherwise, if $\hat{x}$ denotes the maximum likelihood estimate of x, then $\hat{x}$ is the integer equal to or just less than the value of x satisfying the equation

$$\frac{x}{N} = \frac{r}{n} \frac{N + 1}{N}$$

This result shows that as the number of balls in the container increases, the estimate of the proportion black in the container approaches the proportion black in the sample, agreeing with Eq. 3.

Solution (Frederick F. Stephan)

The probability of the observed event is

$$P(r|x) = \frac{\binom{x}{r} \binom{N - x}{n - r}}{\binom{N}{n}} \qquad \text{[P. 113]}$$

The problem is to find what value of x/N makes this a maximum for an observed value of r. Write the difference $P(r|x) - P(r|x - 1)$ and show that it is positive if $x < (N + 1)r/n$ and negative if $x > (N + 1)r/n$. Then show that the maximum likelihood estimate of x is the largest integer $\leq (N + 1)r/n$. If $(N + 1)r/n$ is an integer, show that $(N + 1)r/n - 1$ is also a maximum likelihood estimate of x.

Fixed-interval prediction contrasted with varying-interval prediction.
The broader a prediction, the less useful it is. A prediction that predicts every possibility is of no use whatever. It is no good to say that
it will either rain or not rain tomorrow. Neither is it any good to say
that it will rain unless there is some rational basis for saying so. Out
of 1000 predictions we want as many as possible to be "as nearly right
as possible." But these words have no meaning until a criterion is laid
down by which a particular method of prediction can be tested and found
to conform or not to conform to the requirement "as many as possible
to be as nearly right as possible." There are, as will appear, various
classes of estimates, depending largely on whether prior information can
or can not be used to advantage; and if we are not careful we may work
entirely within one class of estimates, finding how to get a minimum
variance within that class, but ignoring "better" estimates obtainable
from another class of procedures.

All this probably makes little sense to the reader at this point. It
may be helpful to state a few problems. Let us think of an ideal-bowl
experiment, carried out in such manner that binomial theory applies.
Suppose that 100 chips have been drawn with replacement [4] and shuffling; 30 of the 100 chips drawn were red and 70 were white. Questions
like the following arise in practice:

> Can we find from such results a close approximation to p (the
> actual ratio of red to total in the bowl)?

Once "close approximation" is defined, statistical theory helps us to
answer questions of this type with a measurable amount of confidence,
where *confidence* is defined as the proportion of correct answers that the
statistical method provides in a series of random experiments. Amongst
many other forms of questions, statistical theory has been developed to
answer the two following questions, with the confidence required
therein-

> *i.* How can we calculate from past data good approximations
> to a pair of fixed limits p_1 and p_2 that will contain closely 99 percent
> of the values of $\bar{x}$ or r/n in 1000 future samples?
>
> *ii.* How can we compute most efficient intervals,[5] one interval for

---

[4] Nonreplacement requires the use of the hypergeometric series in place of the
binomial series in what follows, but this is only a formality and does not have a
bearing on the argument.

[5] "Efficient" is intended here in the Neyman sense; see "Outline of a theory of
statistical estimation based on the classical theory of probability," *Phil. Trans.
Royal Soc.*, vol. 236A, 1937: pp. 333–80. The reader may also wish to consult Neyman's *Lectures and Conferences in Mathematical Statistics* (The Graduate School,
Department of Agriculture, Washington, 1938; now out of print), p. 154.

each sample, and the intervals varying from sample to sample, so that closely 99 percent of these intervals will overlap the true value μ or p in the next 1000 experiments?

It should be noted that statistical theory deals, not with one experiment, but with a sequence of experiments that theoretically could be repeated, giving rise to random variables like r, r/n, x, X, etc., which have probability-distributions. Two useful kinds of prediction are now explained.[6]

i. Fixed-interval prediction. There may be evidence on hand (as from a series of previous experiments) from which there can be calculated in advance a pair of numbers p_1 and p_2 within which closely 99 per cent of the values of r/n will fall in a future 1000 experiments. If then the inequality

$$p_1 \leq \frac{r}{n} \leq p_2 \qquad \begin{array}{l}\text{[Here the ends are fixed} \\ \text{but the middle is a} \qquad (7) \\ \text{random variable]}\end{array}$$

be written for every one of the 1000 experiments, then closely 99 percent of the inequalities will be true. Now p_1 and p_2 form an interval, and as they remain fixed for all the 1000 experiments, this is *fixed-interval prediction.*

ii. Varying-interval prediction. We may be content to find a method for calculating a pair of limits $\underline{p}$ and $\bar{p}$ (read "p lower" and "p upper") from each experiment, one after another, such that if the inequality

$$\underline{p} \leq p \leq \bar{p} \qquad \begin{array}{l}\text{[Here the middle is fixed,} \\ \text{but the ends are ran-} \qquad (8) \\ \text{dom variables]}\end{array}$$

be written for every one of the 1000 experiments, then closely 99 percent of the inequalities will be true. Because the limits $\underline{p}$ and $\bar{p}$ are calculated anew from each experiment, they are variables, and this is *varying-interval prediction.*

In the second type of prediction, $\underline{p}$ and $\bar{p}$ are functions of n and r. The procedure is to draw a sample of a stated size n, note the number r of red chips, and then calculate $\underline{p}$ and $\bar{p}$. Both they and the interval between them vary from sample to sample. Yet we shall find a method of calculating $\underline{p}$ and $\bar{p}$ so that the Ineq. 8 will be found true in 99 percent or any desired percentage of samples; and the method will work just as well for small samples as for large ones; indeed the prediction is just as valid for $n = 2$ as for $n = 100$. Moreover, we need know nothing in advance about the contents of the bowl or of any previous samples

[6] See Fig. 84 on page 563 and the accompanying text. See also pp. 59–61 in Shewhart's *Statistical Method from the Viewpoint of Quality Control* (The Graduate School, Washington, 1939).

drawn from it; it is only required that the procedure of drawing be random with known probabilities.

On the contrary, any attempt at fixed interval requires previous experiments on which to base the calculations, such as a series of 25 or more small samples, or one big sample that is known to be random.

Remark 1. It should be noted that even though the bowl be varied (first one p and then another) throughout the 1000 experiments, approximately 99 percent of the Ineqs. 8 will nevertheless be found correct. This is a very important property of the varying intervals.

Remark 2. If the sampling method is not random in the sense described in the preceding chapter, the methods of this chapter will not apply.

Remark 3. For both the fixed- and varying-interval predictions there will be more than one possible method of calculating limits that satisfy the required probabilities. Neyman in the article cited in the last footnote laid down various criteria that might be chosen as desirable for the varying-interval prediction.

Calculation of an error band. Fixed-interval prediction is almost always desired in statistical problems of design and analysis. A fixed interval is an *error band*, bounded by p_1 below and p_2 above, like the limits on a control chart. When we are able to calculate the fixed limits p_1 and p_2 for a desired probability P, we are in position to answer questions like the following: How big a sample from a particular material is required in order that the error in the mean ($\bar{x}$ or r/n) shall not exceed some permissible maximum error with an average frequency no greater than once in 100 future experiments? Such calculations require an approximation to σ (or to p in the binomial case). In practice, a good approximation can usually be obtained by diligent and resourceful probing into previous experience and data with similar materials and operations. Such an approximation may serve very well until it can be replaced by a sufficiently long series of controlled observations from the same material that is to be sampled in the future (cf. previous comments on p. 108).

The trouble with the varying intervals is that their widths and centers jump around so playfully in small samples that a useful prediction can hardly be made concerning what proportion of the results (e.g., means) of future samples will fall between any pair of limits: in other words, in small samples varying-interval prediction is *not* fixed-interval prediction.

Fortunately the varying intervals become fairly steady as the sample-size is increased and become excellent numerical approximations to fixed limits having the same probability, provided we know how to perform the calculations for "efficient" intervals. The method of calculation depends on the universe and the method of sampling. For random samples from a binomial universe the methods given later on will be found easy and efficient.[7] For most other finite universes, the methods of Chapter 17 are efficient or at least good enough. In social and economic surveys the size of the sample is often large (hundreds or thousands), and excellent approximations to fixed intervals can thus be calculated, although in the small cells of the tabulated results this will not be so.[8] On the other hand, in much agricultural and industrial experimentation the sample-sizes are often necessarily small, although σ is sometimes known from previous series of experiments.

To continue the line of thought in connexion with fixed-interval prediction we may think of a statistical adviser to an administrator who is responsible to the public or the stockholders of his company. Many of the administrator's decisions will depend on statistical studies. As a result of a statistical study the statistician is expected to produce two limits like N_1 and N_2, or p_1 and p_2, within which the "true" result may be presumed to lie. To the statistician this means that the fixed-interval type of prediction is required: in other words, it means to him that approximately 99 or some other percentage of future samples similarly conducted would be expected to show results lying between these limits. (Sometimes only a lower limit like N_1 or p_1 is required, or an upper limit like N_2 or p_2.) The statistician must have a big enough sample so that he can calculate a pretty good approximation to the limits (or limit) associated with a desired degree of probability. Moreover, this interval must not be too wide. But as the costs of a survey mount rapidly as the prescribed allowable width of the interval is decreased for a given probability, the widest possible usable limits or the lowest possible probability should be prescribed.

An example might be described as follows.[9] Suppose that a shortage of proper housing exists, and that a public official, known perhaps as the Director of Housing, needs to know how many dwelling units were started last month in the metropolitan district of Washington. A regular monthly study of a sample of building-permits gives 400 as the estimated number, with a standard error of 40. The survey also shows that a high proportion of the permits that are issued are not followed by actual construction by reason of shortages of certain necessary materials. The director needs to know whether the actual figure (supposedly determinable by a complete canvass) is as high as 320, as it has been decided upon statistical evidence furnished by studies of the population and housing characteristics of the city that if the figure is below 320 he will be

[7] Cf. the reference to Neyman in footnote 5 and the books mentioned in the preface.

[8] Moreover, the size of the sample as measured by number of households is not the sole determining factor: *how the sample was drawn* is equally important. This point was emphasized in Chapter 5.

[9] The example is actual except for the place, date, number, title, etc.

compelled to allocate to essential needs building materials that are in
short supply. A vendor or manufacturer of building materials should
have an equal interest in the figures. The statistician, we may suppose,
decides that one-sided odds of 49:1 corresponding to 2 sigma give sufficient protection, and he therefore declares that the number of starts
was indeed not less than 320, with the result that no allocation of building
materials is made. He knows that he is taking a risk, but he feels that
this risk is preferable to the risk of the director making a wrong decision.
In the long run the statistician is able to minimize the losses arising from
incorrect decisions based on samples.

The administrator only wants to know *what to do*. He only wants the
numbers N_1 and N_2, p_1 and p_2, or perhaps merely N_1 or N_2 alone, or
p_1 or p_2; his decision will be based on them, not the probabilities associated with them. In my experience he is not interested in probabilities,
even though he may be very dependent on his statistician and respectful
of the statistician's knowledge of probability. It is up to the statistician
to judge what level of probability to use, and this judgment will depend
on the cost of a mistake in case he gives a wrong answer. Of course,
if the statistician can never be caught, then any survey and any answer
are good enough. The statistician's usefulness and his reputation will
depend on how sharp his judgment is regarding the levels of probability
that are needed for the various problems that he encounters. Obviously
he must acquire deep understanding of these problems as they arise.

The statistician will of course speak and write to his colleagues in
technical terms, but in my experience administrators are not interested
in such discourse: they only want to know *what to do*. (See also p. 562.)

Practical, efficient, and simple error bands (fixed limits) for large
samples from ordinary universes are expressible as 1, 2, or 3 times an
estimated standard error. An example of 3-sigma limits would be
$\hat{p} \pm 3\sigma_{\hat{p}}$, or simply $\hat{p} - 3\sigma_{\hat{p}}$ or $\hat{p} + 3\sigma_{\hat{p}}$ for a single lower or upper limit,
$\hat{p}$ being an estimate of p, having standard error $\sigma_{\hat{p}}$. For large samples
from "ordinary" finite universes the 1-, 2-, and 3-sigma limits correspond
roughly to two-sided odds of 2:1, 24:1, and practical certainty against
a wrong decision occurring from failure of the limits to include the results
of future samples. One-sided odds would be 5:1, 49:1, and practical
certainty against a wrong decision occurring from failure of a lower (or
upper) limit to fall below (or above) the results of future samples.
These figures follow from the approach to normality of the distributions
of X and $\bar{x}$ in large samples from a discrete universe. (See the table,
p. 552.)

In practice, when the sample-size is large, p in the expression
$\pm 3\sqrt{q/np}$ is replaced by its observed value r/n, where r as before is
the number of red chips found in the sample of n. The results of the

survey, along with the estimated 3-sigma error band, are then expressed as

$$\frac{r}{n}\left\{1 - 3\sqrt{\left(\frac{1}{r} - \frac{1}{n}\right)\frac{N-n}{N-1}}\right\} < \hat{p} < \frac{r}{n}\left\{1 + 3\sqrt{\left(\frac{1}{r} - \frac{1}{n}\right)\frac{N-n}{N-1}}\right\} \quad (9)$$

or as

$$\hat{p} = \frac{r}{n}\left\{1 \pm 3\sqrt{\left(\frac{1}{r} - \frac{1}{n}\right)\frac{N-n}{N-1}}\right\} \quad (10)$$

The interpretation of these equations is a *prediction*—viz., that if the survey were repeated a great number of times with the same operation of testing or interviewing, practically all of the values of $\hat{p}$ so obtained would fall between the limits so estimated. This prediction has little validity, however, unless n is large. Soon we shall learn how to calculate varying intervals and we shall see that they have an operationally verifiable meaning, even for small samples, and that the varying intervals merge into fixed intervals as the size of sample increases.

For estimating the 2-sigma limits, the factor 3 is to be replaced by 2 throughout.

Remark 1. It is important, when presenting either an estimated fixed interval (Eq. 9 or 10) or a varying interval (next sections), to state the sample-size n on which the calculations are based; also the formula and the justification for the use of the formula. Without this information the reader is not able to judge whether the interval so calculated represents a reliable or unreliable prediction of an interval supposed to contain a certain percentage of the results of future samples.

Remark 2. The width of the standard error and the 3-sigma error band are known more accurately than p itself, because of the square root. To illustrate, suppose that p is expected to lie between 8 and 12 percent, and a survey is to be taken to discover more precisely what p is. The sample-size is to be 900, no larger sample being possible for lack of funds and time. The questions are: Will this size of sample yield the required reliability? Can a satisfactory answer be obtained in advance without knowing p any more accurately? The design of this particular sampling procedure is such that calculations by the binomial expansion will apply, wherefore the width of the 3-sigma error band is $\pm 3\sqrt{q/np}$. It will be seen that a satisfactory approximation to the width of the error band can be obtained in advance, even though p is known only within broad limits.

With $p = 8$ percent,

$$\text{The error band} = p\left(1 \pm 3\sqrt{\frac{q}{np}}\right) = p\left(1 \pm 3\sqrt{\frac{0.92}{900 \times 0.08}}\right)$$

$$= p(1 \pm 0.34)$$

With $p = 12$ percent,

$$\text{The error band} = p\left(1 \pm 3\sqrt{\frac{0.88}{900 \times 0.12}}\right) = p(1 \pm 0.27)$$

Thus, although p is not known very closely in advance (if it were, there would be no use proposing a sample), still the error band is known in advance closely enough, for the calculations just made tell us that p will be determined within about 30 percent of p.

Note that whereas the two values of p used in the calculations differ by 50 percent (12 against 8 percent), the error band is determined within about 20 percent (27 against 34). Evidently if p lies anywhere in the neighborhood of 8 or 12 percent, it will be determined within about 30 percent.

Remark 3. If 30 percent were considered too wide for a 3-sigma error band, a bigger sample would be required. Indeed, as already seen in Chapter 4, the 3-sigma error band is often used the other way around to determine the sample-size for a required degree of reliability (width of error band).

Confidence intervals. The interval extending from $\underline{p}$ to $\bar{p}$ in the Ineqs. 8 is a *confidence interval*, and the limits $\underline{p}$ and $\bar{p}$ themselves are *confidence limits*. The figure 99 percent, which represents the "expected" fraction that will be found correct for 1000 future experiments, is called the *confidence coefficient* of the method of calculating $\underline{p}$ and $\bar{p}$. One might desire that the method of computing $\underline{p}$ and $\bar{p}$ should place them as close together as possible, on the average, and in more complicated sampling they should satisfy certain other desirable requirements that will not be brought up here (cf. the reference to Neyman).

The confidence coefficient can be varied. It can be made 90 percent or 95 percent, or indeed 50 percent. If the confidence coefficient is 99 percent, the interval from $\underline{p}$ to $\bar{p}$ will be broader than if the confidence coefficient were 95 percent. Thus one sacrifices narrowness to gain a greater proportion of correct statements.

The varying-interval type of prediction has been largely developed by Neyman and Egon Pearson. Some fundamental ideas were nevertheless first expressed by Millot [10] in 1923. He took $\underline{p}$ and $\bar{p}$ to be the two roots of the quadratic

$$\left(\frac{r}{n} - p\right)^2 = \frac{\lambda^2 pq}{n} \tag{11}$$

the interpretation being that Ineq. 8 will be true in the fraction P_λ of a large number of experiments of n drawings each.

[10] Stanislas Millot, "Sur la probabilité à posteriori," *Comptes rendus, Paris*, vol. 176, 1923; p. 30: *Théorie nouvelle de la probabilité des causes* (Gauthier-Villars, Paris, 1925). Millot suggested the normal integral as one way of evaluating P_λ. E. B. Wilson, in the *J. Amer. Stat. Assoc.*, vol. 22, 1927: pp. 209–12, suggested the use of the Tchebycheff inequality for evaluating P_λ. An excellent treatment of the distinction between the fixed and varying intervals, with references to Millot, is provided by Oskar Anderson in his *Einführung in die mathematische Statistik* (Julius Springer, Wien, 1935): pp. 114–16. I am indebted to Prof. A. Hald of Copenhagen for the references to Millot and Anderson.

Suppose that for a given value of p (meaning the limit of r/n) the chance distribution of the number of red chips in 10 trials is represented by varying the sizes of 11 beads stretched along a horizontal weightless thread, the weight of the beads being in proportion to the numerical values of the 11 terms of the binomial expansion $(q + p)^{10}$. For purposes of illustration let $p = \frac{1}{4}$; then what we get is the horizontal string of beads at the ordinate $p = \frac{1}{4}$ in Fig. 24. Now place a pair of arrows in the figure so that no more than 2.5 percent of the total weight of the beads is external to each arrow, yet so that the arrows are as close to-

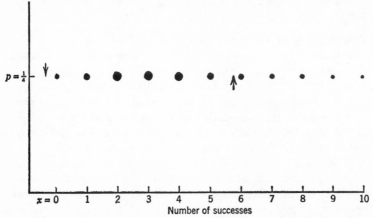

$p = \frac{1}{4}$

$x = 0$ 1   2   3   4   5   6   7   8   9   10

Number of successes

Fig. 24. One way of representing the probabilities of the possible number of successes in 10 trials when $p = \frac{1}{4}$. Here probabilities are represented by varying the areas of eleven beads stretched along a horizontal line according to the terms of the expansion of $(\frac{3}{4} + \frac{1}{4})^{10}$. The arrows are placed so that not more than one-fortieth of the weight of the beads is external to each arrow.

gether as possible. To do this we take the terms of the binomial expansion

$$[\tfrac{3}{4} + \tfrac{1}{4}]^{10} = (\tfrac{3}{4})^{10} + 10(\tfrac{3}{4})^9(\tfrac{1}{4}) + \cdots + 10(\tfrac{3}{4})(\tfrac{1}{4})^9 + (\tfrac{1}{4})^{10}$$

$$= 0.0563 + 0.1877 + 0.2816 + 0.2503 + 0.1460 + 0.0584$$

$$+ 0.0162 + 0.0031 + 0.0004 + 0.0000 + 0.0000 \quad (12)$$

to see where the arrows should go. The 0-th term is 0.0563, which is already greater than 0.025; hence the left-hand arrow is placed to the left of the 0-th bead. The sum of the last six terms is 0.0781, and the sum of the last five is 0.0197; hence the right-hand arrow must be placed between the 5th and 6th beads from the right.

Suppose now that we elaborate this plan a little by constructing beads for several other values of p, say for $p = 0, \frac{1}{6}, \frac{1}{2}, \frac{3}{4}, \frac{5}{6}, 1$, and that we put them all on the same diagram to obtain Fig. 25. In this figure the stepped boundary accomplishes for every possible value of p precisely what the arrows would do if they were shown for every value of p. At least 95 percent of the weight of the beads that might be constructed for any and hence for *all* values of p will fall inside the stepped boundary,

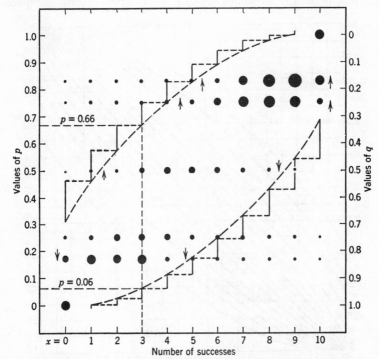

FIG. 25. Chart for reading off varying intervals of probability 0.95 for samples of 10. At least 95 percent of the weight of the beads for any and hence for all values of p will fall inside the stepped boundaries. The dashed curves drawn through the inside corners of the stepped boundary are more convenient to use than the stepped boundary. For $x = 3$, the interval is $\underline{p} = 0.06$ and $\bar{p} = 0.66$.

and obviously inside the dashed curves. This statement holds regardless of whether the total weight of all the beads for one value of p is equal to the total weight of all the beads for some other value of p, since, if the weights of all the beads for any value of p are doubled with respect to the other strings of beads, then every bead on that horizontal line is doubled, but the *fractional* part of this weight that falls externally to the stepped boundary is left unaltered. It follows that the results for $\underline{p}$ and $\bar{p}$ will hold in Ineq. 8 regardless of the distribution of p. The stepped boundary or the dashed curve in Fig. 25 defines what Neyman calls a *confidence belt* of coefficient 0.95. One could also construct a confidence belt of coefficient 0.99, or of any other desired fraction (*vide infra*).

The interpretation is this. Every sample of 10 must be represented by a point in the x, p plane of Fig. 25. This diagram is therefore known as a *sample space*. No sample can possibly fall outside the sample space,

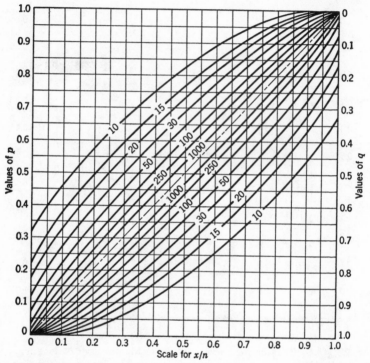

FIG. 26. Confidence belts of coefficient 0.95 for various values of n (Clopper-Pearson).

since p must lie between 0 and 1 inclusive, and the number of white chips can not be fewer than 0 nor more than 10. If we go from one bowl to another, we may deal with one value of p and then another, but whatever be the values of p (whatever be the prior distribution of p), and whether we know them or not, we do know that at least 95 percent of our sample points will lie inside the confidence belt of coefficient 0.95.

The interpretation of the confidence limits for the point binomial. Suppose that 10 chips have been drawn with replacement and shuffling; 3 turn out to be red, and 7 white. Turning now to Fig. 25, we erect a vertical line at $x = 3$ and read $\underline{p} = 0.06$ from the lower boundary of the confidence belt, and $\bar{p} = 0.66$ from the upper boundary. These are the confidence limits, and the interval from p to $\bar{p}$ is the confidence interval. If one were to calculate p and $\bar{p}$ in this way and write the Ineq. 8 for every one of a large number of such experiments, even if p and n were to change from one experiment to another (i.e., even if the contents of the bowl and the size of the experiment were to change), we could still feel safe in saying that at least 95 percent of the Ineqs. 8

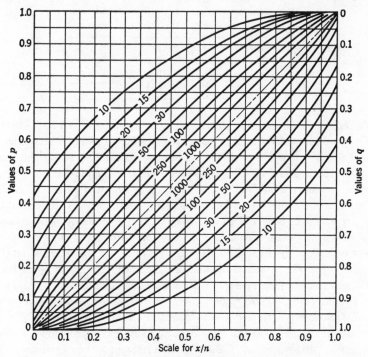

FIG. 27. Confidence belts of coefficient 0.99 for various values of n (Clopper-Pearson).

so obtained would be correct. This is the interpretation of confidence limits. It is a varying-interval type of prediction. By calculating p and $\bar{p}$ with a confidence coefficient of 0.99 one would find that at least 99 percent of his inequalities were correct.

The Clopper-Pearson charts. Clopper and Pearson [11] devised a scheme for presenting confidence belts for several values of n on one chart, as seen in Figs. 26 and 27, which are reproduced here with the kind permission of the authors. Fig. 26 shows confidence belts of coefficient 0.95, and Fig. 27 shows confidence belts of coefficient 0.99.

The double square-root paper of Mosteller and Tukey, soon to be described, may be used in place of the Clopper-Pearson charts.

[11] C. J. Clopper and Egon S. Pearson, *Biometrika*, vol. 26, 1934: pp. 404–13.

An experiment in confidence intervals for the binomial. In this section the results of several class experiments are presented. The exact contents of the bowls are described in the two columns on the right in the table below. Each sample was 20 chips, drawn with replacement.

TABLE 1.   CONTENTS OF THE BOWLS

p	$p:q$	Red	White
0.25	1:3	100	300
.3	3:7	105	245
.4	4:6	104	156
.5	1:1	200	200

The color of each chip was noted, and the chip was returned to the bowl. The number of red chips in each sample of 20 was recorded under r in Table 2. r and r/n vary from sample to sample; so do the confidence limits $\underline{p}$ and $\bar{p}$, which were read from the Clopper-Pearson charts with confidence coefficient = 0.95 (Fig. 26). The student should verify the results by the use of double square-root paper. The inequality $\underline{p} < p < \bar{p}$ was observed to be wrong twice out of the 70 experiments. Of course, 70 experiments are too few for discovering experimentally just how often the Ineq. 8 will be incorrect, but it can be asserted on the basis of experience and mathematics that it will be correct in at least 95 percent of the samples of any long series of experiments.

The Mosteller-Tukey paper. A transformation called the *arc sine transformation* is extremely useful in simplifying the calculation of probabilities associated with the binomial and many other problems. It was originally discovered by Fisher,[12] and, although it has been used by many people, its extreme utility in graphical form seems not to have been appreciated before the issuance of a double square-root paper designed by Mosteller and Tukey [13] and an article by the same authors. Let

$$\theta = \sin^{-1} \sqrt{\frac{r}{n}} \quad \text{or} \quad \left.\begin{array}{c} \dfrac{r}{n} = \sin^2 \theta \\[2mm] \dfrac{n-r}{n} = \cos^2 \theta \end{array}\right\} \tag{13}$$

[12] R. A. Fisher, "On the dominance ratio," *Proc. Royal Soc. Edinburgh*, vol. 42, 1922: pp. 321–41.

[13] Frederick Mosteller and John W. Tukey, "The uses and usefulness of probability paper," *J. Amer. Stat. Assoc.*, vol. 44, 1949: pp. 174–212. The double square-root paper is manufactured by the Codex Book Company of Norwood, Mass.

TABLE 2. AN EXPERIMENT IN CONFIDENCE INTERVALS

Sample-size, $n = 20$

Chips drawn from a bowl, with replacement

Sample number	r = number of red chips in sample	$\dfrac{r}{n}$	Confidence limits read from Fig. 26 (Probability = 0.95)		$\underline{p} < p < \bar{p}$	
			$\underline{p}$	$\bar{p}$	Right	Wrong
		$p = 0.25$ (12 March 1945)				
1	5	0.25	0.08	0.49	✓	
2	7	.35	.15	.60	✓	
3	4	.20	.06	.44	✓	
4	5	.25	.08	.49	✓	
5	7	.35	.15	.60	✓	
6	3	.15	.03	.38	✓	
7	4	.20	.06	.44	✓	
8	3	.15	.03	.38	✓	
9	7	.35	.15	.60	✓	
10	3	.15	.03	.38	✓	
11	9	.45	.23	.69	✓	
12	3	.15	.03	.38	✓	
13	5	.25	.08	.49	✓	
14	9	.45	.23	.69	✓	
15	3	.15	.03	.38	✓	
16	6	.30	.12	.55	✓	
17	3	.15	.03	.38	✓	
18	4	.20	.06	.44	✓	
19	5	.25	.08	.49	✓	
20	5	.25	.08	.49	✓	
$n = 400$	100	.25	.208 *	.292 *	✓	
		$p = 0.3$ (12 Feb. 1944)				
21	5	.25	.08	.49	✓	
22	5	.25	.08	.49	✓	
23	10	.50	.27	.73	✓	
24	7	.35	.15	.60	✓	
25	7	.35	.15	.60	✓	
26	6	.30	.12	.55	✓	
27	8	.40	.19	.64	✓	
28	4	.20	.06	.44	✓	
29	1	.05	0	.26		✓
30	5	.25	.08	.49	✓	
31	3	.15	.03	.38	✓	
32	4	.20	.06	.44	✓	
33	5	.25	.08	.49	✓	
$n = 260$	70	.27	.216 *	.324 *	✓	

TABLE 2. AN EXPERIMENT IN CONFIDENCE INTERVALS (*Continued*)

Sample-size, $n = 20$

Chips drawn from a bowl, with replacement

Sample number	r = number of red chips in sample	$\dfrac{r}{n}$	Confidence limits read from Fig. 26 (*Probability = 0.95*)		$\underline{p} < p < \bar{p}$	
			$\underline{p}$	$\bar{p}$	Right	Wrong
		$p = 0.4$ (12 Feb. 1944)				
34	6	.30	.12	.55	✓	
35	10	.50	.27	.73	✓	
36	10	.50	.27	.73	✓	
37	10	.50	.27	.73	✓	
38	8	.40	.19	.64	✓	
39	8	.40	.19	.64	✓	
40	5	.25	.08	.49	✓	
41	8	.40	.19	.64	✓	
42	11	.55	.32	.77	✓	
43	8	.40	.19	.64	✓	
44	9	.45	.23	.69	✓	
45	5	.25	.08	.49	✓	
46	8	.40	.19	.64	✓	
47	8	.40	.19	.64	✓	
48	6	.30	.12	.55	✓	
49	7	.35	.15	.60	✓	
50	9	.45	.23	.69	✓	
$n = 340$	136	.40	.348 *	.452 *	✓	
		$p = 0.5$ (12 March 1945)				
51	7	.35	.15	.60	✓	
52	11	.55	.32	.77	✓	
53	11	.55	.32	.77	✓	
54	13	.65	.41	.85	✓	
55	10	.50	.27	.73	✓	
56	10	.50	.27	.73	✓	
57	9	.45	.23	.69	✓	
58	11	.55	.32	.77	✓	
59	12	.60	.36	.81	✓	
60	5	.25	.08	.49		✓
61	12	.60	.36	.81	✓	
62	7	.35	.15	.60	✓	
63	7	.35	.15	.60	✓	
64	9	.45	.23	.69	✓	
65	14	.70	.46	.87	✓	
66	9	.45	.23	.69	✓	
67	10	.50	.27	.73	✓	
68	11	.55	.32	.77	✓	
69	11	.55	.32	.77	✓	
70	11	.55	.32	.77	✓	
$n = 400$	200	.50	.451 *	.549 *	✓	

* These limits are calculated from the normal integral. The variance of r/n is pq/n, and the $2\frac{1}{2}$ percent limit each side of the mean is 1.96σ in the normal integral, as may be observed from the table on page 552.

Then approximately

$$\sigma_\theta{}^2 = \frac{1}{4n} \tag{14}$$

and

$$\sigma_\theta = \frac{1}{2\sqrt{n}} \tag{15}$$

where σ_θ denotes the standard error of θ or the standard error of r in angular measure. This result was proved on page 132. It is true only when θ is in radians, when $\dfrac{d}{d\theta}\sin\theta = \cos\theta$. For measurement in degrees,

$$\mathrm{Var}\ \theta^\circ = \frac{821}{n}\ (\mathrm{deg.})^2$$

Although the above result is an approximation, as for simplicity it was derived with differentials instead of finite increments and the second and higher powers of $\delta\theta$ were neglected, nevertheless it fortuitously gives excellent results even for small sample-sizes.

Now let $\sqrt{r}$ be plotted as in Fig. 28 on the vertical axis, and $\sqrt{n-r}$ on the horizontal to get the position P of a random sample. Then

$\theta = \sin^{-1}\sqrt{r/n}$ is the angle made by OP with the horizontal axis. Moreover, the length of OP is $\sqrt{n}$ by the law of the hypotenuse. With $\sqrt{n}$ as a radius and 0 as center, draw an arc. Repeated samples of size n will give sample-points on this arc.

In random sampling the "expected" positions of these points will lie not far from $\sqrt{n-Er},\ \sqrt{Er}$. Their standard deviations will correspond to a curvilinear distance of $\frac{1}{2}$ along the arc, regardless of n. In other words, the standard deviation of the arc is always $\frac{1}{2}$. This is so because $\sigma_\theta = 1/2\sqrt{n}$, and as the radius is $\sqrt{n}$, σ_{arc} must be $\frac{1}{2}$.

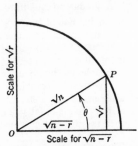

Fig. 28. Relations in the arc sine transformation.

To plot the standard error and 3-sigma limits for binomial sampling it is only necessary to perform the operation described in Fig. 29, wherein it will be observed that the calculation of bands of sampling variability is reduced to the barest simplicity.

Remark 1. On the double square-root paper, a small auxiliary scale gives the distances of 1, 2, 3, and 4 sigma so that probabilities can be gauged quickly with a ruler or dividers (closely 5 mm. for each sigma). The article by Mosteller and Tukey shows tables giving deviations in

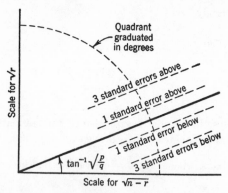

FIG. 29. Illustrating some principles of the Mosteller-Tukey double square root paper. The sample-points that represent binomial samples from a universe containing red and white chips in the proportions $p:q$ will fall around a radial line making the angle $\tan^{-1}\sqrt{p/q}$ with the axis of $\sqrt{n-r}$. The standard error or the 1-sigma limits in either direction will be lines parallel to this radial line and distant $\frac{1}{2}$ unit therefrom. The 3-sigma limits will be parallel at a distance of $\frac{3}{2}$ unit.

millimeters on the double square-root paper, corresponding to percentage points on the distributions of a normal deviate, χ^2 (p. 514), and normal ranges.

A graduated quadrant with radius corresponding to $n = 100$ enables the radial line of $\tan^{-1}\sqrt{p/q}$ to be drawn instantly for any pair of values of p and q, and the angle θ to be read off.

Remark 2. It is suggested that at this point the class draw with replacement 10 or 20 samples of 5 chips each from a mixture of red and white, to be followed by about the same number of samples of 20 chips each. The sample points should be plotted one by one on the M-T paper, and a table made up to show r and the angle θ for each sample. The variance of the experimental values of r and θ should be computed and compared with the theoretical values npq and $1/4n$.

To see whether a sample differs by 1, 2, or 3 sigma from a particular value of p, it is only necessary to lay off the radial line at the angle $\tan^{-1}\sqrt{p/q}$, then plot the observed sample point $\sqrt{n-r}$, $\sqrt{r}$, and note the separation of the sample point and radial line. Their separation in units of sigma is quickly read off by means of the auxiliary scale or by using a millimeter-scale, on which 2 sigma = 10 mm., 3 sigma = 15 mm.

An important step in statistical inference is accomplished by Mosteller and Tukey by plotting the observational triangle as shown in Fig. 30. For large values of n and r, the three points are indistinguishable, but for small numbers they are not. The probabilities (or "sigma distances") of points B and C are important in interpreting a sample. If the probability changes greatly from B to C, as will happen with small samples, or with very small r or very small $n - r$, it will not be wise to stick to exact "levels of significance." One of Mosteller and Tukey's examples follows.

Fisher and Mather [14] described a genetical experiment in which the individuals in 32 litters of 8 mice each were observed for straight hair and wavy hair. The Mendelian theory predicts that half the mice would have straight hair. It was observed that

$$r = 139, \text{ straight hair}$$
$$n - r = 117, \text{ wavy hair}$$

$$n = 256, \text{ total}$$

Did this discrepancy arise by chance?

The observed numbers are plotted as the triangle $A, B, C = (117, 139)$, $(118, 139)$, $(117, 140)$. The theoretical 50:50 proportion is plotted as the radial line of $\tan^{-1} 1$. The distance to Point B (the closest point) is 6.7 mm. or about 1.3σ, which is "not significant," as such deviations in both directions from 50:50 could occur by chance in about 1 in 5 of such samples. And the distance to Point C is 7.5 mm. or 1.5σ, corresponding to chances of about 1 in 7, which again is hardly significant. It can only be concluded that simple Mendelian genetics is not to be rejected by this experiment.

The observed percentage of straight-haired mice is quickly read from the quadrant as 54.5 percent.

To lay off 95 percent confidence limits (*vide supra*) for a binomial sample, one merely *i.* plots the sample-point and measures off 2 estimated standard errors each way, *ii.* reads off the corresponding values of $\sqrt{n - r}$ and $\sqrt{r}$, and *iii.* converts them to r and $n - r$. Or, in Step ii he may read off p and $\bar{p}$ on the graduated quadrant by passing radii through 0 and the 2-sigma points which were located in Step i. Three standard errors may be used for the confidence limits if preferred.

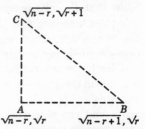

Fig. 30. The observation triangle. r red chips are observed in a sample of n. The observed point is $\sqrt{n - r}$, $\sqrt{r}$. If one more chip is drawn, the sample-point A must move to one or the other corner of the triangle shown. The corners B and C furnish probabilities which are to be compared with the probability associated with the observed point A.

[14] R. A. Fisher and K. Mather, "A linkage test with mice," *Annals Eugenics*, vol. 7, 1936: pp. 265–280.

A host of other sampling problems, both in design and analysis, are solved on the double square-root paper by Mosteller and Tukey in their article. In Chapter 8 it was seen that a single-sampling plan may be developed almost instantly with the aid of this paper. Curiously enough, although designed for binomial probabilities, the paper is advantageous for many other types of problems as well, e.g., the comparison of two variances, as in the analysis of variance (Ch. 17 of this book).

Exercise 1. A sample of 1500 poles was drawn from the maps of a public utility company. Before commencing the field-work of examining the condition of the poles and the equipment on the poles (p. 95), a tally was made of various characteristics (age, type of wood, creosote-treatment, length) of the poles drawn into the sample. This tally was possible through information given on the maps: it was made in order to discover any bad blunders that might have been made in drawing the sample. The tally showed 90.5 percent of a type called CP (for creosoted pine), whereas the property record in the accounting department showed 93.5 percent. The actual plan was 10 systematic subsamples (p. 353). Show that the two proportions are definitely in disagreement. (It is safe to assume here that the variance of $\hat{p}$ was equal to or smaller than the binomial variance pq/n. The cause of the disagreement, it was demonstrated, lay in unposted work-orders, in which many old poles had lately been replaced by CP poles. These changes were already reflected in the accounting records, but not yet posted to the maps. The unposted orders were thereupon sampled in the same manner as the maps, and the field-work was permitted to commence. Such checks are extremely important.)

Exercise 2. According to present regulations of the Civil Aeronautics Administration in the United States, the total weight of the passengers boarding an aeroplane need not be determined by weighing them individually but optionally by substituting for each person an average weight. Suppose that the average weight of an adult passenger is 160 lbs. and that the standard deviation of the distribution of the weights of adult passengers is 20 lbs.

a. Show that under these assumptions the upper 3-sigma limit of the passenger-load will be

$$\text{For 21 passengers, } X = 21\left(160 + \frac{60}{\sqrt{21}}\right)$$

$$= 3360 + 275 = 3635 \text{ lbs.}$$

$$\text{For 52 passengers, } X = 52\left(160 + \frac{60}{\sqrt{52}}\right)$$

$$= 8320 + 433 = 8753 \text{ lbs.}$$

b. Would you prefer to specify that the upper limit should be calculated with 3 sigma or 2 sigma?   Give reasons.

c. Suppose that you were an official of the air line, desirous of increasing the revenue on each trip, and that as a rule there is air-cargo waiting to be put aboard whenever the aeroplane is not loaded to capacity. Suppose furthermore that it costs 25¢ to weigh a passenger, and that each pound of cargo pays 50¢ for a particular trip.   Would you prefer to weigh every passenger or to substitute the average weight and allow the margin of 2 sigma or 3 sigma for safety?

d. Why is no finite multiplier used in these calculations?

e. Suppose that the average weights of men and women passengers differ by 40 lbs. and that the standard deviation of the weights of each sex separately is 15 lbs. Suppose that 2 out of 3 passengers are on the average male: do you think it would pay to keep separate tallies by sex in computing the load?

NOTE. The figures and calculations used here were not suggested by the Civil Aeronautics Administration.

Difference between two binomial proportions. As a special case in the propagation of error (Eq. 48, p. 130) one may write

$$f = \bar{x} - \bar{y}$$

wherein $\bar{x}$ is the mean of n_x observations drawn at random from a universe of mean μ_x and variance $\sigma_x{}^2$, $\bar{y}$ being similarly defined.   Then because $\bar{x}$ and $\bar{y}$ are uncorrelated it may be seen from Exercise 3b on page 133 that

$$E(\bar{y} - \bar{x}) = \mu_y - \mu_x$$

$$\text{Var}\,(\bar{y} - \bar{x}) = \frac{\sigma_x{}^2}{n_x} + \frac{\sigma_y{}^2}{n_y} \qquad \begin{array}{l}\text{[Factors for sampling}\\ \text{without replacement}\\ \text{may be introduced}\\ \text{if desired]}\end{array}$$

In practice one has only estimates of σ_x and σ_y. Sometimes it may be asserted from the nature of the problem that $\sigma_x = \sigma_y$, in which case the variances of the sample are pooled for the purpose of estimating σ_x (Ch. 17). If σ_y is a known multiple of σ_x, the problem is not more difficult, but if σ_x and σ_y are merely supposed to be different with no known factor of proportionality, the problem presents many hazards.[15] If n_1 and n_2 are large, however, an answer believed by the author to be satisfactory is given in Exercise 11 of Chapter 17 (p. 585).

[15] B. L. Welch, "The significance of the difference between two means when the population variances are unequal." *Biometrika*, vol. xxix, 1938: pp. 350–62. See also M. G. Kendall, *The Advanced Theory of Statistics*, vol. II (Griffin, 1946), pp. 111 ff.

The binomial case is interesting, particularly because of the simple solution that is possible with the Mosteller-Tukey double square-root paper which was described on pages 306–12. In the binomial case let r_1 be the number of red chips observed in a random sample of n_1 from a certain universe containing red and white chips, and let r_2 be observed in a sample of n_2 drawn from the same universe. Also let $\hat{p}_1 = r_1/n_1$ and $\hat{p}_2 = r_2/n_2$. Then

$$\text{Var}\,(\hat{p}_2 - \hat{p}_1) = \frac{p_1 q_1}{n_1} + \frac{p_2 q_2}{n_2} \qquad (16)$$

To test the hypothesis that p_1 and p_2 are equal, place

$$\hat{p} = \frac{r_1 + r_2}{n_1 + n_2} \qquad \hat{q} = 1 - \hat{p} \qquad (17)$$

Then the question is whether the fraction

$$\frac{\hat{p}_2 - \hat{p}_1}{\sqrt{\hat{p}\hat{q}\left(\dfrac{1}{n_1} + \dfrac{1}{n_2}\right)}}$$

is "significant." The distribution of this fraction is the distribution of t with $k = n_1 + n_2 - 1$ degrees of freedom. This distribution will be

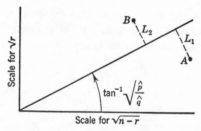

FIG. 31. Two binomial samples give the points A and B. Is there statistical evidence that they differ "significantly," indicating that the two samples came from different universes ($p_1 \neq p_2$)? An answer is found in the length of the line-segments L_1 and L_2: if either is more than a whole unit of length, such evidence exists. $\hat{p} = (r_1 + r_2)/(n_1 + n_2); \; A = \sqrt{n_1 - r_1},\; \sqrt{r_1}; \; B = \sqrt{n_2 - r_2},\; \sqrt{r_2}.$

encountered in Eq. 9 of Chapter 16, where the t-test may be made with the nomograph of Nekrassoff (Fig. 81, p. 555). At the moment we shall be content to observe whether the fraction exceeds 3, a crude test that is sufficient, especially if $n_1 + n_2 - 1$ is 10 or 15. When it is important to reduce the risks of error of the second kind (concluding that $p_1 = p_2$ when $p_1 \neq p_2$), 2 may be used as a critical ratio in place of 3.

On the Mosteller-Tukey double square-root paper it is only necessary to draw the radial line of $\tan^{-1}\sqrt{\hat{p}/\hat{q}}$; then to plot the two points $A = \sqrt{n_1 - r_1}$, $\sqrt{r_1}$ and $B = \sqrt{n_2 - r_2}$, $\sqrt{r_2}$, and measure the perpendicular distances from A and B to the radial line, as illustrated in Fig. 31. These distances, expressed in units of σ_{arc} (5 mm. on the M-T paper), measure the respective deviations in standard units, i.e., as multiples of their standard errors. If one point is distant more than 3 standard deviations (15 mm.) from the radial line $\tan^{-1}\sqrt{\hat{p}/\hat{q}}$, there is statistical evidence that the two samples came from different universes ($p_1 \neq p_2$). Again, 2-sigma limits may be used to indicate significance, if preferred.

Mosteller and Tukey's Table 4 makes possible the comparison of two or more sets of observations on binomial frequencies. One may plot the observed points and draw a radial line from the origin through the center of gravity of the points, which may often be estimated well enough by eye; or if desired, the center may be calculated by pooling the frequencies in the manner of calculating $\hat{p}$ in Eq. 17. If the sum of the greatest perpendicular distances above and below the radial line is smaller than

2.77	standard errors for		2	points
3.34	"	" "	3	"
3.87	"	" "	5	"
4.48	"	" "	10	"
5.01	"	" "	20	"

then the range between these extremes fails to achieve the 5 percent level of significance, and there would be no evidence, on this test, to doubt that the observed frequencies came from the same universe (same p).[16]

Exercise. The variance of $L_1 + L_2$ in Fig. 31 is $\frac{1}{2}$. (HINT. The variance of L_1 is $\frac{1}{4}$, and likewise the variance of L_2 is $\frac{1}{4}$.)

Tchebycheff's inequality. This is a theorem that is very important in connexion with error bands. It gives an upper limit to the proportion (probability) of sample sums or means that can fall outside the 2- or 3-sigma limits for any universe. Unfortunately the limits are so broad that in practice it pays when at all permissible to assume some particular shape for the universe in order to calculate narrower limits (as by the methods of this chapter and Ch. 17).

Let a universe consist of n strata, all containing an equal number of chips. For convenience in writing, n will be illustrated as 3, but the number of strata can be extended at will at any stage. The three strata

---

[16] The article by Mosteller and Tukey (cited in footnote 13) contains many and divers other examples and is highly recommended as a magnificent abbreviated treatise in theoretical statistics.

will be referred to by the letters X, Y, and Z. Their descriptions can be set down in Table 3.

TABLE 3.   DESCRIPTION OF THE THREE STRATA

Symbol	Labels on the chips and their proportions			Mean	Standard deviation
X	x_1 p_1	x_2 p_2	x_3 ⋯ p_3 ⋯	μ_x	σ_x
Y	y_1 q_1	y_2 q_2	y_3 ⋯ q_3 ⋯	μ_y	σ_y
Z	z_1 r_1	z_2 r_2	z_3 ⋯ r_3 ⋯	μ_z	σ_z

The general mean of all the strata will be

$$\mu = \frac{1}{n}(\mu_x + \mu_y + \mu_z) \qquad [n = 3 \text{ here}]$$

and the average variance within the strata will be

$$\sigma^2 = \frac{1}{n}(\sigma_x{}^2 + \sigma_y{}^2 + \sigma_z{}^2)$$

Tchebycheff began by taking the sum of all possible values of

$$[x_i - \mu_x + y_j - \mu_y + z_k - \mu_z]^2 p_i q_j r_k$$

Here x_i, y_j, and z_k are any numbers (chips) drawn from the three strata, and p_i, q_j, and r_k are the chances of drawing them. To perform this summation, it is well first to square the bracket as indicated, thus getting

$$[(x_i - \mu_x)^2 + (y_j - \mu_y)^2 + (z_k - \mu_z)^2 + 2(x_i - \mu_x)(y_j - \mu_y)$$
$$+ 2(x_i - \mu_x)(z_k - \mu_z) + 2(y_j - \mu_y)(z_k - \mu_z)]p_i q_j r_k$$

then to make a partial summation by holding i and j constant while k runs through all its values; k disappears in the process, and there is left

$$[(x_i - \mu_x)^2 + (y_j - \mu_y)^2 + \sigma_z{}^2 + 2(x_i - \mu_x)(y_j - \mu_y) + 0 + 0]p_i q_j$$

To go ahead, we make a second partial summation by holding i constant and letting j run through all its values; j then disappears, and there results

$$[(x_i - \mu_x)^2 + \sigma_y{}^2 + \sigma_z{}^2 + 0 + 0 + 0]p_i$$

The final summation over i leaves merely $\sigma_x{}^2 + \sigma_y{}^2 + \sigma_z{}^2$, which is just $3\sigma^2$ or, in general, $n\sigma^2$. Accordingly we write

$$\Sigma \, [x_i - \mu_x + y_j - \mu_y + z_k - \mu_z]^2 p_i q_j r_k = n\sigma^2 \qquad (18)$$

The remaining step is to divide Eq. 18 through by $n\lambda^2\sigma^2$ and get

$$\sum \frac{[x_i - \mu_x + y_j - \mu_y + z_k - \mu_z]^2}{n\lambda^2\sigma^2} \, p_i q_j r_k = \frac{1}{\lambda^2} \qquad (19)$$

then to distort this summation by retaining only the terms that satisfy the inequality

$$\frac{[x_i - \mu_x + y_j - \mu_y + z_k - \mu_z]^2}{n\lambda^2\sigma^2} > 1 \qquad (20)$$

From Eq. 18 it is clear that if λ is chosen greater than unity, there will be terms that fail to satisfy this inequality.

The equality (19) is immediately destroyed by any rejections from the summation on the left, the $=$ sign changing to $<$. The disparity between the right and left members will be still further increased if all the coefficients of $p_i q_j r_k$ satisfying the Ineq. 20 are counted as unity instead of something greater than unity as they actually are. Both operations performed on Eq. 19 will leave the distorted summation

$$\Sigma' \, p_i q_j r_k < \frac{1}{\lambda^2} \qquad (21)$$

the stroke indicating a summation of only those products $p_i q_j r_k$ for which the Ineq. 20 is satisfied. Ineq. 21 is the Tchebycheff inequality, which is now ready for discussion.

Statement of the theorem. Because $p_i q_j r_k$ is the probability of drawing chips labeled x_i, y_j, and z_k with replacement, it follows from Ineq. 21 that

i. If P denotes the probability that the sum $x_i + y_j + z_k$ of the sample falls outside the limits

$$\overbrace{\mu_x + \mu_y + \mu_z + \cdots}^{n \text{ strata}} \pm \lambda\sigma\sqrt{n}$$

then

$$P < \frac{1}{\lambda^2}$$

Stated otherwise,

ii. The probability $P < 1/\lambda^2$ that the mean of the sample falls outside the limits

$$\mu \pm \frac{\lambda\sigma}{\sqrt{n}}$$

When strata contain unequal numbers of chips, they can be broken up to form strata containing equal or nearly equal numbers. One drawing per stratum or an equal number of draws with replacement from each stratum then corresponds to *proportionate sampling* (Ch. 6) from the original universe; hence, more generally,

iii. In proportionate stratified sampling with replacement, the probability that the mean of the sample lies outside the limits $\mu \pm \lambda\sigma/\sqrt{n}$ must be less than $1/\lambda^2$, n being the size of the sample.

Any of these statements is the Tchebycheff inequality, the last form being perhaps the most useful. Stated in words, it says that of all possible samples that can be drawn with replacement from a given universe, each stratum thereof contributing in proportion to its size, the means of fewer than $1/\lambda^2$ of them will differ from the mean μ of the universe by so much as $\lambda\sigma/\sqrt{n}$.

Some consequences of the theorem. Actually, Tchebycheff stated several theorems, all closely related. They are reworded in the corollaries that follow.

Corollary 1. The probability is less than $1/\lambda^2$ that the mean of n chips drawn with replacement from any universe of standard deviation σ will differ from the mean μ of this universe by so much as $\lambda\sigma/\sqrt{n}$.

This follows at once from the theorem by taking the universe itself to be a single stratum, and the sample to be unrestricted. Or—what is the same thing—the universe may be divided into n equal and like strata, each contributing one chip to the sample.

By making n ever bigger and bigger, we have the *law of large numbers*, which can be stated as:

Corollary 2. The probability that the mean of n observations drawn with replacement from any universe of standard deviation σ will differ from the mean μ of this universe by so much as Δ, can be made less than any preassigned number ϵ, no matter how small Δ and ϵ be.

A rigorous proof of this corollary by the use of Tchebycheff's theorem is more difficult than might appear at first. No proof will be attempted here.

This theorem is illustrated for the binomial by Figs. 44 and 45, pages 416 and 417.

Corollary 3. Of n given numbers, not all equal, of mean μ and standard deviation σ, not more than n/λ^2 of them will differ from μ by the amount $\lambda\sigma$.

This is simply a restatement of Corollary 1. The student should test this corollary by observing that no matter how he chooses 10 numbers, not more than 1 of them can be made to fall outside a range of 3σ.

More powerful theorems, giving narrower limits than those of Tcheby-cheff for special universes that satisfy certain very general conditions, have been derived. For example, Gauss showed that for a continuous unimodal distribution, the probability P of theorem i above is $< 4/9\lambda^2$ (cf. Harald Cramér, *Mathematical Methods of Statistics*, Princeton, 1946: pp. 183, 256).

Historical note. Tchebycheff (1821–1894) was one of Russia's celebrated mathematicians. In statistics he is known for the theorem here presented and for his no less brilliant researches in orthogonal polynomials. His original paper on the inequality was published in French in *Liouville's Journal de mathematiques*, 2d series, vol. 12, 1867: pp. 177–84. Tchebycheff's article was translated into English by Helen M. Walker for David Eugene Smith's *Source Book in Mathematics* (McGraw-Hill, 1929), pp. 580–87.

B. INVERSE PROBABILITY [17]

The meaning of inverse probability. So far in this chapter we have dealt with likelihood and with two kinds of prediction by the use of intervals, fixed and varying. The calculation of the "most likely" value of p (p. 293) for a particular sample result tells what proportion of red chips is most favorable or optimum to the production of the one sample that was observed; this optimum value of p, when sampling with replacement,[18] turned out to be nothing more nor less than the proportion r/n of red chips in the sample.

There is a third problem, now to be tackled. It is the problem of discovering the frequency distribution of the different contents of bowls that when sampled produce samples of a specified kind (such as 4 red and 1 white; next section). This problem is called *inverse probability*, in contrast with direct probability, in which the problem is to discover the frequency distribution of the samples that are drawn from a given bowl. We shall discover that nothing can be done with small samples in inverse probability without the initial or prior distribution of the contents of the different bowls.

In industry, the bowl is a lot of product. Lots of product are presented for sampling one after another, and on the basis of tests on these samples the lots are rejected or accepted. The lots are not all alike but are variable. Moreover, from any one lot, many different samples can be drawn. By rejecting a lot that gives rise to a sample worse than some specified fraction defective, the frequency distribution of the quality of the lots as manufactured is altered to another frequency distribution

[17] In a course in the sampling of human populations, the remainder of this chapter may be omitted at the discretion of the instructor.

[18] For samples drawn without replacement the maximum likelihood proportion is given in the exercise on page 294.

in which poor quality is presumably less in evidence. The calculation of the shape of this new distribution can be carried out for any rule of acceptance if we know the frequency distribution of the quality as manufactured. The problem is one in inverse probability.

An example in inverse probability.[19] A supply holds equal numbers of similar black and white balls. In each experiment, a blindfolded man picks 10 balls out of the supply, one at a time, and places them in a "lot" container. The contents of the supply are kept constant by an assistant who watches and immediately, at every draw, puts into the supply a ball of the same color as the one just drawn, and stirs. The blindfolded man then draws a sample of 4 balls without replacement from the lot-container and places them in the "sample" container. Their colors are recorded on a card by the assistant, who thereupon empties the lot-container and the sample-container in preparation for the next experiment. The experiment is repeated again and again. The arrangement is depicted schematically in Fig. 16 on page 255.

> *Problem.* For those samples having 3 black and 1 white ball, find by calculation the "expected" number of times that the lot-container has had in it 0, 1, 2, 3, 4, 5, 6, 7, 8, 9, and 10 black balls.

On each card the assistant records the information shown in Table 4, either by punching or by encircling the proper number in each column.

TABLE 4.   RECORD OF THE EXPERIMENTS

The particular card shown in this table records the data on Experiment 347, in which the lot-container showed 6 black and 4 white balls, and the sample showed 3 black and 1 white.

Number of the experiment	Lot-container Black	Lot-container White	Sample Black	Sample White
0 0 0 0 0	0	0	0	0
1 1 1 1 1	1	1	1	●
2 2 2 2 2	2	2	2	2
3 3 ● 3 3	3	3	●	3
4 4 4 ● 4	4	●	4	4
5 5 5 5 5	5	5	5	5
6 6 6 6 6	●	6	6	6
7 7 7 7 ●	7	7	7	7
8 8 8 8 8	8	8	8	8
9 9 9 9 9	9	9	9	9

After 32,000 experiments have been recorded, the cards are sorted into 5 piles, according to the number of black balls (r) drawn into the sample.

---

[19] Borrowed with modifications from an example in Fry's *Probability* (Van Nostrand, 1928), p. 123. This same example, varied, served as the basis for Chapter 7.

The "expected" results are depicted in Fig. 32, for which we shall now make the calculations. As the supply is held constant by the assistant, the probability is $P(x, n) = \binom{n}{x} q^{n-x}p^x$ that x of the n balls placed in the lot-container are black, this being the xth term in the binomial expansion of $(q + p)^n$ (Ch. 4, p. 113). Now, having put the n balls into the lot-container, the blindfolded man draws therefrom a sample of m (= 4) balls, *without replacement*. What is the probability of r having

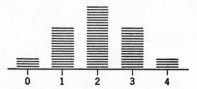

FIG. 32. Showing the relative theoretical frequencies of occurrence of 0, 1, 2, 3, and 4 black balls in the 32,000 samples of 4.

the value 0, 1, 2, $\cdots$, m? As there are initially x black balls and $n - x$ white balls in the lot-container, this probability is

$$P(r|x) = \frac{\binom{x}{r}\binom{n-x}{m-r}}{\binom{n}{m}} \qquad \text{[P. 113]} \quad (22)$$

The probability that there will be x black balls in the lot-container *and* r black balls in the sample is obtained by compounding the two separate probabilities by multiplication, which gives

$$P(r, m; x, n) = \frac{n!}{x!(n-x)!}p^x q^{n-x} \frac{\binom{x}{r}\binom{n-x}{m-r}}{\binom{n}{m}} \qquad (23)$$

$$= \left\{ \frac{m!}{r!(m-r)!} q^{m-r}p^r \right\}$$

$$\times \left\{ \frac{(n-m)!}{(x-r)!(n-m-x+r)!} q^{\overline{n-m}-\overline{x-r}}p^{x-r} \right\} \quad (24)$$

Remark 1. This interesting result consists of the product of two binomial terms, in spite of the fact that the sample is drawn from the lot-container without replacement. The first brace is the rth term of the expansion of $(q + p)^m$, and the second is the $(x - r)$th term of the expansion of $(q + p)^{n-m}$. Thus the probability of the compound event of:

i. Getting x black balls into the lot-container by drawing n balls from a constant supply holding black and white balls in the ratio $p:q$, and then

ii. Getting r black balls in a sample of m balls drawn without replacement from the lot-container;

is equal to the probability of the compound event of:

i. Getting r black balls in a sample of m drawn with replacement from a constant supply of proportions p and q, and then

ii. Getting $x - r$ black balls in another sample of $n - m$ drawn also with replacement from the same supply.

Remark 2. You can not draw more black or white balls out of the lot-container than there are in it; hence Eq. 24 for the probability of the compound event should give 0 for $r > x$ or for $(m - r) > (n - x)$, as in fact it does. This is brought about automatically by the factors $(x - r)!$ and $(n - m - x + r)!$ in the denominators (p. 487).

Now suppose we ask what is the probability of getting r black balls in the sample no matter how many black balls are in the lot-container. To answer this question we take the sum of the probabilities given by Eq. 24 for $x = 0, 1, 2, \cdots, n$ and get

$$
\begin{aligned}
P(r, m) &= \binom{m}{r} p^r q^{m-r} \sum_{x=0}^{n} \frac{(n - m)!}{(x - r)!(n - m - x + r)!} p^{x-r} q^{\overline{n-m}-\overline{x-r}} \\
&= \binom{m}{r} p^r q^{m-r} \sum_{s=0}^{n'} \frac{n'!}{s!(n' - s)!} p^s q^{n'-s} \qquad \begin{bmatrix} s = x - r \\ n' = n - m \end{bmatrix} \\
&= \binom{m}{r} p^r q^{m-r} \qquad\qquad\qquad\qquad\qquad\qquad (25)
\end{aligned}
$$

for the chance that the sample of m will contain r black balls, regardless of how many were in the lot-container.

Remark 3. This result, being merely the rth term of the binomial expansion of $(q + p)^m$, shows that the distribution of black balls in samples drawn from the lot-container *without* replacement is the same as if the samples had been drawn *with replacement directly from the supply* (Case B, p. 254).

Remark 4. As the result in Eq. 25 is merely the first brace in Eq. 24, it follows that *i.* the first brace of Eq. 24 is precisely the proportion of experiments showing r black balls in the sample regardless of the contents of the lot-container, and that *ii.* the second brace shows how these particular experiments are further classified by the number of black balls in the lot-container.

Having made our theoretical calculation, we proceed to answer the question set forth in the problem by putting $m = 4$ and $p = q = \frac{1}{2}$.

r may take on the values 0, 1, 2, 3, 4, the chances for which by Eq. 25 are merely the five terms of the binomial expansion

$$(\tfrac{1}{2} + \tfrac{1}{2})^4 = (\tfrac{1}{2})^4 \{1 + 4 + 6 + 4 + 1\} \tag{26}$$

Thus, if 32,000 experiments had been performed, the expected frequencies in Fig. 32 would be

$$2000; \; 8000; \; 12{,}000; \; 8000; \; 2000$$

In each pile the cards are alike with respect to r, the number of black balls in the sample, but not with respect to x, the number of black balls in the lot-container. In the pile of cards showing no black balls in the sample (at the extreme left of Fig. 32), for example, some of the cards show that the lot-container had no black balls in it; other cards right in the same pile show that the lot-container had 1 black ball and 9 white in it; etc. (But there is no card in that pile corresponding to 10 black and no white balls. Why?)

We are now ready to attack the problem that was posed. Suppose that the pile of 8000 cards made up from those experiments that show 3 black balls ($r = 3$) in each sample of 4 ($= m$) is itself sorted into piles corresponding to the contents of the lot-container which might have had 3, 4, 5, 6, 7, 8, or 9 black balls in it (but not 0, 1, 2, or 10; why?). The answer is contained in the second brace of Eq. 24. We need only insert the numbers $n = 10$, $p = q = \tfrac{1}{2}$, $m = 4$, and $r = 3$, and find that

$$P(3, 4; x, 10) = \left\{ \frac{4!}{1!3!} \frac{1}{2^4} \right\} \left\{ \frac{6!}{(x-3)!(9-x)!} \frac{1}{2^6} \right\} \tag{27}$$

The first brace has the value $4/2^4$, or $\tfrac{1}{4}$. It is to be identified with the 3d term in the expansion of $(\tfrac{1}{2} + \tfrac{1}{2})^4$ in the preceding equation. When multiplied by 32,000 this term gives 8000, which is the "expected" number of experiments showing 3 black and 1 white (as was stated before). The second brace is 0 for $x = 0$, 1, 2, and 10 (see Remark 2) but takes on the values of the terms of the expansion of $(\tfrac{1}{2} + \tfrac{1}{2})^6$ for the seven values of x between $x = 3$ and $x = 9$, as shown in Table 5. These seven values of the second brace tell how the 8000 cards are divided up amongst the seven possible contents of the lot-container. Table 5 contains the answers to the problem as stated.

Table 6 shows a continuation of the problem. In samples of 4 there can be 0, 1, 2, 3, or 4 black balls. We have just worked out the problem of the distribution of the lot-container for samples of 3 black balls, and in Table 6 the results are tabulated for the other possibilities as well. As an exercise the student should work out all the values shown.

TABLE 5.   DISTRIBUTION OF THE CONTENTS OF THE LOT-CONTAINERS THAT GIVE
FORTH SAMPLES OF 4, SHOWING 3 BLACK AND 1 WHITE

(1)	(2)	(3)	(4)
x (Number of black balls in the lot-container)	Value of the second brace in Eq. 27	$P(3, 4; x, 10)$ in Eq. 27	Expected number of experiments
All values of x	1	$\frac{1}{4}$	8000
0	0	0	0
1	0	0	0
2	0	0	0
3	$\frac{1}{64}$	$\frac{1}{256}$	125
4	$\frac{6}{64}$	$\frac{6}{256}$	750
5	$\frac{15}{64}$	$\frac{15}{256}$	1875
6	$\frac{20}{64}$	$\frac{20}{256}$	2500
7	$\frac{15}{64}$	$\frac{15}{256}$	1875
8	$\frac{6}{64}$	$\frac{6}{256}$	750
9	$\frac{1}{64}$	$\frac{1}{256}$	125
10	0	0	0

NOTE.   8000 is the expected number of samples showing 3 black and 1 white.
Col. 2 shows the expected proportions of the 8000 samples coming from the various
contents of the lot-container.   Col. 3 shows these same proportions referred to the
entire 32,000 experiments.   The last column is obtained by multiplying Col. 3 by
32,000.

A larger sample from the lot-container.   Suppose that 4 more balls
are drawn, making 8 all together.   Suppose that 6 are black and 2
white; that, in other words, the sample is twice as big as before but still
shows the same proportion black.   What are the relative frequencies with
which samples like this come from lot-containers holding 2 black balls?
3? 4, etc.?   We return to Eq. 24 and insert $n = 10$, $p = q = \frac{1}{2}$, $m = 8$,
$r = 6$, and find that

$$P(6, 8; x, 10) = \left\{ \frac{8!}{2!6!} \frac{1}{2^8} \right\} \left\{ \frac{2!}{(x - 6)!(8 - x)!} \frac{1}{2^2} \right\} \tag{28}$$

The first brace is the 8th term in the expansion of the binomial $(\frac{1}{2} + \frac{1}{2})^8$.
The second brace is 0 for $x = 0, 1, 2, 3, 4, 5, 9, 10$, but takes on the
values of the expansion of $(\frac{1}{2} + \frac{1}{2})^2$ for $x = 6, 7, 8$.   The product of the
two braces is shown in Table 7 in the column $r = 6$.   This table also
shows the distribution of the lot-container for all other possible numbers
of black balls in samples of 8.   The student should verify these calculations.

Table 7 should be compared carefully with Table 6.   See how much more decisive the probabilities are in Table 7 because of the larger sample-size.

Influence of the initial probabilities.   The reader should be reminded that the results obtained in Tables 6 and 7 would have been different if

TABLE 6.   THE "EXPECTED" PROPORTIONATE DISTRIBUTION OF THE CONTENTS OF THE LOT-CONTAINER AND THE SAMPLES DRAWN THEREFROM

Sample-size, $m = 4$.   Probabilities worked out from Eq. 24

x (Number of black balls in the lot-container)	r, the number of black balls in the sample						
	0–4 incl.	0	1	2	3	4	
0–10 incl.	16	1	4	6	4	1	All by $\frac{1}{16}$
0	1	1	0	0	0	0	
1	10	6	4	0	0	0	
2	45	15	24	6	0	0	
3	120	20	60	36	4	0	
4	210	15	80	90	24	1	All by 2^{-10}
5	252	6	60	120	60	6	
6	210	1	24	90	80	15	
7	120	0	4	36	60	20	
8	45	0	0	6	24	15	
9	10	0	0	0	4	6	
10	1	0	0	0	0	1	

there had been different proportions of black and white balls in the supply.   The influence of the initial probabilities is particularly great for small samples, such as samples of 4.   But the initial probabilities (the supply) have less and less influence on the final results as the size of the sample increases; and in the limit, for large samples the final result is practically independent of the prior probabilities.   It follows that for large samples from the lot-container it matters little whether we have precise knowledge regarding the supply and the way in which the lot-container was filled.   It follows, too, that in the absence of precise knowledge of the supply, to be sure of the contents of the lot-container it is necessary to go to large samples—large enough to overwhelm our ignorance of the supply.

The reader should be aware of the fact that the calculations made for Tables 6 and 7 could not have been made without knowledge of the supply and the manner in which the lot-container was filled. This fact is observed in the presence of p and q in all the formulas that were used.

There have been attempts in the history of probability to substitute the doctrine of "the equal distribution of ignorance" (Boole) for positive

TABLE 7. THE "EXPECTED" PROPORTIONATE DISTRIBUTION OF THE CONTENTS OF THE LOT-CONTAINER AND THE SAMPLES DRAWN THEREFROM

Sample-size, $m = 8$. Probabilities worked out from Eq. 24

x (Number of black balls in the lot-container)	0–8 incl.	r, the number of black balls in the sample									
		0	1	2	3	4	5	6	7	8	
0–10 incl.	256	1	8	28	56	70	56	28	8	1	All by 2^{-8}
0	1	1	0	0	0	0	0	0	0	0	
1	10	2	8	0	0	0	0	0	0	0	
2	45	1	16	28	0	0	0	0	0	0	
3	120	0	8	56	56	0	0	0	0	0	
4	210	0	0	28	112	70	0	0	0	0	
5	252	0	0	0	56	140	56	0	0	0	All by 2^{-10}
6	210	0	0	0	0	70	112	28	0	0	
7	120	0	0	0	0	0	56	56	8	0	
8	45	0	0	0	0	0	0	28	16	1	
9	10	0	0	0	0	0	0	0	8	2	
10	1	0	0	0	0	0	0	0	0	1	

knowledge regarding the probabilities of the composition of the lot-container. If nothing is known regarding the way the lot-container is filled, it has been argued, then black is as likely as white, and all compositions are equally probable. Such reasoning is of course fallacious and will lead to disrepute of both the user and the method unless he is rescued either by big samples or just plain good luck.

Exercise 1. A bag contains 3 balls, each of which is either black or white, the 4 possible combinations of black and white all having the same probabilities. (The equal probabilities could be produced by filling the bag one time after another from a set of random numbers. Then 3, 2, 1, or 0 black balls would be placed in the bag according as the random number turned up 3, 2, 1, or 0; other digits would be counted as blanks.) Two balls are drawn at random from the bag and are found to be white. What are the odds that the remaining 1 is white? (*Answer:* 3:1.)

Solution

Let x be the number of black balls in the bag, and $P(x)$ the probability of x before any drawings are made from it. Then $P(x) = \frac{1}{4}$ for $x = 0, 1, 2, 3$, because the stipulation is that the bag be filled in such manner that all combinations of black and white are equally probable. The probability of selecting a bag that contains x black balls and $3 - x$ white balls, and then drawing therefrom 2 white balls running is

$$P(2, x) = P(x)P(2|x) = \frac{1}{4} \times 2 \times \frac{3-x}{3} \frac{3-x-1}{3-1}$$

If the remaining ball in the bag is white, $x = 0$; if it is black, $x = 1$. What we need to do is to take the ratio of $P(2, x)$ for $x = 0$ to $P(2, x)$ for $x = 1$; in other words, $P(2, 0):P(2, 1)$.

For $x = 0$, $P(2, x) = \frac{1}{4} \times 2 \times \frac{3}{3} \times \frac{2}{2} = \frac{1}{2}$

For $x = 1$, $P(2, x) = \frac{1}{4} \times 2 \times \frac{2}{3} \times \frac{1}{2} = \frac{1}{6}$

The ratio is 3:1, as stated. Note that $P(2, x) = 0$ if $x = 2$ or 3. Why?

Remark. The interpretation of the results is as follows. If one selects for consideration only those experiments that yielded 2 white balls in 2 drawings, in three-quarters of them the remaining ball would be white, and in one-quarter of them it would be black.

One might plot 2 curves (or sets of points, rather), one showing $P(x)$, and the other showing ordinates proportional to $P(2, x)$, for $x = 0, 1, 2, 3$. The former is often spoken of as the *prior* probabilities, and the latter as the *posterior* probabilities. The points are plotted in Fig. 33.

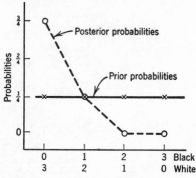

Fig. 33. Showing the distribution of the four different combinations of black and white balls in the bag:

× Solid curve, the prior probabilities, or the distribution of the contents of the bag as filled repeatedly, regardless of what was drawn out of it.

○ Dashed curve, the posterior probabilities, or the distribution of the contents of the bags from which samples of 2 were both white.

Exercise 2. Suppose that the bag of Exercise 1 contained originally 1,000,001 balls, the bag being filled in such manner that all 1,000,002 combinations of black and white balls are equally probable. 1,000,000 balls are drawn from the bag, and all are found to be white. What are the odds that the remaining ball is white? [*Answer:* $(10^6 + 1):1$.]

Remark. The last two exercises bring out important links in the history of the understanding of inverse probability. Chrystal (*Trans. Actuarial Soc. Edinburgh*, vol. ii, 1891: p. 421) made the statement, "Anyone who knows the definition of mathematical probability, . . . will not hesitate for a moment to say that the chance is half, that is to say, that the third ball is just as likely to be white as black." The fallacy of his argument was disclosed by Whittaker (*Trans. Faculty Actuaries Scotland*, vol. viii, 1920: pp. 163–74), who in the same article also proposed and gave the answer to the second of the two preceding problems.

Exercise 3. Suppose that the 3 balls in the bag were drawn from a very large supply containing equal numbers of black and white balls, after careful shuffling. Two balls are drawn at random from the bag and found to be white. What are the odds that the remaining one is white?

Exercise 4. Solve the preceding exercise if the supply contained initially 6 black and 4 white balls.

CHAPTER 10. ESTIMATION OF THE PRECISION OF A SAMPLE

Fortunately I learned a good many years ago that mathematicians are not always the strictly accurate creatures of popular belief. Being in Paris, I developed a cold, and, fearing a high temperature, bought a thermometer, which was naturally marked on the Centigrade scale, assuming that my companion, who had taken a first class in the Mathematical Tripos, would be able to translate it into Fahrenheit for me. When I asked him to do so he looked a little dubious, but sat down with pencil and paper. At the end of half an hour, he presented for my choice a variety of results, ranging from 93 deg. to 108 deg. F.—Edward Shanks, in the *Sunday Times* (London), 10 February 1946; a review of H. McKay's, *The World of Numbers* (Cambridge, 1946).

Need for appraisal of precision. The first two chapters described some of the problems encountered in arriving at a specification of the precision required for some particular characteristic that is to be observed in a proposed survey. A sample-design is a plan by which the specified precision will supposedly just about be met at lowest cost. After the returns of the survey are in comes the preparation of the estimates, and along with it, *the appraisal of the precision of these estimates.* It is the purpose of this chapter to learn how to make this appraisal.

The design of any sample is predicated on the basis of certain assumptions regarding the variance σ^2 of the universe, or its coefficient of variation, γ. After the survey is finished comes the question, what *is* the precision of the results? It ought to be $\gamma/\sqrt{n}$, but were our advance calculations of the precision carried out with the right value of γ? And were the instructions followed faithfully? Did the mistakes and necessary or willful departures from the specified procedures affect γ?

It has been found empirically (never mathematically) that certain mathematical rules correspond pretty closely to certain ideal but realizable procedures of sampling. Departures from instructions in a probability-sample may be presumed to alter the advance estimates of the variance. A statistical appraisal not only sheds light on a particular survey but also provides another link in a long chain of information that is extremely useful in the design of the next sample. By continuous appraisal of surveys and by keeping careful records of costs, the statistician acquires experience by which he may predict the effects of various alternative procedures. Without the theory of probability such studies are impossible.

Regardless of any wrong assumptions that were made in arriving at the necessary advance estimates of the variances and proportions that enter into the formulas for the sampling error, after the survey is completed there need be no guess-work concerning the precision of the final results. A statistical appraisal will tell what the precision really was, regardless of what it ought to have been. The purpose of this chapter is to learn how to make this appraisal. It will be made by studying the returns—usually by studying a *subordinate sample* of the returns. We must learn how to draw the subordinate sample.

The cost of measuring the reliability of a survey is a legitimate part of the cost of the survey.

It should be noted that what we need to do is to make an appraisal of precision from the consistency of the results. There is no other way to do it. Close agreement between the results of two surveys (e.g., comparison of a sample with census figures or other bench-mark data) is often cited as proof that one or both samples must be good. In the author's judgment, such comparisons are useful mainly for psychological purposes; they prove little or nothing, but it must be admitted that they are impressive to the layman, who has often been misled.

It is not conclusive to cite good agreement as proof that a sample is good, any more than it is conclusive to seize upon poor agreement as proof that sampling is and always will be unreliable and a delusion. In good sample design, agreement within 1 standard error is to be expected in about 2 samples out of 3; within 2 standard errors in about 95 samples out of 100; and within 3 standard errors practically always; and when samples are professionally designed and executed, this is inevitably what happens. Thus, when quality is built into a sample, good results will occur frequently, and wide discrepancies will also occur but only rarely, and the statistician knows how often to expect them. Disagreement beyond the 3-sigma limits points to differences in definitions, training, or auspices, to failure to follow instructions, or to some other cause *not connected with the procedure for selecting the sampling units*. In actual practice one does not always need to know the sampling error very accurately. It is usually sufficient to estimate it with a coefficient of variation of 10 or 20 or even 30 percent.

Throughout this chapter we must bear in mind that an estimate of precision is *not* a measure of the errors of response or nonresponse, nor a measure of the reliability of a forecast (p. 18).

Examples of published appraisals of precision occur on page 14 of the book *A Chapter in Population Sampling* (Bureau of the Census, 1947), and in the two examples of application given in Chapters 11 and 12 of this book. A more extensive appraisal was cited in a footnote on page 21.

Purpose of the chapter. The purpose of this chapter is to indicate how to draw a sample of the returns (questionnaires, preferably transferred to punched cards for numerical analysis) in order to estimate the variance of X, the estimated total population of the universe. The procedure will depend on the plan of sampling and the form of estimate used.

The theory to be given here measures only the sampling errors. Measurement of the nonsampling errors and studies of other indexes of reliability must be carried out by proper experimental design, as was mentioned in Chapters 1 and 2.

Suppose that the n elements of the sample have been drawn independently at random. Then if the universe is unstratified, the unbiased estimate of the total population will be

$$X = N\bar{x} \qquad \text{[Ch. 4]}$$

It is then only necessary to estimate the standard deviation σ from the sample and to use this estimate in the formula

$$\text{Var } X = \frac{N - n}{N - 1} N^2 \frac{\sigma^2}{n}$$

[N is the number of elements in the universe; n the number in the sample, supposedly drawn independently (not in clusters)]   [Ch. 4]

The question is then how to estimate σ. Results soon to be derived will show that if a subordinate random sample of size n' be drawn from the returns, and if its variance s'^2 be computed for the particular population-characteristic under discussion, then the

$$\left.\begin{array}{l} \text{Est'd Var } X = \dfrac{N - n}{N} \dfrac{N^2}{n} \dfrac{n's'^2}{n' - 1} = (N - n) \dfrac{N}{n} \dfrac{\displaystyle\sum_1^{n'} (x_i - \bar{x}')^2}{n' - 1} \\[4ex] \text{Est'd Var } \bar{x} = \dfrac{N - n}{N} \dfrac{1}{n} \dfrac{n's'^2}{n' - 1} = \dfrac{N - n}{N} \dfrac{1}{n} \dfrac{\displaystyle\sum_1^{n'} (x_i - \bar{x}')^2}{n' - 1} \end{array}\right\} \quad (1)$$

Here $n' =$ the size of the subordinate sample (the sample to be drawn from the returns). $n' - 1 =$ the degrees of freedom (p. 352).

$$\bar{x}' = \frac{1}{n'} \sum_1^{n'} x_i = \text{the average population per element in the subordinate sample.}$$

As will be seen, the estimate of Var X made from a subordinate sample of size n' (each element of which is supposed to be drawn independently at random) is subject to a coefficient of variation of $\sqrt{(\beta_2 - 1)/(n' - 1)}$, and σ_X is subject to half as big a coefficient of variation. (β_2 is, as usual, μ_4/σ^4 for the universe.) Hence if (e.g.) $\beta_2 = 8$ and $n' = 80$, then the estimate of the standard error of X has a coefficient of variation of about 15 percent, which is close enough for most uses. If by good fortune β_2 is only 5, so much the better. β_2, like σ, is sometimes known fairly well in advance, but the subordinate sample of n' returns will give new estimates of both σ and β_2, and from the latter a new value of n' may be indicated. It may turn out that the original subordinate sample should be augmented. Some of the theory that will be given later on in this chapter will be of assistance in controlling β_2 to reasonable levels.

We proceed now to derive and study the theoretical relation between the variance of a sample and the variance of the universe, under the assumption that the elements are drawn individually (not in clusters). This relation is needed in the foregoing equations and in those that are to follow later.

The "expected" variance in random sampling from a bowl-universe. Mathematically, the problem of estimating a variance in the simplest problem of sampling, viz., sampling at random from a single class (exemplified by drawing chips from a bowl), can be worked out in the following manner. As before, x_1 will denote the population recorded for the first chip of the sample, x_2 will denote the population for the second chip, etc. x_1, x_2, etc., are random variables, but regardless of randomness the following algebraic identities hold:

$$
\left.
\begin{aligned}
x_1 - \mu &= (x_1 - \bar{x}) + (\bar{x} - \mu) \\
x_2 - \mu &= (x_2 - \bar{x}) + (\bar{x} - \mu) \\
&\;\;\vdots \\
x_n - \mu &= (x_n - \bar{x}) + (\bar{x} - \mu)
\end{aligned}
\right\}
\tag{2}
$$

$\bar{x}$ is, as before, the mean of the n values x_1, x_2, $\cdots$, x_n. The left-hand member $x_i - \mu$ is the *error* in x_i. The first term in parenthesis on the right is the *residual* in x_i, and the second term is the error in $\bar{x}$, i.e., the *error in the mean*, $\bar{x}$. Now square each side of the n identities, add, and divide by n: the result is

$$
\frac{1}{n} \sum_1^n (x_i - \mu)^2 = \frac{1}{n} \sum_1^n (x_i - \bar{x})^2 + (\bar{x} - \mu)^2 + \frac{2}{n} (\bar{x} - \mu) \sum_1^n (x_i - \bar{x})
$$

The last term vanishes by definition of $\bar{x}$, and what is left can be written as the right-triangular relationship

$$
\delta^2 = s^2 + u^2
\tag{3}
$$

wherein by definition

$$
\delta^2 = \frac{1}{n} \sum_1^n (x_i - \mu)^2 = \text{the mean square } error \text{ in}
\tag{4}
$$
$$
\text{the sample}
$$

$$s^2 = \frac{1}{n} \sum_1^n (x_i - \bar{x})^2 = \text{the mean square } residual \qquad (5)$$
$$\text{or the } variance \text{ of the}$$
$$\text{sample}$$

$$u = \bar{x} - \mu \qquad = \text{the } error \text{ in the mean, } \bar{x} \qquad (6)$$

δ, s, and u vary from one sample to another, and by supposition they are random variables whose distributions have "expected" values and variances. From Eq. 4 the "expected" value of δ^2 must be

$$E\delta^2 = E \frac{1}{n} \sum_1^n (x_i - \mu)^2$$

$$= \frac{1}{n} \sum_1^n E(x_i - \mu)^2 = E(x_i - \mu)^2 = \sigma^2 \quad \begin{matrix} \text{[By definition} \\ \text{of } \sigma^2] \end{matrix} \quad (7)$$

Also,

$$Eu^2 = E(\bar{x} - \mu)^2 = \text{Var } \bar{x} \qquad \text{[Because } E\bar{x} = \mu]$$

$$= \frac{\sigma^2}{n} \qquad \text{with replacement} \qquad (8)$$

$$\qquad\qquad\qquad\qquad\qquad\qquad\qquad \text{[Ch. 4]}$$

$$= \frac{N - n}{N - 1} \frac{\sigma^2}{n} \quad \text{without replacement} \qquad (9)$$

It is now a simple matter to calculate Es^2, for by Eq. 3, $Es^2 = E\delta^2 - Eu^2$. Separate calculations will be required, depending on whether the sample is drawn with or without replacement.

With replacement

$$Es^2 = E\delta^2 - Eu^2$$

$$= \sigma^2 - \frac{\sigma^2}{n}$$

whence

$$Es^2 = \frac{n - 1}{n} \sigma^2 \qquad (10)$$

Thus one may write

$$\sigma''^2 = \frac{n}{n - 1} s^2 \qquad (11)$$

as an unbiased estimate of σ^2. This is the Gauss estimate of σ^2; compare with pages 522 and 537.

Without replacement

$$Es^2 = E\delta^2 - Eu^2$$

$$= \sigma^2 - \frac{N - n}{N - 1} \frac{\sigma^2}{n}$$

whence

$$\frac{n}{n - 1} Es^2 = \frac{N}{N - 1} \sigma^2 \qquad (10')$$

Thus one may write

$$\sigma''^2 = \frac{n}{n - 1} \frac{N - 1}{N} s^2 \qquad (11')$$

as an unbiased estimate of σ^2. The reader will recognize this estimate as the same one obtained in Eq. 71 of Chapter 5, and designated by the same symbol.

It follows from Eq. 11′ that for a subordinate sample of size n' and variance s'^2, drawn without replacement from a sample of n returns of variance s^2, which in turn were drawn independently (not in clusters) and without replacement from a universe of variance σ^2,

$$\frac{n'}{n'-1} E s'^2 = \frac{n}{n-1} s^2 \tag{12}$$

$$\frac{n'}{n'-1} E E s'^2 = \frac{n}{n-1} E s^2 = \frac{N}{N-1} \sigma^2 \tag{13}$$

where EE denotes the "expected" value of Es'^2 arising from supposed repetitions of the double sampling procedure, viz., i. draw the sample of n returns from the universe; then ii. draw the subordinate sample of n' returns from the first sample, both without replacement. Thus

$$\frac{n's'^2}{n'-1} \quad \text{or} \quad \frac{\sum_1^{n'} (x_i - x')^2}{n'-1}$$

is an unbiased estimate of $\dfrac{N}{N-1} \sigma^2$, and this substitution gives Eq. 1, the use of which is thus justified.

Simple illustration of some of the preceding theory. It will be useful now to perform a simple theoretical experiment to illustrate some of the theory in the last section and previous sections. Take a universe consisting of only 4 chips, labeled 1, 2, 3, 4, as shown in Fig. 34. List all

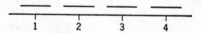

$$\underset{1}{\rule{2em}{0.4pt}} \quad \underset{2}{\rule{2em}{0.4pt}} \quad \underset{3}{\rule{2em}{0.4pt}} \quad \underset{4}{\rule{2em}{0.4pt}}$$

FIG. 34. The universe for an experiment. It consists of four chips labeled 1, 2, 3, 4.

Mean, $\mu = 2.5$

Variance, $\sigma^2 = 1.25$

This is a *rectangular universe* because all the proportions are equal.

possible samples of 2, to see what results would be obtained for the average mean square error, the average standard deviation, and the variance of the sample mean, in samples drawn with and without replacement. The resulting probability-distribution is displayed in the accompanying table.

Take samples of 2. For sampling with replacement use all figures. For sampling without replacement delete those that are starred. $(n = 2)$

Sample	$s^2 = \dfrac{1}{n}[(x_1 - \bar{x})^2 + (x_2 - \bar{x})^2]$	$\delta^2 = \dfrac{1}{n}[(x_1 - \mu)^2 + (x_2 - \mu)^2]$	$(\bar{x} - \mu)^2$	$\bar{x}$
1, 1 *	0 *	9 ÷ 4 *	$\frac{9}{4}$ *	1 *
2	$\frac{1}{4}$	5	1	1.5
3	1	5	$\frac{1}{4}$	2
4	$\frac{9}{4}$	9	0	2.5
2, 1	$\frac{1}{4}$	5	1	1.5
2 *	0 *	1 *	$\frac{1}{4}$ *	2 *
3	$\frac{1}{4}$	1	0	2.5
4	1	5	$\frac{1}{4}$	3
3. 1	1	5	$\frac{1}{4}$	2
2	$\frac{1}{4}$	1	0	2.5
3 *	0 *	1 *	$\frac{1}{4}$ *	3 *
4	$\frac{1}{4}$	5	1	3.5
4, 1	$\frac{9}{4}$	9	0	2.5
2	1	5	$\frac{1}{4}$	3
3	$\frac{1}{4}$	5	1	3.5
4 *	0 *	9 *	$\frac{9}{4}$ *	4 *
Totals				
Starred nos. incl.	10	20	10	40
Starred nos. excl.	10	15	5	30
Averages				
Starred nos. incl.	$\frac{5}{8}$	$\frac{5}{4}$	$\frac{5}{8}$	2.5
Starred nos. excl.	$\frac{5}{6}$	$\frac{5}{4}$	$\frac{5}{12}$	2.5

The conclusions to be drawn are listed below.

Sampling with replacement

$$\begin{cases} \dfrac{n}{n-1}\, \text{Av } s^2 = \dfrac{2}{1} \times \dfrac{5}{8} = 1.25, \text{ which is } \sigma^2 \\ \text{This illustrates } E\sigma''^2 = \sigma^2. \end{cases}$$ [Eq. 11]

$$\begin{cases} \text{Var } \bar{x} = \text{Av } (\bar{x} - \mu)^2 = \frac{5}{8} \\ \dfrac{\sigma^2}{n} = \dfrac{1.25}{2} = \dfrac{5}{8} \\ \text{This illustrates } \text{Var } \bar{x} = \sigma^2/n. \end{cases}$$ [Eq. 8]

$$\begin{cases} \text{Av } \delta^2 = 1.25 \\ \text{This illustrates } E\delta^2 = \sigma^2. \end{cases} \qquad \text{[Eq. 7]}$$

$$\begin{cases} \text{Av } \bar{x} = 2.5 \\ \text{This illustrates } E\bar{x} = \mu. \end{cases} \qquad \text{[P. 74]}$$

Sampling without replacement

$$\begin{cases} \dfrac{n}{n-1}\dfrac{N-1}{N}\text{ Av } s^2 = \dfrac{2}{1}\dfrac{3}{4}\dfrac{5}{6} = \dfrac{5}{4} = 1.25 \\ \text{This illustrates } E\sigma''^2 = \sigma^2. \end{cases} \qquad \text{[Eq. 11']}$$

$$\begin{cases} \quad \text{Var } \bar{x} = \text{Av }(\bar{x} - \mu)^2 = \tfrac{5}{12} \\ \dfrac{N-n}{N-1}\dfrac{\sigma^2}{n} = \dfrac{4-2}{4-1}\dfrac{1.25}{2} = \dfrac{2}{3}\dfrac{5}{4}\dfrac{1}{2} = \dfrac{5}{12} \\ \text{This illustrates } \text{Var } \bar{x} = \dfrac{N-n}{N-1}\dfrac{\sigma^2}{n}. \end{cases} \qquad \text{[Eq. 9]}$$

$$\begin{cases} \text{Av } \delta^2 = 1.25 \\ \text{This illustrates } E\delta^2 = \sigma^2. \end{cases} \qquad \text{[Eq. 7]}$$

$$\begin{cases} \text{Av } \bar{x} = 2.5 \\ \text{This illustrates } E\bar{x} = \mu. \end{cases} \qquad \text{[P. 74]}$$

Exercise 1. Turn to the standard deviations in Tables 1, 2, and 3 on pages 400 ff, which illustrate the multinomial, and show that they obey the equation

$$Es^2 = \frac{n-1}{n}\sigma^2$$

HINT: Es^2 is to be calculated as the average s^2 in the distributions shown in the aforementioned tables. This can be done by summing the products of s^2 by their respective term-values or, easier, by going to the summary tables and forming the products there. Thus, for samples of 2 the products of s^2 by the respective term-values would be those shown in the following table. The Roman numerals I, II, and III refer to the universe (Fig. 47, p. 439). Thus the theoretical relation between Es^2 and σ^2 is illustrated. The student should perform similar calculations for $n = 3$ and $n = 5$. The variances of s, δ, s^2, and δ^2 will be developed in the following exercises.

s^2	I	II	III
0	0	0	0
0	0	0	0
0	0	0	0
0.25	0.06	0.09	0.09
1	0.08	0.06	0.12
0.25	0.06	0.03	0.015
$Es^2 = 0.20$		0.18	0.225
$\sigma^2 = 0.40$		0.36	0.45
$\dfrac{n-1}{n}\sigma^2 = 0.20$		0.18	0.225

Exercise 2. Find general expressions for the variance and coefficient of variation of δ^2 (and hence of any estimate of σ^2 that is a multiple of δ^2) for samples drawn with replacement from any universe whatever that possesses 2d and 4th moments. (δ^2 is the mean square error, as defined on p. 332.)

Answer:
$$\left. \begin{array}{l} \text{Var } \delta^2 = \dfrac{1}{n}\,(\mu_4 - \sigma^4) = \dfrac{\beta_2 - 1}{n}\,\sigma^4 \\[3ex] \text{C.V. } \delta^2 = \sqrt{\dfrac{\beta_2 - 1}{n}} \end{array} \right\} \tag{14}$$

wherein μ_4 and σ^2 are the 4th moment coefficient and variance of the original distributions, and $\beta_2 = \mu_4 : \sigma^4$.

Solution

A sample of n is drawn, with replacement. Let each x be measured from the mean of the universe to simplify the calculations. Then

$$\delta^4 = (\delta^2)^2 = \left(\frac{1}{n}\,\Sigma\,x_i{}^2\right)^2 \qquad [\delta^2 \text{ defined on p. 332}]$$

$$= \frac{1}{n^2}\,(\Sigma\,x_i{}^2)^2 \qquad \begin{array}{l}[\text{All sums run from 1 to } n \text{ unless} \\ \text{otherwise stated}]\end{array}$$

$$= \frac{1}{n^2}\left(\Sigma\,x_i{}^4 + \sum_{i=1}^{n}\sum_{j=1}^{n} x_i{}^2 x_j{}^2\right) \qquad [j \neq i]$$

By definition, $Ex_i{}^4 = \mu_4$ (Ch. 3) and $E\,\Sigma\,x_i{}^4 = \Sigma\,Ex_i{}^4 = n\mu_4$ because there are n terms in the summation. $Ex_i{}^2 = \sigma^2$ by definition, and $E\,\Sigma\,x_i{}^2 = n\sigma^2$. Also

$$E\sum_{i=1}^{n}\sum_{j=1}^{n} x_i{}^2 x_j{}^2 = n(n-1)Ex_i{}^2 Ex_j{}^2$$
$$= n(n-1)\sigma^4$$

because there are $n(n-1)$ terms in the double sum. Moreover, by supposition the drawing is done with replacement, so x_i and x_j are independent and the "expected" value of $x_i{}^2 x_j{}^2$ is the product of $Ex_i{}^2$ by $Ex_j{}^2$, both of which are equal to σ^2. Therefore,

$$E\delta^4 = \frac{\mu_4}{n} + \frac{n(n-1)}{n^2}\sigma^4$$

$$\text{Var } \delta^2 = E\delta^4 - (E\delta^2)^2$$

$$= \frac{1}{n}(\mu_4 - \sigma^4) = \frac{1}{n}(\beta_2 - 1)\sigma^4$$

$$(\text{C.V. } \delta^2)^2 = \frac{\text{Var } \delta^2}{(E\delta^2)^2} = \frac{\text{Var } \delta^2}{\sigma^4} \quad \begin{bmatrix}\text{Because } E\delta^2 = \sigma^2, \text{ by} \\ \text{definition of } \sigma^2\end{bmatrix}$$

$$= \frac{1}{n}(\beta_2 - 1) \qquad\qquad Q.E.D.$$

It is to be noted that for samples from a normal universe $\mu_4 = 3\sigma^4$ (p. 64), and the formulas obtained here reduce to

$$\left.\begin{array}{l}\text{Var } \delta^2 = \dfrac{2}{n}\sigma^4 \qquad \text{[Normal universe]} \\[2ex] \text{C.V. } \delta^2 = \sqrt{\dfrac{2}{n}} \qquad [\quad \text{``} \qquad \text{``} \quad]\end{array}\right\} \quad (15)$$

as will be found also on pages 529 and 531 from normal distribution theory.

Exercise 3. Find a general expression for the variance and coefficient of variation of s^2 (and hence of any estimate of σ^2 that is a multiple of s^2) for samples drawn with replacement from any universe that possesses 2d and 4th moments.

Answers:

$$\text{Var } s^2 = \sigma^4\left\{\frac{\beta_2 - 1}{n} - \frac{2}{n^2}(\beta_2 - 2) + \frac{\beta_2 - 3}{n^3}\right\} \rightarrow \sigma^4\frac{\beta_2 - 1}{n} \qquad (16)$$

$$\text{C.V. } s^2 = \text{C.V. } \sigma''^2$$

$$= \sqrt{\frac{\beta_2 - 1}{n - 1} - \frac{\beta_2 - 3}{(n-1)^2}\frac{n+1}{n}} \rightarrow \sqrt{\frac{\beta_2 - 1}{n - 1}} \quad [\text{where } \sigma'' = \lambda s] \quad (17)$$

$$\text{C.V. } s = \text{C.V. } \sigma'' \rightarrow \frac{1}{2}\sqrt{\frac{\beta_2 - 1}{n - 1}} \qquad\qquad [\text{For large } n] \quad (18)$$

Solution

A sample of n is drawn, with replacement. Let each x_i be measured from the mean of the universe to simplify the calculations. Then (subscripts omitted except where necessary)

$$s^4 = (s^2)^2 = \left(\frac{1}{n}\Sigma\,x^2 - \bar{x}^2\right)^2$$

$$= \left(\frac{\Sigma\,x^2}{n}\right)^2 - \frac{2\,\Sigma\,x^2}{n}\left(\frac{\Sigma\,x}{n}\right)^2 + \left(\frac{\Sigma\,x}{n}\right)^4 \quad \text{[All sums run from 1 to } n \text{ unless otherwise indicated]}$$

$$= \frac{1}{n^2}\left[\Sigma\,x^4 + \sum_{i=1}^{n}\sum_{j=1}^{n} x_i^2 x_j^2\right] \qquad\qquad [j \neq i]$$

$$-\frac{2}{n^3}\left[\Sigma\,x^4 + \sum_{i=1}^{n}\sum_{j=1}^{n} x_i^2 x_j^2\right]$$

$$+\frac{1}{n^4}\left[\Sigma\,x^4 + 3\sum_{i=1}^{n}\sum_{j=1}^{n} x_i^2 x_j^2\right]$$

+ other terms in odd powers whose "expected" values are zero

Then

$$Es^4 = \frac{1}{n}\mu_4 + \frac{n(n-1)}{n^2}\sigma^4 - \frac{2}{n^2}\mu_4 - \frac{2n(n-1)}{n^3}\sigma^4 + \frac{1}{n^3}\mu_4 + \frac{3n(n-1)}{n^4}\sigma^4$$

$$= \sigma^4 + \frac{\mu_4 - 3\sigma^4}{n} - \frac{2\mu_4 - 5\sigma^4}{n^2} + \frac{\mu_4 - 3\sigma^4}{n^3}$$

Proceeding,

$$\text{Var } s^2 = Es^4 - (Es^2)^2$$

$$= Es^4 - \left(\frac{n-1}{n}\sigma^2\right)^2$$

$$= Es^4 - \left(1 - \frac{2}{n} + \frac{1}{n^2}\right)\sigma^4$$

$$= \frac{\mu_4 - \sigma^4}{n} - 2\frac{\mu_4 - 2\sigma^4}{n^2} + \frac{\mu_4 - 3\sigma^4}{n^3}$$

which is equivalent to Eq. 16. Going on,

$$(\text{C.V. } s^2)^2 = \frac{\text{Var } s^2}{(Es^2)^2} = \frac{\text{Var } s^2}{\left(\dfrac{n-1}{n}\sigma^2\right)^2} = \frac{\text{Var } s^2}{\sigma^4}\left(\frac{n}{n-1}\right)^2$$

$$= \frac{n(\beta_2 - 1)}{(n-1)^2} + \frac{4 - 2\beta_2}{(n-1)^2} + \frac{\beta_2 - 3}{n(n-1)^2}$$

$$= \frac{\beta_2 - 1}{n-1} - \frac{\beta_2 - 3}{(n-1)^2}\,\frac{n+1}{n}$$

which is equivalent to Eq. 17. Eq. 18 follows from the fact that C.V. $s =$ $\frac{1}{2}$ C.V. s^2 very closely (from Ex. 3f on p. 133).

Note that *for a normal universe* $\beta_2 = 3$ (p. 64) and any term containing $\beta_2 - 3$ drops out regardless of n, wherefore

For a normal universe

$$\text{Var } s^2 = \frac{2\sigma^4}{n}\left(1 - \frac{1}{n}\right) \qquad \text{[No approximation]}$$

$$\text{C.V. } s^2 = \text{C.V. } \sigma''^2 = \sqrt{\frac{2}{n-1}} \quad \text{[No approximation]}$$

$$\text{C.V. } s = \frac{1}{\sqrt{2(n-1)}} \qquad \begin{array}{l}\text{[Slight approximation,}\\ \text{p. 133]}\end{array}$$

Size of sample required for an estimate of a standard error. The theoretical results just reached are of great importance. As noted in the above exercise, C.V. $\sigma'' = $ C.V. s, provided only that σ'' is some multiple of s. Hence a discussion of the subordinate sample that will be required to provide an estimate of σ with a coefficient of variation of x percent may be confined to the coefficient of variation of s. The quantity β_2 is obviously the important characteristic of the universe in this discussion. The higher β_2, the larger the sample required to give a desired C.V. σ''. For a normal universe β_2 has the value 3, but universes are never normal. In practice, β_2 may be as low as 2 and easily as high as 8 or 10 or much higher (*vide infra*, two sections ahead).

For a numerical illustration of the foregoing theory, suppose that a sample of households has been interviewed, and that an estimate of Var $\bar{x}$ is to be made, where $\bar{x}$ is the estimated number of rooms per household. Suppose that the original sample has been drawn at random one dwelling unit at a time, i.e., not in clusters (Ch. 5), and not systematically. Let us compute the required size n' for the subordinate sample.

The subordinate sample will not only provide an estimate of σ but of β_2 as well, but to commence with it will be necessary to assume some likely value of β_2. It will be supposed, for purposes of illustration, that on the basis of previous experience β_2 is thought to be not greater than 8. The standard error of $\bar{x}$ is desired with a coefficient of variation of about 15 percent. The size n' of the subordinate sample is then computed by using the results of the last exercise. The symbol n appearing there is now our n'; wherefore

$$\text{C.V. } s = \text{C.V. } \sigma'' = \frac{1}{2}\sqrt{\frac{\beta_2 - 1}{n' - 1}}$$

is to be set equal to 15 percent, with $\beta_2 = 8$. A mental computation shows that n' is to be about 80; and, if the size n of the original sample were (say) $n = 1000$, the interval for drawing the subordinate sample would be $n/n' = 1000/80 = 12$. The rule might be, "Start with the 6th household; take it and every 12th thereafter." Eq. 1 would then be used for estimating Var $\bar{x}$. If the estimate of β_2 turns out to lie between 20 and 40, the subordinate sample should be quadrupled. The easiest way to augment the first one is to take another subordinate sample by starting with the 3d return and taking every 12th thereafter. The new 2d and 4th moment coefficients would be computed by making use of Eq. 12 in Chapter 3, by which the effort put into the first set of calculations need not be lost.

In stratified sampling the above procedure would apply to any stratum for which a separate estimate is to be prepared. A strict calculation of the subordinate sample that would be required for an estimate of Var X for the whole universe (and not for any one stratum individually) would call for advance estimates of both σ and β_2 for each stratum, and an advance estimate of the variance in each stratum. The law for the propagation of variance [p. 130] would then give an advance estimate of the variance of Var X for any proposed set of subordinate samples to be taken from the several strata. However, the relation, though not profound, is certainly cumbrous, as it involves in a complicated way σ^4, β_2, N_i, and n_i and the finite multiplier $(N_i - n_i)/(N_i - 1)$ for every stratum. For a simple way out, the following procedure will be found satisfactory. Take a total subordinate sample of 150 or 200 returns, distributed proportionately amongst the several strata. This can be done by using a random start and a constant interval throughout the whole lot of returns. Then use Eq. 1 for each stratum to find that

$$\text{Var } X = \text{Var } X_1 + \text{Var } X_2 + \cdots + \text{Var } X_M$$

$$= \sum_{i=1}^{M} \frac{N_i}{n_i} (N_i - n_i) \frac{\displaystyle\sum_{j=1}^{n_i'} (x_{ij} - \bar{x}_i')^2}{n_i' - 1} \tag{19}$$

Here M is the number of strata.

n_i' is the size of the subordinate sample in Stratum i.

N_i is the number of elements (chips) in Stratum i.

n_i is the size of the original sample in Stratum i.

x_{ij} is the population of the element drawn into the subordinate sample at the jth draw from Stratum i; $\quad j = 1, 2, \cdots, n_i'$.

If the original sample was proportionate (Ch. 6), N_i/n_i is a constant and may be placed in front of the summation. The work-sheet at the end of this chapter for estimating Var X in multistage sampling may easily be adapted to stratified sampling.

If it is suspected that β_2 for any stratum is particularly high, say 20 or more, a larger subordinate sample may be taken from this stratum to get a better estimate of Var X_i. The cause of any particularly high value of β_2 should be investigated, as the results may throw considerable light on the procedure of stratification and indeed may have a bearing on the utility of the whole survey. Some of the theory given further on in this chapter is helpful in controlling β_2.

The above procedure will give an unbiased estimate of Var X whether disproportionate sampling was used or not (Ch. 6).

For cluster-sampling (Ch. 5) the average variance $\sigma_w{}^2$ within the clusters must be estimated by studying a subordinate sample of clusters, which will also provide an estimate of the external variance $\sigma_e{}^2$. No special formulas other than those already given in Chapter 5 are needed, but it is important to note that the subordinate sample is to consist of clusters, not of individual returns (as of d.us.) selected at random. In other words, in repetition, the subordinate sample must simulate the procedure of the original sample; otherwise a false picture of the variance will be obtained.

The problem of small cells in tabulation plans. In considering the tabulation plans of a census or some other survey, any particular cell may be looked upon as occupying a proportion p of the total population to be tabulated, and what does not belong in that cell is simply the remainder, which is of proportion q. Let $n\hat{p}$ be the number of chips (people, households, farms, etc.) of the sample that fall into a particular cell; then if N/n is the sampling interval, $(N/n)n\hat{p}$ or $N\hat{p}$ is the estimated total in this cell. If the sampling distribution of $\hat{p}$ is binomial, the C.V. $\hat{p} = \sqrt{q/np}$ (Ch. 4).

One of the perplexing questions that always arise in planning the tabulations of a survey, sample or complete census, is to decide how small a cell is worth tabulating for analytic purposes.[1] My advice is to tabulate finer than is allowed by considerations of standard error in order to permit combinations and comparisons of cells, which frequently turn out to be of great importance in the analysis of the data. Combinations and comparisons can not be made unless there are smaller cells to build them of. The theory that has just been derived applies to the standard error of a combination in the same manner that it applies to the standard error of a single cell, and the tabulation plans should be based on the combinations of cells that the users may wish to make. If a coefficient of variation of 15 percent is permissible, the minimum value

[1] A distinction between analytic and enumerative problems was explained in Chapter 7.

of np in a random sample of n individual households or people turns out to be about 45, as is found by setting

$$0.15 = \sqrt{\frac{q}{np}} \qquad \text{[P. 111]} \quad (20)$$

first putting $q = 1$, which is usually a respectable approximation. It must not be assumed that 45 is an absolute minimum. There are certainly circumstances under which tabulations would be justified for sample numbers of even 10 or 15.

Remark 1. It should be borne in mind that the symbol np refers to the "expected" content of the sample, not to an estimate made from the sample, which may be many times as large. Thus, if a 5 percent sample were being planned, an "expected" 45 households or individuals in some particular cell of the sample would correspond to 900 in the corresponding cell of the universe.

Remark 2. In a complete count the symbol np and the estimated number are identical, and the last equation tells us that for analytic purposes (Chapter 7) it is illusory to tabulate a complete count in cells smaller than about 45. It used to be supposed that a complete count might be tabulated to any degree of cross-classification. This is of course true if no thought is given to the meaning of the figures, and the same may be said for a sample. For analytic purposes the tabulations arising from a complete census possess sampling errors at least equal to those that would be calculated from the last equation, just as if it were a sample, as indeed it is.

Remark 3. It should be borne in mind when planning the tabulations from a sample that is to be taken by clusters or in several stages that the minimum allowable cell will be larger than would be computed by the binomial theory.

We are now ready to ask and answer the question of whether in the sampling of individual elements an estimated standard error of 15 percent is dependable in the probability sense. In other words, what is the standard error of the estimated standard error of a cell having a particular sample-size? It will be recalled that from a previous exercise

$$\beta_2 = \frac{p^3 + q^3}{pq} \qquad \text{[P. 121]}$$

for a two-celled universe. We are here concerned with very small values of p, hence shall be satisfied with the approximation

$$\beta_2 = \frac{1}{p} - 2 + \text{powers of } p \qquad (21)$$

which the student should derive from the previous equation as an exercise. Now by dropping all terms on the right beyond $1/p$ it may be seen from Eq. 18 that the coefficient of variation of the coefficient of variation as estimated from the sample in a cell having an "expected" size of 45 is

$$\text{C.V. } \sigma'' \doteq \frac{1}{2} \sqrt{\frac{\beta_2 - 1}{n - 1}} \doteq \frac{1}{2} \sqrt{\frac{1}{np}}$$

$$= \frac{1}{2\sqrt{45}} = 7\tfrac{1}{2} \text{ percent} \quad \begin{array}{c} \text{[For an "expected"} \\ \text{sample of 45]} \end{array} \quad (22)$$

Thus for an "expected" sample of 45, we now know not only that the coefficient of variation is 15 percent, but that an estimate of the coefficient of variation is itself fairly reliable, being subject to a coefficient of variation of only $7\tfrac{1}{2}$ percent. It follows that when the elements of a sample are drawn independently at random (not in clusters) there is a sound basis for drawing up tabulation plans with an eye to the coefficient of variation of the numbers that are found in the cells or combinations thereof, because the coefficients of variation themselves are fairly reliable even for pretty small cells.[2]

Causes of high values of β_2. A high value of β_2 often originates from bad luck in attempting to classify the areas, households, or firms that go into the various strata. Graphically, it is ripples in the far tail of a distribution that cause β_2 to be high. (Ripples in the near tail are usually harmless.) An example may be seen in a sample of farms to determine the total acreage of alfalfa. If most farms contain around 20 acres of alfalfa, say from 5 to 100 acres, with an average of 20, the presence of some huge farms of 160, 320, and particularly 640 acres will cause trouble, not only in raising β_2 but—what is worse—in raising σ and lowering the precision of the estimated total acreage. Every effort should be made to find out beforehand where these huge farms are and to list them and sample them separately, perhaps 100 percent, leaving the remaining smaller farms to be sampled. However, in spite of honest efforts, mistakes will occur, causing increases in the values of σ and β_2, necessitating larger samples than would otherwise be required.

For the present discussion the distribution of the populations (people employed, acres, inventory, etc.) of the chips (households, areas, farms, business firms, etc.) within any class or stratum may be supposed to fall into two parts, a main part and a wiggly tail. The chips of a universe are stratified, i.e., sorted into piles on the basis of information that was valid at some previous date. In some instances this information is now highly inaccurate: a firm suddenly expands; had 60 employees 2 years ago, now has 600: a farm that last year or 3 years ago contained 20 acres of alfalfa now has 200. Moreover, mistakes are made in the listings

---

[2] This conclusion was first called to my attention by my colleague Morris H. Hansen in 1947.

through inadvertence or failure to understand the definitions; figures are miscopied; some areas are dropped into the wrong pile; in short, for various reasons, the distribution of any stratum, instead of being entirely "well behaved," possesses a main part plus a wiggly tail. The effect of these unwanted wiggles is to increase both σ and β_2, and it is necessary to keep such possible effects in mind when designing the sample and in appraising its precision. The increase in σ requires a bigger sample to meet a prescribed precision, and the increase in β_2 requires a bigger sample to estimate σ.

For ease in computation it will be assumed that there is but one wiggle in one tail. Let the main part and the wiggle contain respectively the proportions q and p (see Fig. 35).

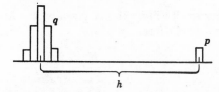

FIG. 35. The distribution of any stratum can be expected to possess one or more wiggles in the tails. These wiggles increase σ and β_2 and require consideration in the design of the sample and in the appraisal of its precision. For simplicity in calculation it will be assumed that there is only one wiggle in one tail, at a distance h, of proportion p.

Let the symbols σ_A, μ_{4A}, and β_2 refer to the main part of the distribution. (We shall not need corresponding symbols for that part of the distribution contained in the wiggle.) If by good fortune the information on hand regarding the sizes of most of the sampling units is accurate, σ_A will be fairly small, p perhaps as low as 0.01, and h not large. With less accurate information, σ_A may be broad, and—what is more unfortunate—p may be as high as 0.05 or higher and h may be a long distance. By Eqs. 21 on page 65 the first four moment coefficients of the entire distribution in Fig. 35 taken about the mean of the main part will be

$$
\left.
\begin{aligned}
\mu_4' &= q\mu_{4A} + ph^4 = \mu_{4A} + p(h^4 - \mu_{4A}) \\[4pt]
\mu_3' &= 0 + ph^3 \quad \text{[Assuming symmetry of the main part of} \\
&\qquad\qquad\qquad \text{Fig. 35, for simplicity, so that } \mu_3 = 0] \\[4pt]
\mu_2' &= q\sigma_A{}^2 + ph^2 \\[4pt]
\mu' &= 0 + ph
\end{aligned}
\right\} \quad (23)
$$

Transferred to the mean of the combined distribution these moment coefficients take the values

$$
\left.\begin{aligned}
\mu_4 &= \mu_4' - 4\mu'\mu_3' + 6\mu'^2\mu_2' - 3\mu'^4 \\
&= \mu_{4A} + p(h^4 - \mu_{4A}) \\
&\quad - p^2h^2\{4h^2 - 6[\sigma_A^2 + p(h^2 - \sigma_A^2)]\} - 3p^4h^4 \\
\sigma^2 = \mu_2 &= \mu_2' - \mu'^2 \\
&= q\sigma_A^2 + pqh^2
\end{aligned}\right\} \qquad (24)
$$

as may be seen by applying Eqs. 22, page 66. For the combined distribution

$$
\beta_2 = \frac{\mu_4}{\sigma^4} \qquad \text{[By definition]}
$$

which if p is very small will be closely given by the first two terms of a power series in p in the form

$$
\beta_2 = \beta_{2A}\left\{1 + p\left(1 + \frac{h^4}{\mu_{4A}} - \frac{2qh^2}{\sigma_A^2}\right) + \cdots\right\} \qquad (25)
$$

which the student may wish to derive from the values of σ^2 and μ_4 just found above. Comparison of σ^2 with σ_A^2 along with inspection of this series discloses the fact that some wild maverick units falling in the tail at a far distance h from the main part of the distribution can literally work havoc with σ and β_2, particularly the latter because of the factor $(h/\sigma_A)^4$ multiplying p. On the other hand, a number of moderate mavericks (say $h/\sigma_A = 4$ or 5) do not do so much damage. As an exercise the student should make a calculation of β_2 with, for example,

$$
p = 0.02 \qquad \text{[1 in 50 units being mavericks and forming a wiggle in the tail]}
$$

$$
h = 10\,\sigma_A
$$

$$
\beta_{2A} = 6 \qquad \text{[i.e., } \mu_{4A} = 6\sigma_A^4]
$$

Calculations show that $\sigma = 1.7\sigma_A$, and thus σ is increased 70 percent by the mavericks. As $\sigma^2 = 2.9\sigma_A^2$, the required sample-size in this stratum (formed by the combined distribution) is practically three times what would have been needed if the mavericks had not been present and the stratum consisted of the main part only. The calculations also show that $\beta_2 = \mu^4:\sigma^4 = 22$, which is nearly four times β_{2A}, wherefore the size of sample of returns that is required to estimate the variance of an estimate made from a sample drawn from the combined distribution is four times the size that would be required if the stratum consisted of the main part only.

A larger value of h will increase β_2 even further, but a larger value of p (more of the same grade of mavericks) will raise σ and β_2 only by a proportionate amount, approximately.

An important lesson in sampling has just been learned. Wild mavericks must be avoided, and it will pay to go to some expense to eliminate them. It may not pay, however, to go to much expense to eliminate the moderate and small ones. As has been stated, mavericks arise from blunders in classification, occasionally from illegibility; but oftener from inaccurate or complete lack of recent information regarding changes and shifts in population, acreage, employment, inventories, etc. Inaccuracies in the present information may be avoided or at least greatly diminished by subscribing to some routine device for receiving reports concerning large shifts in population, acreage, number of employees, inventories, construction-projects, or whatever is the basis of stratification. Checks and controls will help to decrease blunders.

Estimating the external variance in two-stage sampling. The preceding part of the chapter deals only with a single stage of sampling within a universe or stratum. When there are two stages, the equations so far derived will apply directly to the problem of estimating an internal variance σ_j^2 but not to an estimate of the external variance σ_e^2 of the universe. If the external variance of the sample be defined as

$$s_e^2 = \frac{1}{m} \sum_{j=1}^{m} \left[X_j - \frac{X}{M} \right]^2 \tag{26}$$

then s_e^2 is a random variable and will, on the average, differ from σ_e^2 for two reasons: it varies with the particular m primary areas that are chosen for the sample, and it also varies with the subsamples that are drawn from these primary units. The notation to be used will be that used in Part A of Chapter 5, and the student may wish to refresh his memory by glancing back at it. The exact relationship between σ_e^2 and Es_e^2 can be derived by rewriting s_e^2 as

$$s_e^2 = \frac{1}{m} \sum_{1}^{m} X_j^2 - \frac{X^2}{M^2} \tag{27}$$

Then

$$Es_e^2 = E_j \frac{1}{m} \sum_{1}^{m} E_k X_j^2 - \frac{EX^2}{M^2}$$

$$= \frac{1}{M} \sum_{1}^{M} E_k X_j^2 - \frac{EX^2}{M^2} \tag{28}$$

$E_k X_j^2$ will now be worked out. Let

$$X_j = [X_j - A_j] + A_j \tag{29}$$

Square this to get

$$X_j^2 = [X_j - A_j]^2 + A_j^2 + 2[X_j - A_j]A_j$$

The "expected" value of the term on the extreme right is 0, and the "expected" values of the remaining terms give

$$\underset{k}{E}X_j^2 = \underset{k}{E}[X_j - A_j]^2 + A_j^2$$

$$= \underset{k}{\text{Var}}\, X_j + A_j^2$$

$$= \left(\frac{N_j}{n_j}\right)^2 \frac{N_j - n_j}{N_j - 1} n_j \sigma_j^2 + A_j^2 \qquad \text{[P. 103]} \quad (30)$$

This last equation evaluates the first term on the right of Eq. 27. For the second term, the procedure is similar. Thus

$$X = [X - A] + A$$

$$X^2 = [X - A]^2 + A^2 + 2[X - A]A$$

$$EX^2 = E[X - A]^2 + A^2 \qquad \begin{array}{l}\text{[Because } EX = A \text{ and the last} \\ \text{term drops out]}\end{array}$$

$$= \text{Var}\, X + A^2 \qquad\qquad\qquad (31)$$

Then

$$Es_e^2 = \frac{1}{M} \sum_1^M \underset{k}{E}X_j^2 - \frac{EX^2}{M^2} \qquad\qquad \text{[Eq. 28]}$$

$$= \frac{1}{M} \sum_1^M \left\{ \left(\frac{N_j}{n_j}\right)^2 \frac{N_j - n_j}{N_j - 1} n_j \sigma_j^2 + A_j^2 \right\}$$

$$- \left(\frac{1}{m}\right)^2 \left\{ \frac{M - m}{M - 1} m\sigma_e^2 + \frac{m}{M} \sum_1^M \left(\frac{N_j}{n_j}\right)^2 \frac{N_j - n_j}{N_j - 1} n_j \sigma_j^2 \right\} - \frac{A^2}{M^2}$$

$$= \left[1 - \frac{1}{m} \right] \frac{1}{M} \sum_1^M \left(\frac{N_j}{n_j}\right)^2 \frac{N_j - n_j}{N_j - 1} n_j \sigma_j^2$$

$$+ \left[\frac{1}{M} \sum_1^M A_j^2 - \left(\frac{A}{M}\right)^2 \right] - \frac{M - m}{M - 1} \frac{\sigma_e^2}{m} \qquad (32)$$

Or

$$\frac{m}{m - 1} Es_e^2 = \frac{M}{M - 1} \sigma_e^2 + \frac{1}{M} \sum_1^M \left(\frac{N_j}{n_j}\right)^2 \frac{N_j - n_j}{N_j - 1} n_j \sigma_j^2 \qquad (33)$$

If σ_j were 0 in all primary units, as would be true if all the d.us. within any primary unit had the same population, the second term on the right would vanish, leaving an equation analogous to Eq. 13' on page 102 for a single stratum. The same effect is obtained by making a complete coverage in each of the primary areas that comes into the sample; the second term then vanishes because $N_j - n_j = 0$. An estimate of σ_e^2 is obtained from the above equation by writing

$$\hat{\sigma}_e^2 = \frac{M-1}{M} \frac{m}{m-1} s_e^2 - \frac{M-1}{M^2} \sum_1^M \left(\frac{N_j}{n_j}\right)^2 \frac{N_j - n_j}{N_j - 1} n_j \sigma_j^2 \quad (34)$$

then introducing the estimate of σ_j^2 found in Eq. 11' on page 333, thus getting

$$\hat{\sigma}_e^2 = \frac{M-1}{M} \frac{m}{m-1} s_e^2 - \frac{M-1}{mM} \sum_1^m N_j(N_j - n_j) \frac{s_j^2}{n_j - 1} \quad (35)$$

The student may wish to prove that this estimate is unbiased; i.e.,

$$E\hat{\sigma}_e^2 = \sigma_e^2 \quad (36)$$

The variance of the estimate σ_e^2 is difficult, but to a first approximation it must depend on β_2 for the universe of primary units through a relation similar to Eq. 18.

The estimate of σ_e^2 just found will hold within any stratum.

Sampling the returns for estimating σ_e and σ_j. To see how to use a subordinate sample of the returns it is well to rewrite Var X so as to involve the expected variances of samples. Thus,

$$\text{Var } X = \left(\frac{M}{m}\right)^2 \left\{ \frac{M-m}{M-1} m\sigma_e^2 + \frac{m}{M} \sum_1^M \left(\frac{N_j}{n_j}\right)^2 \frac{N_j - n_j}{N_j - 1} n_j \sigma_j^2 \right\}$$

$$= \left(\frac{M}{m}\right)^2 \left\{ \frac{M-m}{M-1} m \left[\frac{M-1}{M} \frac{m}{m-1} Es_e^2 \right. \right. \qquad \text{[By inserting the "ex-}$$
$$\text{pected" values of } s_e^2$$
$$\left. - \frac{M-1}{M^2} \sum_1^M N_j(N_j - n_j) \frac{Es_j^2}{n_j - 1} \right] \qquad \text{and } s_j^2]$$

$$\left. + \frac{m}{M} \sum_1^M \left(\frac{N_j}{n_j}\right)^2 \frac{N_j - n_j}{N_j - 1} n_j \frac{N_j - 1}{N_j} \frac{n_j}{n_j - 1} Es_j^2 \right\}$$

$$= M \left(\frac{M}{m} - 1\right) E \frac{ms_e^2}{m-1} + \frac{M}{m} \sum_1^m \frac{N_j}{n_j} (N_j - n_j) E \frac{n_j s_j^2}{n_j - 1} \quad (37)$$

Estimates of $Ems_e^2/(m-1)$ and $En_j s_j^2/(n_j - 1)$ are to be formed from a subordinate sample of the returns, to consist of a total of n' returns, taken as follows:

m' primary units drawn at random from the m primary units of the sample

$n_j{'}$ returns chosen at random from the returns in primary unit j

$n' = \displaystyle\sum_{1}^{m'} n_j{'}$, the total number of returns in the subordinate sample

The simplest procedure for drawing the subordinate sample is to take a random start and with a constant interval go right through the returns in the m' primary units. The interval is to be adjusted so that the desired total n' is obtained. This procedure automatically allocates the subordinate sample in proportion to the sizes of the primary units. For actual numbers I suggest

$m' = 50$, or take $m' = m$ if $m < 100$.

$n' = 200$, or take $n' = 400$ if the average β_2 for the secondary units is estimated as 25 or higher.

Let x_{jk} be the population of the kth unit of the subordinate sample from primary unit j. Then define:

$$\bar{x}_j{'} = \frac{1}{n'} \sum_{1}^{n_j{'}} x_{jk}, \quad \text{the estimated average population} \tag{38}$$
$$\text{per secondary unit}$$

$$x_j{'} = N_j \bar{x}_j{'}, \qquad \text{the total estimated population for} \tag{39}$$
$$\text{primary unit } j$$

$$\bar{x}' = \frac{1}{m'} \sum_{1}^{m'} x_j{'}, \quad \text{the estimated average population} \tag{40}$$
$$\text{per primary unit}$$

Then in the formula for the Var X,

$$E \frac{ms_e{}^2}{m-1} \text{ is to be estimated by } \frac{\displaystyle\sum_{j=1}^{m'} (x_j{'} - \bar{x}')^2}{m'-1} \quad \text{[Unbiased if } n_j{'} = n_j]$$

$$E \frac{n_j s_j{}^2}{n_j - 1} \;\text{``}\;\text{``}\;\text{``}\qquad\text{``}\qquad\text{``}\quad \frac{\displaystyle\sum_{k=1}^{n_j{'}} (x_{jk} - \bar{x}_j{'})^2}{n_j{'} - 1}.$$

In terms of the variances calculated from the subordinate sample,

$$\text{Est'd Var } X = M\left(\frac{M}{m} - 1\right) \frac{\displaystyle\sum_{j=1}^{m'} (x_j{'} - \bar{x}')^2}{m'-1} \quad \text{[Assume } n_j{'} = n_j]$$

$$+ \frac{M}{m'} \sum_{j=1}^{m'} \frac{N_j}{n_j} (N_j - n_j) \frac{\displaystyle\sum_{k=1}^{n_j} (x_{jk} - \bar{x}_j)^2}{n_j - 1} \tag{41}$$

The above procedure will usually give an estimate of Var X with a coefficient of variation of 15 to 30 percent, which is close enough (for it should be remembered that if Var X is estimated with 20 percent error, then σ_X is estimated with 10 percent error). High values of β_2 amongst the primary units (which would affect the first term on the right) can be pretty well avoided by careful planning. High values of β_2 within the primary units are more difficult to control.

If $n' = 200$ and $m' = 50$, the procedure outlined above allows for $200 - 50 = 150$ "degrees of freedom" in the second term on the

FORM FOR ESTIMATING THE INTERNAL AND EXTERNAL VARIANCES OF SAMPLING UNITS, FOR USE IN ESTIMATING VAR X IN 2-STAGE SAMPLING

The notation is explained on page 350, but the strokes are omitted from x here for convenience.

Sampling units by number and other identification		N_j	x_{jk}	$\bar{x}_j$	$(x_{jk} - \bar{x}_j)^2$	$x_j = N_j\bar{x}_j$	$(x_j - \bar{x})^2$
$j = 1$	$k = 1$	N_1	x_{11}		$(x_{11} - \bar{x}_1)^2$		
	2		x_{12}		$(x_{12} - \bar{x}_1)^2$		
	.		.		.		
	.		.		.		
	.		.		.		
	n_1		x_{1n_1}		$(x_{1n'} - \bar{x}_1)^2$		
				$\bar{x}_1$	$\dfrac{1}{n_1 - 1} \sum_1^{n_1} (x_{1k} - \bar{x}_1)^2$ *	$x_1 = N_1\bar{x}_1$	$(x_1 - \bar{x})^2$ **
$j = 2$	$k = 1$	N_2	Filled				
	2				in		
	.						
	.				likewise		
	n_2						
				$\bar{x}_2$	$\dfrac{1}{n_2 - 1} \sum_1^{n_2} (x_{2k} - \bar{x}_2)^2$ *	$x_2 = N_2\bar{x}_2$	$(x_2 - \bar{x})^2$ **
.	.	.	.		.	.	.
.	.	.	.		.	.	.
.	.	.	.		.	.	.

* ** The starred quantities are to be summed and used in Eq. 41 for estimating Var X.

right [3] (the term that deals with the estimate of the internal variances), and this should be sufficient provided the average β_2 for the secondary units is not higher than 25.

Work-sheets can be ruled to show the n_j populations x_{jk} for each of the m' primary units in the subordinate sample. A form is shown on page 351. x_j' and $\bar{x}_j'$ are to be computed for every j; then $\bar{x}'$; and the work of squaring proceeds. Each sum of squares marked with a single asterisk (*) is to be multiplied by $(N_j/n_j)(N_j - 1)$; then all are to be summed to get the second term in the above equation for Est'd Var X. The m' squares marked with a double asterisk (**) are to be summed and multiplied by $M(M/m - 1)/(m' - 1)$; the result is the first term in the Est'd Var X. However complicated this procedure appears to be, actually only a relatively small amount of labor is involved for this final appraisal, and the survey should be considered incomplete until it is carried out.

The student is referred to pages 42–53 of *A Chapter in Population Sampling* (Bureau of the Census, 1947) for an application to stratified population samples in urban areas. Actual calculations of the precision and of β_2 are made there from samples of the returns.

Special procedure for evaluating the precision of a systematic or patterned sample. In the preceding paragraphs the sampling units were supposedly drawn independently at random. If they were drawn systematically (Chs. 4 and 5), the universe was supposedly thoroughly mixed, in which case a systematic drawing is equivalent to a random drawing so far as probabilities are concerned. A systematic or patterned selection of the subordinate sample would then also be permissible. It is necessary, however, to be prepared with a new method for evaluating the precision of a systematic selection under circumstances where such assumptions are violated.[4]

Before going on it should be noted that it is not alone the drawing of the sample that causes a systematic or patterned selection to differ in character from an independent random selection; the difference really starts with the systematic listing of the universe, and the fact that there may be hidden periodicities and serial correlations between sampling units as listed. A systematic listing is often a natural and necessary

---

[3] "Degrees of freedom" is a term denoting the proper divisor (example, $n - 1$) required under a 2d moment of a sample drawn without replacement to give an unbiased estimate of a variance.

[4] A fundamental theoretical treatment of the variance of a systematic sample was published by the Madows, reference to which was made on page 83. The simple procedures suggested here will suffice for the empirical evaluation of the variance of a systematic sample after the results are in, but the theoretical papers of the Madows and others should be studied for insight into the causes of losses and gains over random sampling.

routine. In listing d.us., for example, a lister will start at one corner of an area and proceed in a systematic manner up and down the streets, roads, and corridors until the job is finished; then he moves into an adjacent area and continues likewise. A map showing small areas numbered in serpentine fashion within a country, county, or city is, in effect, a systematic list. A field or forest is crossed by real or imaginary rows, bands, or lines, numbered serially from north to south or east to west; and these rows, bands, or lines constitute a systematic listing. The variance between samples drawn systematically from any such list may be different, much, not at all, or little, from the variance of samples drawn independently at random.

One simple way of evaluating the precision of a systematic sample is to plan it as 10 systematic subsamples, obtained by using 10 independent random starts. This is the Tukey plan (p. 96). The variability between the results of the 10 subsamples gives an estimate of the standard error of the result obtained from the whole 10 combined. Let $X^{(1)}$, $\cdots$, $X^{(10)}$ be the 10 estimates of the population bearing some particular characteristic: then

$$\text{Est'd } \sigma_X{}^2 = \frac{1}{10} \frac{1}{9} \sum_1^{10} [X^{(i)} - \overline{X}]^2 \tag{42}$$

The sampling and tabulating plans may be so laid out that the identity of the 10 subsamples is maintained, and the estimates of the variances of the chief characteristics are obtained automatically.

Thus, in an actual example, 10 independent subsamples each of 150 poles and their attachments (wire, cable, crossarms, terminals, etc.) gave the following results for the weighted average physical condition of the property in the aerial account.[5]

$x = 67.5\%$	68.4%
65.9	69.6
67.1	69.4
67.9	69.0
70.8	68.0

whence the estimated average physical condition of the aerial account was $\bar{x} = 68.4$ percent, with an estimated standard error of 0.44 percent, which the student should calculate.

For consideration of this one account it should be noted that as each subsample covered 150 poles and attachments, the 10 results may cer-

[5] W. Edwards Deming, "On the sampling of physical materials," a paper read at a meeting of the International Statistical Institute held in Berne, 5–10 September 1949; *La Revue de l'Institut International de Statistique*, vol. 50, 1950.

tainly be treated as 10 normally distributed random variables, to which Fisher's t-test may be validly applied, either with the aid of tables or with the nomograph of Nekrassoff (p. 555), whence it is found that the odds are 99:1 that the value $\bar{x} - 0.44t_1 = 68.4 - 0.44 \times 2.82 = 67.2$ does not fall below the result that would have been obtained by examining all the poles and attachments (in this example, 973,000 poles) with the same care as was exercised on the samples, were such performance possible. The student may also wish to observe from the results of Exercise 9 on page 127 that the probability is only $(\frac{1}{2})^{10}$ or $1/1024$ that the lowest of the 10 results, viz., 65.9, lies above the median of the universe of physical conditions of the entire aerial account.

In this example, as in many of its kind, interest centered in the weighted mean of 14 accounts, so the t-test was not actually used. Although only 9 degrees of freedom were provided by each account, the standard error of the weighted mean of the 14 accounts was estimated with considerable precision. (For the optimum allocation of the sample in such problems, see p. 219.)

Obviously the Tukey plan may be used with any procedure of sampling—systematic, 2-stage, or any other. In essence, it is only 10 repetitions of the sampling procedure, whatever it is, which produces 10 independent normal variates which are to be compared with each other.

More than 10 subsamples would be preferable, to get an improved estimate; but to make much of a gain the number would require to be doubled, and any such recommendation would under most circumstances lead to administrative panic.

Although the Tukey plan, just described, is strongly recommended in design, one sometimes discovers, after the fact, that a systematic sample was not laid out in independent subsamples. The variance of an estimate X may still be obtained satisfactorily in many problems by the following device, which I shall refer to as the *loop* plan. Imagine loops to be thrown around successive pairs of sampling units. These loops form hidden strata in the universe, and two successive units of the sample have come from one of these hidden strata. The device about to be described reflects the gains arising from geographic stratification but is at fault in assuming that the two units of a pair were drawn at random. Each pair will nevertheless be used to provide the estimate of the variance $\sigma_i{}^2$ of the hidden stratum whence they were drawn. In practice, 50 or 100 pairs of successive returns will be selected systematically from the stack of completed returns left in the order in which they came in from the survey. It might be slightly better to arrange them in the order of listing but this is not necessary. For an estimate of the variance of X we note first that

$$\sigma_X^2 = \sum_{i=1}^{n/n_i} N_i^2 \frac{N_i - n_i}{N_i - 1} \frac{\sigma_i^2}{n_i}$$

$$\doteq a^2 \left(1 - \frac{1}{a}\right) n_i \Sigma \sigma_i^2$$

$$= a^2 \left(1 - \frac{1}{a}\right) n \operatorname{Av} \sigma_i^2 \qquad \cdot(43)$$

The subscript i in the summation runs over the successive hidden strata that are created by the imaginary loops, and $a = N\!:\!n$, the original counting interval. Now let R_i denote the range between the two values of the pair of returns drawn from the ith hidden stratum. Then $\frac{1}{2}R_i^2$ is an estimate of σ_i^2 and

$$\text{Est'd } \sigma_X^2 = \frac{1}{2} na^2 \left(1 - \frac{1}{a}\right) \operatorname{Av} R_i^2 \qquad (44)$$

This equation is easy to apply, but the user thereof must be on guard for the fault already referred to.

Remark. It is to be noted that in both the Tukey plan and the loop plan the samples for any two characteristics are correlated, wherefore variances are not additive. This is so because the estimates are all derived from the same set of sampling units. A sampling unit (d.u., or cluster of d.us., for example) that shows high rent would on the average also show high expenditures for clothing. As the correlations are unknown or are too much trouble to compute, it is highly desirable to decide in advance of the tabulations just what standard errors are going to be needed. If the standard error of the expenditure for rent plus clothing combined is going to be needed, it should be computed by one plan or the other, along with any other standard errors that are contemplated.

PART IV. APPLICATIONS OF SOME OF THE FOREGOING THEORY

CHAPTER 11. INVENTORIES BY SAMPLING [1]

To slip upon the pavement is better than to slip with the tongue.—
Ecclesiasticus, 20:18.

Statement of the problem. Administration of tire-rationing during
the war required current and reliable knowledge of the numbers of
passenger, truck, and other civilian-grade tires held by dealers, mass
distributors, and manufacturers. To fill this need, inventories of tire
dealers were taken each quarter, which together with monthly reports
from tire manufacturers and mass distributors gave fairly complete
information concerning stocks of tires available for civilian consumption.
In the interest of speed, accuracy, guaranteed reliability, and economy,
the quarterly inventories were eventually taken by sampling. These
advantages of sampling gave rise to increased usage and to the development of new theory, but it should be stated that the sampling plan
described here depends only on comparatively simple theory.

The immediate aim of this chapter is to describe the sample of dealers
used in estimating the tire-inventory for March 1945; more specifically,

   i. To state the reliability desired.

   ii. To describe how this reliability was achieved through the application of basic principles and simple procedures.

   iii. To describe how available knowledge concerning dealers' stocks
was used to minimize the number of dealers in the sample and
the workload of weighting the results.

   iv. To describe the estimation of the precision actually attained, and
to compare the estimate of the actual precision with the aimed-
at precision. This was done by examining a subordinate sample
of the returns.

   v. To estimate the gains in efficiency accomplished by stratification.

The methods herein described may be profitably used in administrative
problems of business and government relating to inventories, sales, employment, or traffic, wherever conditions are similar—i.e., where a complete list of possible respondents exists, along with other information

[1] Copied almost verbatim from an article by the author and Willard R. Simmons,
J. Amer. Stat. Assoc., vol. 41, 1946: pp. 16–33.

such as size or type, which may serve as useful criteria for stratification.[2]

Definition of the universe. The universe to be covered was by definition the list of dealers on record with the OPA. This list contained the original registrations at the beginning of rationing (October 1942), plus dealers authorized subsequently ("new authorizations"), minus deletions that were made when notice was received that a business was defunct. Changes in name and address and unreported defunct businesses created the usual problems in interpreting nonresponses.

Evils of nonresponse; advantages of sampling. Nonresponses would be harmless if it were certain that a nonresponding dealer is out of business or has no tires. But unfortunately a nonresponse may mean other things as well—moved, change in name, or business as usual at the same address, coupled with inadvertence or inability to fill out the questionnaire. Every effort should therefore be made to keep the list trimmed of dead wood and to evaluate nonresponses by personal calls on a subsample.

The minimum in nonresponses can be reached much easier with a sample than with a complete coverage because there are fewer of them, and the district offices are not bogged down with more calls on dealers than is humanly possible to make; moreover, the sample almost always, by the principle of Neyman sampling (Ch. 6), contains a preponderance of big dealers employing bookkeepers able to supply the figures and even taking pride in doing so. In illustration, the "complete count" of September 1944 resulted in 24,015 nonresponses or 17 percent of the 140,989 questionnaires mailed out, whereas in the December 1944 and March 1945 samples the nonresponses were only 1.2 percent and 3.9 percent respectively of the 16,000 questionnaires mailed out, and be it noted that both these samples included a sample of the nonresponses of September. Incidentally, the average number of tires per dealer was 50 percent higher in both December and March for this group than the average inventory of all other dealers, thus pretty well deflating the possible interpretation in which refuge is so often taken, that nonresponses can be ignored on the pretext that they are average, or are

---

[2] Information on size and type is not necessary for sampling, but ordinarily such information will make possible a considerable reduction in the size of sample required to attain the reliability desired (Ch. 6). Stratification by area alone can be expected to show some small gains in sampling efficiency. Separation of dealers into groups widely different in type and size as of the last inventory will bring further gains, provided the inventories of dealers on the date of the sample still differ widely in type and size, in which case the additional efficiency gained by applying the principle of optimum allocation (*vide infra*) of the sample to the various strata is often striking. Fortunately, any large list must be arranged in some systematic fashion, such as by city, type, size, and order of receipt, whatever it is to be used for; and little or no additional labor may be required to form strata that are ideal for sampling.

composed mostly of small dealers, or represent no stock at all. These results bear out an interesting observation that had been made in tabulating the returns from the September complete count, viz., that the last 3000 dealers that reported subsequently to the second follow-up letter sent to delinquents, actually had on hand an average of over 10 tires per dealer against an overall average of 6 tires per dealer for the 114,000 dealers that had already responded.

Some history of the quarterly inventories. The first inventory of tires was accomplished simultaneously with registration, following which quarterly inventories were taken, at first by the "cut-off" method by which small dealers were not asked to respond. Because the list of dealers showed not only names and addresses, but also the number of tires by size and type on hand and consigned as of the latest complete inventory, it was possible to cut off any class of dealer, or to classify the dealers for efficient sample-design. In June 1943 a complete count was attempted, following two inventories by cut-off that had been taken in December 1942 and March 1943, subsequently to the first registration (October 1942). The quarterly cut-off was repeated in September and December of 1943, but samples were taken in March and June 1944, followed by the attempted complete count of September 1944. At this time, the number of nonresponses was huge (24,015), and proper follow-up was impossible because of the sheer enormity of the task. Worrisome discrepancies appeared between the complete count of September 1944 and the earlier samples taken in March and June, the explanation of which turned out to be the *more energetic and successful follow-up of the nonresponses in the samples.* This explanation was not at first accepted, and indeed was not demonstrated until the sample of December 1944 was taken. Meanwhile, a study of these discrepancies had led to a re-examination of the entire reporting system, as a result of which sampling was recommended again for the inventories of December 1944 and March 1945.

Preliminary considerations and the stratification of dealers. This chapter will describe only the sampling plan as revised for March 1945. The sample for the previous quarter (December 1944) was planned along the same lines and will therefore be mentioned only as its results were used in the plans for March.

It is important to bear in mind that data were needed concerning several types and sizes of tires (of which there are new passenger tires, new truck and bus tires, motorcycle tires, used passenger tires, and tractor-implement tires). Now a dealer in the sample was expected to report his stock of all types of tires, but owing to the differing dispersions of the various types of tires amongst dealers, some handling (e.g.) motorcycle tires, and some not, some handling truck and bus tires and some not, and some handling all types, a sample of dealers that would provide adequate precision for one type of tire might not do so for another. Analysis of the dispersions of tire-stocks demonstrated that a sample producing adequate results for new passenger tires would probably suffice for other types of tires with a few minor adjustments.

The particular kind of sampling to be described here has these characteristics:

> *i.* Stratified sampling.
>
> *ii.* Optimum allocation of sample to provide the aimed-at precision with the minimum number of questionnaires.
>
> *iii.* Simple systematic selection of dealers within strata.
>
> *iv.* A definite precision-requirement, consistent with the uses that were to be made of the data.
>
> *v.* A predetermined procedure for weighting the results to get unbiased estimates.
>
> *vi.* A determination of the precision that was actually attained, based on a sample of the returns.

The stratification. Study of past quarterly inventories disclosed two main groups of tire dealers: Group A, containing dealers reporting stocks of new truck and bus tires in September 1944; and Group B, containing dealers reporting no new truck and bus tires but reporting stocks of new passenger tires in September 1944. This division was very effective in reducing the variance in the estimated total of new truck and bus tires. Similarly, it was effective for new passenger tires, because the larger dealers were automatically thrown into one group (Group A). This is so because dealers that handle new truck and bus tires usually aim to keep good stocks of all types of tires; actually, these dealers held about 85 per cent of the new passenger tires, as well as practically all the new truck and bus tires. On the other hand, the dealers in Group B consisted mainly of small stores and service stations handling relatively few popular-size passenger tires usually incidental to the main business.

Besides Groups A and B there were five special groups, described later. Samples were drawn from these special groups, but they were relatively unimportant; the main consideration was Groups A and B, which contained practically all the tires.

Dealers in each of these two groups were classified according to the numbers of tires that they held in September 1944 in intervals of 10 tires, i.e., 1–9 tires, 10–19 tires, 20–29 tires, and so forth. For Group A this classification was based on new truck and bus tires, and for Group B, on new passenger tires.

Method of estimation. From Class i, consisting of N_i dealers, a sample of n_i dealers was selected (cf. Remark 3). N_i was known; n_i was determined from considerations of optimum allocation as discussed on page 218. If ξ denotes the total number of (e.g.) new passenger tires on hand in March, then

$$\xi = N_1\mu_1 + N_2\mu_2 + \text{similar terms for the other sub-} \atop \text{classes through all groups} \qquad (1)$$

where μ_i is the average number of new passenger tires held per dealer in Class i, which consists of N_i dealers. The sample gives, not μ_i, but $\bar{x}_i$ as an unbiased estimate thereof, as is indicated in Remark 1. The comparison is disclosed by their definitions:

$$\mu_i = \frac{1}{N_i}(x_{i1} + x_{i2} + \text{similar terms through all } N_i \atop \text{dealers of Class i}) \qquad (2)$$

$$\bar{x}_i = \frac{1}{n_i}(x_{i1} + x_{i2} + \text{ similar terms through all the } n_i \atop \text{sample dealers of Class i}) \qquad (3)$$

x_{ij} being the number of new passenger tires held by the jth dealer in the ith class. Because $\bar{x}_i$ is an unbiased estimate of μ_i, it follows that

$$X = N_1\bar{x}_1 + N_2\bar{x}_2 + \text{similar terms for the dealers of the sample} \qquad (4)$$

is an unbiased estimate of ξ.

Remark 1. For the estimate X in Eq. 4,

$$EX = E(N_1\bar{x}_1 + N_2\bar{x}_2 + \text{similar terms through all classes})$$
$$= N_1E\bar{x}_1 + N_2E\bar{x}_2 + \text{similar terms} \qquad (5)$$

Now from Eq. 3 it follows that

$$E\bar{x}_i = E\frac{1}{n_i}(x_{i1} + x_{i2} + \text{similar terms through all } n_i \atop \text{sample dealers in Class i}) \qquad (6)$$

The probability of including any particular dealer in the sample is n_i/N_i, wherefore the sum of all possible sample-values of inventories multiplied by their probabilities of occurrence is

$$E\bar{x}_i = \frac{1}{n_i}\left(\frac{n_i}{N_i} \cdot x_{i1} + \frac{n_i}{N_i} \cdot x_{i2} + \text{similar terms through all} \atop N_i \text{ dealers in Class i}\right)$$

$$= \frac{1}{N_i}(x_{i1} + x_{i2} + \text{similar terms through all } N_i \text{ dealers in Class i})$$

$$= \mu_i \qquad (7)$$

Substitution of μ_i for $E\bar{x}_i$ in Eq. 5 gives Eq. 1. Thus the "expected" value of the estimate X is equal to ξ, which is the sum of the inventories of passenger tires of all dealers.

The sample in groups A and B. There are two fundamental principles for guidance:

$$\frac{n_i}{N_i} = \frac{\sigma_i}{k} \qquad \text{[Eq. 21, p. 220]} \quad (8)$$

$$\sigma_{\bar{x}_i}{}^2 = \frac{N_i - n_i}{N_i - 1} \frac{\sigma_i{}^2}{n_i} \qquad \text{[Ch. 4]} \quad (9)$$

The first is the principle of Neyman sampling, which says that to obtain the greatest efficiency in sampling for (e.g.) passenger tires, the proportion (n_i/N_i) of dealers to be drawn into the sample from the ith class should be proportional to the standard deviation σ_i of the passenger tires held by the dealers of this class on the date of the sample. The word "class" is used here in the general sense. It might mean a size-class within a group, or it might mean a whole group. The symbol k in Eq. 8 is a factor of proportionality and will be determined later by conditions fixed by the precision that is aimed at (see Eqs. 19 and 21).

Remark 2. Sometimes the total sample-size $(n = \Sigma\, n_i)$ is fixed in advance by considerations of cost (and not, as here, by an aimed-at precision). For such problems it is convenient to know that the symbol k in Eq. 8 has the value

$$k = \frac{\Sigma\, N_i \sigma_i}{n} \qquad \text{[Eq. 23, p. 221]} \quad (10)$$

in which case Eq. 8 gives

$$\frac{n_i}{n} = \frac{N_i \sigma_i}{\Sigma\, N_i \sigma_i} \qquad (11)$$

This equation fixes the sample n_i within a class in terms of the total sample n, supposed known. In the problem of this chapter, the approach was from the other direction—a desired precision was aimed at, and k determined accordingly (cf. Eqs. 19 and 21). The sample sizes n_i were then computed from Eq. 22. They of course satisfy Eq. 11, but n was not known until the n_i were computed and added up.

In symbols, the standard deviation σ_i in Eqs. 8 and 9 is defined by the equation

$$\sigma_i{}^2 = \frac{1}{N_i} \sum_{j=1}^{N_i} (x_{ij} - \mu_i)^2 \qquad (12)$$

Eqs. 8 and 9 are intended to refer to conditions existing at the date of inventory. Actually, σ_i is known only for some earlier date (in this case, the preceding September and December), so it is necessary to approximate σ_i for the date of the sample (cf. Eq. 14). Eq. 9 gives the variance of the mean number of tires per dealer as would be obtained in an ideal experiment in which repeated samples of n_i dealers are drawn at random and without replacement from the N_i dealers. Each class of dealer will contribute a term (see Eq. 13 ahead).

Remark 3. The n_i dealers from any class were selected systematically by starting at a random point in the first interval in the file and taking out one card at a time at a constant interval throughout the file for that class. A systematic selection is much easier to carry out than any serious attempt at random selection. Moreover, systematic selection in this type of problem may introduce slight gains over a random selection, for the reason that systematic selection assures nearly proportionate representation of dealers from each city and state, because of the arrangement of the files. This may have been of some importance because the inventories in one broad locality usually differ from those in another both qualitatively and quantitatively. On the other hand, it must also be recognized that systematic selection will sometimes introduce losses.

The Tukey plan for evaluating the precision of a systematic sample (p. 353) was not used here.

An assumption for arriving at usable values of σ_i. The symbol X in Eq. 4 denotes the total number of new passenger tires as estimated from the sample. As in Eq. 6 of Chapter 6 the variance of X will be

$$\sigma_X^2 = N_1^2 \sigma_{\bar{x}_1}^2 + N_2^2 \sigma_{\bar{x}_2}^2 + \text{similar terms for all other classes} \\ \text{through all groups}$$

$$= N_1^2 \frac{N_1 - n_1}{N_1 - 1} \frac{\sigma_1^2}{n_1} + N_2^2 \frac{N_2 - n_2}{N_2 - 1} \frac{\sigma_2^2}{n_2} + \text{similar terms}$$

$$\doteq N_1^2 \left(1 - \frac{n_1}{N_1}\right) \frac{\sigma_1^2}{n_1} + N_2^2 \left(1 - \frac{n_2}{N_2}\right) \frac{\sigma_2^2}{n_2} + \text{similar terms} \quad (13)$$

In the planning of the sample, suitable values of σ_1, σ_2, etc., must be settled upon from considerations of possible trends and dispersions of tire inventories. For the March sample there was the experience of the December sample to fall back on, which helped considerably.

It seems reasonable to suppose that the changes in dispersion took place in the inventories of a group of dealers over a 3- or 6-month period will be roughly proportional to their average inventory of the base date (September 1944). This is only saying that the stock of a large dealer will usually change more in either direction than the stock of a small dealer, and about in proportion to size as measured by initial inventory. This is the assumption that is contained in Eq. 29 (p. 223) of Chapter 6. It of course breaks down in the small size-classes (cf. the treatment of the smallest size-class in Table 4 and the treatment of "0-blocks" at the end of the chapter). With this in mind, it was perceived that in the returns from the December sample σ_i/μ_i was roughly constant in the different size-classes of Group A and hardly exceeded 1.25. Similarly, it hardly exceeded 0.75 in Group B. To allow for still further dispersion during the additional 3 months between December and March it was assumed that

$$\left.\begin{array}{l} \dfrac{\sigma_i}{\mu_i} = 2 \text{ in Group A} \\[2em] \dfrac{\sigma_i}{\mu_i} = 1 \text{ in Group B} \end{array}\right] \quad \begin{array}{c} \text{[Corresponding to} \\ \text{Eq. 29, p. 223]} \end{array} \quad (14)$$

These approximations are admittedly rough, as is only too obvious from Tables 1 and 2.

TABLE 1. COEFFICIENTS OF VARIATION FOR SEPTEMBER AND DECEMBER FOR GROUP A

Group A: Dealers that in September had new truck and bus tires, but

fewer than 40 { new passenger tires used truck and bus tires used passenger tires

μ_i and σ_i were estimated from a subsample of about 100 returns taken systematically from each size-class

| Number of new truck and bus tires in September | New truck and bus | | | | | | New passenger | | | | | |
| | September | | | December | | | September | | | December | | |
	μ	σ	σ/μ	μ	σ	σ/μ	μ	σ	σ/μ	μ	σ	σ/μ
1- 9	3.5	2.3	0.66	4.1	5.9	1.44	8.9	9.3	1.04	14.8	18.2	1.23
10-19	13.5	2.7	.20	13.0	9.6	0.74	11.1	10.6	0.95	21.0	26.3	1.25
20-29	24.1	2.8	.12	25.0	13.2	.53	13.6	11.2	.82	34.2	40.6	1.19
30-39	34.0	2.5	.07	38.2	17.9	.47	18.5	12.1	.65	34.2	28.2	0.82

Remarks regarding the assumptions. It is appropriate to recall three remarks. As was seen in Chapter 6, Neyman sampling only requires that the *ratios* $\sigma_1 : \sigma_2$, etc., between the standard deviations σ_i be assigned, and not their absolute magnitudes. Now very often it is possible to make good guesses at these ratios even though the value assigned to each σ_i is too high or too low. Of course, if one consistently assigns too high a value to every σ_i and computes the size of sample required to give the precision aimed at, the precision actually attained will be needlessly high, which is not desirable as it entails a heavier burden of response and heavier cost than are actually necessary. On the other hand, if one consistently assigns too low a value to each σ_i, the precision actually attained will not be as good as the precision aimed at, though it may nevertheless be good enough. In the second place, it is a fact that the efficiency of the sample (as measured by the inverse of $\sigma_X{}^2$) is only feebly altered by moderate departures from the optimum ratios between n_1, n_2, etc., as given by Eq. 8, wherefore a sampling plan may still be very efficient even if some of the assumptions and approximations are crude.

In the third place, regardless of what assumptions and approximations were made, or why, there need be no guess-work in the final results, because an analysis of the returns will show what precision was actually attained.

TABLE 2. COEFFICIENTS OF VARIATION FOR SEPTEMBER AND DECEMBER FOR GROUP B

Group B: Dealers without new truck and bus tires in September and with

fewer than 40 $\begin{cases} \text{used passenger tires} \\ \text{used truck and bus tires} \end{cases}$

μ_i and σ_i were estimated from a subsample of about 100 returns taken systematically from each size-class

Number of new passenger tires in September	New truck and bus						New passenger					
	September			December			September			December		
	μ	σ	σ/μ	μ	σ	σ/μ	μ	σ	σ/μ	μ	σ	σ/μ
0	...	...	...	0.2	0.9	4.5	....	...	....	1.0	3.6	3.60
1- 9	...	...	...	0.5	1.8	3.6	4.0	2.4	0.60	6.7	8.2	1.22
10-19	...	...	...	1.0	3.3	3.3	13.1	3.1	.24	13.0	9.9	0.76
20-29	...	...	...	1.2	4.9	4.1	24.0	2.7	.11	24.7	11.4	.46
30-39	...	...	...	0.8	2.7	3.4	33.9	2.9	.09	32.0	12.4	.39

Evaluation of k. To find k we note that Eq. 8 gives

$$\frac{\sigma_i^2}{n_i} = \frac{k\sigma_i}{N_i} \tag{15}$$

which when used in Eq. 13 shows that

$$\tau_X^2 = k \left\{ \frac{\sigma_1}{\mu_1} N_1 \mu_1 \left(1 - \frac{n_1}{N_1}\right) + \frac{\sigma_2}{\mu_2} N_2 \mu_2 \left(1 - \frac{n_2}{N_2}\right) + \text{similar terms} \right\} \tag{16}$$

Now suppose that we restrict consideration to Groups A and B, fixing the aimed-at precision so that it should be attained from these two groups alone. The relatively small contributions of passenger tires from other groups of dealers will only slightly increase the precision. This procedure is not strictly Neyman allocation over all strata, but is not far from it. One further simplification will be made: the finite multiplier $1 - n_i/N_i$ in each term of Eq. 16 will be dropped (i.e., assumed to be near unity, as it would be if the sampling ratio n_i/N_i were small). Actually, the calculations for k (*vide infra*) should perhaps have been repeated to get a second approximation by retaining the finite multipliers in Eq. 16 with rough values of n_i/N_i. The coefficients of ξ_A and ξ_B in Eq. 17 would thereby have been decreased, but it was supposed that the effect was slight and not worth the effort. The next step, therefore, was to ignore the finite multipliers and factor out σ_i/μ_i for Groups A and B separately in Eq. 16 to find that

$$\sigma_X{}^2 = k \left\{ \left(\frac{\sigma}{\mu}\right)_A (N_1\mu_1 + N_2\mu_2 + \text{similar terms in Group A}) \right.$$

$$\left. + \left(\frac{\sigma}{\mu}\right)_B (N_1\mu_1 + N_2\mu_2 + \text{similar terms in Group B}) \right\}$$

$$= k\{2\xi_A + \xi_B\} \tag{17}$$

because $\sigma_i/\mu_i = (\sigma/\mu)_A = 2$ for Group A and $\sigma_i/\mu_i = (\sigma/\mu)_B = 1$ for Group B according to a previous decision (Eq. 14). It follows that the

$$(\text{C.V. } X)^2 = \frac{\sigma_X{}^2}{\xi^2} = k \frac{2\xi_A + \xi_B}{\xi^2} \tag{18}$$

where C.V. X denotes the coefficient of variation of X, or its standard error expressed in units of ξ itself. Solved for k, this equation gives

$$k = \frac{\xi^2(\text{C.V. } X)^2}{2\xi_A + \xi_B} \tag{19}$$

It is now necessary to anticipate the division of the total number of tires ξ between Groups A and B. For simplicity it was assumed that the division in March would be what it was in December, viz.,

$$\xi_A : \xi_B = 8:1 \tag{20}$$

Moreover, it was necessary to decide on a suitable value for ξ as of March 1945, and after some deliberation it was decided that 1.8 million could not be far wrong. This number fixed the values of ξ_A and ξ_B as 1.6 and 0.2 million respectively, whereupon k could be obtained from Eq. 19 as soon as the aimed-at precision C.V. X was decided upon (next section).

The aimed-at precision. It was decided that a coefficient of variation of 1.5 percent would be desirable. This corresponds to a 3-sigma error band of 4 or $4\frac{1}{2}$ percent, outside of which errors of sampling practically never fall. This band of error seems to be small enough in view of the uses to be made of the data, and in view of other errors and biases that arise from difficulties in counting tires on hand and consigned (Ch. 2). Also there was the stock of unknown magnitude held by unauthorized dealers. Altogether, further refinement of sampling error seemed out of place, because it would entail additional burden of response on the part of dealers and increased costs to the government, and would moreover enhance liability to errors arising from nonresponse and other sources, most of which, as surveys are actually carried out, become more troublesome as the sample increases, and reach their maximum in attempts at complete counts.

With the aimed-at coefficient of variation of X set at 1.5 percent, Eq. 19 gave

$$k = \frac{1.8^2 \times 10^{12} \times 1.5^2 \times 10^{-4}}{2 \times 1.6 \times 10^6 + 0.2 \times 10^6} = 214 \qquad (21)$$

If the finite multipliers in Eq. 16 had not been neglected, k would have turned out to be a little greater than 214. Nevertheless, for ease in computation k was fixed at 200, whereupon from Eqs. 8 and 14

$$\left. \begin{aligned} \frac{n_i}{N_i} &= \mu_i \quad \text{directly in percent for Group A} \\ &= \tfrac{1}{2}\mu_i \quad \text{directly in percent for Group B} \end{aligned} \right\} \qquad (22)$$

And thus the sample was allocated in Groups A and B, which account for most of the new passenger tires (truck and bus also). Tables 3 and 4 show the calculations and recommended sampling ratios.

TABLE 3.   SIZE OF SAMPLE FOR GROUP A

(Group A is defined in Table 1)

Number of new truck and bus tires in September	For new passenger tires; assumed for March μ_i	Size of sample computed (Eq. 22)	Actual size of sample decided upon	Number of dealers in the	
				Universe N_i	Sample n_i
1– 9	15	15%	1 in 6	19,850	3,308
10–19	22	20	1 in 5	3,250	650
20–29	30	30 ⎱	1 in 3	1,613	538
30–39	35	35 ⎰			
40–49	45	45 ⎱	1 in 2	894	447
50–59	55	55 ⎰			
60 and over	..	100	All	1,662	1,662
Total				27,269	6,605

Special groups of dealers. *The nonresponses of September.* In the September attempted "complete count" 24,015 questionnaires were not returned, even after two follow-up notices. In December a 4 percent sample of these names was drawn: in number this sample was 997. Here is what happened:

Number of questionnaires mailed out as a 4 percent sample of the September nonresponses.. 997

Failed to return by mail and personal visit or could not be located or no personal follow-up made... 310

Returned.. 687
  Marked out of business.................................. 217
  Marked 0 passenger tires, or completely blank (interpreted as
    having no stock)....................................... 134 ⎱
  Reports showing 1 or more new passenger tires.............. 336 ⎰ 470

For March, a 5 percent sample of the September nonresponses was recommended. The coefficient of variation (37.9/11.9 or about 3) of this group perhaps called for a slightly larger sample, but it was deemed better not to risk overloading the district offices with this group (which was obviously inclined not to respond) and to accept a little larger sampling error at the gain of decreased bias of nonresponse.

TABLE 4. SIZE OF SAMPLE FOR GROUP B

(Group B is defined in Table 2)

Number of new passenger tires in September	For new passenger tires; assumed for March μ_i	Size of sample computed (Eq. 22)	Actual size of sample decided upon	Number of dealers in the	
				Universe N_i	Sample n_i
0	1	$\frac{1}{2}\%$ *	1 in 20	6,340	317
1– 9	7	$3\frac{1}{2}$	1 in 25	23,650	946
10–19	13	$6\frac{1}{2}$	1 in 12	6,500	542
20–29	25	$12\frac{1}{2}$	1 in 8	1,833	229
30–39	35	$17\frac{1}{2}$ ⎱	1 in 5	1,136	227
40–49	45	$22\frac{1}{2}$ ⎰			
50–69	60	30	1 in 3	358	119
70–99	85	$42\frac{1}{2}$	1 in 2	163	82
100 and over	..	100	All	168	168
Total				40,148	2,630

* The sample decided upon was larger than computed for this size-class not only because its dispersion was much higher than the other size-classes (see Table 2) but because this class of dealer is subject to more important relative shifts over time.

For the 470 dealers returning blanks or 0 or 1 or more tires, the mean number of new passenger tires per dealer was 11.9, with a standard deviation of 37.9. When expanded by 25 this group gave

$$11.9 \times 470 \times 25 = 140{,}000 \text{ new passenger tires}$$

This figure is to be interpreted as a minimum in the number of passenger tires held by the nonresponses of September because the 310 that failed to return likely also held tires. Anyhow, this minimum of 140,000 was a twelfth of all the new passenger tires in the hands of dealers in December, and if these dealers held anything like the same stocks in September it is easy to see why there should have been discrepancies between the September complete count and the two quarterly samples in the preceding March and June (as was mentioned earlier).   The ability of the staff of the district offices to call on the smaller number of delinquent dealers in a sample is responsible for the better response in December and March.

Blanks.   In September, 29,133 dealers returned blank questionnaires. A sample of 3 percent of these was used in December, which when expanded on the basis of the returns accounted for 71,000 new passenger tires.   One might conjecture that this group of dealers held enough tires in September to aggravate the trouble of balancing the books. A 3 percent sample was recommended again for March.

New authorizations.   There were 1000 new authorizations between September and December, and another 1000 were allowed for over the interval between December and March.   A sample of 200 was recommended as being adequate and easy to draw, being 10 percent.

Manufacturers' outlets.   There were about 2000 manufacturers' outlets, averaging for September about 75 new passenger tires per dealer.   A 100 percent sample was recommended, but a 1 in 2 or 1 in 4 sample might have been good enough because these outlets were afterward observed to have fairly uniform stocks.   No calculation of their coefficient of variation (σ/μ) was made.

Dealers holding in September 40 or more used tires of any kind but fewer than 40 new passenger or new truck or bus tires.   A 25 percent sample was recommended: every 4th card starting with the 2d, first classifying the dealers in intervals of 10 tires (40–49, 50–59, etc.).

Motorcycle tires.   Of the dealers reporting no new passenger tires and no new truck and bus tires in September, and fewer than 40 used passenger or truck and bus tires, 296 reported new motorcycle tires, and 91 reported used motorcycle tires.   After inspecting the reports of these 387 dealers the sampling plan shown below was recommended:

Number of motorcycle tires in September	Sample
1– 4	25% or 1 in 4
5– 9	25% " "
10–14	25% " "
15–19	25% " "
20 and over	100% " all

Of course there were motorcycle tires held by other dealers, but stocks of motorcycle tires were well dispersed, and the precision in this category should be good enough.

The contributions to the variance from these special groups were small because they were sampled adequately (approximately according to optimum allocation) and also because Groups A and B contained nine-tenths of the new passenger tires (and even more of the new truck and bus tires). In particular, the manufacturers' outlets, like the dealers with heavy stocks in Groups A and B, contributed no variance, being sampled 100 percent.

Size of sample—all groups. The overall sample from all groups, as calculated, turned out to be 14,750 dealers, which would have been about 1 dealer in 10. However, partly by mistaken carry-over of instructions from previous samples, and partly to gain simplicity in operation, bigger samples than specified were taken out of some of the special classes, and the actual number of questionnaires mailed out was close to 16,000. As mentioned earlier, the response from the March sample (with two follow-up letters) was 96.1 percent, and this sample included a 5 percent sample of the dealers that did not respond in September.

Results: comparison of the precision attained with the precision aimed at. *The precision attained: the effect of stratification.* After the returns were in, a subsample of about 100 cases was drawn systematically from every group for the purpose of estimating σ_i and μ_i for use in Eq. 13 whence the precision of the results was obtained. For comparison, the coefficients of variation were also computed for the precision as it would have been if no stratification had been introduced. In both investigations the assumption was made that the equations of random sampling can be applied to estimate the precisions of our systematic samples, as in Remark 3. The apparent coincidence between the precision aimed at for the new passenger tires (1.5 percent) and the precision actually attained (again 1.5 percent) is partly an illusion, because the precision aimed at applies strictly only to Groups A and B, whereas the precision actually attained and shown in the table was computed for all groups combined. However, as was explained earlier, it was known that the contributions from the other groups could not have much effect on the overall precision, so it can be concluded that the agreement was excellent—unfortunately too good for illustrative purposes. The outcome does nevertheless show qualitatively the kind of results that will be obtained from plans that are drawn up with reasonable care, even though some of the assumptions and simplifications that one is forced to make often appear crude at the time.

The precision for the new truck and bus tires (1.4 percent; see Table 5) turned out to be entirely satisfactory, as was expected, even though the criterion for fixing k and hence for fixing the size of the sample was based on new passenger tires. The reason is that the truck and bus tires were heavily concentrated in Group A.

TABLE 5. ACTUAL COEFFICIENTS OF VARIATION ATTAINED OVER ALL DEALERS

Plan	New truck and bus tires	New passenger tires
Stratified with optimum allocation	1.4%	1.5%
Unstratified (for comparison)	4.4	2.5

Interpretation of sampling error. The observed coefficient of variation, measured after the returns are in, tells us that the number of new passenger tires held by dealers on March 31st, as determined by the sample, can hardly differ from the results of a "complete count" by more than $4\frac{1}{2}$ percent (three coefficients of variation). The error band for truck and bus tires happens to be about the same width or smaller. But to interpret these error bands it must be understood that the theoretical complete count with which the sample is compared must be taken with the same thoroughness as the sample, including pressure on the non-responses. Moreover, the complete count must cover the same universe of dealers and not include (e.g.) unauthorized dealers. Thus, while the sample contains sampling errors and is therefore less precise than a complete count, it is undoubtedly more accurate than a complete count would have been in practice.

Note in regard to stratification and conservation of information from complete counts. It is obvious from Table 5 that the stratification based on the complete count of September increased the efficiency of sampling enormously, even though considerable dispersion of inventories took place between September and December and still more between December and March. Unstratified, the required sample for new passenger tires would have been $2\frac{1}{2}$ times as large; and for the new truck and bus tires it would have been 5 times as large. It should be explained that the gains in efficiency attributed here to stratification were actually the gains of stratification with Neyman allocation. The gains that would have been made by stratification with proportionate sampling would not have been as great—perhaps two-thirds as great in this survey (cf. the preceding chapter).

This experience seems to be typical in many sample surveys and teaches an important lesson in regard to conserving whatever information is available concerning the universe. Thus, a census of population providing statistics by blocks, tracts, and city size-groups, even though now out of date in many spots because of heavy population shifts, may nevertheless still be helpful for purposes of stratification, even for sampling blocks within cities, as experience and theory demonstrate. Some of the reasons are found in the section entitled "Remarks regarding assumptions." For example, Neyman allocation will indeed be obtained even though the census or latest available complete count indicates only rough *relative* sizes (of firms, blocks, farms, etc.) and is entirely wrong on absolute values. But this is not all; a complete count, although out of date, will often go a long way toward segregating areas that may have suffered important changes and that should be sampled in higher proportion. Thus, in the footnote to Table 4 it is explained that the 0-class was sampled heavier than was indicated by the rule that was used for the other classes. Similarly, in the sample censuses described in the booklet *A Chapter in Population Sampling*, the blocks were stratified into classes according to the number of dwelling units in 1940, and every *n*th block was taken out for the sample. But the "0-blocks" (those that had no dwelling units in 1940) were thrown into a separate class for special investigation as a safeguard against the enormous changes that can and do take place when a large housing development is built on land not previously occupied. Another illustration of the conservation of available information is found in the best practice in sampling in industry, wherein the lot, shipment, shift, or even order of receipt is usually effective for subgrouping even though it may appear that the product must have become thoroughly mixed after production. It never is thoroughly mixed unless there has been a deliberate and skillful attempt to mix it.

It is not to be inferred that promiscuous stratification will get results. After the main assignable causes of variability have been removed through stratification, further stratification will usually be too ineffective to repay the cost.

CHAPTER 12. A POPULATION SAMPLE FOR GREECE [1]

In this so great a diversity of opinions concerning the true measure of the earth's circumference, let it be free for every man to follow whomsoever he please.—Robert Hues, *Tractatus de Globis* (London, 1592).

The purposes of the samples. In the summer of 1946 a combined mission of American and British observers was sent to Greece by their governments to examine the electoral lists which were being prepared for a plebiscite on the issue of whether George II was to be retained as king of the Hellenes. To be exact, this was the second mission, and was designated as AMFOGE II. An earlier mission had been sent to Greece for observations on the election of 31 March 1946; its work was completed and a report was issued by the Department of State, 1946: publication 2522. Both missions were formed to carry out the provisions of the Varkiza agreement. This agreement, signed 12 February 1945, a day after publication of the Yalta Declaration, required that Greece would hold both parliamentary elections and a plebiscite, and stated that "representatives of both sides agree that for the verification of the genuineness of the popular will, the great allied powers shall be requested to send observers."

In carrying out its observations the mission made a number of statistical surveys to obtain not only first-hand information on registration but also considerable background information on general population characteristics. Only the population aspects of the mission's statistical activities will be presented here.

The problem of investigating the electoral lists. Any electoral list may fail in two ways:

  a. It may contain surplus names—that is, names of men who have died before a certain date and should either not be on the list at all or should be subtracted out by entry on a subsequent negative register; fictitious names; multiple registrations; names of men who have lost their right to vote by conviction of certain felonies, desertion from the army, etc.

  b. It may fail to contain names that should be on it.

The two types of error are independent; i.e., a list may be entirely free of the first type, yet not of the second, containing no surplus names but failing to contain the names of some citizens who are entitled to register. Likewise it may be free of the second type, yet not of the first. The purpose of the mission was to determine the extent and variability of each type of error, and the reasons therefor, and to decide whether the electoral rolls could safely be used in a national referendum.

[1] Copied almost verbatim from an article by Jessen, Blythe, Kempthorne, and Deming, *J. Amer. Stat. Assoc.*, vol. 42, 1947: pp. 357–84.

The sample of names from the rolls. In order to determine the extent of the first type of error, samples of names were drawn with known probabilities from all the electoral lists in Greece, positive and negative, basic and supplemental.[2] Every name in the sample was investigated to see whether it was a valid entry. This investigation included not only information from the man himself (when possible) and his neighbors, but also a further examination of the lists to determine whether his positive (or negative) entry was annulled, properly or improperly, by a negative (or positive) entry. The sample of names was not valid as a household sample; it was, as intended, a sample of names from the rolls. This sample will therefore not be mentioned again here except in brief.

The sample of households. This is the sample that forms the subject of this article. It was taken primarily to investigate the second type of error in the rolls. It was a sample of households drawn with known probability from all of Greece, and it could thus be used for obtaining data on the population as well as for measuring the extent of the second type of error in the electoral rolls. One very important population characteristic in the investigation of the electoral lists was the total number of male Greek citizens 21 and over, which determined an upper limit to the potential number of electors.

The population sampled included those inhabitants who were reported as belonging to households. A household was defined as one person living separately or a group of people living together as a unit (in general, a family). Observers were required to include as members of households "all persons associated with it who will not be included as members of some other household as defined here." People temporarily absent from a household, such as members of the armed forces, were to be included. It should be noted that people who would not come into this enquiry could not in practice be taken into account for election purposes, no matter how perfect the electoral procedure. In the opinion of the authors the total number of people thus excluded could not materially affect the estimates given here.

The form of questionnaire that was used for eliciting the information from the households contained spaces for sex, age, literacy, employment, class of worker, and occupation, aside from the spaces required for information concerning registration. The form used is shown at the close of the chapter. The information obtained from the household concerning the registration, citizenship, and residence of every male 21 and over was checked against the local electoral rolls, positive and negative, to

---

[2] For an explanation of the nature of the Greek electoral lists see Jessen, Kempthorne, Daly, and Deming, "Observations on the 1946 elections in Greece," *Amer. Sociol. Rev.*, vol. 14, 1949: pp. 11–16.

determine whether this information was correct—more precisely, to determine the extent of the second type of error (Type b as described above).

This paper presents the method of selecting the household sample, the method of estimating the various population characteristics, the calculation of the sampling errors, and the results on some of the population characteristics of Greece.

Outline of the plan for the sample of households. The household sample was so designed that every household had a known probability of being drawn into the sample. It was moreover drawn in such manner that the bias of selection was eliminated, and the standard errors were computable. These features were considered absolutely necessary because of the importance of the decisions to be based on the data. Some of the towns and villages selected for the sample were accessible only with difficulty, but no substitutions of either village or household were permitted. Ofttimes a village could be reached only after a day's journey by jeep, supplemented by transportation by burro or on foot. Boats furnished transportation to the islands and to coastal towns inaccessible otherwise because of swamp or mountains. Aeroplanes were used occasionally for dispatching observers and for supervision of the work.

The field-work required the services of 65 observer-teams for 3 weeks. Each observer-team consisted of an observer, an interpreter, a driver, and the indispensable jeep. The British observers were officers in His Majesty's and Dominions' Forces; the American observers were civilians (mostly instructors and graduate students in statistics, government, political science, and economics) dispatched to Greece for that purpose.

In regard to the time required for the entire job, it should be noted that the first contingent of the sampling staff arrived in Athens about 23 June, and the report of the mission was made to the press 19 August— a total elapsed time of about 8 weeks.

On the basis of estimates of the variability of certain important characteristics of the population from one area to another, and in the light of information obtained on a similar mission in the spring of 1946, a sample of the following description was indicated: about 200 cities, towns, and villages drawn from strata suitably created on the basis of geographical divisions and the population in 1940, and including 1/500th of the population of Greece.

The primary sampling unit, which we shall refer to as a "place," was a village, town, or city, along with its satellite villages and any households in the open country (rare in Greece) within a carefully delimited area whose combined population appeared in the 1940 Greek census as a single entry or "place." Several stages of subsampling were usually used within a primary area. The ultimate unit was the house-hold.[3] The sampling ratios in the various stages were adjusted so that the overall sampling ratio in every stratum was uniformly 1/500. Thus the sample was self-weighting. This self-weighting feature was extremely important because of the simplicity introduced into the tabulations which were necessarily carried out by hand under great pressure for speed.

For obtaining the total population of Greece, a regression estimate was used. More explicitly, the sample was used to obtain estimates of *i*. the change in population since 1940; and *ii*. the proportions of the population in various subclasses. The mathematical expressions for the regression estimates and their standard errors will be given further on.

The sample of villages, towns, and cities. The basis for the sample design was the 1940 population census of Greece, which, although out of date in many respects, nevertheless contained information valuable for controlling and diminishing the errors of sampling. The census publication distinguished two types of area: first, the *koinotis* (plural, *koinotetes*), which is a small community or village; and second, the *demos* (plural, *demoi*), which is usually a town or city with more than 10,000 population, or else the capital of a *nomos* (plural, *nomoi*), which represents a province. The census of 1940 gave the population of each of the 5690 *koinotetes* and *demoi* covering the whole of Greece. Many of the *koinotetes* and *demoi* include more than one community or population group. The census publication showed the name of only the main community in an area, but a list and a map were obtained from the Ministry of Interior which showed the names and locations of all the populated centers within each *koinotis* and *demos*. These smaller villages and towns were termed "satellites" and for purposes of sampling were regarded as parts of the listed *koinotis* or *demos*. Thus every part of Greece had a chance to be included in the sample.

At the time of the 1940 census Greece was divided into 38 *nomoi*. These in turn were divided into *eparchies*, and the *eparchies* into *koinotetes* and *demoi*. Thus, every square foot of Greece was in one or another *koinotis* or *demos* as listed in 1940. The Dodecanese Islands were not included.

The first step in the sampling plan was to subdivide the *koinotetes* into 4 classes and the *demoi* into 2, according to their 1940 populations. Table 1 shows the plan of classification and the sampling ratios that were used for selecting the *koinotetes* and *demoi*, and for selecting the sample households from within them. Attention is called to the fact that the product of any of the two sampling ratios standing side by side in the fourth and fifth columns is always 1/500. Had straight multi-

[3] See Schemes 2a and 2b on page 380; also the appendixes at the end of this chapter.

plication by 500 been intended, instead of a ratio-estimate, probably 12 or 15 size-classes of community would have been used, instead of 6.

The sample *koinotetes* in Class 1 were selected by starting with a random number between 1 and 100, such as 37, and marking off in the census publication the 37th, 137th, 237th, ⋯ *koinotis* of the size specified (0–499). The sample *koinotetes* and *demoi* in the other classes were drawn in a similar manner. All 22 cities having 25,000 or more inhabitants in 1940 were in the sample.

It was necessary to consider in the overall sample-design the needs of both aspects of the mission's work, that is, to investigate the two kinds of error in the electoral lists (*vide supra*). In order that a sample-design may be made as efficient as possible, it is essential that the designer know, or have estimates of, the nature and amount of variability existing in the universe under consideration, and the costs, in terms of time and equipment, of performing the various possible sampling operations. It was possible to use experiences acquired in the United States, supplemented by experience gained from the first mission and from trial runs, to provide useful information on a number of these points. The principal items of importance from the standpoint of design are listed below.

1. *Travel.* The average distance that could be covered to or from a randomly selected sample place in Greece by jeep, burro, or boat was estimated at 75 miles per day.

2. *Drawing the sample of households.* The average time required to make lists or maps and to designate the sample households thereon according to the prescribed rules was estimated to be 1 sample place per day.

3. *Interviewing.* The rate of interviewing was estimated at 15 sample households per day for 1 observer.

4. *Locating and examining the electoral lists.* The observer was required to find the local electoral lists (which sometimes required traveling to another city or town) and to draw off a sample of names according to prescribed rules, along with certain identifying information, so that he could investigate errors of Type *a* (next step). He was also required to examine the information appearing on the electoral lists concerning the men previously interviewed in the household sample, to investigate errors of Type *b*. It was estimated that the observer would require a whole day to do this work.

5. *Investigating the registrants.* It was estimated that the investigation of the errors of Type *a*, which required interviewing the registrants drawn from the electoral lists found in Step 4, and making the necessary enquiries of neighbors, would proceed at an average rate of 20 names per day.

With these advance estimates of cost in mind, it was possible to draw up a sampling plan that would produce the reliability required and which at the same time would approach the optimum use of resources. In consideration of the above costs of operation, the number of observers (65), and the period of 3 weeks allotted to the field-work, it was decided that about 15 sample households or 1 average day's

interviewing in 1 sample place (including satellite villages), and about
200 sample places, would constitute a good workable plan. Each observer would then have 3 sample places to complete in the 3 weeks, or
1 sample place per week, the time being allocated as follows:

                                                                    Days
1. Travel to and from the sample place................. 2
2. Preparing the household sample..................... 1
3. Interviewing the sample households.................. 1
4. Locating and using the electoral lists................ 1
5. Investigating names drawn from electoral lists......... 1
                                                            —
    Total.. 6

The decision was made, on the above considerations, to adopt as a
sampling unit, an "expected" 15 households within a sample area, this
number being based on the 1940 census of Greece. For example, in
a village of 750 inhabitants in 1940 (150 households with 5 people per
household), a sampling ratio of 1 in 10 would result in an "expected"
15 households (see size-class 2 in Table 1). For places in this class with
populations other than 750 the "expected" number of households would
of course be different from 15. It would have been possible to make use
of the 1940 population data for each place separately (instead of by
size-class) to determine the particular within-place sampling ratio that
would be required to obtain a constant "expected" size of sampling
unit, but this would have required too much time for the selection of the
sample places and moreover, under the circumstances, would have been
too cumbersome in the field. Such a plan offers greater accuracy than
the one used if the method of estimating totals is the simple unbiased
estimate, obtained by multiplying the sample totals by the reciprocal
of the overall sampling fraction (500), but it was believed that the loss
of information resulting from this variation in size of sampling unit
(caused by varying size of village) could mostly be recovered in the plan
that was followed, wherein a ratio-estimate [4] was used (*vide infra*).

A sample of 200 areas with 15 households per area contains a total
of 3000 households. On the assumption of a total population of $7\frac{1}{2}$
million people, or $1\frac{1}{2}$ million households (5 people per household), the
overall sampling ratio would be 3000 divided by $1\frac{1}{2}$ million, or 1 in 500.
It was decided to keep the sampling ratio constant over the whole of
Greece, regardless of size of community. A preponderant justification
for the constant ratio was simplicity of tabulation, enforced by the
rigid requirements for speed.

[4] The ratio-estimate studied in Chapter 5 is one type of regression-estimate, but
the type to be used here is more specific to an assumed model; see the footnote on
page 165

The actual procedure for drawing the sample of villages, towns, and cities was described above. This was merely the means of putting into effect certain theoretical considerations of sample allocation which will now be described. The procedure of drawing the sample areas can be looked upon as creating, within each size-class, geographic-alphabetic strata of 500 sampling units, 15 households to the sampling unit as measured in 1940—geographic-alphabetic because the *demoi* and *koinotetes* were listed in the 1940 census volumes alphabetically by *nomoi*. In size-class 1 each stratum consisted of 100 places, each of 5 sampling

TABLE 1. SUMMARY OF THE SAMPLE-DESIGN

(1) Size-class code	(2) Population in 1940	(3) Assumed average population in 1940	(4) Sampling ratios For selection of sample places	(5) Sampling ratios For selection of names and households within a sample place	(6) Number of places In size-class	(7) Number of places In sample
			For koinotetes			
1	0– 499	350	1/100	1/5	2147	20
2	500– 999	750	1/50	1/10	2049	40
3	1000–4999	2500	1/20	1/25	1366	70
4	5000 and over	7000	1/5	1/100	54	10
Totals					5616	140
			For demoi			
5	Under 25,000	17,000	1/2	1/250	52	26
6	25,000 and over	....	1/1	1/500	22	22
Totals					74	48

NOTE. The total number of sample places is perhaps better counted as 199, and not as 188(140 + 48) shown in Table 1. Every *demos* of 25,000 or more inhabitants was counted only once in Table 1, but the large ones contributed several sample parishes; 7 parishes from the city of Athens, for instance, were in the sample. Actually, data were obtained from 199 places and "semiplaces" (parishes) referred to in the text as "about 200."

units; in size-class 2 each stratum consisted of 50 places, each of 10 sampling units, and so on (cf. Table 1).

One sampling unit (15 "expected" households) was drawn per stratum. The constant sampling ratio of 1 in 500, sufficiently justified on the grounds of simplicity as mentioned above, happens to accord fairly well with the principle of optimum allocation, according to which

$$\frac{n_i}{N_i} = \frac{\sigma_i}{\lambda \sqrt{c_i}} \qquad \text{[Eq. 15, p. 218]} \quad (1)$$

where λ is a proportionality constant, n_i is the number of sampling units drawn from Stratum i in the sample (1 in this problem), N_i is the number of sampling units originally there (500), c_i is the cost of carrying out the enquiry in 1 sampling unit, and σ_i is the standard deviation of the desired characteristic defined by the equation

$$\sigma_i{}^2 = \frac{1}{N_i} \sum_j (x_{ij} - \bar{x}_i)^2 \qquad (2)$$

wherein x_{ij} represents the value of the desired characteristic in sampling unit j, and $\bar{x}_i$ represents the average value of this characteristic

TABLE 2. SIZES OF STRATA AND ESTIMATED NUMBER OF INHABITANTS TO BE
SAMPLED FROM EACH, BY COMMUNITY SIZE-CLASS

Size-class code	Number of strata	Population (estimated from 1940 census) per stratum	Estimated number of inhabitants in sample per stratum	Estimated number of households * in sample per stratum
1	20	35,000	70	14
2	40	37,500	75	15
3	70	50,000	100	20
4	10	35,000	70	14
5	26	34,000	68	14
6	33 †	37,500 †	75	15

* A household was assumed to consist of 5 inhabitants.   † See note in Table 1.

in all N_i sampling units of Stratum i. As a great many characteristics were to be measured (population change since 1940, number of Greek male citizens 21 and over, number registered, and many others) and as the strata created were of approximately equal size, it could only be assumed for purposes of allocation that σ_i was the same for all strata. It has already been explained that c_i is about the same in city and village; hence the right-hand member is practically constant and demands the constancy of n_i/N_i. But regardless of the principle of allocation, the constancy of n_i/N_i would probably have been required in order to gain time and eliminate complications in the tabulation. Tables 1 and 2 show the sample plan in summary.

Experience has shown that even crude approximations to optimum allocation are highly preferable to no guide at all. The worst to be expected is some loss in efficiency and some loss in advance control of the sampling error.

It should be noted that no bias is introduced by failure of any of the above assumptions and, moreover, that the actual sampling errors as calculated from the returns are entirely independent of the above assumptions that were made in the planning.

The selection of the ultimate sampling unit, the household. A sample of households was selected from every place that was drawn into the sample. The procedure in every instance was to select a prescribed proportion of the households of the place. The procedure varied. In the smaller villages, principally in the first two community size-classes, it was sometimes possible to procure a list of all the families, which with some effort could be made complete. When this was the case, the drawing of the sample families was very simple; a starting number and a sampling interval were applied to the household lists (this being, incidentally, the plan of drawing names from the electoral lists for examining errors of the first type). In some cities and towns, however, special procedures were designed, depending on the size of place, whether it was scattered or concentrated, whether readily divisible into its aliquot parts or parishes, and particularly on whether a reliable map could be found, or whether one had to be constructed on the spot. In summary, several plans for drawing the samples of households were used, depending on the circumstances met. These plans are described briefly below, and they will be referred to as Schemes 1a, 1b, 2a, and 2b.

1. A sample of households drawn directly from accurate and up-to-date lists of households covering an area (village, parish, or other area, even an entire *koinotis*). The method of drawing was to use a random start with the appropriate sampling interval. The list of households could be of one or another type:

    a. A list of households by name and address. An example is furnished by the priest's list of the families in his parish which could easily be converted into a list of households. However, any list had to be carefully checked and brought up-to-date before it could be used. In small communities this revision was accomplished by viewing the area in company with the priest or other prominent citizens who knew practically all the families and who would assist in finding any errors or omissions.

    b. A "map-list," identifying the families by location on a map which was usually of necessity made on the spot (see Fig. 10 on p. 139). The households were numbered serially on the map, and the sampling interval applied as before.

2. A sample of blocks, followed by a further sampling of households from within the sample blocks. To carry out this scheme, a map had to be pro-

vided or made which showed the street pattern and definitely identified the boundaries of the blocks or any conveniently small parcels of land within the area to be sampled. There were two methods of drawing the sample of households:

a. Blocks chosen for the sample with probabilities in proportion to their estimated sizes (size being measured by the number of households within the block, not by area). The households within the sample block were then carefully map-listed and numbered serially. The sample of households was drawn with a random start and constant interval within any one sample block, but the interval was adjusted from block to block to give a constant number of sample households as calculated from the estimated sizes. Block sizes were estimated only approximately before sampling, as by cruising the area and making quick eye-estimates. The drawing of blocks with probabilities proportionate to sizes was easy to carry out by applying the proper sampling interval to the cumulative totals of sizes. (See the example in Appendix A of this chapter.)

b. Blocks chosen for the sample with equal probabilities. The households within the sample blocks were then map-listed and numbered serially throughout, and a sample of them was drawn by applying a constant sampling interval over the entire list. (See the example in Appendix B of this chapter.) In both Schemes 2a and 2b, strong geographic control was achieved by numbering the blocks in a serpentine fashion.

Completeness of the sample. A series of controls by maps and census information served to ensure that every village and every household within a primary area had its proper chance of inclusion in the sample, and also—equally important—to ensure that no outside area could come in. Excellent maps were obtained from the British Forces in Greece and more detailed ones from the Greek government.

The completeness of the sample was remarkable, and the cooperation of the Greek people unforgettable. Actually, information was obtained concerning all but 10 of the 3052 households in the sample. (These 10 were allocated to the tabulations by the simple device of assigning to each of them the characteristics of the preceding family.) Nine of the 10 families could not be found at home after repeated trials. The 10th family was a refusal, the only one; it had recently moved in from a foreign country and contained no eligible voters. There was, of course, an occasional omission of an item of information, but the entire field-job was remarkably free of imperfections.

Estimation of total population. Two methods were compared for the estimation of the total population of Greece in 1946. The simpler estimate would be that obtained by multiplying the sample total by 500, the reciprocal of the sampling fraction. As pointed out above in the discussion of the sample-design, however, it was possible to utilize the information concerning the population in 1940 for a regression-estimate, which should and did have a lower sampling error.   For each sample place there were two numbers, x and y:

x: The "expected" number of people in 1940, which is the number of people in the area in 1940 multiplied by the within-place sampling ratio for that place.

y: The number of people counted in the sample households in 1946.

A plot of the points (x, y) is given in Fig. 36.   The 199 sampling units represented in Fig. 36 may for the present purpose be regarded as a

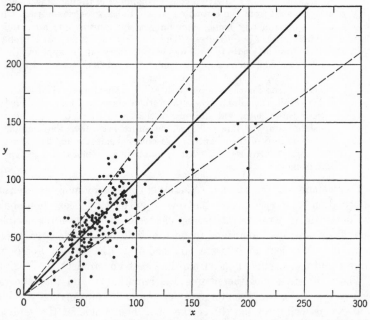

Fig. 36.   Regression of the observed population in sampling units against the "expected" population.   The broken lines indicate the extent of variability within x-arrays.

random sample of the total number of sampling units in the whole population (this total number is of the order of $7\frac{1}{2}$ millions divided by 500, that is, 15,000).   It is clear from the chart that the relationship between x and y is approximately linear and that the line may be passed through the origin.   The problem is then to estimate the parameter β in the equation

$$y = \beta x \tag{3}$$

If the estimate of β is b, the total population in 1946 may be estimated as b multiplied by the total population in 1940—the sum of x over all possible sampling units. It was thought that the best linear unbiased estimate b of β would be given by the equation

$$b = \frac{\Sigma\, wxy}{\Sigma\, wx^2} \qquad (4)$$

where w is the weight of the observation y and is the reciprocal of σ_y^2. The variance σ_y^2 of y within x-arrays clearly increases with x, and in order to estimate the relationship the y-values were divided into groups according to the x-values, the range of x within a group being 5 units. The within-group variance of y will be a near approximation to the

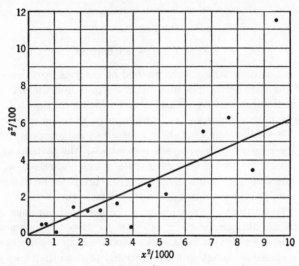

FIG. 37. The relationship between x^2 and the variance s^2 within x-arrays.

variance of y in the x-arrays of the group, because the within-group correlation of y and x will be small. Examination of these estimates of variance indicated that they were approximately proportional to the square of x, and Fig. 37 contains a plot of s^2, the estimated variance, against x^2. The variance σ_y^2 of y was therefore assumed to be equal to x^2 multiplied by the variance per unit weight. With this relationship the coefficient of variation of y is constant because the mean value of y for a given x is proportional to x. In this manner the estimate

$$b = \frac{\sum \dfrac{1}{x^2}\,xy}{\sum \dfrac{1}{x^2}\,x^2} = \frac{1}{n}\sum \frac{y}{x} \qquad (5)$$

of β was obtained, where n is the number of sample pairs of values of x and y. The variance per unit weight may then be estimated as

$$s_0{}^2 = \frac{1}{n-1} \sum \frac{(y-bx)^2}{x^2}$$
$$= \frac{1}{n-1} \left[\sum \left(\frac{y}{x}\right)^2 - nb^2 \right] \tag{6}$$

and the variance of b is equal to $s_0{}^2/n$. It was found that

$$b = 0.988$$
$$s_0{}^2 = 0.0861$$
$$\sigma_b{}^2 = 0.00043$$
$$\sigma_b = 0.021 = 2.1 \text{ percent}$$

It may be noted that the above method of analysis provides two estimates of the variance $s_0{}^2$ per unit weight: Eq. 6, and the one provided by the relationship between s^2 and x^2 (for which a maximum likelihood estimate may be obtained if a distribution of y within x-arrays is assumed. These two estimates are of course not independent, but if the distribution of y within x-arrays is normal with the stated variance, a comparison of the two estimates gives a rough indication whether a straight line through the origin adequately represents the data. In the present case, the data are not sufficiently numerous to make this comparison. To indicate the reliability of the assumed variance relationship, there are also given on Fig. 36 lines to show the assumed variability of y in the x-arrays; because the variance is assumed proportional to x^2, the standard deviation is proportional to x, so that lines of slope $b \pm s_0$ through the origin should contain about two-thirds of the y-values in each x-array, and this is seen to be about so. There are a few anomalous points, the existence of which may in part at least be ascribed to the use of the less accurate sampling schemes in a few places or to large changes in population between 1940 and 1946.

The population of Greece in 1940 was 7,344,860, so that the estimate of the population in 1946 was 7,344,860 × 0.988 = 7,257,000, with a standard error of 2.1 percent, or 152,000. This estimate may be compared with that obtained by multiplying the sample total by 500, namely, 6,975,000, with a standard error of 3.7 percent. The efficiency of the regression-estimate is therefore 3 times that of the simple estimate.

Remark.[5] The reader should be warned that Eq. 5 for b is in a form that, while presumably justified here from the observation that the variance of

---

[5] This remark was not included in the article published in the *Journal of the American Statistical Association*.

y appears to be proportional to x^2, may be a hazardous one in another example. An accidentally low value of x without a correspondingly low value of y in Eq. 5 will send the estimate b sky high, and a huge error may be the result. The estimate used here comes about by assuming a particular model for the relation between the departures of x and y from strict proportionality, so as to evaluate the variance of y, or the w in Eq. 4. When this or any other model is justifiable, it should certainly be used, because under proper conditions the equation corresponding to the correct model gives unbiased estimates, often with much greater precision (smaller variance)

TABLE 3.   ESTIMATED STANDARD ERRORS FOR VARIOUS CHARACTERISTICS

Characteristic	Estimate in thousands	Standard error as percentage of estimate
Population by sex and age		
15–19 male and female	825	3.4
20–29 male and female	1197	3.2
50–59 male and female	534	3.9
20–29 male	560	3.9
50–59 male	258	4.7
Literacy by sex and age		
14–19 female, literate	432	4.6
14–19 male, illiterate	47	13.6
20–39 male and female, literate	1757	3.4
20–39 male, literate	966	3.0
20–39 female, literate	791	4.9
20–39 male, illiterate	90	10.5
20–39 female, illiterate	358	6.4
Labor force by sex and age		
20–39 male, in labor force, at work	926	4.4
20–39 male, in labor force, not at work	78	13.9
20–39 male and female, in labor force, at work	1150	4.3
20–39 male and female, in labor force, not at work	101	14.4
40–59 male, in labor force, at work	618	4.6
40–59 male, in labor force, not at work	40	17.7
20–39 male, employer of others	39	20.6
40–59 male, employer of others	38	26.5
14–19 male, employed by others	175	8.0
14–19 female, employed by others	69	14.0
20–39 male, employed by others	452	6.6
40–59 male, employed by others	214	9.4
Sex-ratio of the age group 0–4	1.18 *	2.0

* The heading over the column does not apply to this figure.

than will be obtained with the wrong model. Improperly used, however, an assumed model that leaves x in the denominator may do severe damage by introducing heavy bias. The form of estimate $f = \Sigma X_i/\Sigma Y_i$ found in Chapter 5 is recommended whenever a particular model for departures from a proportionate relationship between x and y can not be strongly justified.

Sampling errors of other results. The main use of the figures is for administrative purposes for which approximate measures of precision will suffice. Evaluation of the exact sampling errors for most of the results would be prohibitively complicated. In the first place, various schemes of sampling households within sample places were used. Second, the results were obtained by multiplying the proportions estimated from the sample by the estimated total population, and both the proportions and the estimated total population are subject to sampling error. Approximate measures of accuracy, however, are essential and obtainable.

TABLE 4.   POPULATION OF GREECE BY SEX AND AGE

July 1946

(All figures in thousands)

(1) Age-group	(2) Total	(3) Male	(4) Female
All ages	7257	3599	3658
0– 4	634	343	291
5– 9	721	372	349
10–14	842	434	408
15–19	825	392	433
20–29	1197	560	637
30–39	1008	496	512
40–49	813	425	388
50–59	534	258	276
60 and over	683	319	364

Approximate standard errors for several characteristics are shown in Table 3. For other characteristics it may be surmised that there is a fairly close relationship between the standard error of an estimated figure and the magnitude of that figure. Such relationships are usual with survey data. Thus, from the magnitude of the errors listed in Table 3, standard errors of many other characteristics may be qualitatively inferred.

The results: some population characteristics of Greece. Six tables are presented herewith (Tables 4–9). The figures in them were derived from hand tabulations carried out in Athens. All tables except Table 9 include the armed forces of Greece.

Table 4 shows the population by sex- and age-classes. The standard error of sampling of the estimate of the total population of Greece (7.26 million) is 2.1 percent, which fixes the total population with 95 percent probability within about 4 percent. The decrease in population between 1940 and 1946 is indicated by the sample to be 1.2 percent.

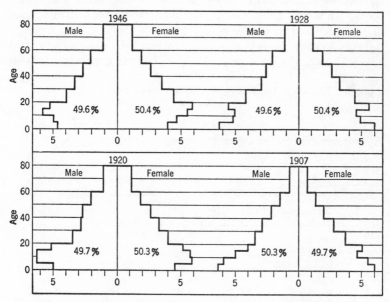

Fig. 38.   The population of Greece by age-groups at different dates.

This indicated decrease, however, is subject to the aforementioned sampling error of 2.1 percent, wherefore it can be concluded that the population change since 1940 must have been slight, lying with 95 percent probability between a decrease of 5 percent and an increase of 3 percent.

Fig. 38 shows the age-sex pyramids for 1946, 1928, 1920, and 1907 (unfortunately the census of 1940 was never tabulated by age). Greece's population of age 0–9 shows a serious deficit. This deficit can be laid to two factors: *i.* the rapid decline of the birth-rate, which fell from an average of 29.5 per thousand in 1931–35 to 23.5 per thousand in 1939; and *ii.* the high mortality-rate of children during the war.[6] The importance

[6] The author is indebted to Dr. Dudley Kirk of the Office of Population Research in Princeton for a number of helpful suggestions, such as pointing out the decline in the birth-rate of Greece, and the sex composition of the immigrants and emigrants (mentioned in a later paragraph).

of public health measures to conserve the population of children now aged 0–9 is apparent.

Calculations made by Dr. T. Nicholas Panay who assisted the authors in Athens indicate that if the logarithmic rate of growth determined by the censuses of 1928 and 1940 had been maintained through 1946, the population of Greece in July 1946 would have been 8 million, or $\frac{3}{4}$ million higher than the population determined by the sample. This estimated $\frac{3}{4}$ million may be largely ascribed to losses due to the war and occupation both in actual deaths, military and civil, from all causes, and in deficit arising from the decreased number of births.

Table 5 shows the sex-ratio in the various age-classes and comparisons with 1928, 1920, and 1907. It should be remembered that between

TABLE 5. THE SEX-RATIO (MALE : FEMALE) BY AGE-GROUPS IN GREECE AT VARIOUS CENSUSES

July 1946

(1) Age-class	(2) 1946 AMFOGE	(3) 1928 census	(4) 1920 census	(5) 1907 census
All ages	0.98	0.98	0.99	1.01
0– 4	1.18	1.04	1.09	1.06
5– 9	1.07	1.05	1.08	1.08
10–14	1.06	1.09	1.10	1.07
15–19	0.91	0.98	0.95	0.80
20–29	0.88	0.93	0.87	0.93
30–39	0.97	0.90	0.86	1.03
40–49	1.10	0.96	1.04	1.09
50–59	0.93	1.04	1.08	1.11
60 and over	0.88	0.94	0.94	1.07

the censuses of 1920 and 1928 huge interchanges of population took place between Greece, Turkey, and Bulgaria. The net result of these transfers was to increase the population of Greece by nearly a third. There are at least three factors operating to bring about a low sex-ratio in the young adult classes: *i.* war, *ii.* the refugees from Asia Minor had a lower sex-ratio than native Greeks, *iii.* emigration, which has included a disproportionately large number of young men. As in censuses in most countries, an undercount of young children must be presumed in the observations reported here. Moreover, it appears by comparing the figures for male and female children, or by observing from Table 5 that the sex-ratio in the age-group 0–4 is 1.18, that the undercount must have been more pronounced for female children than for male. The phenomenon may be exaggerated by the sampling errors, but the pattern by sex- and age-classes for 1928, 1920, and 1907 (Table 5) indicates that relative under-reporting of female children has existed in previous censuses as well. A possible alternative, not to be thrown out

TABLE 6. LITERACY OF THE GREEK POPULATION

July 1946

(Absolute figures in thousands)

(1)	(2)	(3)	(4)	(5)	(6)	(7)
	Total population		Literate		Illiterate	
Age-group and sex	Number	Percent	Number	Percent	Number	Percent
8 and over	6219	100	4470	71.9	1749	28.1
Male	3042	100	2570	84.5	472	15.5
Female	3177	100	1900	59.8	1277	40.2
8–13	984	100	774	78.6	210	21.4
Male	505	100	410	81.2	95	18.8
Female	479	100	364	76.0	115	24.0
14–19	1000	100	864	86.4	136	13.6
Male	479	100	432	90.2	47	9.8
Female	521	100	432	82.9	89	17.1
20–39	2205	100	1757	79.7	448	20.3
Male	1056	100	966	91.5	90	8.5
Female	1149	100	791	68.8	358	31.2
40 and over	2030	100	1075	53.0	955	47.0
Male	1002	100	762	76.0	240	24.0
Female	1028	100	313	30.4	715	69.6

summarily, is that perhaps because of special conditions of nutrition in Greece, there has been unusually high mortality of female children in the past several years and perhaps for many years. Unfortunately, the birth and death records are incomplete, and the degree of under-reporting of young children can not be ascertained.

This was a household sample, but all members of the household not having permanent residence elsewhere were to be counted at home, whether actually living at home or not at the time of enumeration. This approach gave a count of the armed forces and obviated the need for special counts of armed bands and prisons, which would have been impossible. Monasteries came into the sample in the regular way. The number of homeless children in asylums has since been determined and

TABLE 7.   EMPLOYMENT STATUS OF THE GREEK POPULATION 14 YEARS OF AGE AND OVER

July 1946

(Absolute figures in thousands)

(1)	(2)	(3)	(4)	(5)	(6)	(7)	(8)	(9)
		In the labor force		Not in the labor force			Percent of labor force	
Age-group and sex	Total population	At work or with a job	Not at work but seeking work	Normally in school	House-wives	Other	At work or with a job	Not at work but seeking work
14 and over	5235	2466	197	282	1391	899	92.6	7.4
Male	2537	2032	147	175	. . .	183	93.3	6.7
Female	2698	434	50	107	1391	716	89.7	10.3
14–19	1000	360	40	241	34	325	90.0	10.0
Male	479	262	20	143	. . .	54	92.9	7.1
Female	521	98	20	98	34	271	83.0	17.0
20–39	2205	1150	101	41	639	274	91.9	8.1
Male	1056	926	78	32	. . .	20	92.2	7.8
Female	1149	224	23	9	639	254	90.7	9.3
40–59	1347	705	45	. .	531	66	94.0	6.0
Male	683	618	40	. .	. . .	25	93.9	6.1
Female	664	87	5	. .	531	41	94.6	5.4
60 and over	683	251	11	. .	187	234	95.8	4.2
Male	319	226	9	. .	. . .	84	96.2	3.8
Female	364	25	2	. .	187	150	92.6	7.4

turns out to be under 14,000 of which roughly two-thirds are male. This information was obtained by Dr. Panay, mentioned earlier, who kindly consulted the proper government officials in Athens and even ascertained some figures directly from heads of institutions in various parts of Greece where it seemed desirable to do so.

Tables 6–9 show the results obtained for various other population characteristics. In all instances the answers given in the household were accepted.[7] In regard to literacy, for example, if the information obtained was that a particular person could read and write in any language, there was no resort to any test.

Likewise a man claiming to be employed was not asked for any proof; in particular, he was not asked how many hours he worked last

TABLE 8. NATURE OF EMPLOYMENT OF THE GREEK POPULATION OF AGE 14 AND OVER

July 1946

(All figures in thousands)

(1)	(2)	(3)	(4)	(5)	(6)	(7)	(8)	(9)	(10)
	Employer of others			*Employed by others*			*Self-employed*		
Age-group	Total	Male	Female	Total	Male	Female	Total	Male	Female
	114	106	8	1127	877	250	1225	1049	176
14–19	3	2	1	244	175	69	113	85	28
20–39	43	39	4	590	452	138	517	435	82
40–59	39	38	1	251	214	37	415	366	49
60 and over	29	27	2	42	36	6	180	163	17

week. There was thus no measure of underemployment. The employment figures by themselves do not give an adequate picture of the degree of employment. Under the conditions existing, there could very well be some upward distortion in the number of people claiming to be employed and self-employed. Many people, especially men and boys, undoubtedly reported themselves as employed when in fact they were only working part-time as peddlers. On the other hand, some of them were doing this sort of work from choice rather than to accept some regular form of employment at the prevailing low wages. Unfortunately it was impossible to ask for the number of hours worked per week; this information, however valuable, would have crowded the schedule and complicated the work beyond the allowable tolerances.

[7] This statement applies only to the population characteristics, not to information obtained concerning registration. As already stated, information obtained in the household regarding registration was compared with information from the registers.

In regard to employment status (Table 7) it should be remarked that the classification was decided upon after conferences with officials in the Greek government. The figures given in Table 9 are not exhaustive —that is, the Army, Navy, and certain miscellaneous and unclassified industries have been omitted. The figures for "Normally in school" in Table 7 are to be interpreted as the number of males and females 14

TABLE 9. EMPLOYMENT OF THE GREEK POPULATION IN CERTAIN INDUSTRIES

July 1946

(All figures in thousands)

(1)	(2)	(3)	(4)	(5)	(6)	(7)	(8)	(9)	(10)	(11)	(12)
	Both sexes, 14 and over	Male					Female				
Industry		14 and over	14–19	20–39	40–59	60 and over	14 and over	14–19	20–39	40–59	60 and over
Agriculture	1436	1183	187	507	332	157	253	64	117	55	17
Woodswork	40	38	2	18	17	1	2	1	1	..	..
Fishing	25	25	5	10	7	3	..	..	..	..	..
Mining, including salt mining	13	13	1	5	6	1	..	..	..	..	..
Manufacturing	141	94	14	48	28	3	47	14	27	4	2
Construction	60	59	4	23	23	9	1	..	1	..	..
Transportation	79	79	3	37	35	4	..	..	..	..	..
Business	341	307	23	137	108	39	34	7	20	6	1

and over intending to go back to school in the fall. Some of them were, of course, at work at the time the observations were made, but they were, nevertheless, classed as normally in school. No one was given a double assignment.

No family characteristics were tabulated, but a count of heads by sex showed 83 percent of the heads to be male. The number of people in the sample of all ages was 13,827, and the number of households was 3052; accordingly the average size of household was 4.5 people, with an insignificant sampling error (not actually calculated). This figure, of course, includes in-laws and friends and lodgers living under the same roof, sharing expenses, food, and household facilities.

APPENDIX A: BLOCKS DRAWN WITH PROBABILITIES IN PROPORTION TO SIZE (SCHEME 2a)

The Sample for Parish Ayios Panteleemous in the Demos of Ayios Georgios Keratsiniou

This appendix shows the actual drawing of the household sample in a sample parish in the Athens-Piraeus metropolitan district. Blocks were drawn from the sample parish with probability proportionate to their estimated sizes (Scheme 2a), and households were then drawn from the sample blocks. The procedure of selection is illustrated in the next sections.

The *demos* Ayios Georgios Keratsiniou is part of the metropolitan area of Athens-Piraeus. Because of its size (Class 6) it had a probability of unity (certainty) of coming into the sample. There were five parishes in this *demos*, and one was drawn at random (Parish Ayios Panteleemous). The sampling rate within the sample parish was assigned as 1 household in 100, satisfying the requirement that the overall probability of a household being selected from the *demos* should be 1 in 500, as expressed by the equation

$$\tfrac{1}{1} \times \tfrac{1}{5} \times \tfrac{1}{100} = \tfrac{1}{500}$$

The first fraction is the probability (unity) of the *demos* being in the sample. The second fraction is the probability of the parish being in the sample, and the third fraction is the probability for the selection of households from within the sample parish. Had sizes of parishes (as of 1940) been known at the time the sample was drawn, it might have been tempting to draw the parishes from within *demoi* with probability proportionate to their 1940 sizes, but the figures were not on hand till later, and moreover, there were often complications arising from changes in boundaries of parishes since 1940, rendering this suggestion a difficult one to carry out.

The first step was to obtain or make a map of the sample parish and to number the blocks in serpentine fashion. The second step was to cruise the parish with a jeep to estimate the number of households in each block, making stops only where necessary to get fair approximations of the actual number of households, resulting in a table of estimated households (Table 10).

It will be noted that Blocks 54, 55, and 56 were tied together and came into the sample as a unit. With this type of sampling scheme, small blocks should be tied with other small ones or with a large one, for two reasons: first, because it is desirable that a sample block (or group of tied blocks) contain enough households to provide the required number for the sample (Col. 3 of Table 11); second, because there is usually high correlation between households within a small block. A decrease in the standard error of sampling can be brought about by enlarging the block, although it should be noted that failure to heed this advice does

TABLE 10. ESTIMATED HOUSEHOLDS BY BLOCKS: PARISH AYIOS PANTELEEMOUS

(This table was brought in by the observer)

(1) Block number	(2) Estimated households	(3) Cumulative total	(1) Block number	(2) Estimated households	(3) Cumulative total
1	14	14	50	20	555
2	20	34	51 }	20	575
3	10	44	52	3	578
4	13	57	53	25	603
5 }	12	69	54	3	606
6	5	74	55 } (s)	12	618
7	20	94	56	3	621
8	15	109	57 }	0	621
9	23	132	58	0	621
10	16	148	59	0	621
11	29	177	60	22	643
12	15	192	61 }	11	654
13	18	210	62	8	662
14	15	225	63	17	679
15	22	247	64	19	698
16	17	264	65	14	712
17	20	284	66	12	724
18(s)	23	307	67	8	732
19	12	319	68	8	740
20 }	1	320	69	9	749
21	0	320	70	8	757
22	2	322	71	8	765
23	0	322	72	6	771
24	14	336	73	9	780
25	19	355	74	7	787
26 }	0	355	75	8	795
27	11	366	76	7	802
28	19	385	77 }	6	808
29	1	386	78	0	808
30	8	394	79	0	808
31	7	401	80 }	0	808
32	5	406	81	0	808
33 }	2	408	82	0	808
34	4	412	83	7	815
35 }	4	416	84	37	852
36	1	417	85 }	6	858
37	0	417	86	2	860
38 }	7	424	87	2	862
39	2	426	88	0	862
40	0	426	89	17	879
41	0	426	90	12	891
42	38	464	91 }	7	898
43 }	0	464	92	9	907
44 }	0	464	93	11	918
45	16	480	94 }	10	928
46	16	496	95	3	931
47	20	516	96(s)	13	944
48 }	19	535	97	12	956
49	0	535	98	10	966

not result in bias. The tying should be done before the sample blocks are selected; otherwise bias can be expected.

The sample blocks are designated by the letter s in Table 10. They were selected by applying the constant interval 322 to the cumulative totals, starting with the random number 288 (selected between 1 and 322) according to the tabular scheme shown below:

THE SAMPLE OF BLOCKS

Sampling rate within the sample parish (1/500) (5/1)	1/100
Cumulative total of estimated households	966
Estimated number of households in the sample, 966/100	10
Take 3 blocks (a convenient workload)	
Sampling interval for blocks, 966/3	322
Random start	288

The sampling plan for the parish is summarized in Table 11, which constitutes a set of instructions actually handed to the observer, who was instructed to map-list the sample blocks, include and identify

TABLE 11.   THE SAMPLE OF HOUSEHOLDS

(1) Block number	(2) Estimated households Total	(3) In sample	(4) Sampling interval	(5) Households in the sample
18	23	4	6	3, 9, 15, 21, $\cdots$
54, 55, 56	18	3	6	2, 8, 14, 20, $\cdots$
96	13	3	4	4, 8, 12, 16, $\cdots$

every household with a number, starting afresh with 1 in each block. The estimated 10 sample households were assigned to the 3 blocks with approximate equality (4, 3, 3). The numbers in the last column designate the sample households, which were obtained by using a random start in each block and applying the designated sampling interval. The sampling interval to be applied in any sample block was simply the total estimated households in that block divided by the estimated number of households in the sample. If the block turned out to contain twice as many households as were estimated in advance, the actual number of sample households would also be twice the estimated number. The sample was thus self-correcting.

The actual number of sample households may be more or less than the number designated in the column "Estimated households in sample" of Table 11, depending on the accuracy of the preliminary estimates. The specified sampling interval was to be applied to the actual number of households found in the block, even if the block were found to contain twice as many households as estimated (last column of Table 11). The sampling interval is fixed by the preliminary estimates, which, because they also determine the probabilities of selection, leave the sample unbiased.

APPENDIX B: BLOCKS DRAWN WITH EQUAL PROBABILITIES (SCHEME 2b)

With this scheme no table of estimated households for each block is required. In the above sample parish, a workload of 10 blocks might be considered adequate for this plan. There being 98 blocks in this parish, the sampling interval for blocks would be 98/10 = 10. With a random start of 4, blocks 4, 14, 24, 34, $\cdots$ would be in the sample. They would then be map-listed, and the households numbered continuously from the first through to the last household in each sample block. The sampling interval for households would be 100/10 = 10. With 7 as a random start, households numbered 7, 17, 27, 37, $\cdots$ would be the sample households for block 4, for example. The sample household for the other selected blocks would be drawn in the same way with different starting points.

NOTE IN REGARD TO SCHEMES 2a AND 2b

In the preliminary work of cruising the area preparatory to drawing blocks with probability in proportion to size (Scheme 2a), the amount of cruising, and hence the cost of cruising, can theoretically range all the way from very little to a high amount when an overmeticulous estimate is made of the size of every block. In the former case the blocks are merely assigned equal sizes, and drawing them with probability in proportion to size is then identically the same operation as drawing them at random. As a matter of fact, drawing blocks with equal probabilities is usefully regarded as a modification of drawing them with probabilities in proportion to estimated sizes.

A limited amount of preliminary cruising will often net considerable saving by eliminating the need for a large sample of blocks. The fewer the blocks in the sample, the easier the job, because every sample block must be carefully map-listed, even if only one sample household is to be drawn out of it. Hence, in place of cruising to make a fair estimate of the size of every block, and to draw blocks with probability in proportion to size, one may well prefer to make a quicker cruise for the purpose of tying small but not necessarily contiguous blocks together on the map with blue pencil to build artificial blocks of approximately equal size and then to draw these artificial blocks with equal probabilities. It is also well to mark the 0-areas on the map (those that obviously contain no households), and to exclude them from the sample by skipping them in the numbering of the blocks, which takes place at this stage. Whether one designates this plan as 2a or 2b is unimportant.

In case of doubt, it is probably advisable to cruise the area for the purpose of making quick estimates for every block, and to apply Scheme 2a. This plan has the advantage of being definite and requiring less supervision than the plan of partially equalizing the sizes of blocks.

It should be pointed out that the plan to be recommended may well depend on whether there is to be only one survey or a continuing series of monthly or quarterly surveys. In the latter event, the greater first-cost of Scheme 2a should be regarded as an investment that may pay big dividends if it is used often enough, because of the smaller samples required. Under either Scheme 2a or 2b the household listings can be sampled over and over until exhausted. Here appears another reason for tying small blocks together or to large ones in Scheme 2a: the resulting larger sampling interval for households within the block (last column of Table 11) permits more samples to be drawn for future surveys, without drawing and listing a new sample of blocks.

ALLIED MISSION TO OBSERVE THE GREEK ELECTIONS II

Household interview

Form D
July 10, 1946

AMPOGE District
1 2 3 4 5 6 7

Observer _____
Cluster No. _____

Nomos _____
Place _____ Serial No. _____
Observer's Serial No. _____

AB Date _____
Household No. _____

| Population Characteristics | | | | | | | | | Registration Status | | | |
Members of household (1)	Sex (2)	Age (3)	Literacy (4)	Occupa-tion (5)	Worker class (6)	Labor status (7)	Res. status (8)	Citizen (9)	If now registered, where? (10)	If not, reason (11)	If registered in 1936, where? (12)	Ver. of reg. (13)
1												
2												
3												
4												
5												
6												
7												

CODE SYMBOLS

(4) Y—Literate
    N—Illiterate

(5) Industry code
    1—Agriculture
    2—Woodswork
    3—Fishing
    4—Mining, salt work
    5—Industry
    6—Construction
    7—Transportation (including sailors, except Navy)
    8—Business, commerce
    9—School
    10—No occupation, retired
    11—Army and Navy
    12—All others

(6) E—Employer of others
    L—Employed by others
    S—Works for self

(7) E—Employed
    NE-S—Not employed, seeking work
    NE-N—Not employed, not seeking work

(11) Personal
    1—Indifference
    2—Too busy or sickness
    3—Residence requirements not met
    4—Birth certificate unobtainable
    5—Police ident. card unobtainable
    6—Bread-ration card lacking
    7—Criminal record disqualifies
    8—Governmental authorities refuse registration, although legal requirements believed met
    9—Intimidation by police and gendarmerie
    10—Intimidation by political opponents
    11—Other (describe)

PART V. SOME FURTHER THEORY FOR DESIGN AND ANALYSIS

CHAPTER 13. DETAILED STUDY OF SOME BINOMIAL AND RELATED DISTRIBUTIONS

. . . there was little information available as to the strength and direction of currents, and as the log was heaved only once every two hours, the position of a sailing ship fixed by such crude means must have been almost worthless. We actually read of navigators looking out for land a week or more before they expected to make it, lying by at night for fear they might run aground before morning. . . . Pepys probably did not increase his popularity in the *Grafton* by getting Dartmouth to call for the dead reckoning from twelve different persons on board, especially as this was done before they sighted land. Their errors were subsequently found to be very considerable—one was as much as 70 leagues out! . . . the inference drawn from these discrepancies was that the chart must be wrong, and it was corrected accordingly.—Edwin Chappell, Introduction to *The Tangier Papers of Samuel Pepys* (The Navy Records Society, London, 1935).

A. THE BINOMIAL AND POISSON SERIES

The binomial expansion. In Chapter 4 the binomial series was encountered when a two-celled universe was sampled with replacement. Here we shall examine the binomial series with more thoroughness.

Suppose that in the binomial $q + p$ the two quantities q and p are complementary. Then $q + p = 1$, and if the binomial be raised to any power whatever, the sum of all the terms will be unity; that is

$$(q + p)^n = q^n + nq^{n-1}p + \frac{n!}{(n-2)!\,2!} q^{n-2}p^2 + \cdots$$

$$+ \frac{n!}{2!(n-2)!} q^2 p^{n-2} + nqp^{n-1} + p^n = 1 \quad (1)$$

It is customary to denote the rth binomial coefficient by $\binom{n}{r}$. In this symbol, n (invariably an integer) denotes the exponent of the binomial, and r the ordinal number of the term, counting the term q^n as the *zeroth*, $nq^{n-1}p$ the *first*, etc. In terms of factorials,

$$\binom{n}{r} = \frac{n!}{(n-r)!\,r!} = \binom{n}{n-r} \quad (2)$$

If r assumes the values $0, 1, 2, \cdots, n$ successively, $\binom{n}{r}$ assumes the values of the binomial coefficients, thus:

With $r = 0$ or n, $\dbinom{n}{r} = \dfrac{n!}{(n - 0)!\,0!} = 1$ [Since $0! = 1$; see p. 487]

With $r = 1$ or $n - 1$, $\dbinom{n}{r} = \dfrac{n!}{(n - 1)!\,1!} = n$

With $r = 2$ or $n - 2$, $\dbinom{n}{r} = \dfrac{n!}{(n - 2)!\,2!} = \dfrac{n(n - 1)}{2!}$

With $r = 3$ or $n - 3$, $\dbinom{n}{r} = \dfrac{n!}{(n - 3)!\,3!} = \dfrac{n(n - 1)(n - 2)}{3!}$

Etc.

The law of formation is evident. It should be noted that the coefficients of the binomial are symmetrical from either end, as expressed by Eq. 2; that is, the coefficient of q^n is the same as the coefficient of p^n; the coefficient of $q^{n-1}p$ is the same as that of qp^{n-1}; etc. But although the coefficients are symmetrical from either end, the terms themselves are not, save for the case $p = q = \frac{1}{2}$, in which event $q^{n-r}p^r = 1/2^n$ no matter what r may be. So when p and q are equal, the expansion is

$$\left(\frac{1}{2} + \frac{1}{2}\right)^n = \frac{1}{2^n}\left\{1 + n + \frac{n!}{(n - 2)!\,2!} + \cdots + \frac{n!}{2!(n - 2)!} + n + 1\right\} \quad (3)$$

This is the only possible symmetrical expansion, for when p and q are unequal the value of $q^{n-r}p^r$ will vary from term to term and can not be factored out like the $1/2^n$ in Eq. 3.

By introducing the summation operator Σ, the expansion written out in Eq. 1 may be put in the compact form

$$(q + p)^n = \sum \binom{n}{r} q^{n-r}p^r \quad (4)$$

In the process of summation, r is to take the values 0, 1, 2, $\cdots$, n successively. But, if $\dbinom{n}{r}$ vanishes when r is negative or greater than n, as it will do when $x!$ is identified with $\Gamma(x - 1)$ as in Chapter 14 ahead, the summation may as well be written over all integral values of r, positive and negative, from $-\infty$ to $+\infty$: only the integers 0, 1, 2, $\cdots$, n will contribute to the summation.

The $n + 1$ terms on the right of Eq. 1 when plotted one by one against the index r from $r = 0$ to $r = n$ are called a *point binomial* (Fig. 39).

Remark 1. The student should examine the expansions $(q + p)^1$, $(q + p)^2$, $(q + p)^3$, $(q + p)^4$.

$(q + p)^0 = 1$	1
$(q + p)^1 = q + p$	1 1
$(q + p)^2 = q^2 + 2qp + p^2$	1 2 1
$(q + p)^3 = q^3 + 3q^2p + 3qp^2 + p^3$	1 3 3 1
$(q + p)^4 = q^4 + 4q^3p + 6q^2p^2 + 4qp^3 + p^4$	1 4 6 4 1
$(q + p)^5$	1 5 10 10 5 1

At the right are written the binomial coefficients in order, to make up a "Pascal pyramid." Any coefficient may be obtained by adding to the one directly above it that just to the left and above, or by adding downward in the column just to the left to a point just above (see the exercise following).

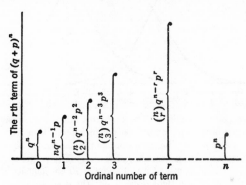

FIG. 39.   The point binomial.

Remark 2. The physical aspects of Eq. 1 should be noted for mnemonic reasons. As the exponent of q runs from n to 0, the exponent of p runs from 0 to n, and the sum of the exponents is always n. These exponents are identical with the two factorials in the denominators of the binomial coefficients; thus the coefficient of $q^n p^0$ (i.e., of q^n) is $\binom{n}{0} = \dfrac{n!}{(n-0)!\,0!}$; the coefficient of $q^{n-1}p^1$ is $\binom{n}{1} = \dfrac{n!}{(n-1)!\,1!} = n$; etc.

Remark 3. Because $p + q = 1$, the sum of the terms in Eq. 1 is unity, and the point binomial is hence naturally normalized. In Chapter 3 a curve was said to be normalized when the area under it is unity. Strictly speaking, in dealing with the point binomial we deal not with area but with discrete points. As a matter of expediency these points are usually joined by lines to make a *polygon* as in Fig. 40 or are used for making a *histogram* as in Fig. 42. The area under either the polygon or the histogram will be unity.

Remark 4. Tables of binomial coefficients show $\binom{n}{r}$ tabulated against n and r. Thornton C. Fry, in his *Probability* (Van Nostrand, 1928), shows such a table as far as $n = 100$.   Max Sasuly, in his *Trend Analysis of Statistics* (Brookings, 1934), shows $\binom{n}{r}$ with n running through 132 and r through 12.

Reference should be made to Degen's tables of 1824, in which he published $\log n!$ to 18 decimals, n stepping by unity from 1 to 1200.

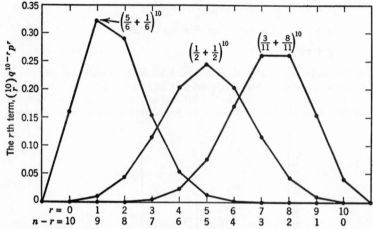

FIG. 40. Three point binomials. The ordinates are the successive terms of the binomial expansion of $(q + p)^n$ as written out in Eq. 1. Here $n = 10$ for all three sets of points. q and p have the values marked in the figure. The sum of the ordinates is in every case unity because $p + q = 1$. The points are joined by lines for identification. If the points at $r = 0$ and $r = 10$ are joined with the zero ordinates one unit to the left and right respectively, the resulting polygons each enclose unit area. Another style of polygon that also encloses unit area is used in later figures. (The two styles are sometimes differentiated by calling this figure a *polygon*, and Figs. 42–45 *histograms*.)

Exercise. By noting that in the array of binomial coefficients shown in the Pascal pyramid on page 401 any coefficient is the sum of all those in the column next on the left as far as the place next above, deduce the Bernoulli sums—

$$1 + 2 + 3 + \cdots + r = \tfrac{1}{2}r^2 + \tfrac{1}{2}r = \tfrac{1}{2}r(r + 1)$$

$$1^2 + 2^2 + 3^2 + \cdots + r^2 = \tfrac{1}{3}r^3 + \tfrac{1}{2}r^2 + \tfrac{1}{6}r$$

$$1^3 + 2^3 + 3^3 + \cdots + r^3 = \tfrac{1}{4}r^4 + \tfrac{1}{2}r^3 + \tfrac{1}{4}r^2$$

$$1^4 + 2^4 + 3^4 + \cdots + r^4 = \tfrac{1}{5}r^5 + \tfrac{1}{2}r^4 + \tfrac{1}{3}r^3 - \tfrac{1}{30}r$$

Etc.

Bernoulli also wrote the following general equation for the sum of the

tth powers of the first r integers—

$$1^t + 2^t + 3^t + \cdots + r^t$$

$$= \frac{r^{t+1}}{t+1} + \frac{1}{2}r^t + \frac{t}{2!}B_1 r^{t-1} - \frac{t(t-1)(t-2)}{4!}B_2 r^{t-3}$$

$$+ \frac{t(t-1)(t-2)(t-3)(t-4)}{6!}B_3 r^{t-5} - + \cdots \text{ ending in } r \text{ or } r^2$$

wherein $B_1 = \frac{1}{6}$, $B_2 = \frac{1}{30}$, $B_3 = \frac{1}{42}$, $B_4 = \frac{1}{30}$, etc., these numbers now being called the Bernoulli numbers, after their discoverer.

Solution [1]

Take the coefficient 6 in the expansion of $(q + p)^4$. It may be formed by adding coefficients downward in the column just to the left, to a point just above, thus—

$$1 + 2 + 3 = 6$$

In terms of binomial coefficients this can be written

$$\binom{1}{1} + \binom{2}{1} + \binom{3}{1} = \binom{4}{2}$$

Applying this same rule in the 1st column (counting the left-hand column of 1's as the 0-th column) we find that

$$\binom{1}{1} + \binom{2}{1} + \binom{3}{1} + \cdots + \binom{r}{1} = \binom{r+1}{2}$$

or

$$1 + 2 + 3 + \cdots + r = \tfrac{1}{2}r(r + 1)$$

which is a well-known result. Now add coefficients in the 2d column likewise. The same rule gives

$$\binom{1}{2} + \binom{2}{2} + \binom{3}{2} + \binom{4}{2} + \cdots + \binom{r}{2} = \binom{r+1}{3}$$

That is

$$\frac{1(1-1) + 2(2-1) + 3(3-1) + 4(4-1) + \cdots + r(r-1)}{2!}$$

$$= \frac{(r+1)r(r-1)}{3!}$$

whence

$$1^2 + 2^2 + 3^2 + 4^2 + \cdots + r^2 - (1 + 2 + 3 + 4 + \cdots + r)$$

$$= \frac{2!}{3!}(r+1)r(r-1)$$

[1] This is the procedure followed by Jacques Bernoulli himself in his *Ars Conjectandi* (posthumous 1713), pp. 90–7. This book is rare but is available in many libraries in the Ostwald Classics, No. 107 (Library of Congress No. QA 273. B52).

The series in parenthesis on the left we already know to be $\frac{1}{2}r(r+1)$, whence

$$1^2 + 2^2 + 3^2 + 4^2 + \cdots + r^2 = \frac{2!}{3!}(r+1)r(r-1) + \frac{1}{2}r(r+1)$$

$$= \frac{1}{3}r^3 + \frac{1}{2}r^2 + \frac{1}{6}r$$

as stated above. The student should continue to the sum of cubes and higher powers. For example, for the sum of cubes, summation in the 3d column gives

$$\binom{1}{3} + \binom{2}{3} + \binom{3}{3} + \binom{4}{3} + \binom{5}{3} + \cdots + \binom{r}{3} = \binom{r+1}{4}$$

$$\frac{\left\{\begin{array}{c}1(1-1)(1-2) + 2(2-1)(2-2) + 3(3-1)(3-2) \\ + 4(4-1)(4-2) + \cdots + r(r-1)(r-2)\end{array}\right\}}{3!}$$

$$= \frac{(r+1)r(r-1)(r-2)}{4!}$$

which gives

$$\frac{\left\{\begin{array}{c}1^3 + 2^3 + 3^3 + 4^3 + \cdots + r^3 - (2+1)(1^2 + 2^2 + 3^2 + 4^2 + \cdots + r^2) \\ + 1 \cdot 2(1 + 2 + 3 + 4 + \cdots + r)\end{array}\right\}}{3!}$$

$$= \frac{(r+1)r(r-1)(r-2)}{4!}$$

which will reduce to the result stated for the sum of the cubes of the first r integers.

The sums of the 4th and higher powers may be derived likewise.

The mode of the point binomial. The *mode* of the point binomial is the abscissa of the point having the maximum ordinate. Stated otherwise, it is the value of r in Eq. 5 that corresponds to the term that is greater than any other term of the expansion. A slight complication occurs when there are two equal modal values of the point binomial, as occasionally happens; see, for example, the polygon labeled $(\frac{3}{11} + \frac{8}{11})^{10}$ in Fig. 40.

The position of the mode (or of the pair of modes) will now be investigated. Let R denote for the moment the ratio of the rth term to the preceding term. This ratio in terms of n, r, p, and q is

$$R = \frac{n!}{(n-r)!\,r!}\,q^{n-r}p^r \div \frac{n!}{(n-r+1)!\,(r-1)!}\,q^{n-r+1}p^{r-1}$$

$$= \frac{n-r+1}{r}\frac{p}{q} \tag{5}$$

from which

$$R = \frac{n - r + 1}{n - np}\frac{np}{r} \tag{6a}$$

or

$$\frac{Rq + p}{q + p} = \frac{(n + 1)p}{r} \tag{6b}$$

$R > 1$ signifies that the rth ordinate in Fig. 39, or the rth term of Eq. 5, is greater than the one next on the left; and $R < 1$ means the reverse. If $R = 1$, the rth ordinate is equal to the one next on the left. As the mean of the point binomial falls at np, and since r necessarily takes on integral values, the interpretation of the conditions for $R \gtreqless 1$ is as follows:

a. If np is an integer, then r can take on the values np and $np + 1$; furthermore, as Eq. 6a shows, $R > 1$ if $r \leqq np$ and $R < 1$ if $r \geqq np + 1$; hence *if np is an integer, the mean and mode coincide at np.* An example is the polygon $(\frac{1}{2} + \frac{1}{2})^{10}$ in Fig. 40; another is the expansion of $(\frac{7}{10} + \frac{3}{10})^{10}$, which the student should investigate.

b. If $(n + 1)p$ is an integer, then r can take on this value, and when it does, the ratio R will be unity. In fact, $R \gtreqless 1$ if $r \lesseqgtr (n + 1)p$, as is easiest seen from Eq. 6b. Hence, *if $(n + 1)p$ is an integer, there will be two equal modal ordinates located at the abscissas $np + p$ and $np - q$, and the mean np will lie somewhere between them.* An example is the polygon $(\frac{3}{11} + \frac{8}{11})^{10}$ in Fig. 40.

c. If neither np nor $(n + 1)p$ is an integer, r can take on neither of these values. But from Eq. 6b it is clear that $R \gtrless 1$ if $r \lessgtr (n + 1)p$; hence *if neither np nor $(n + 1)p$ is an integer, the mean and mode will both fall between $np + p$ and $np - q$.* An example is the polygon $(\frac{5}{6} + \frac{1}{6})^{10}$ in Fig. 40.

It should be noted that the mode in every case is the largest integer contained in the fraction $(n + 1)p$; also that the mean and mode will never be separated by a whole unit.

Remark. In practice, frequency distributions having two maxima are occasionally met. Such curves are said to be *bimodal*. It should be noted, however, that a point binomial with a pair of equal modal ordinates (Fig. 40) is not considered to be bimodal because the two top points are always consecutive and there is never a valley between them.

Exercise. If $(n + 1)p$ is an integer, prove that

$$\binom{n}{[n + 1]p - 1} q^{n-[n+1]p+1} p^{[n+1]p-1} = \binom{n}{[n + 1]p} q^{n-[n+1]p} p^{[n+1]p}$$

and there will be two equal modal ordinates.

Numerical illustration of point binomials. In Fig. 40 are shown three point binomials, all for $n = 10$ but for different values of p and hence also of q. The ordinates are worked out as follows:

a. $p = q = \frac{1}{2}$. Here

$$\left(\frac{1}{2} + \frac{1}{2}\right)^{10} = \sum \binom{n}{r} \left(\frac{1}{2}\right)^{n-r} \left(\frac{1}{2}\right)^{r}$$

$$= \frac{1}{2^{10}} \sum \binom{n}{r} = \frac{1}{2^{10}} (1 + 10 + 45 + 120 + 210 + 252$$

$$+ 210 + 120 + 45 + 10 + 1)$$

$$= 0.000977 + 0.009766 + 0.043945 + 0.117187 + 0.205078$$

$$+ 0.246094 + 0.205078 + 0.117187 + 0.043945$$

$$+ 0.009766 + 0.000977 \tag{7}$$

These eleven terms are plotted in Fig. 40. In this case the mean and mode both occur at $np = 5$. As the binomial coefficients are symmetrical from either end, the point binomial is symmetrical when $p = q = \frac{1}{2}$. (See Eq. 3 and the text immediately before and after.)

b. $p = \frac{1}{6}$, $q = \frac{5}{6}$. Here

$$\left(\frac{5}{6} + \frac{1}{6}\right)^{10} = \sum \binom{n}{r} \left(\frac{5}{6}\right)^{n-r} \left(\frac{1}{6}\right)^{r}$$

$$= 0.161506 + 0.323011 + 0.290710 + 0.155045 + 0.054266$$

$$+ 0.013024 + 0.002171 + 0.000248 + 0.000018$$

$$+ 0.000001 + 0.000000 \tag{8}$$

The binomial *coefficients* are the same as they were in Case a where $p = q = \frac{1}{2}$, but the terms themselves are not now symmetrical from either end because p and q are unequal and $q^{n-r} p^r$ varies from one term to another. These eleven successive terms are also plotted in Fig. 40. Here the mean lies at $np = \frac{10}{6} = 1\frac{2}{3}$, and the mode at 1. Note that the mean and mode are separated by $\frac{2}{3}$ unit.

c. $p = \frac{8}{11}$, $q = \frac{3}{11}$. Here

$$\left(\frac{3}{11} + \frac{8}{11}\right)^{10} = \sum \binom{n}{r} \left(\frac{3}{11}\right)^{n-r} \left(\frac{8}{11}\right)^{r}$$

$$= 0.000002 + 0.000061 + 0.000729 + 0.005181$$

$$+ 0.024176 + 0.077362 + 0.171916 + 0.261968$$

$$+ 0.261968 + 0.155240 + 0.041397 \tag{9}$$

These terms, also plotted in Fig. 40, give a polygon with two equal modal (maximum) ordinates at abscissas 7 and 8, as is to be expected whenever $(n + 1)p$ is an integer (*vide supra*). Here the mean lies at $np = \frac{80}{11} = 7\frac{3}{11}$, which is between the pair of modes.

The Type III curve. In the solution of problems of chance it is often necessary to form the sum of a number of consecutive terms of the point binomial. In this day, the computation of such sums is facilitated by computing machines and tables. The basis for some of these tables is the exact relationship between the sum of successive terms of the binomial and the incomplete beta function (next chapter). This exact relationship, along with the approximations afforded by the Type III and the normal curves, to be encountered in a moment, is important for studies of statistical theory and applications thereof in many branches of science.[2]

DeMoivre in 1733 arrived at the normal curve as an approximation to the point binomial, and recognized the need of a table of the normal

[2] Tables of binomial coefficients were mentioned in Remark 4, page 401. Over certain ranges, however, for the computation of binomial terms and sums thereof, binomial coefficients are now outmoded by the new *Tables of the Binomial Probability Distribution* (Applied Mathematics Series, No. 6, edited by Churchill Eisenhart, The National Bureau of Standards, Washington, 1950). These tables show the binomial terms and the sum of such terms from the rth to the nth, by steps of 0.01 in p, and for r and n up to 50.

Abraham DeMoivre (1667–1754) left France at the age of 21 after three years' imprisonment following the revocation of the edict of Nantes by Louis XIV in 1685. He settled in London, where he spent the rest of his life tutoring, writing, and solving problems in chance for professional gamblers. Because of his foreign origin, the influence of neither Newton in England nor of the Bernoullis in Switzerland availed for him a university post.

On November 12, 1733, he printed a seven-page pamphlet in Latin entitled *Approximatio ad Summam Terminorum Binomii $(a + b)^n$ in Seriem expansi* and bound it into the unsold copies of his *Miscellanea Analytica*, which had been published three years previously. A first supplement of 22 pages had already been added in 1731. Copies of the *Miscellanea Analytica* containing the first supplement are rare, but copies containing both the first and second supplement are rarer yet, only two copies having been reported extant—one at University College in London, the other in the Preussische Statsbibliothek in Berlin. DeMoivre's own translation of the second supplement, together with some additions, was included in the second edition (1738) of his *Doctrine of Chances*, pages 235–43. This is the origin of the English translation quoted in this book. DeMoivre's translation, with editorial notes by Helen M. Walker, is given in David Eugene Smith's *Source Book in Mathematics* (McGraw-Hill, 1929). The essential parts of this translation are found on pages 14–7 of Helen M. Walker's *History of Statistical Method* (Williams and Wilkins, Baltimore, 1929). The rare second supplement, the *Approximatio*, was discovered in the library at University College by Karl Pearson in 1924; see *Biometrika*, vol. 16, 1924: pp. 402–4. It has been made generally accessible by a photographic reproduction with a commentary by R. C. Archibald in *Isis*, vol. 8, 1926: pp. 671–83. For more details of DeMoivre's life see an article by Helen M. Walker, *Scripta Mathematica*, vol. 2, 1934: pp. 316–33.

integral for computing binomial sums.   His development, not disclosed,
I opine must have involved the use of the Stirling approximation for large
factorials (p. 520), in the manner found in Fry's *Probability and Its
Engineering Uses* (Van Nostrand, 1928), Art. 82.   Here, Karl Pearson's [3]
approach will be made through the intermediate development of the
Type III curve, which in the limit for large samples approaches the
normal curve.

The rth ordinate of the point binomial is $\binom{n}{r} q^{n-r} p^r$.   Through the
$n + 1$ ordinates of the point binomial draw a smooth curve and put

$$y = \binom{n}{r} q^{n-r} p' \tag{10}$$

We seek an approximate expression for y as a function of r.   The average of the two successive ordinates at r and $r + 1$ is

$$\frac{1}{2}\left[\binom{n}{r} q^{n-r} p^r + \binom{n}{r+1} q^{n-r-1} p^{r+1} \right]$$

which we may say is the ordinate of the proposed curve at the abscissa
$r + \frac{1}{2}$.   The slope of the line joining these two ordinates is simply their
difference

$$\binom{n}{r+1} q^{n-r-1} p^{r+1} - \binom{n}{r} q^{n-r} p^r$$

as they are spaced a unit apart.   This we might require to be the slope
of the curve at the abscissa $r + \frac{1}{2}$.   The problem is then described in
Fig. 41.   The quotient obtained by dividing the slope by the average
ordinate at $r + \frac{1}{2}$ is

$$\frac{\binom{n}{r+1} q^{n-r-1} p^{r+1} - \binom{n}{r} q^{n-r} p^r}{\frac{1}{2}\left[\binom{n}{r+1} q^{n-r-1} p^{r+1} + \binom{n}{r} q^{n-r} p^r \right]} = \frac{2(np - q - r)}{np + q + (q - p)r} \tag{11}$$

To transfer this quotient to the abscissa r, which is $\frac{1}{2}$ unit to the left
of $r + \frac{1}{2}$, replace r in the right-hand side by $r - \frac{1}{2}$.   Then, if y and
dy/dr denote the ordinate and slope of the proposed curve at the abscissa
r,

$$\frac{1}{y}\frac{dy}{dr} = \frac{2(np - q - r + \frac{1}{2})}{np + q + (q - p)(r - \frac{1}{2})} \tag{12}$$

[3] Karl Pearson, *Phil. Trans. Royal Soc.*, vol. A186, 1895: pp. 343–414, pp. 356–57
in particular.

will be the differential equation of the curve. With some rearrangement this can be written

$$\frac{1}{y}\frac{dy}{dr} = -\frac{(r-np)+\frac{1}{2}(q-p)}{npq+\frac{1}{2}(r-np)(q-p)+\frac{1}{4}} \tag{13}$$

Here are three quantities: $r-np$, npq, and $\frac{1}{2}(q-p)$. The first, $r-np$, is simply the abscissa measured from the mean. The second quantity, npq, is the variance σ^2 of the binomial. The third quantity, $\frac{1}{2}(q-p)$, is the *loading*, for which the symbol ϵ will be used.

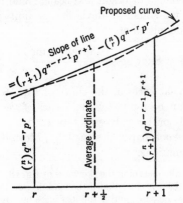

Fig. 41. Seeking an approximation to the point binomial.

It is well to make some changes in the scale before integrating this equation. Introduce the variable t so related to r that

$$\frac{r-np}{\sqrt{npq}} = \frac{r-np}{\sigma} = t\sqrt{1+\frac{1}{4\sigma^2}} \tag{14}$$

t is then proportional to the distance of r from the mean np. Also for convenience let α be introduced by the equation

$$\frac{\frac{1}{2}(q-p)}{\sqrt{npq}} = \frac{\epsilon}{\sigma} = \alpha\sqrt{1+\frac{1}{4\sigma^2}} \tag{15}$$

In these new symbols the differential equation (13) appears as

$$\frac{dy}{y} = -\frac{t+\alpha}{1+\alpha t}dt$$

$$= -\frac{1}{\alpha}dt + \frac{1-\alpha^2}{\alpha^2}\frac{\alpha\,dt}{1+\alpha t} \tag{16}$$

Each term is now ready to be integrated.   The result is

$$\ln y = \ln y_0 - \frac{t}{\alpha} + \frac{1 - \alpha^2}{\alpha^2} \ln (1 + \alpha t) \tag{17}$$

or

$$y = y_0(1 + \alpha t)^{(1-\alpha^2)/\alpha^2} e^{-t/\alpha} \tag{18}$$

y_0 being the constant of integration.

This curve has emerged as an approximation to the point binomial. It is a Pearson Type III curve and has the appearance of Fig. 57, page 468, except for a change of origin and scale. The mode of Eq. 18 lies at $t = -\alpha$, and its standard deviation is unity; hence with Karl Pearson's original definition of

$$\text{Skewness} = \frac{\text{mean} - \text{mode}}{\sigma} \tag{19}$$

it is clear that the skewness of the curve of Eq. 18 must be α. Areas under the curve can be found by the *Tables of the Incomplete Gamma Function* (Ch. 14), or by Salvosa's tables,[4] but for many purposes the double square-root paper designed by Mosteller and Tukey will supplant such calculations.[5]

Remark.   A word concerning the skewness of Eq. 18: From the definition of skewness in Eq. 19 it is clear that if the loading ϵ or $\frac{1}{2}(q - p)$ be zero, the skewness α will also be zero; and if the loading ϵ be not zero, the skewness α will in any case decrease as n increases. If n is so large that $\sqrt{1 + 1/4\sigma^2}$ can be replaced satisfactorily by 1, then to the same approximation α is equal to ϵ/σ; the skewness of the Type III curve is then simply the loading of the point binomial measured in terms of its standard deviation $\sqrt{npq}$.

The normal curve.   Let us now direct our efforts toward the limiting form that Eq. 18 takes for large values of n and for small or moderate loading, that is, for small values of the skewness α. To this end we go back to the logarithmic form in Eq. 17, finding that

$$\ln y = \ln y_0 - \frac{t}{\alpha} + \frac{1 - \alpha^2}{\alpha^2} \ln (1 + \alpha t)$$

$$= \ln y_0 - \frac{t}{\alpha} + \frac{1 - \alpha^2}{\alpha^2} \left\{ \alpha t - \frac{\alpha^2 t^2}{2} + \cdots \right\} \quad \begin{bmatrix} \ln (1 + x) \\ = x - \dfrac{x^2}{2} + \dfrac{x^3}{3} - + \cdots \end{bmatrix}$$

$$= \ln y_0 - \tfrac{1}{2}t^2 - \alpha(t - \tfrac{1}{3}t^3) + \cdots$$

whence

$$y = y_0 e^{-\frac{1}{2}t^2} \{1 - \alpha(t - \tfrac{1}{3}t^3) + \cdots\} \tag{20}$$

----

[4] A reference to Salvosa's tables will be found on page 474.

[5] A reference to Mosteller and Tukey's double square-root paper was given on page 306.

The first power of α is here retained, but no higher powers.[6] If in any case the skewness α is so small that it can be neglected entirely, the limiting reduces to

$$y = \frac{1}{\sqrt{2\pi}} e^{-\frac{1}{2}t^2} \tag{21}$$

which is a normal curve of unit standard deviation.

It will be observed from Eq. 14 that if n is large, t is very closely equal to $(r - np)/\sigma$, and the last equation takes the form

$$y = \frac{1}{\sigma\sqrt{2\pi}} e^{-x^2/2\sigma^2} \qquad \begin{array}{l}\text{[Normal curve of}\\ \text{standard devi-}\\ \text{ation } \sigma, \text{ cen-}\\ \text{tered at 0]}\end{array} \tag{22}$$

Consideration of the series within the braces multiplying the right-hand side of Eq. 20 shows that the neglected terms will not be unimportant unless t, as well as α, is confined to small values. And since, as previously remarked, the skewness α may be small either because the loading ϵ is small or because n is large (or of course from both causes), it follows that *the normal curve must not be expected to approximate the point binomial except near the mean, and then only if n is large and the loading not extreme. The approximation is best when the loading (and hence the skewness also) is zero, but in any case the approximation improves as n increases.*[7]

The normal curve can hardly be expected to give an approximation to the point binomial far in the "tails," for theoretically it is a curve of infinite range in both directions from the mean; the point binomial, on the other hand, cuts off absolutely at 0 and n. It should be noted, too, that the Type III curve has one tail of infinite range (see Fig. 57, p. 468) and hence it must also misrepresent the point binomial in the far tail.

Graphical and numerical illustration of the foregoing. *a.* In Fig. 42 are plotted three point binomials together with normal curves having the same mean and standard deviation. In this style of plotting (histogram), each point is made the center of a rectangular block. The sum of the areas of all the blocks so obtained is unity.

i. $n = 10$, $p = \frac{1}{4}$, $q = \frac{3}{4}$. Here the mean is at $np = 2.5$, and the standard deviation is $\sqrt{npq} = \frac{1}{4}\sqrt{30}$. The approximating normal curve is Eq. 22 written with $\sigma = \frac{1}{4}\sqrt{30}$, the y-axis being placed at the mean of the

---

[6] Eq. 20 will be found written out at length on page 211 of Fry's *Probability and Its Engineering Uses* (Van Nostrand, 1928).

[7] A remarkable plate showing the point binomials $(\frac{1}{2} + \frac{1}{2})^n$ for $n = 2, 4, 10, 50$, together with the limiting normal curve, is shown opposite page 310 of Raymond Pearl's *Medical Biometry and Statistics* (W. B. Saunders, Philadelphia, 1930).

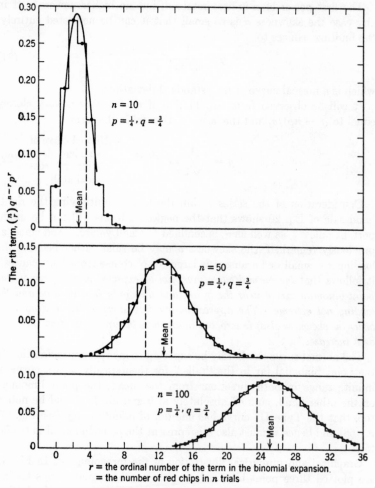

FIG. 42. The point binomial $(\frac{3}{4} + \frac{1}{4})^n$. The successive terms of the binomial $(\frac{3}{4} + \frac{1}{4})^n$ plotted as ordinates against the ordinal number r of the term in the expansion. The ordinates represent the probabilities of getting 0, 1, 2, $\cdots$, n successes in n repeated trials when the probability of success at a single trial is constant and equal to $p = \frac{1}{4}$. The sum of the ordinates in any plot is unity. The means of the polygons lie at np, and their standard deviations are $\sqrt{npq}$. The maximum ordinate always lies close to the mean, and in fact exactly at the mean when np is an integer. As n increases, the polygons become flatter and more symmetrical about the mean, and the area between the dashed lines decreases. A normal curve having the same mean and standard deviation as the polygon is shown for each. The approximation afforded by the normal curve improves as n increases. Figs. 42–45 were constructed for the author's classes out of admiration for similar figures in Fry's *Probability and Its Engineering Uses* (D. Van Nostrand, 1929).

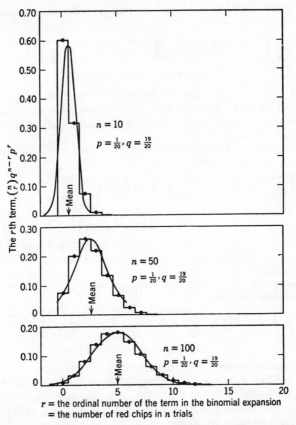

FIG. 43. The point binomial $(\frac{19}{20} + \frac{1}{20})^n$. Comparison with the point binomial $(\frac{3}{4} + \frac{1}{4})^n$ shows that the approach to a symmetrical figure is much slower when p and q are very unequal, that is, when the loading is great. The normal curves are evidently not as good approximations as those in the preceding figure where p and q were more nearly equal; nevertheless the approximation is improving as n increases.

point binomial. Suppose we require the sum of the 2d, 3d, and 4th terms: the true value of this sum is

$$\binom{10}{2}\left(\frac{3}{4}\right)^8\left(\frac{1}{4}\right)^2 + \binom{10}{3}\left(\frac{3}{4}\right)^7\left(\frac{1}{4}\right)^3 + \binom{10}{4}\left(\frac{3}{4}\right)^6\left(\frac{1}{4}\right)^4$$

$$= 0.2816 + 0.2502 + 0.1460 = 0.6778 \quad (23)$$

This may be looked upon as the sum of the three rectangular blocks lying between the abscissas $1\frac{1}{2}$ and $4\frac{1}{2}$, or between -1 and $+2$ if *measured from the mean*, $2\frac{1}{2}$. Measured in units of the standard deviation ($\frac{1}{4}\sqrt{30}$), these abscissas are distant $-4/\sqrt{30}$ and $8/\sqrt{30}$, respectively, from the mean. In order to use a table of the normal integral that starts from the mean, the negative and positive integrations must be performed separately, and the result is

$$\frac{1}{\sigma\sqrt{2\pi}} \int_{-1}^{2} e^{-x^2/2\sigma^2}\, dx = \frac{1}{\sqrt{2\pi}} \int_{0}^{4/\sqrt{30}} e^{-\frac{1}{2}t^2}\, dt + \frac{1}{\sqrt{2\pi}} \int_{0}^{8/\sqrt{30}} e^{-\frac{1}{2}t^2}\, dt$$

$$= 0.2674 + 0.4279 = 0.6953 \qquad (24)$$

ii. $n = 50, p = \frac{1}{4}, q = \frac{3}{4}$. Here the mean is at $np = 12.5$, and the standard deviation is $\sqrt{npq} = \frac{1}{4}\sqrt{150}$. The approximating normal curve is accordingly written with $\sigma = \frac{1}{4}\sqrt{150}$, the y-axis being placed at the mean of the point binomial, as before.

Suppose here that we require the sum of the 10th, 11th, 12th, 13th, and 14th terms: the true value of this sum is

$$\binom{50}{10}\left(\frac{3}{4}\right)^{40}\left(\frac{1}{4}\right)^{10} + \binom{50}{11}\left(\frac{3}{4}\right)^{39}\left(\frac{1}{4}\right)^{11} + \binom{50}{12}\left(\frac{3}{4}\right)^{38}\left(\frac{1}{4}\right)^{12}$$

$$+ \binom{50}{13}\left(\frac{3}{4}\right)^{37}\left(\frac{1}{4}\right)^{13} + \binom{50}{14}\left(\frac{3}{4}\right)^{36}\left(\frac{1}{4}\right)^{14}$$

which turns out to be $0.09852 + 0.11942 + 0.12937 + 0.12605 + 0.11104 = 0.58440$. This is the sum of the areas of the five rectangular blocks lying between the abscissas $9\frac{1}{2}$ and $14\frac{1}{2}$, or between -3 and $+2$ if measured from the mean, 12.5. The area under the approximating normal curve between these limits is

$$\frac{1}{\sigma\sqrt{2\pi}} \int_{-3}^{2} e^{-x^2/2\sigma^2}\, dx = \frac{1}{\sqrt{2\pi}} \int_{0}^{12/\sqrt{150}} e^{-\frac{1}{2}t^2}\, dt + \frac{1}{\sqrt{2\pi}} \int_{0}^{8/\sqrt{150}} e^{-\frac{1}{2}t^2}\, dt$$

$$= 0.33641 + 0.24318 = 0.57959 \qquad (25)$$

which is to be compared with the exact value, 0.58440.

iii. $n = 100, p = \frac{1}{4}, q = \frac{3}{4}$. Here the mean is at $np = 25$, and the standard deviation is $\sqrt{npq} = \frac{1}{4}\sqrt{300}$. The approximating normal curve is accordingly written with $\sigma = \frac{1}{4}\sqrt{300}$, the y-axis being placed at the mean of the point binomial, as before.

Suppose that here we require the sum of the 5 terms from 26 to 30 inclusive. The exact value of this sum is

$$\binom{100}{26}\left(\frac{3}{4}\right)^{74}\left(\frac{1}{4}\right)^{26} + \cdots + \binom{100}{30}\left(\frac{3}{4}\right)^{70}\left(\frac{1}{4}\right)^{30}$$

which after some laborious arithmetic turns out to be 0.34274, the sum of the areas of the 5 rectangular blocks lying between the abscissas $25\frac{1}{2}$ and $30\frac{1}{2}$, or between $+\frac{1}{2}$ and $5\frac{1}{2}$ when measured from the mean, 25. The approximation to this area afforded by the normal curve is

$$\frac{1}{\sigma\sqrt{2\pi}} \int_{\frac{1}{2}}^{5\frac{1}{2}} e^{-x^2/2\sigma^2}\, dx = \frac{1}{\sqrt{2\pi}} \int_{0}^{22/\sqrt{300}} e^{-\frac{1}{2}t^2}\, dt - \frac{1}{\sqrt{2\pi}} \int_{0}^{2/\sqrt{300}} e^{-\frac{1}{2}t^2}\, dt$$

$$= 0.39799 - 0.04596 = 0.35203 \qquad (26)$$

which again is good, but a little in excess.

b. In Fig. 43 we see three more point binomials, this time with $p = \frac{1}{20}$ and $q = \frac{19}{20}$. In spite of such heavy loading, a normal curve may be seen emerging even at $n = 50$.

We shall here try only one numerical example in approximation. Suppose that for $n = 100$ we require the sum of the 4 terms (rectangular blocks) at the abscissas 4, 5, 6, 7. The exact value of this sum is

$$\binom{100}{96}\left(\frac{19}{20}\right)^{96}\left(\frac{1}{20}\right)^{4} + \binom{100}{95}\left(\frac{19}{20}\right)^{95}\left(\frac{1}{20}\right)^{5} + \binom{100}{94}\left(\frac{19}{20}\right)^{94}\left(\frac{1}{20}\right)^{6}$$

$$+ \binom{100}{93}\left(\frac{19}{20}\right)^{93}\left(\frac{1}{20}\right)^{7}$$

After more laborious arithmetic this is found to be 0.614200. This is the sum of the areas of the 4 rectangular blocks lying between the abscissas $3\frac{1}{2}$ and $7\frac{1}{2}$, or between $-1\frac{1}{2}$ and $+2\frac{1}{2}$ when measured from the mean, 5. The approximation afforded by the normal curve is

$$\frac{1}{\sigma\sqrt{2\pi}} \int_{-1\frac{1}{2}}^{2\frac{1}{2}} e^{-x^2/2\sigma^2}\, dx = \frac{1}{\sqrt{2\pi}} \int_0^{3/\sqrt{19}} e^{-\frac{1}{2}t^2}\, dt + \frac{1}{\sqrt{2\pi}} \int_0^{5/\sqrt{19}} e^{-\frac{1}{2}t^2}\, dt$$

$$= 0.25435 + 0.37432 = 0.62867 \qquad (27)$$

which is 2.3 percent high.

It will be interesting to try out the Type III curve here.[8] Instead of using Eq. 18, we may introduce the equivalent form

$$Y = \frac{1}{\Gamma(p+1)} x^p e^{-x} \qquad \begin{array}{l}\text{[In Eq. 18 let } 1 + \alpha t = \alpha^2 x, \\ dt = \alpha\, dx, y\, dt = Y\, dx]\end{array} \qquad (28)$$

which is plotted in Fig. 57 (p. 468) with $p + 1$ replaced by n. This is the form of the integrand of the gamma function; and it will be seen in Exercise 29 of Chapter 14 that its mean is at $x = p + 1$ and its standard deviation is $(p + 1)^{\frac{1}{2}}$. If then we equate the standard deviation of the curve to the standard deviation of the binomial, we shall have $p + 1 = npq = \frac{19}{4}$, or $p = 3.75$. The integration is to be taken from a point $\frac{3}{2}$ to the left of the mean to a point $\frac{5}{2}$ to the right of the mean, as the limits for l in the normal integral indicate. Now the mean of the Type III curve is at $p + 1$, which is 4.75; hence the integral runs from $4.75 - 1.5 = 3.25$ to $4.75 + 2.5 = 7.25$, whereupon the approximation afforded by the Type III curve is

$$\frac{1}{\Gamma(p+1)} \int_{3.25}^{7.25} x^p e^{-x}\, dx = I\left(\frac{7.25}{\sqrt{4.75}}, 3.75\right) - I\left(\frac{3.25}{\sqrt{4.75}}, 3.75\right)$$

$$= I(3.3266, 3.75) - I(1.4912, 3.75) \qquad (29)$$

in the notation of the *Tables of the Incomplete Gamma Function* (Ch. 14). Linear interpolation in these tables gives for the two integrals:

$$I(3.3266, 3.75) = 0.8726$$

$$I(1.4912, 3.75) = 0.2677$$

$$\text{Difference} = 0.6049$$

The difference, 0.6049, is 1.5 percent below the true value 0.6142 and is accordingly a better approximation than that afforded by the normal curve.

[8] The student may prefer to return to this calculation after studying Chapter 14.

The flattening of the point binomial with increasing n. As n increases, the point binomial $(q + p)^n$ becomes flatter and flatter as the standard deviation $\sqrt{npq}$ (or loosely, the "spread") increases. As a consequence,

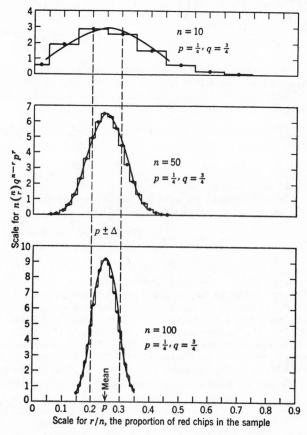

FIG. 44. The probability of the ratio r/n in n random trials as given by the binomial series. This is an alternative form of the point binomials $(\frac{3}{4} + \frac{1}{4})^n$ of Fig. 42. The abscissas have been divided by n and the ordinates multiplied by n. The total area under each polygon is again unity. The means of the polygons lie at p, and their standard deviations are $\sqrt{pq/n}$. As n increases, the polygons become more and more symmetrical and concentrated about the mean, so that the area between the dashed lines increases. A normal curve having the same mean and standard deviation as the polygon is shown for each.

the sum of the terms contained within a prescribed number of integers either side or both sides of the mode decreases. This fact is illustrated by the three point binomials of Figs. 42 and 43, on each of which a pair of vertical lines is drawn equidistant from the mode. As n increases, every ordinate included between these vertical lines decreases, wherefore the included area also decreases, proving the statement. The probability that the number of red chips in n trials will fall exactly on the mode, or that it will be any particular number within a prescribed interval from the mode, becomes less and less as n increases.

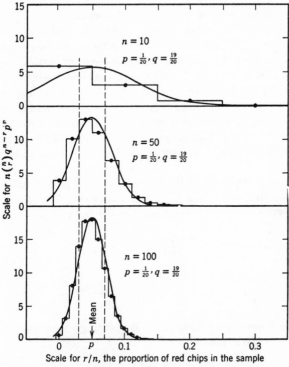

FIG. 45. The probability of the ratio r/n in n random trials as given by the binomial series. This is an alternative form of the point binomials of Fig. 43.

The concentration of the proportion r/n toward p with increasing n. Suppose now that the base line of each of the plots in Figs. 44 and 45 be compressed by dividing the abscissas r by n; the new abscissas are then r/n, which, regardless of the magnitude of n, always run from 0 to 1 while r runs from 0 to n. The probabilities remain unchanged (cf. the table on p. 115). In order that the new plots be normalized (their areas kept equal to unity), the ordinates must at the same time be multiplied by n. The old ordinate scale was $\binom{n}{r} q^{n-r} p^r$, or the rth term of the expansion of $(q + p)^n$; therefore the new ordinate must be n times as great, or $n \binom{n}{r} q^{n-r} p^r$. Let the results be plotted in !Figs. 44 and 45.

As n increases, the plots in Figs. 42 and 43 flatten out; this is because their standard deviations $\sqrt{npq}$ increase with n. In contrast, the plots in Figs. 44 and 45 show more and more concentration toward the mean with increasing n, because their standard deviations are $\sqrt{pq/n}$, which *decrease* with n. The mean of the r/n curves, it should be noted, is always at $r/n = p$, regardless of the sample-size n. More and more of the total area (unity) is included within the pair of vertical lines at the abscissas $r/n = p \pm \Delta$. In Fig. 42 the vertical lines mark off a range for r; in Fig. 44, however, the vertical lines prescribe a range not for r but for r/n, and as n increases they include a greater and greater range of r values, and hence an ever-increasing fraction of the area of the whole plot. It is thus a fact that *the probability that r/n will lie in the interval $p \pm \Delta$ may be increased to as near certainty as desired simply by making n big enough.*

The Poisson exponential limit. In Eq. 1 let $p \to 0$ and $n \to \infty$ in such manner that np remains finite and equal to m. Then for $q = 1 - p$ one may write $1 - m/n$, whereupon the rth term of the binomial is [9]

$$\binom{n}{r} q^{n-r} p^r = \frac{n!}{r!(n-r)!} \left(\frac{m}{n}\right)^r \left(1 - \frac{m}{n}\right)^{n-r}$$

$$= \left(1 - \frac{m}{n}\right)^n \frac{m^r}{r!} \frac{n!}{(n-r)!\, n^r (1 - m/n)^r} \to e^{-m} \frac{m^r}{r!} \quad \text{as } n \to \infty \quad (30)$$

This is the *Poisson probability* of r. By putting r successively equal to $0, 1, 2, 3, \cdots$ the sequence

$$e^{-m} \left\{ 1, m, \frac{m^2}{2!}, \frac{m^3}{3!}, \cdots, \frac{m^r}{r!}, \cdots \right\}$$

is generated. This series is known as the *Poisson exponential limit*, or the *Poisson exponential series*. It may be considered either as a series in its own right or as an approximation term by term to the point binomial for large n and small p.

Let us compare a few of the terms of the Poisson exponential limit with those of the point binomial for $n = 50$, $p = \frac{1}{20}$ (Fig. 43 or 45).

---

[9] The student is referred to C. E. Wetherburn's *Mathematical Statistics* (Cambridge, 1947), p. 64.

For m we use $np = 2.5$ and draw up the comparison below, first calculating or finding from a table of exponentials that $e^{-2.5} = 0.082085$. It

Term	Binomial; $p = \frac{1}{20}$, $n = 50$		Poisson limit; $m = 2.5$	
0th	q^n	$= 0.076945$	e^{-m}	$= 0.082085$
1st	$nq^{n-1}p$	$= .202487$	me^{-m}	$= .205212$
2d	$\binom{n}{2} q^{n-2}p^2 =$	$.261101$	$(m^2/2!)e^{-m} =$	$.256516$
3d	$\binom{n}{3} q^{n-3}p^3 =$	$.219875$	$(m^3/3!)e^{-m} =$	$.213763$
4th	$\binom{n}{4} q^{n-4}p^4 =$	$.135975$	$(m^4/4!)e^{-m} =$	$.133602$

will be observed that the agreement is good. For larger values of n the agreement will be still better for the same value of p.

Since the mean of the point binomial is at np, and its standard deviation is $\sqrt{(npq)}$, the mean and standard deviation of the Poisson exponential limit must be m and $\sqrt{m}$ respectively, as can be proved directly (see the exercises following).

The calculation of probabilities often involves summations of terms of the Poisson series. Molina's tables, cited in the next section, are particularly helpful. In the next chapter it will be shown that any number of successive terms may be summed exactly by an incomplete gamma function (first proved by Molina in 1915).

Another publication which has been found useful for the subject of this chapter is Holbrook Working's *The Binomial and Exponential Distributions* (Stanford University Press, 1943).

Exercise 1. Prove directly that the mean of the Poisson series is m, and that the variance is also m. That is, prove $\sum_{0}^{\infty} r P(r) = m$ and $\sum_{r=0}^{\infty} (r - m)^2 P(r) = m$, where $P(r) = e^{-m}m^r/r!$.

Exercise 2. If two independent variates, x and y, have Poisson distributions with means m_1 and m_2, their sum has also the Poisson distribution with mean $m_1 + m_2$ and variance $m_1 + m_2$.

The student is referred to page 59 of Wetherburn's book, cited at the bottom of the preceding page.

Exercise 3. The sum of any number of Poisson variates has itself a Poisson distribution whose mean is the sum of the means of the separate distributions.

Exercise 4. (*A theorem due to E. C. Molina.*)[10] Define

$$P(r) = \frac{e^{-m}m^r}{r!} \qquad \text{[Poisson probability]}$$

$$P(r \gtreqless s) = \sum_{r=s}^{\infty} P(r)$$

$$S = \sum_{s=c}^{\infty} P(r \gtreqless s)$$

Prove that

$$S = (1 - c)P(r \gtreqless c) + mP(r \gtreqless c - 1)$$

Solution (Molina [10])

By definition,

$$S = e^{-m}\left\{ \frac{m^c}{c!} + \frac{m^{c+1}}{(c + 1)!} + \frac{m^{c+2}}{(c + 2)!} + \cdots \right.$$

[As all terms are positive and the horizontal series all convergent, this two-way series may be added in the vertical]

$$+ \quad `` \quad + \quad `` \quad + \cdots$$
$$+ \quad `` \quad + \cdots$$
$$\left. + \cdots \right\}$$

$$= e^{-m}\left\{ \frac{m^c}{c!} + 2\frac{m^{c+1}}{(c + 1)!} + 3\frac{m^{c+2}}{(c + 2)!} + \cdots \right\}$$

$$= e^{-m}\sum_{i=0}^{\infty} [\overline{c + i} - \overline{c - 1}]\frac{m^{c+i}}{(c + i)!}$$

$$= (1 - c)P(r \gtreqless c) + mP(r \gtreqless c - 1) \qquad\qquad Q.E.D.$$

Exercise 5. *The square-root transformation.* Suppose that r is a random variable having the Poisson distribution—i.e.,

$$P(r) = \frac{m^r}{r!} e^{-m} \qquad r = 0, 1, 2, \cdots$$

Prove that

$$\text{Var } \sqrt{r} = \tfrac{1}{4} \qquad \text{approximately, if } m > \text{about } 10$$

$$E \sqrt{r} = \sqrt{m - \tfrac{1}{4}} \qquad `` \qquad , \quad `` \,`` \,`` \quad `` \quad ``$$

These results are important because in binomial or Poisson sampling they enable tests of significance to be calculated instantly without knowledge of m, provided $m > $ about 10, in which case the standard error of $\sqrt{r}$ is $\tfrac{1}{2}$, 2 sigma is 1, and 3 sigma is $1\tfrac{1}{2}$, regardless of the mean m.   Bartlett [11] and

[10] Private communication from Mr. Molina, 13 June 1948.

[11] M. S. Bartlett, *J. Royal Stat. Soc.*, Suppl., vol. 3, 1936: pp. 68–78.

Irwin [12] have provided tables of Var $\sqrt{r}$ for $m < 15$. The following simple table shows how Var $\sqrt{r}$ varies with m (values copied from Irwin's article):

m	2	3	4	5	8	10	15
Var $\sqrt{r}$	0.3900	0.3400	0.3056	0.2861	0.2630	0.2600	0.2570

The square-root transformation bears the same relation to the Poisson distribution as the arc sine transformation bears to the binomial distribution. As a matter of fact, the square-root transformation may be looked upon as operating on one axis only of the arc sine transformation.

Most economical inventory.[13] A grocer lays in his stock of packages of cream cheese every Monday morning when the driver for the wholesale company makes his regular call. The grocer sells an average of 100 packages per week. (*a*) What is the minimum Monday-morning stock if he is not to be caught short more than 1 week in 100 on the average? (*b*) In order not to miss more than 1 sale in 100, on the average? The grocer recognizes two kinds of error.

Error of the 1st kind: too heavy a stock; packages unsold at the end of the week, to be carried over. The grocer hopes to avoid this, as customers prefer fresh stock. Moreover, too heavy a stock jams his costly refrigerated space; also it represents unnecessary investment of capital.

Error of the 2d kind: too small a stock; caught short; sales lost; customers irritated and go elsewhere.

He can not entirely avoid one error without getting caught in the other: he therefore asks questions *a* and *b* in the hope of minimizing his net losses.

Two assumptions will be made in the solution: *i.* a sale is as probable in one minute as the next; *ii.* sales are made a package at a time. (Later we shall see the effect of a customer purchasing 2 packages at a time.) The sample-size in this problem is the number of intervals of time during the week in each of which a package of cheese might be sold. Let 1 minute be the unit of time, and let the store be open 8 hours per day,

[12] J. O. Irwin, *J. Royal Stat. Soc.*, vol. cvi, 1943: pp. 143–4.

[13] My interest in this problem was stimulated by Fry's dog-biscuit problem on p. 127 of his *Probability and Its Engineering Uses* (Van Nostrand, 1928) and by recent conversations with Dr. Churchill Eisenhart. The reader will recognize Fry's phraseology. I have chosen cream cheese to keep away from Fry's dog biscuits, as his problem deals only with question *b*. The distinction between the two questions *a* and *b* was pointed out to me by Eisenhart, who in a private communication solved them both.

6 days per week. Then $n = 60 \times 8 \times 6 = 2880$ minutes; this is the "sample-size" in the calculations. The probability of a purchase taking place in a 1-minute interval is then

$$p = \frac{np}{n} = \frac{100}{2880} = 0.0347 \tag{31}$$

Let N be the initial stock required. The problem is to find what N should be to control the two errors. The solution, under the simplifying assumptions that were made above, lies in the binomial series. Let r be the number of customers that ask for cream cheese in a particular week (sample); by assumption r is a random variable and may take any value from 0 to n; its probability will be

$$P(r) = \binom{n}{r} q^{n-r} p^r \qquad \text{[Binomial series]} \tag{32}$$

N is to be found from the following inequalities:
For question a,

$$\sum_{r=N+1}^{n} P(r) \not> 0.01 \tag{33}$$

For question b,

$$\sum_{r=N}^{n} (r - N)P(r) \not> 0.01np \tag{34}$$

The labor of using the binomial is uninviting, and for question a we seek refuge in the normal curve as an approximation. This suggestion may appear to violate one of the precepts recently learned about restraining ourselves from going too far out in the tails of the approximating normal curve, but n is a large number here and the approximation may not be bad: fortunately we can check it. First, we need

$$\sigma^2 = npq = 100(1 - 0.0347) = 96.53 \tag{35}$$

$$\sigma = 9.82 \tag{36}$$

From the table of the normal integral on page 552 r/σ for $P = 0.01$ is seen to be 2.33, whence

$$N + \tfrac{1}{2} = 100 + 2.33\sigma \qquad \text{[The integration under the normal curve is to commence at } N + \tfrac{1}{2}\text{]}$$

$$= 100 + 23 = 123 \tag{37}$$

whence $N = 122\tfrac{1}{2}$ or 123 for good measure. This is the answer that the normal integral provides for question a. To verify it with another approximation let us convert binomial terms to Poisson exponential terms, for which Molina's tables [14] supply the answer at a glance.

Molina gives $\sum\limits_{r=c}^{\infty} P(r)$ as a function of c and m. $P(r)$ stands for the

Poisson probability $e^{-m}m^r/r!$, wherein $m = np = 100$. The accompanying brief extraction comes from page 46 of Molina's tables, whence it is

FOR QUESTION a: PROBABILITIES EXTRACTED DIRECTLY FROM MOLINA'S TABLE, PAGE 46

N	$\sum\limits_{r=c}^{\infty} P(r)$
123	0.014306
124	.011244
	← 0.01
125	.008774
126	.006798

seen that if the right-hand side is to be 0.01 or next below, c must be 125, which means that $N = 125 - 1$, in good agreement with the previous result. Thus with 24 percent excess initial inventory the grocer will be caught short only once in 100 weeks or about once in 2 years.

Question b is more difficult. Here,

$$\frac{1}{np} \sum_{r=N}^{n} (r - N)P(r)$$

$$= \frac{1}{np} \sum_{r=N}^{n} r \frac{n!}{(n-r)!r!} q^{n-r}p^r - \frac{N}{np} \sum_{r=N}^{n} P(r)$$

$$= \sum_{r-1=N-1}^{n-1} \frac{(n-1)!}{[n-1-\overline{r-1}]!(r-1)!} q^{\overline{n-1}-\overline{r-1}}p^{r-1} - \frac{N}{np} \sum_{r=N}^{n} P(r)$$

$$= \sum_{r'=N-1}^{n'} \binom{n'}{r'} q^{n'-r'}p^{r'} - \frac{N}{np} \sum_{r=N}^{n} \binom{n}{r} q^{n-r}p^r \quad \begin{array}{l}[r-1 \text{ replaced by } r'; \\ n-1 \text{ replaced by } \\ n'] \end{array}$$

$$\equiv F(N) \tag{38}$$

This is a function of N, as indicated. One way to solve this is by the inverse process of plotting $F(N)$ against N and seeing what value of N makes $F(N)$ equal to 0.01. Again the Poisson exponential will serve. With Poisson terms substituted for binomial terms Eq. 38 gives

$$F(N) = \sum_{r=N-1}^{\infty} e^{-m} \frac{m^r}{r!} - \frac{N}{np} \sum_{r=N}^{\infty} e^{-m} \frac{m^r}{r!} \quad [m = np = 100] \tag{39}$$

[14] E. C. Molina, *Poisson's Exponential Binomial Limit* (Van Nostrand, 1942).

By using Molina's tables one may rapidly fill in Cols. 2 and 3 in the accompanying table, whence it is obvious that N for question b must be 110. Thus, if on the average 100 packages are sold per week, an excess stock of only 10 packages will incur a *loss in sales* of less than 1 percent on the average.

For Question b: Cols. 2 and 3 from Molina's tables

N	$N-1$	$\sum\limits_{r=N-1}^{\infty} e^{-m}\dfrac{m^r}{r!}$	$\dfrac{N}{np}\sum\limits_{r=N}^{\infty} e^{-m}\dfrac{m^r}{r!}$	$F(N)$
			$m = np = 100$	(3)–(4)
(1)	(2)	(3)	(4)	(5)
109	108	0.224408	1.09 × 0.196325	0.010414
				← 0.01
110	109	.196325	1.10 × .170560	.008709

Remark 1 (Variation of results with "expected" sales). The above solutions were obtained on the assumption that the average weekly sales are 100 packages. What if the average were 25 or 10 packages? Let us return to question a and look at that part of Eq. 37 which says that

$$N + \tfrac{1}{2} = np + 2.33\sigma$$

$$= np + 2.33\sqrt{npq} \qquad (40)$$

With the $\tfrac{1}{2}$ dropped for good measure, the excess Monday-morning inventory beyond "expectation" is

giving
$$N - np = 2.33\sqrt{npq} \doteq 2.33\sqrt{np} \quad \text{[As } q \text{ is nearly 1]} \quad (41)$$

$$\frac{N - np}{np} = 2.33\sqrt{\frac{q}{np}} \doteq \frac{2.33}{\sqrt{np}} \qquad (42)$$

Thus, the smaller the weekly sales of a commodity, the greater the proportionate initial stock required. If the average weekly sales were 24 packages, the excess $N - np$ should be closely $2.33\sqrt{24} = 11.4$ or 12 for more good measure, which is 50 percent of the "expected" sales. The grocer's initial stock would be $24 + 12 = 36$. Molina's tables give $C = 37$ and $N = 36$. If the average weekly sales were 10 packages, the excess $N - np$ should be closely $2.33\sqrt{10} = 7\frac{1}{2}$, and the grocer's initial stock should be $10 + 8 = 18$. Molina's tables give C equal to about $18\frac{1}{2}$, whence $N = 18$, in good agreement again.

There is an important lesson here for the operation of retail chains, supply-depots, forms for railroad ticket-offices, and the like. **A moderate number of large stores require a total inventory of much less investment than a large number of small stores selling the same total per week.** This was seen in the fact that 24 percent excess was required when the "expected" sales were 100 per week, 50 percent when the "expected" sales were 24, and almost 100 percent when the "expected" sales were only 10.

Remark 2 (Effect of doubling the unit time-interval). The unit of time, which was arbitrarily chosen as 1 minute, may be varied without practical effect on the results. The unit of time represents one drawing from the bowl. It is the time required for the grocer to go through the motions of selling a package. As an exercise the student should show that the results are practically unaffected if the unit of time be halved or doubled. The reason is seen in the fact that np is unchanged and that q is very small. The assumption that 1 minute is the correct unit of time for selling a package is therefore of no consequence.

Remark 3 (Effect of double and triple purchases). The assumption that packages are sold one at a time is partly vitiated by customers who purchase 2 or 3 or more packages at once. The effect is to compel a bigger inventory than was calculated. This fact can be seen by imagining that every customer buys 2 packages: then np is in effect cut to half, because the total sales are 50 double units instead of 100 single units, and the initial inventory for question a is 66 double units or 132 single packages, in contrast with the 124 packages calculated earlier for the required inventory when sales are made in single units. A few observations taken occasionally on the number of packages of important commodities taken out per customer would provide the necessary information. Studies of this nature could properly be undertaken, I should think, by an association of grocers, as they would be in the nature of a research problem. The cost would be trifling compared with the savings.

Remark 4 (Effect of variable demand). For variable demand, as when 100 packages of a commodity (e.g., raisins) are sold per week at one season of the year against 24 per week at another season, it is only necessary to adjust the initial inventory accordingly.

Remark 5 (Relative importance of commodities). It is one thing in a customer's eyes for a grocer to be out of a commodity like crackers or cream cheese, which all grocers worthy of the name are expected to have on hand, and another to be out of a commodity like tinned stems of mushrooms, sales of which are not large and not vital to the customer. It appears therefore that so far as loss of prestige (and future trade) is concerned, a grocer would need to be most careful of his inventories of the really important items of the trade. A risk of being caught short 1 week in 20 on unimportant items might well be taken; the saving in total excess inventory over "expected" sales would be possibly 30 percent, I should judge from Molina's tables and a rough uncritical appraisal of the relative weights of important and unimportant items commonly carried in grocery stores.

Remark 6. The following considerations may be good reasons for departing from theory (pointed out in a private communication from Professor George W. Brown, then at the Iowa State College).

   a. There may be minimum units in which the grocer may purchase, such as by the dozen, half-dozen, carton of 5.

   b. The Poisson variability may be only a small part of the variability in sales during the week, which may be affected by weather and fluctuating impacts of advertising.

Exercise. Apply the square-root transformation to Eq. 41 and show that an approximate solution to question a is

$$N = (\sqrt{np} + 1.16)^2$$

Solution

$$N = np + 2.33\sqrt{np} \qquad \text{[Eq. 41]}$$

$$= (\sqrt{np} + 1.16)^2 - 1.16^2$$

$$\doteq (\sqrt{np} + 1.16)^2 \quad \text{[Neglecting } 1.16^2] \qquad Q.E.D.$$

Remark 7. The double square-root paper of Mosteller and Tukey (p.306) is admirably adapted to the required calculations. Let the above equation be written

$$\sqrt{N} = \sqrt{np} + 1.16$$

In the variables $\sqrt{N}$ and $\sqrt{np}$ this is a straight line having a slope of 45° and an intercept of 1.16. The vertical distance from this line to the 45° line through the origin is equivalent to 2.33σ for the 1 percent point under the normal integral. For any "expected" average sale (measured along the horizontal), such as $np = 100$ or 24, it is only necessary to read off N on the vertical. With $np = 100$, N is seen to be 125, in close agreement with the results obtained earlier. If the reader does not have a supply of the M-T double square-root paper on hand, he should rule his own, as he may do in a jiffy by (a) measuring off $x = \sqrt{np}$ on the horizontal and $y = \sqrt{N}$ on the vertical, marking off a few points along the scales in terms of np and N, and (b) drawing the line $y = x + 1.16$, along which he should observe that

$$
\begin{array}{rcl}
N = 125 & \text{when} & np = 100 \\
36 & " & 24 \\
19 & " & 10
\end{array}
$$

as obtained earlier, either exactly or very closely.

Exercise 1. By differentiating the identity

$$e^{-npu}(q + pe^u)^n = \sum_{r=0}^{n} P(r)e^{(r-np)u}$$

with respect to u, and again with respect to u, after each differentiation setting $u = 0$, prove that

$$\sum_{r=0}^{n} P(r)(r - np) = 0, \quad \text{and} \quad \sum_{r=0}^{n} P(r)(r - np)^2 = npq$$

wherefore the mean of the point binomial lies at $r = np$, and its variance is npq. Here $P(r)$ denotes $\binom{n}{r} q^{n-r}p^r$, the rth term of the binomial expansion. Students familiar with the use of the characteristic function will recognize this device.

Exercise 2. By differentiating the identity $(q + p)^n = \sum \binom{n}{r} q^{n-r} p^r$ twice with respect to p, treating q as a constant, prove that the mean of the point binomial is at np and its variance is npq.

Solution (Bertrand, Calcul des probabilités, 1889)

Differentiate once, multiply by p, and get

$$n(q + p)^{n-1} p = \sum r \binom{n}{r} q^{n-r} p^r$$

Now put $q + p = 1$ and see that

$$Er = \sum r \binom{n}{r} q^{n-r} p^r = np$$

Now start over, differentiate twice, and multiply by p^2 to get

$$n(n - 1)(q + p)^{n-2} p^2 = \sum r(r - 1) \binom{n}{r} q^{n-r} p^r$$

Now put $q + p = 1$ and see that

$$n(n - 1)p^2 = \sum r(r - 1) \binom{n}{r} q^{n-r} p^r$$

$$= \sum (r^2 - r) \binom{n}{r} q^{n-r} p^r$$

$$= Er^2 - Er$$

whence

$$Er^2 = n(n - 1)p^2 + np \qquad \text{[Because } Er = np]$$

and

$$\sigma_r^2 = Er^2 - (Er)^2$$

$$= n(n - 1)p^2 + np - n^2 p^2 = npq \qquad \text{Q.E.D.}$$

Exercise 3. Prove that

a.
$$E \binom{r}{i} = \sum_r \binom{r}{i} P(r) = \binom{n}{i} p^i$$

b.
$$E \binom{n - r}{i} = \sum_r \binom{n - r}{i} P(r) = \binom{n}{i} q^i$$

Here $P(r)$ denotes $\binom{n}{r} q^{n-r} p^r$, the rth term of the point binomial. i is a constant, and r is a random variable (the number of red chips in a sample of n). Then $\binom{r}{i}$ and $\binom{n - r}{i}$ are random variables, and the results give their "expected" values.

Solution (Dr. Eli S. Marks)

Let

$$F = (q + p)^n = \sum \binom{n}{r} q^{n-r} p^r$$

By differentiating both of these forms of F, it is seen that

$$\frac{\partial^i F}{\partial p^i} = n(n-1)(n-2) \cdots (n-i+1)(q+p)^{n-i}$$

$$= \sum_r r(r-1)(r-2) \cdots (r-i+1) \binom{n}{r} q^{n-r} p^{r-i}$$

whence

$$\frac{p^i}{i!} \frac{\partial^i F}{\partial p^i} = \binom{n}{i} p^i (q+p)^{n-i} = \sum_r \binom{r}{i} P(r)$$

Put $q + p = 1$, and the required result for Part a follows at once. Turn the binomial end for end, and Part b follows by symmetry.

Exercise 4.   Prove that the kth factorial moment coefficient of the point binomial is $n(n-1)(n-2) \cdots [n - (k-1)]p^k$.

Solution

By definition the kth factorial moment coefficient will be

$$\sum r(r-1)(r-2) \cdots [r - (k-1)] \binom{n}{r} q^{n-r} p^r$$

Cancel the k factors involving r into $r!$, and the result follows mentally.

B. THE HYPERGEOMETRIC SERIES

More detailed development of the hypergeometric series.   Consideration will now be given to a more detailed study of the hypergeometric series, which was encountered in Chapter 4, pages 113 ff. First, the probabilities represented by the hypergeometric series will be developed in another way. With Np red chips in the bowl, all having equal P-values, the probability of drawing a red chip therefrom is p. But this draw removes 1 chip, and a red one at that; the P-value of each chip is now not $1/N$ but $1/(N-1)$ because of nonreplacement, and hence the probability of drawing another red chip at the second draw is not p but something less, being $(Np - 1)/(N - 1)$, wherefore the probability of drawing 2 red chips in succession is $\dfrac{Np}{N} \dfrac{Np - 1}{N - 1}$. By symmetry, the probability of drawing 2 white ones in succession is $\dfrac{Nq}{N} \dfrac{Nq - 1}{N - 1}$.

Likewise the probability of getting a white chip at the first draw is q, and the probability of following it with a red one is $Np/(N-1)$. The product is $\dfrac{Nq}{N}\dfrac{Np}{N-1}$, which is also the probability that the first draw is red and the second white. It follows that the probability of drawing 1 red and 1 white with order unspecified in 2 drawings without replacement is

$$P(1, 2, N) = 2\frac{Np}{N}\frac{Nq}{N-1}, \quad \text{or, better,} \quad \binom{2}{1}\frac{Np}{N}\frac{Nq}{N-1}$$

Thus, for samples of 2, the probabilities of the 3 possible results are these:

$$P(2, 2, N) = \frac{Np}{N}\frac{Np-1}{N-1} \qquad = \text{the probability of drawing 2 red chips in succession}$$

$$P(1, 2, N) = \binom{2}{1}\frac{Np}{N}\frac{Nq}{N-1} = \text{the probability of drawing 1 red chip and 1 white in either order}$$

$$P(0, 2, N) = \frac{Nq}{N}\frac{Nq-1}{N-1} \qquad = \text{the probability of drawing 2 white chips in succession}$$

By proceeding in this way, and recalling that a sample of n chips containing r red ones and $n-r$ white ones can be arranged in $\binom{n}{r}$ different ways, it can be seen that, if there are initially Np red chips in the universe, the chance of there being r red ones in the sample of n drawn therefrom without replacement will be

$$P(r, n, N) =$$
$$\binom{n}{r}\overbrace{\frac{Np}{N}\frac{Np-1}{N-1}\cdots\frac{Np-r+1}{N-r+1}}^{r\text{ factors}}\overbrace{\frac{Nq}{N-r}\frac{Nq-1}{N-r-1}\cdots\frac{Nq-(n-r)+1}{N-n+1}}^{n-r\text{ factors}} \quad (43)$$

No matter what be the order of the appearance of the r red and $n-r$ white chips in the sample, the numerator will consist of these same n factors taken in the order corresponding to the appearance of the red and white chips. The order of the factors in the denominator does not change. Going on,

$$P(r, n, N) = \frac{n!}{(n-r)!\,r!}\frac{(N-n)!}{N!}\frac{Np!}{(Np-r)!}\frac{Nq!}{(Nq-\overline{n-r})!}$$

$$= \frac{\dbinom{Np}{r}\dbinom{Nq}{n-r}}{\dbinom{N}{n}} \quad (44)$$

as obtained in Table 5, page 112. This is the rth term of the *hypergeometric series*. As with the binomial series, the sum of all $n + 1$ terms is unity, but term for term the two series differ more or less, depending on how big N is compared with n (see Ex. 5, p. 122). It is helpful to write down the rth term of both series and see why they differ.

Binomial $\quad \dbinom{n}{r} \overbrace{p \quad p \quad \cdots \quad p}^{r \text{ factors}} \overbrace{q \quad q \quad \cdots \quad q}^{n-r \text{ factors}}$

Hypergeometric $\quad \dbinom{n}{r} \dfrac{Np}{N} \dfrac{Np-1}{N-1} \cdots \dfrac{Np-r+1}{N-r+1} \dfrac{Nq}{N-r} \dfrac{Nq-1}{N-r-1} \cdots \dfrac{Nq-(n-r)+1}{N-n+1}$

The fundamental distinction is that in the binomial term q and p are repeated, whereas in the hypergeometric term the factors vary continually as one chip after another is drawn.

The mean of the hypergeometric series is $Er = np$ exactly the same as for the binomial. But the variance of the hypergeometric series is smaller than the variance npq of the binomial by the factor $(N - n)/(N - 1)$. Both of these results have already been derived in Chapter 4.

Remark 1. The rth term of the hypergeometric series can easily be memorized. In the factor $\dbinom{Np}{r}$, Np is the number of red chips initially in the universe, and r the number of red chips in the sample. Similarly, Nq in the factor $\dbinom{Nq}{n-r}$ is the number of white chips initially in the universe, and $n - r$ the number of white chips in the sample. N and n in the denominator represent the totals of red and white chips in universe and sample respectively.

Remark 2. Questions are sometimes asked regarding the name "hypergeometric" series. A general series of the form

$$F(\alpha, \beta, \gamma, x) = 1 + \frac{\alpha \cdot \beta}{\gamma \cdot 1} x + \frac{\alpha \cdot \alpha + 1 \cdot \beta \cdot \beta + 1}{\gamma \cdot \gamma + 1 \cdot 1 \cdot 2} x^2$$

$$+ \frac{\alpha \cdot \alpha + 1 \cdot \alpha + 2 \cdot \beta \cdot \beta + 1 \cdot \beta + 2}{\gamma \cdot \gamma + 1 \cdot \gamma + 2 \cdot 1 \cdot 2 \cdot 3} x^3 + \cdots \quad (45)$$

was encountered by Gauss in the solution of certain differential equations, and because this series includes the ordinary geometric series $\sum\limits_{r=0}^{\infty} x^r$ as a special case, the more general series was described as "hypergeometric." Now, if the hypergeometric probability series be turned end for end and summed with r starting at n and running down to 0, then factored, it can be written as

$$\sum_{r=n}^{0} P(r, n, N)$$

$$= \frac{Np}{N} \frac{Np-1}{N-1} \frac{Np-2}{N-2} \cdots \frac{Np-n+1}{N-n+1}$$

$$\times \left\{ 1 + \binom{n}{1} \frac{Nq}{Np-n+1} + \binom{n}{2} \frac{Nq}{Np-n+1} \frac{Nq-1}{Np-n+2} \right.$$

$$+ \binom{n}{3} \frac{Nq}{Np-n+1} \frac{Nq-1}{Np-n+2} \frac{Nq-2}{Np-n+3} + \cdots$$

$$\left. + \frac{Nq}{Np-n+1} \frac{Nq-1}{Np-n+2} \frac{Nq-2}{Np-n+3} \cdots \frac{Nq-n+1}{Np} \right\} \quad (46)$$

The series in the braces is none other than $F(\alpha, \beta, \gamma, x)$ in which $\alpha = -n$, $\beta = -Nq, \gamma = Np - n + 1, x = 1$. Thus, the theory of sampling without replacement leads to a series that has long been called hypergeometric.

It may be of interest to note that the binomial or Bernoulli series for sampling *with* replacement is also a special form of the Gauss hypergeometric series, as can be seen by writing out the series $F(\alpha, \beta, \gamma, x)$ with $\alpha = -n$, $\gamma = \beta, x = -p/q$. By so doing it will be found that the Bernoulli series

$$(q + p)^n = q^n + nq^{n-1}p + \frac{n \cdot n - 1}{1 \cdot 2} q^{n-2}p^2$$

$$+ \frac{n \cdot n - 1 \cdot n - 2}{1 \cdot 2 \cdot 3} q^{n-3}p^3 + \cdots + p^n$$

$$= q^n F\left(-n, \beta, \beta, -\frac{p}{q}\right) \quad (47)$$

Exercise 1. Prove the theorem

$$\sum_r \binom{Np}{r} \binom{Nq}{n-r} = \binom{N}{n}$$

HINT. The left-hand side is the numerator of $P(r, n, N)$ as given in Eq. 44. In n drawings *some* value of r *must* be obtained; i.e., $\sum_r P(r, n, N) = 1$,

which is equivalent to the theorem.

Another proof, shown to me by Mr. Emil D. Schell, may be had by considering the identity

$$(1 + x)^{Np}(1 + x)^{Nq} = (1 + x)^N$$

and noting that, when the left-hand side is multiplied out, the coefficient of x^n on the left must equal the coefficient of x^n on the right. The equality of the coefficients of x^n demands that

$$\sum_r \binom{Np}{r} \binom{Nq}{n-r} = \binom{N}{n}$$

which is the theorem required.

Exercise 2. *a.* Find the maximum ordinate in the hypergeometric series

$$P(r, n, N) = \frac{\binom{Np}{r}\binom{Nq}{n-r}}{\binom{N}{n}}$$

Solution (due to F. F. Stephan)

$$\frac{P(r, n, N)}{P(r-1, n, N)} = \frac{\binom{Np}{r}\binom{Nq}{n-r}}{\binom{Np}{r-1}\binom{Nq}{n-r+1}}$$

$$= \frac{Np-r+1}{r}\ \frac{n-r+1}{Nq-n+r}$$

$$= 1 + \frac{(Np+1)(n+1) - r(N+2)}{r(Nq-n+r)}$$

Then
$$P(r, n, N) \geq P(r-1, n, N)$$
if
$$r \leq \frac{(Np+1)(n+1)}{N+2}$$

Therefore the maximum (modal) term of $P(r, n, N)$ occurs for that value of r which is the integer next below

$$\frac{(Np+1)(n+1)}{N+2}$$

If the equality holds, there will be two equal modal ordinates at and just preceding the value of r given by equality. This result corresponds to the mode in the point binomial (p. 404).

b. Prove that, as in the binomial, the mean and mode of the hypergeometric series are separated by less than a whole unit.

Solution (Eli S. Marks)

Let r_m be the modal abscissa. Then the following inequalities develop and prove the theorem. From Part *a* of this exercise

$$\frac{(Np+1)(n+1)}{N+2} - 1 \leq r_m \leq \frac{(Np+1)(n+1)}{N+2}$$

whence
$$np - 1 + \frac{(Np+1)(n+1) - np(N+2)}{N+2} \leq r_m$$

$$\leq np + 1 + \frac{(Np+1)(n+1) - (np+1)(N+2)}{N+2}$$

or

$$np - 1 + \frac{p(N - n) + nq + 1}{N + 2} \leq r_m \leq np + 1 - \frac{np + q(N - n) + 1}{N + 2}$$

which gives

$$np < r_m < np + 1 \qquad\qquad Q.E.D.$$

Exercise 3. Find the third moment coefficient of the hypergeometric series about the mean.

Answer:

$$E(r - Er)^3 = npq(q - p)\frac{(N - n)(N - 2n)}{(N - 1)(N - 2)}$$

Thus, in the distribution of r in random samples the

$$\text{Skewness} = \frac{\mu_3}{2\sigma^3} \quad \begin{array}{l}\text{[There are many definitions of skewness.}\\ \text{This is only one of them. See also p. 410.]}\end{array}$$

$$= \frac{\epsilon}{\sqrt{npq}}\frac{N - 2n}{N - 2}\sqrt{\frac{N - 1}{N - n}}$$

where $\epsilon = \frac{1}{2}(q - p)$, the loading of the universe as defined on page 409. The skewness of r is seen to approach the binomial skewness $\epsilon/\sqrt{npq}$ as N increases.

Solution

Expand the binomial $(r - Er)^3$, take the "expected" value of each term, replace Er by np; also add and subtract $3n^3p^3 + 6n^2p^2Er$ and thus get

$$E(r - Er)^3 = Er^3 - 3npE(r - np)^2 - n^3p^3 \tag{1}$$

Then write the identity

$$Er^3 = E[r(r - 1)(r - 2) + 3r^2 - 2r]$$

$$= E[r(r - 1)(r - 2)] + 3E(r - np)^2 + 3n^2p^2 - 2np \tag{2}$$

Substitution of Eq. 2 into Eq. 1 gives

$$E(r - Er)^3 = E[r(r - 1)(r - 2)]$$

$$+ (1 - np)[3E(r - np)^2 - np(2 - np)] \tag{3}$$

But

$$E(r - np)^2 = \text{Var } r = \frac{N - n}{N - 1}npq \tag{4}$$

Also

$$E[r(r - 1)(r - 2)]$$

$$= \sum_{r=0}^{n} r(r - 1)(r - 2) \, P(r, n, N)$$

$$= \sum r(r - 1)(r - 2) \frac{\binom{Nq}{n - r} \binom{Np}{r}}{\binom{N}{n}}$$

$$= Np \cdot Np - 1 \cdot Np - 2 \frac{n \cdot n - 1 \cdot n - 2}{N \cdot N - 1 \cdot N - 2} \sum \frac{\left(\dfrac{Nq}{n - 3 - r - 3}\right) \binom{Np - 3}{r - 3}}{\binom{N - 3}{n - 3}}$$

$$= Np \cdot Np - 1 \cdot Np - 2 \frac{n \cdot n - 1 \cdot n - 2}{N \cdot N - 1 \cdot N - 2} \tag{5}$$

Substitution of these equivalents for $E(r - np)^2$ and $E[r(r - 1)(r - 2)]$ into Eq. 2 will give the desired result.

Exercise 4 (Standard error of a percentile). Show that the standard error of the estimate of a percentile p is

$$\sigma_{\hat{x}_p} = \frac{1}{y_p} \sqrt{\frac{pq}{n}}$$

where $\hat{x}_p$ denotes the estimated percentile, and y_p the ordinate of the distribution at the required percentile. In particular, the standard error of the median is

$$\sigma_{\hat{x}_{50}} = \frac{1}{2y_{50}\sqrt{n}}$$

where $\hat{x}_{50}$ denotes the estimate of the median, and y_{50} denotes the ordinate of the distribution at the median point.

Solution

A universe is divided into two proportions p and q. When p is 0.9, the corresponding abscissa x_p is the 90 *percentile*; when $p = 0.25$ (0.75), the points on the abscissa are the lower (upper) *quartiles*; and when $p = 0.5$, the abscissa is the median. The sample gives an estimate $\hat{p}$ of p. Now from Fig. 46 it may be clear that

$$d\hat{p} = y_p \, dx_{\hat{p}}$$

provided the curve is continuous, whence by the equation on page 130 for the propagation of variance it must be that

$$\sigma_{\hat{p}} = y_p \sigma_{\hat{x}_n}$$

wherein $\sigma_{\hat{x}_p}$ is the standard error of the estimate $\hat{x}_p$ of the percentile $\hat{p}$, and $\sigma_{\hat{p}}$ is the standard error of the proportion of elements in the sample that are drawn from that portion of the universe constituting the percentile p. Then, as $\sigma_{\hat{p}}$ is known to be $\sqrt{pq/n}$ (p. 113) for samples drawn with replacement, the required result follows at once. The law for the propagation of error is valid only for large samples; however, in practice the result is useful for small samples as well. For samples drawn without replacement, the finite multiplier $\sqrt{[(N-n)/(N-1)]}$ would be introduced on the right, giving a slight decrease in the standard error of $\hat{x}_p$.

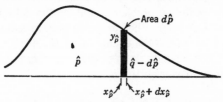

FIG. 46. Curve to assist in the derivation of the standard error of a percentile. $y_{\hat{p}}$ is the ordinate at the abscissa $x_{\hat{p}}$, which is the abscissa which divides the sample into two classes defined by the proportions $\hat{p}$ and $\hat{q}$.

For the normal curve, the midordinate y_{50} is $0.40/\sigma$ (more exactly, $0.3989/\sigma$), wherefore for this curve the standard error of the median of samples of n is $\sigma/2 \times 0.40\sqrt{n}$. On the other hand, the standard error of the mean ($\bar{x}$) is $\sigma/\sqrt{n}$. Let n' be the size of sample required when using the median of a sample as an estimate of the mean of a normal universe, and n the size of sample required when using the mean. The two standard errors will be equal if

$$\frac{1}{2 \times 0.40\sqrt{n'}} = \frac{1}{\sqrt{n}}$$

which requires

$$n = 0.64n'$$

Thus the mean ($\bar{x}$) of a sample of 64 has the same standard error as the median $\hat{x}_{50}$ of a sample of 100, wherefore the median is said to have an efficiency 64 percent of the efficiency of the mean. With a universe other than normal, the comparison may be entirely different. Thus, the median of a sample of n drawn from the Cauchy universe (p. 60) has a probability-distribution with standard deviation $\pi/4\sqrt{n}$, as the student should prove; whereas the mean has no standard deviation and so, by comparison with the median, has zero efficiency.

C. THE MULTINOMIAL SERIES

The multinomial expansion. $q + p$ is a binomial, and $p_1 + p_2 + p_3$ is a trinomial. The binomial and hypergeometric series arose in the theory of sampling from a universe consisting of 2 categories (red and white, vacant and occupied, rejected and accepted). The next step is the multinomial expansion and modifications thereof for the theory of sampling from a universe whose elements are classified into more than 2 categories. For instance, the universe might be a bowl containing red, blue, and green chips, and the sampling problem might be to determine the number and proportions within suitable probability limits. Or, the universe might be the dwelling units of a city, and the sampling problem might be to find the proportions vacant and for rent, vacant and not for rent, and not vacant. The categories might also be numerical magnitudes: thus some chips in the bowl might be labeled 1, some 2, and some 3; and in fact this is just the arrangement in Fig. 47 (p. 439) which will be used for the experiments to be described further on. In such an arrangement the sampling problem might be the determination of the mean of the bowl or some other property, within suitable probability limits. For such problems the expansion of $(p_1 + p_2 + p_3)^n$ is useful. The simplest way to expand it is to separate p_1 from the other terms and apply the binomial expansion to get

$$
\begin{aligned}
(p_1 + p_2 + p_3)^n &= \sum \frac{n!}{(n-r)!\,r!}\, p_1^{\,n-r}(p_2 + p_3)^r \\
&= p_1^{\,n} + np_1^{\,n-1}(p_2 + p_3) \\
&\quad + \frac{n!}{(n-2)!\,2!}\, p_1^{\,n-2}(p_2 + p_3)^2 + \cdots \\
&\quad + (p_2 + p_3)^n
\end{aligned}
\tag{48}
$$

then to apply the binomial expansion again, this time to the powers of p_2 and p_3. The result is

$$
\begin{aligned}
(p_1 + p_2 + p_3)^n &= p_1^{\,n} + p_2^{\,n} + p_3^{\,n} + n(p_1^{\,n-1}p_2 + p_1^{\,n-1}p_3 + p_2^{\,n-1}p_3 \\
&\quad + p_1 p_2^{\,n-1} + p_1 p_3^{\,n-1} + p_2 p_3^{\,n-1}) \\
&\quad + \frac{n(n-1)}{2}(p_1^{\,n-2}p_2^{\,2} + p_1^{\,n-2}p_3^{\,2} + p_2^{\,n-2}p_3^{\,2} \\
&\quad + p_1^{\,2}p_2^{\,n-2} + p_1^{\,2}p_3^{\,n-2} + p_2^{\,2}p_3^{\,n-2} \\
&\quad + 2p_1^{\,n-2}p_2 p_3 + 2p_1 p_2^{\,n-2}p_3 + 2p_1 p_2 p_3^{\,n-2}) + \cdots \\
&= \sum \frac{n!}{n_1!\,n_2!\,n_3!}\, p_1^{\,n_1} p_2^{\,n_2} p_3^{\,n_3}
\end{aligned}
\tag{49}
$$

the summation to cover all possible values of n_1, n_2, and n_3 such that $n_1 + n_2 + n_3 = n$. This result also follows from symmetry. Note that, as in the case of the binomial, the coefficient $\dfrac{n!}{n_1!\,n_2!\,n_3!}$ is a combinatorial expression, being the number of possible combinations of n things, of which n_1 are alike, n_2 are alike, and n_3 are alike. The student should satisfy himself that in this expansion there are 6 terms if $n = 2$, 10 terms if $n = 3$, 15 terms if $n = 4$, 21 terms if $n = 5$. For any n, there are $\binom{n + 3 - 1}{3 - 1} = \binom{n + 2}{2}$ terms.

If there are more than 3 cells in the universe, say M, so that

$$n_1 + n_2 + \cdots + n_M = n \tag{50}$$

$$p_1 + p_2 + \cdots + p_M = 1 \tag{51}$$

the same procedure gives

$$(p_1 + p_2 + \cdots + p_M)^n = \sum \frac{n!}{n_1! n_2! \cdots n_M!} p_1^{n_1} p_2^{n_2} \cdots p_M^{n_M} \tag{52}$$

Here the summation is to be made with all possible values of $n_1, n_2, \cdots, n_M$ compatible with Eq. 50. The term summed is called the general term of the multinomial expansion. The coefficient $\dfrac{n!}{n_1! n_2! \cdots n_M!}$ is the possible number of arrangements of n things of which n_1 are alike, n_2 are alike, $\cdots$, and n_M are alike. The expansion contains $\binom{n + M - 1}{M - 1}$ terms. The student should note that in the simple case $M = 2$, Eq. 52 reduces at once to the binomial expansion

$$(p_1 + p_2)^n = \sum \frac{n!}{n_1! n_2!} p_1^{n_1} p_2^{n_2} = \sum \frac{n!}{(n - r)! r!} p_1^{n-r} p_2^r \tag{53}$$

which was treated in Part A of this chapter.

Remark. The general term of the multinomial expansion may be written as

$$\frac{n!}{n_1! n_2! \cdots n_{M-1}! (n - n_1 - n_2 - \cdots - n_{M-1})!} p_1^{n_1} p_2^{n_2}$$
$$\cdots p_M^{n - n_1 - n_2 - \cdots - n_{M-1}}$$

subject to no restriction whatever on the $M - 1$ integers $n_1, n_2, \cdots, n_{M-1}$, for they can not possibly produce a contribution to the expansion unless they are all positive and their sum is not greater than n (see Ex. 13a at the end of Ch. 14).

An approximation. Let

$$x_i = \frac{n_i - np_i}{\sqrt{np_i}} \tag{54}$$

The following approximation may be derived for the general term of the multinomial expansion: [15]

$$\frac{n!}{n_1! n_2! \cdots n_M!} p_1^{n_1} p_2^{n_2} \cdots p_M^{n_M}$$
$$\rightarrow \frac{1}{(2\pi n)^{\frac{1}{2}(M-1)}(p_1 p_2 \cdots p_M)^{\frac{1}{2}}} e^{-\frac{1}{2}(x_1^2 + x_2^2 + \ldots + x_M^2)} \quad (55)$$

This approximation is good at not too large values of the x_i, provided no one of the np_i is small, which of course implies that n and each of the n_i are large numbers. The approximation has validity similar to that of the normal curve for the point binomial (p. 410); in fact the right-hand member of the approximation 55 may be regarded as the ordinate of a "normal surface" in M dimensions, having unit standard deviation along each of the M coordinate axes $Ox_1, Ox_2, \cdots, Ox_M$.

Remark 1. There is one relation between the x_i; they are not all independent. This relation is seen by summing Eq. 54 to get

$$\Sigma x_i \sqrt{p_i} = \frac{1}{\sqrt{n}} \Sigma n_i - \sqrt{n} \Sigma p_i$$

$$= \sqrt{n} - \sqrt{n} = 0 \quad (56)$$

Remark 2. In the preceding chapter it was found that the mean of the binomial lies at np, or that $Er = np$, so by collapsing $p_2 + p_3 + \cdots + p_M$ it follows that in the multinomial

$$En_1 = np_1$$

But this is a perfectly general result, applying to samples from all cells of the universe; hence

$$En_i = np_i \quad (57)$$

wherefore Eq. 54 defining x_i may be written in the form

$$x_i = \frac{n_i - En_i}{\sqrt{En_i}} \quad (54')$$

Remark 3. The exponential in Eq. 55 is a maximum when $x_1 = x_2 = \cdots = x_M = 0$. The position of the maximum is helpful in picking out the greatest term in the multinomial expansion, which will usually be that for which

n_1 is an integer close to En_1, or np_1
n_2 " " " " " En_2, or np_2
.   .
.   .
.   .
n_M " " " " " En_M, np_M

[15] Thornton C. Fry, *Probability* (Van Nostrand, 1928), Art. 101.

These statements may be tested against the expansion $(0.6 + 0.3 + 0.1)^n$ in Tables 1, 2, and 3 on pages 440 ff.

Remark 4. It is interesting to see what happens to the approximating normal surface in the simple case $M = 2$ (2 cells in the universe). The multinomial then reduces to a binomial, in which

$$n_1 + n_2 = n \quad \text{and} \quad p_1 + p_2 = 1$$

or

$$n_2 = n - n_1 \quad \text{and} \quad p_2 = 1 - p_1$$

With p_1 replaced for convenience by p, and p_2 by q, Eq. 54 makes

$$x_1 = \frac{n_1 - np_1}{\sqrt{np_1}} = \frac{x}{\sqrt{np}} \quad \text{and} \quad x_2 = \frac{n_2 - np_2}{\sqrt{np_2}} = -\frac{x}{\sqrt{nq}} \tag{58}$$

where $x = n_1 - np = -(n_2 - nq)$. x is thus the deviation of n_1 and n_2 from their "expected" values. The approximating normal surface (55) then reduces to a curve of ordinate

$$y = \frac{1}{\sqrt{2\pi npq}} e^{-\frac{1}{2}\left(\frac{x^2}{np} + \frac{x^2}{nq}\right)}$$

$$= \frac{1}{\sqrt{2\pi npq}} e^{-\frac{x^2}{2npq}} \tag{59}$$

But this approximation is just the normal curve of standard deviation $\sqrt{(npq)}$, as found in the preceding chapter.

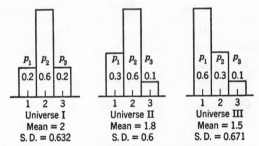

FIG. 47. Three simple discrete universes. Each universe consists of three categories (the headings 1, 2, 3) in proportions p_1, p_2, p_3.

Use of the multinomial in a theoretical experiment. Fig. 47 shows 3 universes, each consisting of 3 cells (wherefore $M = 3$ for each universe). In 1 cell of each universe are chips labeled 1; in another cell are chips labeled 2; and in another are chips labeled 3. The proportions are different in the 3 universes. Universe I is symmetrical; Universe II is skewed; and Universe III is highly skewed. It will be useful to note the similarities and contrasts in the sampling distributions of the means and standard deviations of samples drawn from them.

The compilation for random samples of 2, 3, and 5 follows in Tables 1, 2, and 3. In the third column are written the terms of Eq. 49; opposite each term there are written its numerical values for the 3 universes (Cols. 4, 5, and 6); and to the left, in Cols. 1 and 2, the mean and standard deviation of the sample that corresponds to the term in Col. 3. The results are plotted in Figs. 48 and 49, which also show samples of size $n = 10$, computations for which are omitted.

TABLE 1.   SAMPLES OF 2; NUMERICAL VALUES OF THE TERMS IN THE EXPANSION OF
$$(p_1 + p_2 + p_3)^2$$

Mean $\bar{x}$	Standard deviation s	Term	Term value I	II	III
1	0	$p_1{}^2$	0.04	0.09	0.36
2	0	$p_2{}^2$	.36	.36	.09
3	0	$p_3{}^2$	.04	.01	.01
1.5	$\sqrt{.25} = 0.5$	$2p_1p_2$	.24	.36	.36
2	$\sqrt{1.00} = 1$	$2p_1p_3$	.08	.06	.12
2.5	$\sqrt{.25} = 0.5$	$2p_2p_3$	.24	.12	.06
Total			1	1	1

SUMMARY

Mean $\bar{x}$	Probability I	II	III	Standard deviation s	Probability I	II	III
1	0.04	0.09	0.36	0	0.44	0.46	0.46
1.5	.24	.36	.36	$\sqrt{.25} = 0.5$	.48	.48	.42
2	.44	.42	.21	$\sqrt{1.00} = 1$	.08	.06	.12
2.5	.24	.12	.06				
3	.04	.01	.01	Total	1	1	1
Total	1	1	1				

Remark 1. In the first row in Table 1, the term $p_1{}^2$ is the probability of getting 1 twice in succession. The numerical value of $p_1{}^2$ depends on the universe, i.e., on the value of p_1; hence there are 3 different entries under I, II, and III. But regardless of what universe they come from, the mean and standard deviation of the two numbers 1, 1 are 1 and 0 respectively, as entered in Cols. 1 and 2.

Next consider the term $10p_1{}^2 \cdot 3p_2p_3{}^2 = 30p_1{}^2p_2p_3{}^2$ in the 9th row of Table 3. The coefficient $30 = 5!/2!\ 1!\ 2!$ is the number of ways in which the 1 can occur twice, 2 once, and 3 twice, in a sample of 5 ($n = 5$). The value of $30p_1{}^2p_2p_3{}^2$ depends on p_1, p_2, and p_3, and accordingly will have different numerical entries under I, II, and III. But regardless of order, and of what universe they come from, the mean and standard deviation of the 5 numbers 1, 1, 2, 3, 3, are respectively 2 and 0.89, as given in Cols. 1 and 2.

Remark 2. It is sometimes desirable to increase the number of categories in the universe; but obviously the labor of calculating the probabilities associated with the means and standard deviations of samples therefrom will increase pretty rapidly and get out of hand. The remedy is to go the extreme and use a continuous universe, infinitely fine-grained, for which the infinitesimal calculus becomes applicable and simplifies the labor.

TABLE 2. SAMPLES OF 3; NUMERICAL VALUES OF THE TERMS IN THE EXPANSION OF
$$(p_1 + p_2 + p_3)^3$$

Mean $\bar{x}$	Standard deviation s	Term	Term value I	II	III
1	0	p_1^3	0.008	0.027	0.216
$\frac{4}{3}$	$\sqrt{(\frac{2}{9})} = .47$	$3p_1^2 p_2$	.072	.162	.324
$\frac{5}{3}$	$\sqrt{(\frac{8}{9})} = .94$	$3p_1^2 p_3$	.024	.027	.108
$\frac{5}{3}$	$\sqrt{(\frac{2}{9})} = .47$	$3p_1 p_2^2$	.216	.324	.162
2	$\sqrt{(\frac{2}{3})} = .82$	$6p_1 p_2 p_3$	.144	.108	.108
$\frac{7}{3}$	$\sqrt{(\frac{8}{9})} = .94$	$3p_1 p_3^2$	.024	.009	.018
2	0	p_2^3	.216	.216	.027
$\frac{7}{3}$	$\sqrt{(\frac{2}{9})} = .47$	$3p_2^2 p_3$	.216	.108	.027
$\frac{8}{3}$	$\sqrt{(\frac{2}{9})} = .47$	$3p_2 p_3^2$	.072	.018	.009
3	0	p_3^3	.008	.001	.001
Total			1	1	1

SUMMARY

Mean $\bar{x}$	Probability I	II	III	Standard deviation s	Probability I	II	III
1	0.008	0.027	0.216	0	0.232	0.244	0.244
$\frac{4}{3}$	.072	.162	.324	$\sqrt{(\frac{2}{9})} = .47$	.576	.612	.522
$\frac{5}{3}$	.240	.351	.270	$\sqrt{(\frac{2}{3})} = .82$	.144	.108	.108
2	.360	.324	.135	$\sqrt{(\frac{8}{9})} = .94$	.048	.036	.126
$\frac{7}{3}$	.240	.117	.045	Total	1	1	1
$\frac{8}{3}$	.072	.018	.009				
3	.008	.001	.001				
Total	1	1	1				

Discussion of the sampling results (Figs. 48 and 49). The summary in the lower part of each table shows the total chances of the various means and standard deviations that occurred in Cols. 1 and 2. The standard deviation 0, for instance, will appear whenever all n readings agree exactly. Thus for samples of 2 under universe I, we must add .04 + .36

$+$ .04 to find .44 for the total chance of the standard deviation 0, however occurring. The rest of the summary is made up likewise.

From the summaries of Tables 1, 2, and 3 the polygons of Figs. 48 and 49 are plotted. In statistical parlance these are called the theoretical

TABLE 3.   SAMPLES OF 5; NUMERICAL VALUES OF THE TERMS IN THE EXPANSION OF
$$(p_1 + p_2 + p_3)^5$$

Mean $\bar{x}$	Standard deviation s	Term	Term value I	II	III
1	0	$p_1{}^5$	0.00032	0.00243	0.07776
1.2	$\sqrt{.16} = .4$	$5p_1{}^4p_2$	.00480	.02430	.19440
1.4	$\sqrt{.64} = .8$	$5p_1{}^4p_3$	.00160	.00405	.06480
1.4	$\sqrt{.24} = .49$	$10p_1{}^3p_2{}^2$	.02880	.09720	.19440
1.6	$\sqrt{.64} = .8$	$20p_1{}^3p_2p_3$	.01920	.03240	.12960
1.8	$\sqrt{.96} = .98$	$10p_1{}^3p_3{}^2$	.00320	.00270	.02160
1.6	$\sqrt{.24} = .49$	$10p_1{}^2p_2{}^3$	.08640	.19440	.09720
1.8	$\sqrt{.56} = .75$	$30p_1{}^2p_2{}^2p_3$	.08640	.09720	.09720
2	$\sqrt{.8} = .89$	$30p_1{}^2p_2p_3{}^2$	.02880	.01620	.03240
2.2	$\sqrt{.96} = .98$	$10p_1{}^2p_3{}^3$	.00320	.00090	.00360
1.8	$\sqrt{.16} = .4$	$5p_1p_2{}^4$	.12960	.19440	.02430
2	$\sqrt{.4} = .63$	$20p_1p_2{}^3p_3$	.17280	.12960	.03240
2.2	$\sqrt{.56} = .75$	$30p_1p_2{}^2p_3{}^2$	.08640	.03240	.01620
2.4	$\sqrt{.64} = .8$	$20p_1p_2p_3{}^3$	.01920	.00360	.00360
2.6	$\sqrt{.64} = .8$	$5p_1p_3{}^4$	.00160	.00015	.00030
2	0	$p_2{}^5$	.07776	.07776	.00243
2.2	$\sqrt{.16} = .4$	$5p_2{}^4p_3$	.12960	.06480	.00405
2.4	$\sqrt{.24} = .49$	$10p_2{}^3p_3{}^2$	.08640	.02160	.00270
2.6	$\sqrt{.24} = .49$	$10p_2{}^2p_3{}^3$	.02880	.00360	.00090
2.8	$\sqrt{.16} = .4$	$5p_2p_3{}^4$	.00480	.00030	.00015
3	0	$p_3{}^5$	.00032	.00001	.00001
Total			1	1	1

SUMMARY

Mean $\bar{x}$	Probability I	II	III	Standard deviation s	Probability I	II	III
1	0.00032	0.00243	0.07776	0	0.0784	0.0802	0.0802
1.2	.00480	.02430	.19440	$\sqrt{.16} = .4$	.2688	.2838	.2229
1.4	.03040	.10125	.25920	$\sqrt{.24} = .49$	.2304	.3168	.2952 ⎫ *
1.6	.10560	.22680	.22680	$\sqrt{.40} = .63$	.1728	.1296	.0324 ⎭
1.8	.21920	.29430	.14310	$\sqrt{.56} = .75$	.1728	.1296	.1134 ⎫ *
2	.27936	.22356	.06723	$\sqrt{.64} = .8$	.0416	.0402	.1983 ⎭
2.2	.21920	.09810	.02385	$\sqrt{.80} = .89$	.0288	.0162	.0324
2.4	.10560	.02520	.00630	$\sqrt{.96} = .98$	.0064	.0036	.0252
2.6	.03040	.00375	.00120				
2.8	.00480	.00030	.00015	Total	1	1	1
3	.00032	.00001	.00001				
Total	1	1	1				

* Grouped in plotting.

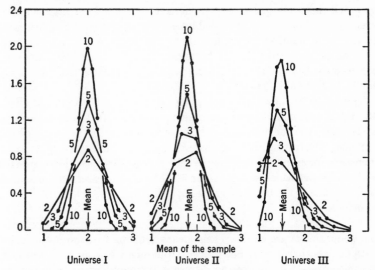

Fig. 48.   The probability-distributions of the means of samples of 2, 3, 5, and 10 from the three discrete universes of the preceding figure.   The arrows indicate the means of the universes and polygons.

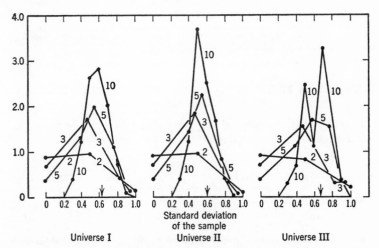

Fig. 49.   The probability-distributions of the standard deviations of samples of 2, 3, 5, and 10 from the three discrete universes of Fig. 47.   The arrows indicate the standard deviations of the universes.

sampling distributions of the means and standard deviations. It is to be noted that these polygons were derived here purely as mathematical exercises in the calculus of probabilities, and not as results of real sampling. It is possible to assimilate theoretical and practical results, however, as we know from Part B, by taking drawings blindfolded and with replacement from a bowl containing physically similar chips, thoroughly shuffled between draws.

The polygons showing the sampling distributions of the means (Figs. 48 and 49) disclose certain fundamental results. In the first place, in every case, the mean of the polygon coincides with the mean of the universe, illustrating the fact that $E\bar{x} = \mu$. In the second place, as n increases, the standard deviation of each polygon decreases in accordance with the theory which says that Var $\bar{x} = \sigma^2/n$. Moreover, as n increases, the skewness of the universe carries over less and less to the distribution of means.

The polygons in Fig. 49 for the standard deviations are more erratic, but some things are clear. Their means (not marked) do not coincide with one another for any one universe, but as n increases the areas seem to concentrate with increasing n toward the standard deviation of the universe, shown by the arrows. In Chapter 10 it was proved that for any universe,

$$Es^2 = \frac{n-1}{n}\sigma^2 \qquad \text{[P. 333]} \quad (60)$$

and the student was asked in Exercise 1 (p. 336) to illustrate this equation for the distributions plotted in Fig. 49.

EXERCISES ON EXTENSION AND APPLICATION OF THE BINOMIAL AND MULTINOMIAL SERIES

Exercise 1 (Correction of the general multinomial term for non-replacement). The general multinomial term was seen to be

$$P(n_1, n_2, \cdots, n_M) = \frac{n!}{n_1!n_2!\cdots n_M!}\, p_1{}^{n_1}p_2{}^{n_2}\cdots p_M{}^{n_M} \qquad \text{[P. 437]}$$

wherein

$$n_1 + n_2 + \cdots + n_M = n$$

Find the corresponding general term for sampling without replacement.

Solution

As with the binomial, nonreplacement requires the following alterations (cf. the comparison on p. 112):

For $p_1{}^{n_1}$ write

$$\frac{Np_1 \quad Np_1 - 1 \quad Np_1 - 2 \quad \cdots \quad Np_1 - \overline{n_1 - 1}}{N \quad N - 1 \quad N - 2 \quad \cdots \quad N - \overline{n_1 - 1}} \quad [n_1 \text{ factors}]$$

For $p_2{}^{n_2}$ write

$$\frac{Np_2 \quad Np_2 - 1 \quad Np_2 - 2 \quad \cdots \quad Np_2 - \overline{n_2 - 1}}{N - n_1 \quad N - n_1 - 1 \quad N - n_1 - 2 \quad \cdots \quad N - \overline{n_1 - 1} - \overline{n_2 - 1}} \quad [n_2 \text{ factors}]$$

Etc. When these alterations are made in the general term of the multinomial, the result is the *generalized hypergeometric* series

$$P(n_1, n_2, \cdots, n_M; n, N) = \frac{\binom{Np_1}{n_1}\binom{Np_2}{n_2}\cdots\binom{Np_M}{n_M}}{\binom{N}{n}}$$

This is the probability of obtaining n_1 chips labeled x_1, n_2 chips labeled x_2, etc., in a sample of n drawn without replacement from a universe containing N chips altogether, the initial proportions being $p_1, p_2, \cdots, p_M$.

Exercise 2. Prove by direct evaluation of the summation involved in the definition of En_i that, when sampling with replacement,

$$En_i = np_i$$

where n_i is the random variable signifying the number of x_i-chips drawn in a sample of n.

Solution (classical)

For the probability of n_i, use the general term of the multinomial expansion, which gives

$$En_i = \Sigma \, n_i \frac{n!}{n_1! \, n_2! \cdots n_M!} p_1{}^{n_1} p_2{}^{n_2} \cdots p_M{}^{n_M} \quad \text{[Definition]}$$

Let $i = 1$ and perform some rearrangements. Then

$$En_1 = np_1 \sum \frac{(n-1)!}{(n_1 - 1)! \, n_2! \cdots n_M!} p_1{}^{n_1 - 1} p_2{}^{n_2} \cdots p_M{}^{n_M}$$

$$= np_1$$

By symmetry, $En_2 = np_2$, etc. $\qquad\qquad$ *Q.E.D.*

Exercise 3. Prove that the preceding result holds also for sampling without replacement.

Solution (classical)

For the probability of n_1, use a term of the generalized hypergeometric series, which by definition gives

$$En_1 = \sum n_1 \frac{\binom{Np_1}{n_1}\binom{Np_2}{n_2}\cdots\binom{Np_M}{n_M}}{\binom{N}{n}}$$

$$= np_1 \sum \frac{\binom{Np_1 - 1}{n_1 - 1}\binom{Np_2}{n_2}\cdots\binom{Np_M}{n_M}}{\binom{N - 1}{n - 1}}$$

$$= np_1 \qquad\qquad\qquad\qquad\qquad\qquad Q.E.D.$$

Exercise 4. A universe of N chips is apportioned as shown.

		Label	
Number	Color	One side	Other side
Np_1	Red	x_1	$u_i = 1, v_i = 0, w_i = 0$ $(i = 1, 2, \cdots, Np_1)$
Np_2	Blue	x_2	$u_i = 0, v_i = 1, w_i = 0$ $(i = Np_1 + 1, \cdots, Np_2)$
Np_3	Green	x_3	$u_i = 0, v_i = 0, w_i = 1$ $(i = Np_2 + 1, \cdots, Np_3)$

Samples of n are drawn at random without replacement. Let $\hat{p}_1$, $\hat{p}_2$, $\hat{p}_3$ be the proportions of red, blue, and green in the sample. Prove that

$$\frac{n}{n - 1} E\hat{p}_1\hat{p}_2 = \frac{N}{N - 1} p_1 p_2 \qquad (1)$$

with similar results for $E\hat{p}_1\hat{p}_3$ and $E\hat{p}_2\hat{p}_3$. Thence discover anew that

$$\text{Var } \bar{x} = \frac{N - n}{N - 1} \frac{\sigma^2}{n} \qquad \text{[As already known]} \quad (2)$$

where

$$\bar{x} = \hat{p}_1 x_1 + \hat{p}_2 x_2 + \hat{p}_3 x_3$$

denotes the mean x-value of a sample.

Solution by William N. Hurwitz

Universe

$$p_1 = \frac{1}{N} \sum_1^N u_i$$

$$p_2 = \frac{1}{N} \sum_1^N v_i$$

$$p_3 = \frac{1}{N} \sum_1^N w_i$$

Sample

$$\hat{p}_1 = \frac{1}{n} \sum_1^n u_i = \frac{n_1}{n}$$

$$\hat{p}_2 = \frac{1}{n} \sum_1^n v_i = \frac{n_2}{n}$$

$$\hat{p}_3 = \frac{1}{n} \sum_1^n w_i = \frac{n_3}{n}$$

$$E\, n_1 n_2 = E \sum_1^n u_i \sum_1^n v_i = E \sum_1^n u_i v_i + E \sum_1^n \sum_1^n u_i v_j \qquad (j \neq i)$$

$$= \sum_1^n E\, u_i v_i + \sum_1^n \sum_1^n E\, u_i v_j \qquad \text{[Theorem on p. 71]}$$

$$= \frac{n}{N} \sum_1^N u_i v_i + \frac{n}{N} \frac{n-1}{N-1} \sum_1^N \sum_1^N u_i v_j$$

$$= \quad `` \quad + \frac{n}{N} \frac{n-1}{N-1} \left\{ \sum_1^N u_i \sum_1^N v_i - \sum_1^N u_i v_i \right\}$$

$$= \frac{n}{N} \left[1 - \frac{n-1}{N-1} \right] \sum_1^N u_i v_i + \frac{n}{N} \frac{n-1}{N-1} N p_1 N p_2$$

The first term on the right is 0 because the ith draw can not be both red and blue; u_i or v_i, one or both, must be 0. The remaining term completes the proof of Eq. 1 above. The proof of Eq. 2 may proceed along the line developed by Mr. Cornfield, below.

Solution by Jerome Cornfield

a. The probability of obtaining exactly n_1 red, n_2 blue, and n_3 green chips in a sample of n is

$$P(n_1, n_2, n_3; n, N) = \frac{\binom{Np_1}{n_1} \binom{Np_2}{n_2} \binom{Np_3}{n_3}}{\binom{N}{n}} \qquad \text{[Ex. 1]}$$

Then

$$E\, \frac{n_1}{n} \frac{n_2}{n} = \frac{1}{n^2} \sum_{n_1=0}^{n} \sum_{\substack{n_2=0 \\ [n_1+n_2+n_3=n]}}^{n} n_1 n_2 \frac{\binom{Np_1}{n_1} \binom{Np_2}{n_2} \binom{Np_3}{n_3}}{\binom{N}{n}} \qquad \text{[By definition]}$$

$$= \frac{N}{N-1} \frac{n-1}{n} p_1 p_2 \sum \sum \frac{\binom{Np_1-1}{n_1-1} \binom{Np_2-1}{n_2-1} \binom{Np_3}{n_3}}{\binom{N-2}{n-2}}$$

$$= \frac{N}{N-1} \frac{n-1}{n} p_1 p_2 \qquad \text{[Eq. 1]} \quad Q.E.D.$$

In the last step, the quantity being summed is the probability of $n_1 - 1$ red, $n_2 - 1$ blue, and n_3 green chips occurring in a sample of $n - 2$ chips drawn without replacement from a universe containing initially $Np_1 - 1$ red, $Np_2 - 1$ blue, and Np_3 green chips. The double sum is unity because *some* set of values of n_1 and n_2 *must* occur.

b. For the variance of $\bar{x}$ we recall from Chapter 3 that

$$\text{Var } \bar{x} = E\bar{x}^2 - (E\bar{x})^2 = E\bar{x}^2 - \mu^2$$

To evaluate $E\bar{x}^2$, square $\bar{x}$ as defined above and take the "expected" value of each term to find that

$$E\bar{x}^2 = \sum_{i=1}^{3} x_i{}^2 E\left(\frac{n_i}{n}\right)^2 + 2\sum_{i<j}^{3} x_i x_j E\frac{n_i}{n}\frac{n_j}{n}$$

The cross-product term on the right was just evaluated in Part a. The first term on the right gives

$$E n_1{}^2 = \text{Var } n_1{}^2 + (E n_1)^2$$

$$= \frac{N-n}{N-1} n p_1 (p_2 + p_3) + (n p_1)^2$$

whence

$$x_1{}^2 E\left(\frac{n_1}{n}\right)^2 = \frac{N-n}{N-1}\frac{1}{n} x_1{}^2 p_1(p_2 + p_3) + p_1{}^2 x_1{}^2$$

$$= x_1{}^2 \left\{\frac{N-n}{N-1}\frac{1}{n} p_1(p_1 + p_2 + p_3) + \frac{N}{N-1}\frac{n-1}{n} p_1{}^2\right\}$$

The parenthesis $p_1 + p_2 + p_3$ is unity and can be dropped as a factor. By permuting the subscripts, similar expressions can be written at once for $x_2{}^2 E\left(\frac{n_2}{n}\right)^2$ and $x_3{}^2 E\left(\frac{n_3}{n}\right)^2$, whereupon

$$E\bar{x}^2 = \frac{N-n}{N-1}\frac{1}{n}\left\{p_1 x_1{}^2 + p_2 x_2{}^2 + p_3 x_3{}^2\right\}$$

$$+ \frac{N}{N-1}\frac{n-1}{n}\left\{p_1 x_1 + p_2 x_2 + p_3 x_3\right\}^2$$

The last brace is μ by definition, wherefore from the identity

$$\text{Var } \bar{x} = E\bar{x}^2 - \mu^2$$

we get

$$\text{Var } \bar{x} = \frac{N-n}{N-1}\frac{1}{n}(p_1 x_1{}^2 + p_2 x_2{}^2 + p_3 x_3{}^2 - \mu^2)$$

$$+ \left[\frac{N}{N-1}\frac{n-1}{n} + \frac{N-n}{N-1}\frac{1}{n} - 1\right]\mu^2$$

The last bracket is zero, leaving

$$\text{Var } \bar{x} = \frac{N-n}{N-1}\frac{\sigma^2}{n} \qquad \text{[Eq. 2]} \quad Q.E.D.$$

c. Show, finally, that if each chip in the universe is of a different color, so that

$$p_1 = p_2 = \cdots = p_N = \frac{1}{N}$$

and if on each is written a number typified by x_i, with

$$\mu = \frac{1}{N}\Sigma x_i$$

and

$$\sigma^2 = \frac{1}{N} \Sigma \, (x_i - \mu)^2$$

then in a sample of n chips drawn without replacement

$$E\bar{x} = \mu$$

and

$$\text{Var } \bar{x} = \frac{N - n}{N - 1} \frac{\sigma^2}{n}$$

just as before. This is a perfectly general result, containing Part b of this exercise as a special case.

d. The sum of a sample may be defined as

$$x = \alpha_1 x_1 + \alpha_2 x_2 + \cdots + \alpha_N x_N$$

where

$$\begin{aligned} \alpha_i &= 1 \quad \text{if } x_i \text{ is in the sample} \\ &= 0 \quad \text{otherwise} \end{aligned} \Bigg\}$$

Then

$$\text{Var } x = \text{Var } n\bar{x} = n^2 \text{ Var } \bar{x}$$

$$= \frac{N - n}{N - 1} n\sigma^2$$

as obtained before.

Remark. If Mr. Cornfield's elegant method of solution is applied to finding Var r for the universe of two categories (which the student should undertake), it will be observed that, although the cross-product term $x_1 x_2 E \dfrac{n_1}{n} \dfrac{n_2}{n}$ does appear, its value can be made zero by putting $x_1 = 0$. The squared terms $x_1{}^2 E \left(\dfrac{n_1}{n}\right)^2 + x_2{}^2 E \left(\dfrac{n_2}{n}\right)^2$ then contain the entire value of $E\bar{x}^2$, and the first of these is 0, leaving

$$E\bar{x}^2 = 0 + x_2{}^2 E \left(\frac{n_2}{n}\right)^2 = \left(\frac{x_2}{n}\right)^2 E n_2{}^2 = \left(\frac{x_2}{n}\right)^2 \left[npq \, \frac{N - n}{N - 1} + n^2 p^2 \right]$$

$$= \frac{N - n}{N - 1} \frac{\sigma^2}{n} + n^2 p^2 \qquad \begin{array}{l} \text{[By the result for } E n_1{}^2 \text{ as} \\ \text{given above]} \end{array}$$

giving

$$\text{Var } r = \frac{N - n}{N - 1} \frac{\sigma^2}{n} \qquad\qquad [\sigma^2 = x_2{}^2 npq]$$

as obtained in Chapter 4.

Exercise 5 (Estimation of the ratio of the sizes of two mutually exclusive classes). A sample of n cards is drawn at random from a file of N cards in such manner that the generalized hypergeometric series may be used for calculating probabilities. Each card represents a d.u., and the universe of cards is made up of the categories and proportions shown in the accompanying table.

Type of equipment for cooking	Proportion in file	Number in sample
Gas	p	x
Electricity	q	y
Coal	r	z
Wood	s	u
Kerosene	t	v

From the sample it is desired to estimate the ratio $p:q$, the ratio of gas to electric equipment. Assume that the categories are mutually exclusive, which means that no card shows more than one type of equipment. Restrict the discussion to nonzero values of y. Let $f = x:y$.

a. Show that the ratio $x:y$ is biased. Find the "expected" value of $x:y$.

b. Find the variance of $x:y$.

Solution to Part a (Frederick F. Stephan)

$$E \frac{x}{y} = \sum_{y=1}^{n} \sum_{x=0}^{n} \frac{x}{y} P(x, y)$$

$$= \frac{1}{1 - P_0} \sum_{y=1}^{n} \sum_{x=0}^{n} \frac{x}{y} \frac{\binom{A}{x}\binom{B}{y}\binom{C}{w}}{\binom{N}{n}} \qquad \begin{array}{l} [w = z + u + v; \\ A = Np, B = Nq, \\ C = N(r + s + t)] \end{array}$$

$$= \frac{1 - P_0'}{1 - P_0} np\, E' \frac{1}{y}$$

where

$$P_0 = \frac{\binom{N - Nq}{n}\binom{Nq}{0}}{\binom{N}{n}} = \left\{ \begin{array}{l} \text{the probability of } y \text{ being 0 in a sample} \\ \text{of } n \text{ from a universe of } N \text{ of which } Nq \\ \text{cook with electricity} \end{array} \right.$$

$$P_0' = \frac{\binom{N - 1 - Nq}{n - 1}\binom{Nq}{0}}{\binom{N - 1}{n - 1}} = \left\{ \begin{array}{l} \text{the probability of } y \text{ being 0 in a sample} \\ \text{of } n - 1 \text{ from a universe of } N - 1 \text{ of} \\ \text{which } (N - 1)q \text{ cook with electricity} \end{array} \right.$$

$$E' \frac{1}{y} = \frac{1}{1 - P_0'} \sum_{y=1}^{n} \frac{1}{y} \frac{\binom{Nq}{y}\binom{N - Nq - 1}{n - y - 1}}{\binom{N - 1}{n - 1}}$$

The student may next show that

$$E \frac{x}{y} \to \frac{p}{q} \quad \text{as} \quad n \to N$$

wherefore $x{:}y$ is a biased but consistent estimate of $p{:}q$.

Solution to Part b (*William N. Hurwitz*)

Each card brought into the sample will show an x_i or a y_i or something else, where

$$\begin{cases} x_i = 1 & \text{if the } i\text{th d.u. cooks with gas} \\ \quad = 0 & \text{otherwise} \end{cases}$$

$$\begin{cases} y_i = 1 & \text{if the } i\text{th d.u. cooks with electricity} \\ \quad = 0 & \text{otherwise} \end{cases}$$

What is wanted is the variance of the estimate

$$f = \frac{\displaystyle\sum_1^n x_i}{\displaystyle\sum_1^n y_i}$$

From Eq. 50 on page 177 it may be seen that the

$$(\text{C.V.} f)^2 \doteq \frac{N-n}{N-1} \frac{1}{n} (c_x{}^2 + c_y{}^2 - 2\rho c_x c_y)$$

wherein

$$c_x{}^2 = \frac{1-p}{p} \qquad\qquad\qquad \text{[As for the binomial]}$$

$$c_y{}^2 = \frac{1-q}{q} \qquad\qquad\qquad [\text{``} \quad \text{``} \quad \text{``} \qquad \text{``} \quad]$$

$$\rho c_x c_y = \frac{1}{pq} E(x_i - p)(y_i - q) \qquad \text{[Definition]}$$

$$= \frac{E x_i y_i - pq}{pq} = \frac{0 - pq}{pq} \quad \begin{array}{l}\text{[Because } x_i y_i = 0 \text{ for} \\ \text{all } i]\end{array}$$

$$= -1$$

Then

$$(\text{C.V.} f)^2 = \left(\text{C.V.} \frac{x}{y}\right)^2 = \frac{N-n}{N-1} \frac{1}{n} \left[\frac{1-p}{p} + \frac{1-q}{q} + 2\right]$$

$$= \frac{N-n}{N-1} \frac{1}{n} \left[\frac{1}{p} + \frac{1}{q}\right]$$

This result is an excellent approximation provided nq is large and zero values of y are excluded from consideration. The student may wish to derive the exact variance and show that it simplifies to this approximation.

Exercise 6 (Mean and variance of the ratio of a subclass to its marginal total). Cards are drawn as in the preceding exercise, but with the aim of estimating what proportion of coal-heated d.us. is equipped for cooking with gas. The proportions in the universe are shown in the table.

Fuel for heating	Equipment for cooking					
	All sources	Gas	Elec-tricity	Coal	Wood	Kero-sene
Coal	p_1	p_{11}	p_{12}	p_{13}	p_{14}	p_{15}
Petroleum	p_2	p_{21}	p_{22}	p_{23}	p_{24}	p_{25}
Gas	p_3	p_{31}	p_{32}	p_{33}	p_{34}	p_{35}

Again assume that the categories are mutually exclusive. Let n_i and n_{ij} be the number of cards of the sample of n which fall into the cells whose subscripts in the table are i and ij. Exclude zero values of n_i.

a. Show that $n_{ij}:n_i$ is an unbiased estimate of $p_{ij}:p_i$.

b. Find the variance of this estimate.

Solution (Frederick F. Stephan)

Before commencing on this problem, the student should prove the following useful lemma: if the n_{ij} are hypergeometric or multinomial, then the marginal totals are likewise. The proof of this lemma may be seen by merely recalling that the cards in row i belong to n_i just as if no separation had been made for j. The same logic that leads to a hypergeometric or multinomial distribution of n_{ij} leads also to a similar distribution of n_i. Or, mathematically,

$$P(n_1, n_{21}, n_{22}, \cdots, n_{35})$$

$$= \sum_{n_{11}} \sum_{n_{12}} \sum_{n_{13}} \sum_{\substack{n_{14} \\ (n_{11}+n_{12}+n_{13}+n_{14}+n_{15}=n_1)}} \frac{\binom{Np_{11}}{n_{11}} \binom{Np_{12}}{n_{12}} \cdots \binom{Np_{35}}{n_{35}}}{\binom{N}{n}}$$

$$= \frac{\binom{Np_1}{n_1} \binom{Np_{21}}{n_{21}} \cdots \binom{Np_{31}}{n_{31}} \cdots \binom{Np_{35}}{n_{35}}}{\binom{N}{n}} \qquad Q.E.D.$$

Part a

No generality is lost if the columns headed electricity, coal, wood, and kerosene be combined into a new heading called "Other than gas," and if the two bottom rows be combined into "Other than coal." Then we need deal only with the array

$$
\begin{array}{c|cc}
n_1 & n_{11} & n_{12} \\[4pt]
n_2 & n_{21} & n_{22}
\end{array}
$$

in which $n_{11} + n_{12} = n_1$ and $n_1 + n_2 = n$. Then by the lemma just proved it follows that for any fixed $n_1 \neq 0$, n_{11} and n_{12} are distributed hypergeometrically with the joint probability $\binom{Np_{11}}{n_{11}} \binom{Np_{12}}{n_{12}} / \binom{Np_1}{n_1}$, which is known to give $E(n_{11}:n_1) = p_{11}:p_1$. As this is true for any $n_1 \neq 0$, the theorem is established, as the proof may be repeated for any subscripts i and ij.

Part b

By the methods used in Part a of the preceding exercise the student may show that

$$
E\left(\frac{n_{11}}{n_1}\right)^2 = \frac{1}{1 - P_0} \sum_{\substack{n_1 = 1 \\ [n_2 = n - n_1; \ n_{12} + n_{11} = n_1]}}^{n} \sum_{n_{11} = 0}^{n_1} \left(\frac{n_{11}}{n_1}\right)^2 \frac{\binom{Np_{11}}{n_{11}} \binom{Np_{12}}{n_{12}} \binom{Np_2}{n_2}}{\binom{N}{n}}
$$

$$
= \frac{Np_{11}(Np_{11} - 1)}{Np_1(Np_1 - 1)} + \frac{p_{11}}{p_1} \frac{N(p_1 - p_{11})}{Np_1 - 1} E\frac{1}{n_1}
$$

wherein P_0 is the probability that n_1 is 0. This result is good for any subscripts i and ij, hence

$$
\operatorname{Var}\frac{n_{ij}}{n_i} = E\left(\frac{n_{ij}}{n_i}\right)^2 - \left(E\frac{n_{ij}}{n_i}\right)^2
$$

$$
= \frac{Np_{ij}(Np_{ij} - 1)}{Np_i(Np_i - 1)} + \frac{p_{ij}}{p_i} \frac{N(p_i - p_{ij})}{Np_i - 1} E\frac{1}{n_i} - \left(\frac{p_{ij}}{p_i}\right)^2
$$

$$
= \frac{Np_{ij}(p_i - p_{ij})}{p_i(Np_i - 1)} \left[E\frac{1}{n_i} - \frac{1}{Np_i} \right]
$$

As np_1 increases toward Np_i, $E(1/n_i) \to 1/np_i$ and $\operatorname{Var} n_{ij}/n_i \to 0$. As Np_i increases,

$$
\operatorname{Var}\frac{n_{ij}}{n_i} \to \frac{p_{ij}(p_i - p_{ij})}{p_i^2} \left[E\frac{1}{n_i} - \frac{1}{Np_i} \right]
$$

If, further, np_i is also large,

$$
\operatorname{Var}\frac{n_{ij}}{n_i} \doteq \frac{N - n}{N} \frac{(p_{ij}:p_i)(1 - p_{ij}:p_i)}{np_i}
$$

This is a very important result, for it gives the sampling variance of the ratio of any cell to a larger class containing this cell. It is easy to recall by noting that it closely duplicates a result derived in Chapter 4, viz.,

$$\text{Var}\,\frac{r}{n} = \text{Var}\,\hat{p} = \frac{N - n}{N - 1}\,\frac{pq}{n}$$

It is only necessary to replace $p:1$ by $p_{ij}:p_i$, q or $1 - p$ by $1 - p_{ij}:p_i$, and n by np_i, which is the "expected" marginal total En_i of the class. Another form of the above result is

$$\left(\text{C.V.}\,\frac{n_{ij}}{n_i}\right)^2 = \frac{N - n}{Nn}\left[\frac{1}{p_{ij}} - \frac{1}{p_i}\right]$$

The student may wish to derive this approximation after the manner of the Hurwitz solution to Part b of the preceding exercise.

Exercise 7 (The random walk in one dimension). A particle starts from 0 and is constrained to take steps of equal length forward or backward at random along the x-axis, the ratio of the probabilities of forward to backward steps being $p:q$. Question: What is the probability that the particle will be at a distance x from 0 after it has made n steps?

Solution [16]

The solution is given by the binomial distribution. It is as if chips marked $+1$ and -1 in the ratio $p:q$ were mixed in a bowl. One chip at a time is

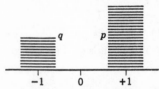

drawn to determine which way the particle is to move; then the chip is replaced, the chips shuffled, and another one drawn. When $+1$ is drawn, the particle is to move 1 step to the right (positive direction), and when a -1 is drawn, the particle is to take 1 step to the left. The problem is really only that of finding the distribution of the sum of the n numbers obtained in n drawings. This problem was solved in Chapter 4. The universe may be pictured as in Fig. 50. A mental calculation shows that the mean and

FIG. 50. The universe for the random walk in one dimension. Chips marked $+1$ and -1 stand in the ratio $p:q$. Here $\mu = p - q$, $\sigma^2 = 2^2\,pq$.

variance of this universe are

$$\mu = p - q \tag{1}$$

$$\sigma^2 = 2^2 pq \tag{2}$$

If x represents the sum of the n numbers, then obviously x gives the position of the particle after n steps. The actual distribution of x will be a binomial, and its mean and variance are known to be

$$Ex = n\mu = n(p - q) \tag{3}$$

$$\text{Var}\,x = n\sigma^2 = 2^2 npq \tag{4}$$

[16] See Exercise 10 for another solution.

The general term of this binomial will be

$$f(x, n) = \begin{pmatrix} n \\ \dfrac{n - x}{2} \end{pmatrix} q^{\frac{1}{2}(n - x)} p^{\frac{1}{2}(n + x)} \tag{5}$$

This can be seen mentally or by noting that if r steps go to the right and $n - r$ to the left, then

$$x = r - (n - r) = 2r - n \tag{6}$$

Solved for r and $n - r$, this relation gives

$$r = \tfrac{1}{2}(n + x), \quad \text{steps to the right} \tag{7}$$

$$n - r = \tfrac{1}{2}(n - x), \quad \text{steps to the left} \tag{8}$$

Substitution into the general binomial term $\dfrac{n!}{r!(n - r)!} q^{n-r} p^r$ gives the result already written for $f(x, n)$.

From Eqs. 3 and 4 it is obvious that the mean of the distribution of x moves to the right with time (more precisely, with n), and that it becomes flatter and flatter. It also approaches the normal distribution. If $p = q = \tfrac{1}{2}$, there is no movement of the mean of the distribution with time; it only flattens out.

If n is even, x may take only the $n + 1$ even values $0, \pm 2, \pm 4, \cdots, \pm n$. If n is odd, x may take only the $n + 1$ odd values $\pm 1, \pm 3, \pm 5, \cdots, \pm n$. Thus, the points on the x-distribution are always spaced 2 units apart, as shown in the table below.

r	$x = 2r - n$	$f(x, n)$
0	$-n$	q^n
1	$-n + 2$	$nq^{n-1}p$
2	$-n + 4$	$\begin{pmatrix} n \\ 2 \end{pmatrix} q^{n-2}p^2$
.	.	.
.	.	.
.	.	.
n	n	p^n

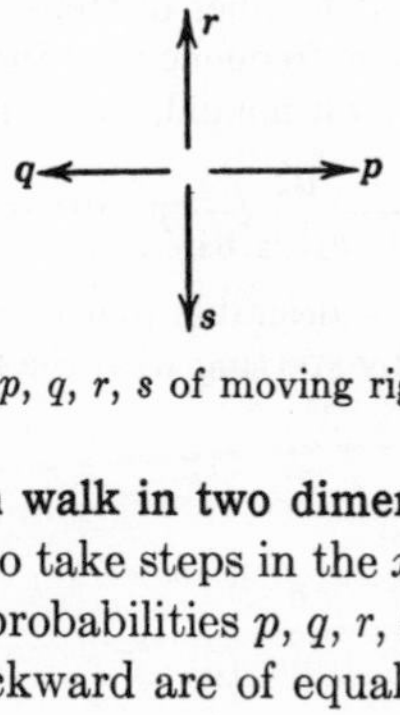

FIG. 51. The probabilities p, q, r, s of moving right and left, up and down.

Exercise 8 (The random walk in two dimensions). A particle starts from 0 and is constrained to take steps in the xy-plane, forward or backward or up or down, with probabilities p, q, r, and s as shown in Fig. 51. The steps forward and backward are of equal length, and the steps up and down are of equal length. The problem is to find the probability-distribution of the point x, y, which denotes the position of the particle after n steps. Figure 52 shows a set of possible terminal positions of the particle.

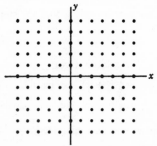

Fig. 52. The possible terminal positions of the particle.

Solution

The solution is given by the multinomial distribution. It is as if red, white, blue, and green chips were mixed in a bowl as shown in the table below. One chip at a time is drawn to determine which way the particle is

Description of chips	How marked	Proportion
Red	One step to the right	p
White	One step to the left	q
Blue	One step up	r
Green	One step down	s

to move; after it is moved the chip is replaced, the chips shuffled, and another one drawn. The problem is to find the distribution of x, y, which denotes the position of the particle after n steps (n drawings). The x-coordinate will be the sum of the numbers on the red and white chips, and the y-coordinate will be the sum of the numbers on the blue and green chips.

Let i be the number of steps taken forward and backward parallel with the x-axis; then $n - i$ is the number of steps taken up and down parallel with the y-axis. The relative frequencies of the red, white, blue, and green chips will be given by the multinomial, whose general term is

$$\frac{n!}{n_1! \, n_2! \, n_3! \, n_4!} \, p^{n_1} q^{n_2} r^{n_3} s^{n_4}$$

x and y are given; hence it is desirable to express n_1, n_2, n_3, and n_4 in terms of x and y. This is done by starting with the relations

$$\left. \begin{array}{l} x = n_1 - n_2 \\ i = n_1 + n_2 \end{array} \right\} \quad (1) \qquad\qquad \left. \begin{array}{l} y = n_3 - n_4 \\ n - i = n_3 + n_4 \end{array} \right\} \quad (2)$$

and solving them to get

$$\left. \begin{array}{l} n_1 = \tfrac{1}{2}(i + x) \\ n_2 = \tfrac{1}{2}(i - x) \end{array} \right\} \quad (3) \qquad\qquad \left. \begin{array}{l} n_3 = \tfrac{1}{2}(n - i + y) \\ n_4 = \tfrac{1}{2}(n - i - y) \end{array} \right\} \quad (4)$$

Substitution into the general term of the multinomial gives the desired result in the form

$$f(x, y; n) = \sum \frac{n!}{\left(\dfrac{i+x}{2}\right)! \left(\dfrac{i-x}{2}\right)! \left(\dfrac{n-i+y}{2}\right)! \left(\dfrac{n-i-y}{2}\right)!}$$

$$\times\ p^{\frac{1}{2}(i+x)} q^{\frac{1}{2}(i-x)} r^{\frac{1}{2}(n-i+y)} s^{\frac{1}{2}(n-i-y)} \quad (5)$$

A summation is required because i may have any value that allows the particle to come to rest at x, y after n steps. The summation is to be taken over all values of i that do not give half integers in the factorials. Stated another way, i may take on the values $|x|$, $|x| + 2$, $|x| + 4$, $\cdots$, n.

Exercise 9 (The random walk in two dimensions, with hesitation).

This is a modification of the preceding exercise. The particle may move forward or backward, up or down, as before, but may also stay put during an interval of time. What is the probability-distribution of the point x, y which denotes the position of the particle after n intervals?

Solution

Chips of a fifth color (yellow), marked "Stay put," are added to the bowl. The proportions of the five colors are now p, q, r, s, t. Let i be the number of steps taken forward and backward, and j the number up and down. Then $n - i - j$ is the number of yellow chips drawn, corresponding to the number of times the particle is to stay put. The solution to the problem is

$$f(x, y; n) = \sum \frac{n!}{\left(\dfrac{i+x}{2}\right)! \left(\dfrac{i-x}{2}\right)! \left(\dfrac{j+y}{2}\right)! \left(\dfrac{j-y}{2}\right)! (n-i-j)!}$$

$$\times\ p^{\frac{1}{2}(i+x)} q^{\frac{1}{2}(i-x)} r^{\frac{1}{2}(j+y)} s^{\frac{1}{2}(j-y)} t^{n-i-j}$$

In this summation i and j are to run over all values that do not give half integers in the factorials. Stated another way,

i takes the values $|x|$, $|x| + 2$, $\cdots$, $n - |y|$ or $n - |y| - 1$

j takes the values $|y|$, $|y| + 2$, $\cdots$, $n - |x|$ or $n - |x| - 1$

Exercise 10 (An iterative solution to the one-dimensional random walk).[17] The problem is as stated in Exercise 7.

Cornfield's solution

Let the probability that the particle be at distance x from 0 after n steps be given by the unknown function $f(x, n)$. The only two positions after the completion of $n - 1$ steps which are compatible with posi-

[17] Charles Jordan, *Calculus of Finite Differences* (Chelsea, 1947), p. 615. Mr. Jerome Cornfield showed me both the solution given here and the reference.

tion x after n steps are $x - 1$ and $x + 1$, the probabilities of which are $f(x - 1, n - 1)$ and $f(x + 1, n - 1)$. If it is at position $x - 1$, the particle can finish at position x only by taking one step to the right, the probability of which is p. If at position $x + 1$, the particle can finish at position x only

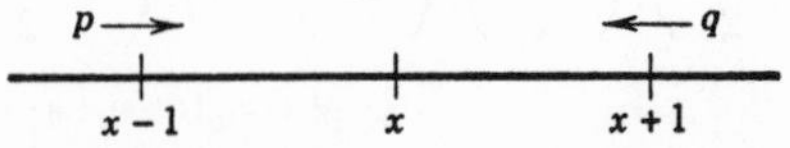

Fig. 53. The random walk in one dimension. The probability is p of moving to the right, q of moving to the left. To arrive at x the particle must have been at either $x - 1$ or $x + 1$ at the preceding step.

by taking one step to the left, the probability of which is q. (A glance at Fig. 53 may be helpful.) Consequently one may write

$$f(x, n) = p f(x - 1, n - 1) + q f(x + 1, n - 1) \tag{1}$$

This is a finite difference equation to be solved for the unknown function $f(x, n)$. We can proceed by first noting that the same process of reasoning yields the additional finite difference equations

$$f(x - 1, n - 1) = p f(x - 2, n - 2) + q f(x, n - 2) \tag{2}$$

$$f(x + 1, n - 1) = p f(x, n - 2) + q f(x + 2, n - 2) \tag{3}$$

Substituting (2) and (3) in (1), we obtain

$$f(x, n) = p^2 f(x - 2, n - 2) + 2pq f(x, n - 2) + q^2 f(x + 2, n - 2)$$

A repetition of this process yields

$$f(x, n) = p^3 f(x - 3, n - 3) + 3p^2 q f(x - 1, n - 3)$$
$$+ 3pq^2 f(x + 1, n - 3) + q^3 f(x + 3, n - 3)$$

By induction it is easy to show that after $k - 1$ substitutions we have

$$f(x, n) = \sum_{i=0}^{k} \binom{k}{i} p^{k-i} q^i f(x - k + 2i, n - k)$$

And finally, if we set $k = n$,

$$f(x, n) = \sum_{i=0}^{n} \binom{n}{i} p^{n-i} q^i f(x - n + 2i, 0) \tag{4}$$

$f(x - n + 2i, 0)$ is the probability that the particle will be at position $x - n + 2i$ after 0 steps. This probability is clearly

$$1, \quad \text{if} \quad x - n + 2i = 0$$
$$0, \quad \text{if} \quad x - n + 2i \neq 0$$

In that case all terms in the summation of Eq. 4 drop out except the term in which $i = \frac{1}{2}(n - x)$, and there remains

$$f(x, n) = \binom{n}{\dfrac{n - x}{2}} p^{\frac{1}{2}(n+x)} q^{\frac{1}{2}(n-x)}$$

as obtained in Exercise 7.

Exercise 11. Make up charts showing the probabilities of arriving at any point x, y after n steps, for $n = 1, 2, 3, 4$, under the fundamental probabilities p, q, r, s of Exercise 8. (See Fig. 54.)

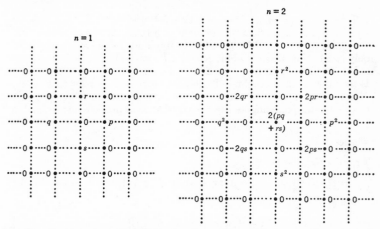

FIG. 54. Showing the probabilities for the random walk in two dimensions, for $n = 1$ and $n = 2$. The problem is to draw up similar charts for $n = 3$ and $n = 4$. Note that there will be a check on the numerical coefficients, because they must add up to 4^n.

Exercise 12. Make up charts showing the probabilities of arriving at any point x, y after n intervals, for $n = 1, 2, 3, 4$, under the fundamental probabilities p, q, r, s, t of Exercise 9. (See Fig. 55.)

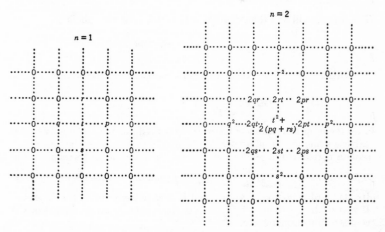

FIG. 55. Showing the probabilities for the random walk in two dimensions, with hesitation, for $n = 1$ and $n = 2$. The problem is to draw up similar charts for $n = 3$ and $n = 4$.

Exercise 13a (Rayleigh's approximate solution of the random walk in one dimension).

Solution

This will be an approximate solution patterned after Lord Rayleigh's solution in his paper, "On the resultant of a large number of vibrations," *Phil. Mag.*, vol. xlvii, 1899: pp. 246–51; also in his *Theory of Sound*, 2d ed. only (1894), Sec. 42a, and in his *Scientific Papers*, vol. iv, p. 370.

Let $f(x, n)$ have the same significance as assigned to it heretofore; $f(x, n)$ is thus the function sought. (The student may wish to refer to Fig. 50.) Now in order to arrive at x in n steps the particle must have been at $x - 1$ or $x + 1$ at the preceding step, no matter what be its previous history, and so

$$p f(x - 1, n - 1) + q f(x + 1, n - 1) = f(x, n)$$

whence

$$f(x, n) - f(x, n - 1)$$

$$= p[f(x - 1, n - 1) - f(x, n - 1)] + q[f(x + 1, n - 1) - f(x, n - 1)]$$

$$= q\{[f(x + 1, n - 1) - f(x, n - 1)] - [f(x, n - 1) - f(x - 1, n - 1)]\}$$

$$+ (p - q)[f(x - 1, n - 1) - f(x, n - 1)]$$

This is a difference equation. It may be solved by methods already demonstrated in previous exercises, but Lord Rayleigh, apparently unaware of such possibilities, found an approximate solution by passing to the corresponding differential equation. In his problem, p and q were equal, but a similar approach can be used here where p and q may be unequal, in which case the differential equation is

$$\frac{\partial f}{\partial n} = q \frac{\partial^2 f}{\partial x^2} - (p - q) \frac{\partial f}{\partial x}$$

The student will find this equation to be satisfied by the function

$$f = \frac{1}{\sqrt{n}} e^{-[x - (p - q)n]^2/an}$$

provided $a = 4q$, wherefore an approximate solution of the problem is

$$f \, dx = \frac{1}{\sqrt{4\pi nq}} e^{-[x - (p - q)n]^2/4nq} \, dx$$

after the proper normalizing factor is introduced, for the probability of the particle being in the interval $x \pm \frac{1}{2}dx$ after n steps. This function represents a normal curve centered at $x = (p - q)n$ and hence traveling to the right with speed $(p - q) \dfrac{dn}{dt}$ where $\dfrac{dn}{dt}$ is the number of steps per unit of time. Its standard deviation is $\sqrt{2nq}$, and it thus flattens out as time goes on. If $p = q$ (the case solved by Lord Rayleigh), there is no movement of the center of the normal curve.

b. The random walk in two dimensions. Here the particle is confined to the xy-plane, and the problem is to find the probability that after n steps it lies at a distance r from the starting point.

Solution (again after Lord Rayleigh)

In order to arrive at x, y in n steps the particle at the end of $n - 1$ steps must have been somewhere on a circle of radius equal to one step and centered at x, y. Let x', y' be a point on this circle. Introduce the angle φ connecting the coordinates by the following equations:

$$\left.\begin{array}{l} x' - x = \cos \varphi \\ y' - y = \sin \varphi \end{array}\right\} \quad \text{[1 step = 1 unit of length]}$$

Then

$$f(x, y, n) = \int_0^{2\pi} f(x', y', n - 1) \frac{d\varphi}{2\pi}$$

$$= \int_0^{2\pi} \frac{d\varphi}{2\pi} \left\{ f(x, y, n - 1) + \frac{\partial f}{\partial x} \cos \varphi + \frac{\partial f}{\partial y} \sin \varphi \right.$$

$$+ \frac{1}{2} \frac{\partial^2 f}{\partial x^2} \cos^2 \varphi + \frac{\partial^2 f}{\partial x \, \partial y} \sin \varphi \cos \varphi$$

$$\left. + \frac{1}{2} \frac{\partial^2 f}{\partial y^2} \sin^2 \varphi + \cdots \right\}$$

$$= f(x, y, n - 1) + \frac{1}{4} \left\{ \frac{\partial^2 f}{\partial x^2} + \frac{\partial^2 f}{\partial y^2} \right\} + \cdots$$

The approximating differential equation is

$$\frac{\partial f}{\partial n} = \frac{1}{4} \left\{ \frac{\partial^2 f}{\partial x^2} + \frac{\partial^2 f}{\partial y^2} \right\}$$

a solution for which is

$$f = \frac{1}{n\pi} e^{-(x^2 + y^2)/n}$$

Thus the probability of the particle lying in the area $dx \, dy$ at x, y is

$$f \, dx \, dy = \frac{1}{n\pi} e^{-(x^2 + y^2)/n} \, dx \, dy$$

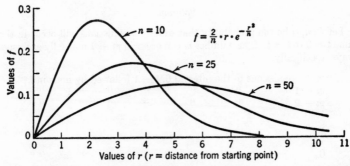

FIG. 56. Probabilities for the random walk in two dimensions derived by Lord Rayleigh's approximation.

Integration over the ring of radius $r \pm \frac{1}{2}dr$, where

$$r^2 = x^2 + y^2$$

yields $2\pi r \, dr$ times $\dfrac{1}{n\pi} e^{-(x^2+y^2)/n}$, which is constant over the ring. Thus

$$f \, dr = \frac{2}{n} re^{-r^2/n} \, dr$$

is the probability of the particle lying in the ring distant $r \pm \frac{1}{2}dr$ from 0 at the end of n steps. This curve has a shape like that shown in Fig. 56 for 3 values of n (not drawn to scale). The student may wish to look ahead in Chapter 14 on the gamma function to note that—

The mean of this curve lies at $\frac{1}{2}\sqrt{n\pi}$

Its standard deviation is $\sqrt{n(1 - \frac{1}{4}\pi)}$

Thus the mean moves outward and flattens out as the number of steps increases.

Exercise 14 (Disintegration of radioactive substances). Suppose that over 1 unit interval of time (1/1000th or 1 second perhaps), the probability is p that an atom of a certain radioactive substance will disintegrate and emit an α or other particle. If there are N_0 atoms present at time $t = 0$, there will be some number $N_t \le N_0$ which have not yet emitted at a later time t. It will be assumed that p is the same for all atoms and that it remains constant with time. The wave-mechanical picture of nuclei as potential barriers and the customary statistical interpretation lead to the statement that the probability of decay in unit time is constant.[18] By means of an observation-screen or other counting device, the number of emissions over any interval can be counted, and in fact cumulated, beginning at time $t = 0$. The problem is to find the probability that exactly m emissions will occur in the time interval t elapsing after $t = 0$.

The student will perceive that the results apply to many problems in insurance.

Solution

Let $P(m, t)$ be the probability that exactly m emissions will occur in the time $t = 0$ to $t = t$. This number m can occur in $m + 1$ mutually exclusive ways, specifically,

m	emissions in the interval 0, $t - 1$ followed by none in t
$m - 1$	" " " " " " " " 1 " "
$m - 2$	" " " " " " " " 2 " "
.	
.	
.	
0	" " " " " " " " m " "

[18] This sentence is quoted from Arthur E. Ruark, "The exponential law of radioactive disintegration," *Phys. Rev.*, vol. 44, 1933: pp. 654–56.

and it follows that

$$P(m, t) = \sum_{r=0}^{m} \binom{N_0 - r}{m - r} q^{N_0 - m} p^{m - r} P(r, t - 1) \tag{1}$$

Let $t = 1$ to see what has happened at the end of the first interval of time.

$$P(m, 1) = \sum_{r=0}^{m} \binom{N_0 - r}{m - r} q^{N_0 - m} p^{m - r} P(r, 0) \tag{2}$$

But as there can be no disintegrations in a zero time-interval,

$$\left. \begin{array}{ll} P(r, 0) = 0 & \text{if } r > 0 \\ = 1 & \text{if } r = 0 \end{array} \right\} \tag{3}$$

wherefore in Eq. 2 only the term in the summation for $r = 0$ is to be used, and

$$P(m, 1) = \binom{N_0}{m} q^{N_0 - m} p^m \tag{4}$$

Thus m takes the binomial distribution at $t = 1$, the successive probabilities of 0, 1, 2, 3, $\cdots$ disintegrations in the time-interval 0, 1 being accordingly

$$q^{N_0}, \quad N_0 q^{N_0 - 1} p, \quad \binom{N_0}{2} q^{N_0 - 2} p^2, \quad \binom{N_0}{3} q^{N_0 - 3} p^3, \quad \cdots$$

Now let $t = 2$.   Eq. 1 combined with Eq. 4 then gives

$$P(m, 2) = \sum_{r=0}^{m} \binom{N_0 - r}{m - r} q^{N_0 - m} p^{m - r} \binom{N_0}{r} q^{N_0 - r} p^r$$

$$= q^{N_0} \binom{N_0}{m} q^{N_0 - m} p^m \sum_{r=0}^{m} \binom{m}{r} q^{-r} \qquad \text{[By rearrangement of factorials]}$$

$$= \binom{N_0}{m} [q^2]^{N_0 - m} p^m \sum_r \binom{m}{r} q^{m - r} 1^r$$

$$= \binom{N_0}{m} [q^2]^{N_0 - m} p^m (1 + q)^m \sum_r \binom{m}{r} \left[\frac{q}{1 + q} \right]^{m - r} \left[\frac{1}{1 + q} \right]^r$$

$$= \binom{N_0}{m} [q^2]^{N_0 - m} [p(1 + q)]^m \qquad \begin{array}{l} \text{[The summation in the} \\ \text{above line is 1]} \end{array} \tag{5}$$

This again is binomial but the shape is different from the binomial obtained for $t = 1$ for now q^2 has replaced q and $p(1 + q)$ has replaced p.

Now let $t = 3$.   Eq. 1 combined with the last result gives

$$P(m, 3) = \sum_{r=0}^{m} \binom{N_0 - r}{m - r} q^{N_0 - m} p^{m - r} \binom{N_0}{r} [q^2]^{N_0 - r} [p(1 + q)]^r$$

$$= \binom{N_0}{m} [q^3]^{N_0 - m} p^m \sum_r \binom{m}{r} [q^2]^{m - r} [1 + q]^r$$

$$= \binom{N_0}{m} [q^3]^{N_0 - m} p^m (1 + q + q^2)^m$$

$$\times \sum_r \binom{m}{r} \left[\frac{q^2}{1 + q + q^2} \right]^{m - r} \left[\frac{1 + q}{1 + q + q^2} \right]^r$$

$$= \binom{N_0}{m} [q^3]^{N_0 - m} [p(1 + q + q^2)]^m \tag{6}$$

which is another binomial. It is now more or less obvious, as the student may wish to prove by induction, that

$$P(m, t) = \binom{N_0}{m} [q^t]^{N_0-m}[p(1 + q + q^2 + \cdots + q^{t-1})]^m$$

$$= \binom{N_0}{m} [q^t]^{N_0-m}[1 - q^t]^m \qquad (7)$$

Thus, after any interval t, the distribution of m is binomial, and the successive probabilities of 0, 1, 2, 3, $\cdots$ disintegrations are

$$q^{tN_0}, \quad N_0 q^{t(N_0-1)}(1 - q^t), \quad \binom{N_0}{2} q^{t(N_0-2)}(1 - q^t)^2, \quad \cdots$$

Eq. 7 might be written down at once on consideration of the fact that the probability that any particular atom will come through t intervals without disintegrating must be q^t if q is the probability of coming through one interval without disintegrating. Then $1 - p^t$ will be the probability of this atom disintegrating in t intervals, and Eq. 7 must therefore be the probability that any m atoms will disintegrate.

Let N_t be the number of atoms remaining (not disintegrated) at the end of the interval t. N_t is a random variable; it has an "expected" value and a variance. We have just found in Eq. 7 the distribution of $N_0 - N_t$; it is binomial, and from theory already given in this chapter we know that the mean of this binomial is $N_0(1 - q^t)$ and that its variance is $N_0(1 - q^t)q^t$ (compare with np and npq for the binomial whose rth term is $\binom{n}{r} q^{n-r}p^r$; p. 113). Therefore

$$E(N_0 - N_t) = N_0(1 - q^t)$$

or

$$EN_t = N_0 q^t \qquad (8)$$

Moreover, by what has just been said,

$$\begin{aligned}\text{Var } N_t &= EN_t^2 - (EN_t)^2 \\ &= E(N_0 - N_t)^2 - [E(N_0 - N_t)]^2 \\ &= \text{Var } (N_0 - N_t)^2 \\ &= N_0(1 - q^t)q^t = (1 - q^t)EN_t \qquad (9)\end{aligned}$$

The coefficient of variation of N_t is found to be

$$\begin{aligned}\text{C.V. } N_t &= \sqrt{\frac{\text{Var } N_t}{(EN_t)^2}} = \sqrt{\frac{1 - q^t}{EN_t}} \\ &= \sqrt{\frac{1 - q^t}{N_0 q^t}} \qquad (10)\end{aligned}$$

Thus the C.V. N_t will be quite small in the first stages of decomposition, provided N_0 is large, but with large values of t (i.e., in the latter stages of decomposition) the C.V. N_t will be large and N_t may vary considerably from its "expected" value.

Next it is of interest to note from Eq. 8 that

$$\frac{EN_t}{EN_{t-1}} = q \tag{11}$$

and that

$$\frac{E(N_t - N_{t-1})}{EN_{t-1}} = -p \tag{12}$$

Thus p is the ratio of the "expected" number of atoms that will disintegrate in the time-interval $t-1$, t to the number present (not yet disintegrated) at $t-1$.

The problem of disintegration is usually treated as one in the infinitesimal calculus, and some treatments tacitly deal only with "expected" values. Thus, if in the last equation $E(N_t - N_{t-1})$ be replaced by dN as representing an increment in the "expected" number of atoms over the time interval dt, and if EN_{t-1} is replaced by N, and p by $\lambda\, dt$, there arises the differential equation

$$\frac{dN}{N} = -\lambda\, dt \tag{13}$$

This is the usual starting point. It merely says that the amount of disintegration (dN) will be proportional to N and to dt. On a probability basis we know, however, that there will be fluctuations above and below any mean value, as our previous equations have shown. Integration of the differential equation is easy, giving

$$N = N_0\, e^{-\lambda t} \tag{14}$$

as the student may wish to verify. N is the "expected" number of atoms remaining at any time t, and N_0 is the number at the start ($t = 0$). The "mean life" of an atom is

$$Et = \frac{\displaystyle\int_0^\infty Nt\, dt}{\displaystyle\int_0^\infty N\, dt}$$

$$= \frac{1}{\lambda} \tag{15}$$

Historical note. In 1910 Bateman [19] explored the probabilities of radioactive disintegration, allowing for single but not multiple emissions during the period of time between $t-1$ and t, which he called δt. His starting point was thus similar to Eq. 1, but with r taking only the values m and $m-1$ in the summation. His results give a Poisson exponential series in place of the binomial series derived here for $P(m, t)$. The student may wish to prove that if pt is small, the binomial probabilities for 0, 1, 2, 3, $\cdots$ disintegrations in time t are reasonably well approximated by the Poisson series

$$e^{-N_0 pt} \left\{ 1,\quad N_0 pt,\quad \frac{1}{2!}(N_0 pt)^2,\quad \frac{1}{3!}(N_0 pt)^3,\quad \cdots \right\}$$

[19] H. Bateman, *Phil. Mag. (London)*, vol. 20, 1910: p. 704.

An exhaustive treatment of the probability theory of radioactive transformation, including results for the amounts of radioactive products formed by successive disintegrations, all on the assumption that one or no disintegrations of any product take place in the time dt, was published by Ruark and Devol, *Phys. Rev.*, vol. 49, 1936: pp. 355–67.

Exercise 15. In the last exercise let p and t be so small that p^2t^2 and higher powers can be neglected in comparison with pt. Then prove that under this assumption

$$EN_t = N_0(1 - pt + \cdots) \tag{16}$$

$$\text{Var } N_t = N_0 pt(1 - \cdots) \tag{17}$$

$$\text{C.V. } N_t = \sqrt{\frac{pt + \cdots}{N_0 pt(1 + \cdots)}} = \frac{1}{\sqrt{N_0}} \tag{18}$$

Also prove that the binomial distribution of m at time t, as given in Eq. 7, reduces to the approximation

$$P(m, t) = \binom{N_0}{m} (1 - pt + \cdots)^{N_0 - m}(pt + \cdots)^m \tag{19}$$

Neglecting the higher powers of pt, this is the same as $P(m, 1)$ in Eq. 4, except that p is now t times as large as it was.

CHAPTER 14. THE GAMMA AND BETA FUNCTIONS

Sir,—You report that at a whist table at a function at the Royal Hotel, Plymouth, a player was dealt all thirteen trumps. The odds against this happening are 635,013,559,599 to 1. The odds against all four players holding complete suits are 2,235,197,406,895,366,368,301,559,599 to 1.

A.

Sir,—The letter from A. stating that the odds against four players holding the complete suits in cards are 2,235,197,406,895,366,368,301,559,599 to 1 is too ridiculous for words, as it would be humanly impossible for anyone to work out such odds. In any case, he might have made the last 599 an even 600.

B.

Sir,—The odds against all four players at whist holding complete suits are 2,235,197,406,895,366,368,301,559,599 to 1. These figures are worked out by combinations and permutations, and have been checked by the Royal Statistical Society and other eminent mathematicians. They are undoubtedly correct, and may be accepted as such.

A.

—Letters to the editor of *The Western Morning News*, Plymouth.

Definition of the gamma function. The gamma function for positive values of n may be defined by the integral [1]

$$\Gamma(n) = \int_0^\infty x^{n-1} e^{-x} \, dx \tag{1}$$

$\Gamma(n)$ is read "the gamma function of n." The integral converges for all positive values of n; hence it truly defines a function of n. With the upper limit infinite, as here written, the gamma function is said to be *complete*, but this adjective is usually omitted except when it is desired to distinguish a complete from an incomplete gamma function. The *incomplete* gamma function is defined by [2]

$$\Gamma_x(n) = \int_0^x x^{n-1} e^{-x} \, dx \tag{2}$$

---

[1] The integral given here as the definition of the gamma function was studied by Euler, mainly in the form $\Gamma(\lambda + 1) = \int_0^1 (\ln 1/x)^\lambda \, dx$ [vol. 4 of his *Institutionis Calculi Integralis* (St. Petersburg, 1770)] and is sometimes referred to as "Euler's second integral." An excellent presentation of the Euler integrals is given by Joseph Edwards in his *Integral Calculus*, vol. 2, Chapter 24 (Macmillan, 1922). In the theory of the functions of a complex variable, other definitions of the gamma function are also given, all of which are equivalent to the Euler integral in Eq. 1 when the real part of n is positive.

[2] An integral with limits is a function of those limits, and not a function of the variable used in the integrand. In Eq. 2 the subscript x in $\Gamma_x(n)$ is to be identified

which is the same integral except that the upper limit is a variable. The incomplete gamma function $\Gamma_x(n)$ is thus not only a function of n but also of x. For consistency the complete function $\Gamma(n)$ should have the subscript ∞, but for brevity and by convention the subscript is generally omitted.

It is instructive to plot the integrand of Eqs. 1 and 2 for a given value of n, for positive values of x. The area under the whole curve to infinity

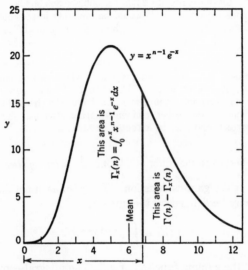

FIG. 57.   The curve $y = x^{n-1}e^{-x}$, this being the integrand in the definition of the gamma function.   The curve is drawn for $n = 6$.   The area under the whole curve to infinity is the *complete* gamma function $\Gamma(n)$, and the area to the finite abscissa x is the incomplete gamma function $\Gamma_x(n)$.

is by definition the complete function $\Gamma(n)$. The area included between 0 and the finite abscissa x is $\Gamma_x(n)$. The exact shape of the curve varies with the parameter n, but the relation between the complete and incomplete functions is always as shown in Fig. 57. The reader may agree that the adjective "incomplete" is aptly applied here; the gamma function is incomplete when the integration extends only part way, i.e., to some finite distance.[3]

with the upper limit of the integral, not with the x in the integrand. The student may find it helpful to rewrite Eq. 2 as $\Gamma_x(n) = \int_0^x t^{n-1}e^{-t}\,dt$. Any letter whatever may be used in the integrand without affecting the value of the integral. Similar remarks apply to Eq. 14.

[3] Several ways of evaluating the incomplete gamma function by series and continued fractions will be found in Exercises 27 and 28.

Other forms of the gamma integral. The above integrals can take a variety of other forms [4] by change of variable. In the theory of errors an important form is obtained by setting $x = u^2$; then $dx = 2u\,du$, and

$$\Gamma_x(n) = \Gamma_{u^2}(n) = 2\int_0^{u=\sqrt{x}} u^{2n-1}e^{-u^2}\,du \tag{3}$$

The student should take careful note that in the incomplete gamma function the subscript of Γ always refers to the upper limit[2] of the integral in Eq. 2; thus the subscript x or u^2 in Eq. 3 is the upper limit of the integral in Eq. 2 and not of the new form Eq. 3. With $n = \frac{1}{2}$, Eq. 3 gives

$$\Gamma_x(\tfrac{1}{2}) = 2\int_0^{\sqrt{x}} e^{-x^2}\,dx \tag{4}$$

whence we perceive that the normal probability integral may be regarded as a table of the incomplete function $\Gamma_x(\frac{1}{2})$.

Several other forms of the integral for the gamma function that are important in application will be found in the examples at the end of this chapter.

The recursion formula. The most important property of the gamma function is a recursion formula. It can be found by integrating the right-hand side of Eq. 1 by parts, thus:

$$\Gamma(n) = \int_0^\infty x^{n-1}e^{-x}\,dx = \frac{1}{n}\int_0^\infty e^{-x}\,dx^n \quad (n > 0)$$

$$= \frac{1}{n}\{[x^n e^{-x}]_0^\infty + \int_0^\infty x^n e^{-x}\,dx\}$$

$$= \frac{1}{n}\{0 + \Gamma(n+1)\}$$

Accordingly, there follows the *recursion formula*

$$\Gamma(n+1) = n\,\Gamma(n) \tag{5}$$

Thus $\Gamma(n+1)$ is simply related to $\Gamma(n)$. Eq. 5 may be regarded as a difference equation, and the gamma function as a solution thereof.

By repeated applications of the recursion formula it is seen that

$$\Gamma(n+1) = n\,\Gamma(n) = n\,\Gamma(n-1+1) = n(n-1)\,\Gamma(n-1)$$

$$= n(n-1)(n-2)\cdots(n-k)\,\Gamma(n-k) \tag{6}$$

Hence if the gamma function is tabled over a unit interval of the argu-

[4] See, for example, E. B. Wilson, *Advanced Calculus*, Chapter 14 (Ginn and Company, 1912).

ment, the function for arguments outside this range can be found; for example

$$\Gamma(3.37) = \Gamma(2.37 + 1) = 2.37 \times \Gamma(2.37) = 2.37 \times \Gamma(1.37 + 1)$$

$$= 2.37 \times 1.37 \ \Gamma(1.37) \tag{7}$$

so that, if $\Gamma(1.37)$ is known, then $\Gamma(3.37)$ can be evaluated by a simple multiplication.

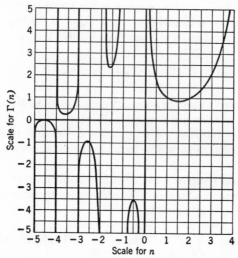

FIG. 58. Graph of the gamma function. (From Jahnke und Emde's *Tafeln.*)

The recursion formula was here derived from the definition contained in Eq. 1, and it was necessary to postulate that $n > 0$ and real; otherwise the term $[x^n e^{-x}]_0^\infty$ would not have dropped out in the integration by parts. But the recursion formula, it so happens, is a general property of the gamma function otherwise defined, and it holds for negative and complex values of the argument as well as for positive real values.

The relation of the gamma function to factorials. With $n = 1$, Eq. 1 gives

$$\Gamma(1) = \int_0^\infty e^{-x} \, dx = 1$$

Hence if n is a positive integer, and if Eq. 6 is extended, the recursion formula leads to the following simple and important relation with the factorial:

$$\Gamma(n + 1) = n(n - 1)(n - 2) \cdots 5 \cdot 4 \cdot 3 \cdot 2 \cdot 1 \ \Gamma(1) = n! \tag{8}$$

For values of n not integral, $\Gamma(n + 1)$ may be regarded as a generalized $n!$; the practice is in fact fairly common. Thus $\frac{3}{4}!$ is to be identified with $\Gamma(\frac{7}{4})$; $-\frac{1}{2}!$ with $\Gamma(\frac{1}{2})$; 0! with $\Gamma(1) = 1$.

The graph of the complete gamma function. It is easy to evaluate the gamma function at any positive integer with Eq. 8. For example, $\Gamma(4) = 3! = 6$ and $\Gamma(5) = 4! = 24$. The curve is continuous on the positive side of the vertical axis, as can be proved; [5] hence a graph like Fig. 58 is not difficult to construct for positive values of n. The minimum between $n = 1$ and $n = 2$ is interesting and requires special analysis. We have already noticed that $\Gamma(1) = 1$, and by Eq. 8 it follows that $\Gamma(2) = 1! = 1$ also; hence there must be a proper minimum somewhere between. Calculations have located the position of the minimum at $n = 1.461632144968 \cdots$, at which point the gamma function has the value $0.8856031944 \cdots$.

For negative values of n, the integral in Eq. 1 is not used to define the gamma function; instead, other definitions are used such as Gauss's or Weierstrass's. Gauss defined his infinite product $\pi(n)$ by the equation

$$\pi(n) = \operatorname*{Lim}_{m \to \infty} \frac{m!\, m^n}{(n + 1)(n + 2)(n + 3) \cdots (n + m)}$$

When n is real and positive, Gauss's infinite product $\pi(n)$ has the same numerical value as $\Gamma(n + 1)$ as the gamma function is defined in Eq. 1 (not proved here); hence $\pi(n)$ is taken as the Gauss definition of the gamma function whether n be real, negative, or complex. In Exercise 4 the student is asked to prove that the recursion formula (5) holds for the Gauss definition regardless of whether n be positive, negative, or a complex number; also that Gauss's infinite product $\pi(n) = n!$ in the ordinary factorial sense when n is a positive integer.

The student should plot the four curves [6]

$$y = \frac{1}{n + 1}$$

$$y = \frac{2!\,2^n}{(n + 1)(n + 2)}$$

$$y = \frac{3!\,3^n}{(n + 1)(n + 2)(n + 3)}$$

$$y = \frac{4!\,4^n}{(n + 1)(n + 2)(n + 3)(n + 4)}$$

and note that successively they appear to give better and better approximations to the graph of the gamma function $\Gamma(n + 1)$ plotted against n (same as Fig. 58 with the horizontal scale displaced one unit to the right).

[5] See, for example, H. S. Carslaw, *Fourier's Series and Integrals*, pp. 132 and 133 in the 1930 edition (Macmillan, 1906, 1921, 1930).

[6] These curves are shown in Edwards's *Integral Calculus* (Macmillan, 1922), Chapter XXIV in vol. 2, pages 77 and 78.

Gauss's $\pi(n)$ or $\Gamma(n+1)$ has infinite discontinuities at $n = -1$, -2, -3, $\cdots$, as appear in Fig. 58. Moreover, in Exercise 5 the student is asked to prove that the infinite discontinuities in the gamma function $\Gamma(n+1)$ are alternatively positive and negative on the two sides of a negative integer.

Tables of the incomplete gamma function. By transformation of variable, as called for in the exercises, several important curves (the Pearson Types III and V and others) can be changed into the form $y = x^{n-1}e^{-x}$, the integrand of Eq. 1, wherefore the incomplete gamma function constitutes an important probability integral. The need of tables of this integral was recognized by Karl Pearson, who, with the assistance of his staff at the Biometric Laboratory, turned out in 1921, after many years' labor, the *Tables of the Incomplete Gamma Function*.[7] A sample page is shown in Fig. 59. Therein the function $I(u, p)$ is tabled to seven decimals against the arguments u and p, along with second and fourth differences with respect to both u and p. The function $I(u, p)$ is related to u and p, and through them to the incomplete gamma function, by the definition

$$I(u, p) = \frac{\Gamma_x(p+1)}{\Gamma(p+1)} = \frac{\int_0^x x^p e^{-x}\, dx}{\int_0^\infty x^p e^{-x}\, dx} \tag{9}$$

wherein

$$u = \frac{x}{\sqrt{p+1}}$$

Evidently for any fixed value of p, $I(u, p)$ varies from 0 to 1 as x varies from 0 to ∞; $I(u, p)$ is in fact just the fractional part of the area under the curve of Fig. 57 between 0 and the abscissa x, p being identified as $n - 1$, and u as $x/\sqrt{p+1} = x/\sqrt{n}$. The advantage of tabling against $x/\sqrt{n}$ instead of against x directly arises from the fact that, as n increases, the curve of Fig. 57 flattens out, and one is accordingly compelled to move to higher and higher values of x if he would include always a specified fraction of the area. The device of using $x/\sqrt{n}$ amounts to a change of scale that keeps the table within bounds; when n is high, a moderate value of $x/\sqrt{n}$ corresponds to a high value of x. $x/\sqrt{n}$ is in fact just the upper limit x of the incomplete gamma function (Eq. 2) expressed in units of the standard deviation $\sqrt{n}$ of the curve. (Exs. 1 and 29 will illuminate this point. It may be noted here that the mean of the curve of Fig. 57 lies at abscissa n, and the standard deviation of the curve is $\sqrt{n}$.) The introductory material in the *Tables of the Incom-*

---

[7] *Tables of the Incomplete Gamma Function*, published in 1922 by His Majesty's Stationery Office, and reissued in 1934 by the Office of Biometrika, University College, London W.C. 1.

18 $u = 0{\cdot}0$ to $6{\cdot}2$ TABLES OF THE INCOMPLETE Γ-FUNCTION $p = 3{\cdot}0$ to $3{\cdot}5$

u	$p = 3{\cdot}0$: $I(u, p)$	δ_u^2	δ_u^4	δ_p^2	δ_p^4	$p = 3{\cdot}1$: $I(u, p)$	δ_u^2	δ_u^4	δ_p^2	δ_p^4	$p = 3{\cdot}2$: $I(u, p)$	δ_u^2	δ_u^4	δ_p^2	δ_p^4
·0	·0000000	—	—	0		·0000000	—	—	0		·0000000	—	—	0	
·1	·0000568	+6626	—	+40	+1	·0000436	+5494	—	+30		·0000334	+4549	—	+24	
·2	·0007762	+18625	+775	+298	+10	·0006366	+16304	+1417	+247	+9	·0005217	+14237	+1934	+204	+8
·3	·0033581	+31399	−2308	+832	+16	·0028600	+28531	−1648	+713	+14	·0024337	+25859	−1050	+618	+13
·4	·0090799	+41865	−3606	+1563	+21	·0079365	+39110	−3148	+1384	+19	·0069316	+36431	−2682	+1225	+17
·5	·0189882	+48725	−3830	+2331	+19	·0169240	+46541	−3594	+2128	+18	·0150726	+44321	−3332	+1924	+17
·6	·0337690	+51755	−3466	+3080	+16	·0305656	+50378	−3404	+2835	+15	·0276457	+48879	−3310	+2609	+14
·7	·0537253	+51319	−2816	+3660	+13	·0492450	+50811	−2880	+3420	+12	·0451067	+50127	−2909	+3192	+11
·8	·0788135	+48067	−2083	+4046	+8	·0730055	+48364	−2215	+3829	+9	·0675804	+48466	−2321	+3620	+9
·9	·1087084	+42733	−1367	+4224	+3	·1016024	+43702	−1531	+4044	+3	·0949007	+44484	−1633	+3866	+3
1·0	·1428765	+36030	−744	+4207	−1	·1345695	+37509	−914	+4069	+0	·1266694	+38819	−1066	+3930	0
1·1	·1806476	+28584	−220	+4022	−4	·1712875	+30402	−375	+3927	−3	·1623200	+32088	−536	+3829	−3
1·2	·2212771	+20918	+177	+3705	−4	·2110457	+22920	+44	+3654	−3	·2011794	+24821	−81	+3592	−3
1·3	·2639984	+13429	+474	+3292	−4	·2530959	+15482	+367	+3276	−4	·2425209	+17473	+250	+3253	−3
1·4	·3080626	+6414	+666	+2823	−5	·2966943	+8411	+588	+2837	−5	·2856097	+10375	+506	+2843	−5
1·5	·3527682	+65	+789	+2329	−7	·3411338	+1928	+734	+2366	−7	·3297360	+3785	+673	+2395	−6
1·6	·3974803	−5495	+839	+1839	−7	·3857661	−3821	+803	+1889	−7	·3742408	−2132	+766	+1936	−6
1·7	·4416429	−10216	+840	+1369	−5	·4300163	−8767	+832	+1427	−5	·4185324	−7283	+809	+1486	−5
1·8	·4847839	−14097	+812	+934	−4	·4733898	−12881	+807	+1000	−4	·4620957	−11625	+805	+1062	−4
1·9	·5265152	−17166	+755	+545	−3	·5154752	−16188	+766	+613	−3	·5044965	−15162	+771	+676	−4
2·0	·5665299	−19480	+683	+209		·5559418	−18729	+697	+274		·5453811	−17928	+711	+336	
2·1	·6045966	−21111	+601	−73		·5945355	−20573	+627	−15		·5844729	−19983	+640	+44	
2·2	·6405522	−22141	+520	−305		·6310719	−21790	+540	−252		·6215664	−21392	+564	−199	
2·3	·6742937	−22651	+437	−488		·6654293	−22467	+465	−442		·6565207	−22237	+486	−395	
2·4	·7057701	−22724	+361	−627		·6975400	−22679	+386	−586		·6892513	−22596	+409	−547	

u	$p = 3{\cdot}3$: $I(u, p)$	δ_u^2	δ_u^4	δ_p^2	δ_p^4	$p = 3{\cdot}4$: $I(u, p)$	δ_u^2	δ_u^4	δ_p^2	δ_p^4	$p = 3{\cdot}5$: $I(u, p)$	u
·0	·0000000	—	—	0		·0000000	—	—	0		·0000000	·0
·1	·0000256	+3759	—	+18		·0000196	+3102	—	+14		·0000150	·1
·2	·0004271	+12407	+2320	+169	+6	·0003494	+10788	+2608	+139	+5	·0002855	·2
·3	·0020693	+23375	−494	+533	+12	·0017580	+21082	−9	+457	+11	·0014924	·3
·4	·0060490	+33849	−2238	+1082	+16	·0052748	+31367	−1798	+956	+14	·0045961	·4
·5	·0134136	+42085	−3043	+1737	+16	·0119283	+39854	−2741	+1565	+15	·0105995	·5
·6	·0249867	+47278	−3183	+2395	+13	·0225672	+45600	−3038	+2200	+12	·0203677	·6
·7	·0412876	+49288	−2912	+2976	+10	·0377661	+48308	−2875	+2770	+10	·0345216	·7
·8	·0625173	+48386	−2401	+3416	+10	·0577958	+48141	−2467	+3223	+10	·0533966	·8
·9	·0885856	+45083	−1813	+3691	+4	·0826396	+45507	−1920	+3519	+4	·0770454	·9
1·0	·1191622	+39967	−1216	+3790	0	·1120341	+40953	−1354	+3651	+1	·1052710	1·0
1·1	·1537355	+33635	−681	+3729	−2	·1455239	+35045	−821	+3626	−2	·1376749	1·1
1·2	·1916723	+26622	−218	+3530	−3	·1825182	+28316	−352	+3464	−3	·1737105	1·2
1·3	·2322713	+19391	+142	+3224	−2	·2223441	+21235	+31	+3192	−1	·2127361	1·3
1·4	·2748094	+12302	+418	+2844	−5	·2642935	+14185	+323	+2842	−4	·2540618	1·4
1·5	·3185777	+5631	+606	+2420	−6	·3076614	+7458	+538	+2440	−6	·2969891	1·5
1·6	·3629091	−434	+726	+1977	−6	·3517751	+1269	+678	+2015	−6	·3408426	1·6
1·7	·4071971	−5773	+781	+1539	−6	·3960157	−4242	+751	+1589	−6	·3849932	1·7
1·8	·4509078	−10331	+799	+1122	−5	·4398321	−9002	+783	+1178	−5	·4288742	1·8
1·9	·4935854	−14090	+770	+740	−4	·4827483	−12979	+773	+798	−4	·4719910	1·9
2·0	·5348540	−17079	+724	+397		·5243666	−16183	+730	+457		·5139249	2·0
2·1	·5744147	−19344	+662	+99		·5643666	−18657	+676	+159		·5543344	2·1
2·2	·6120410	−20947	+587	−146		·6025009	−20455	+607	−95		·5929514	2·2
2·3	·6475726	−21963	+507	−348		·6385897	−21646	+530	−301		·6295767	2·3
2·4	·6809079	−22473	+432	−506		·6725139	−22309	+457	−466		·6640733	2·4

FIG. 59. A sample page from Karl Pearson's *Tables of the Incomplete Gamma Function.*

plete Gamma Function gives a history of the work and instructions for the use of the tables, together with illustrative examples.

In 1930 L. R. Salvosa [8] published an extensive table of the areas and ordinates of the Pearson Type III curve

wherein

$$y = y_0(1 + \tfrac{1}{2}\alpha t)^{4/\alpha^2 - 1} e^{-2t/\alpha}$$

$$y_0 = \frac{\left(\dfrac{4}{\alpha^2}\right)^{4/\alpha^2 - \frac{1}{2}}}{e^{4/\alpha^2} \Gamma\left(\dfrac{4}{\alpha^2}\right)} \qquad (10)$$

the arguments being α and t. $\tfrac{1}{2}\alpha$ is one measure of the "skewness" of the curve, and t is the abscissa measured from the mean.

Tables of "chi-square" can also be used to evaluate the incomplete gamma function. Some of these tables show

$$P(\chi) = \frac{1}{\Gamma(\frac{1}{2}k)2^{\frac{1}{2}k}} \int_{\chi^2}^{\infty} (\chi^2)^{\frac{1}{2}k - 1} e^{-\frac{1}{2}\chi^2} \, d\chi^2 \qquad (11)$$

tabulated against χ^2 and k as arguments. Other tables, following Fisher, show χ^2 tabulated against $P(\chi)$ and k as arguments. It can readily be shown by a change of variable that

$$P(\chi) = 1 - \frac{\Gamma_{\frac{1}{2}\chi^2}(\frac{1}{2}k)}{\Gamma(\frac{1}{2}k)} \qquad (12)$$

(See Exs. 34 and 42.)

Another notation for the incomplete gamma function, used occasionally by continental writers, is $(n, x)!$ for $\Gamma_x(n + 1)$. In this symbolism $(n, \infty)!$ means $n!$

Besides supplying tables for integrals under the Type III curve and other curves reducible thereto, the incomplete gamma function affords a method of summing any number of terms of the Poisson exponential limit without approximation, as is demonstrated in a later section (p. 482).

Definition of the beta function. The beta function will here be defined by the integral [9]

$$B(m, n) = \int_0^1 x^{m-1}(1 - x)^{n-1} \, dx \qquad (13)$$

---

[8] L. R. Salvosa, *Ann. Math. Stat.*, vol. 1, 1930: pp. 191–225.

[9] The integral given here as the definition of the beta function was studied by Euler mainly in the form $\int_0^1 x^{m-1}(1 - x^n)^{1-k/n} \, dx$, n fixed [Art. 345 in vol. 1 of his *Institutiones Calculi Differentialis* (St. Petersburg, 1768)] and is sometimes referred to as "Euler's first integral." The relation between the gamma and beta functions, shown here as Eq. 24, was given by him in Art. 27 of vol. 4.

$B(m, n)$ is read "the beta function of m and n." It is truly a function of m and n; that is, it depends on both m and n for its value. With the upper limit unity, as here written, the beta function is said to be *complete*, but, as with the gamma function, the adjective is usually omitted except when there is a possibility of confusing the complete with the incomplete beta function. The *incomplete* beta function is defined by [2]

$$B_x(m, n) = \int_0^x x^{m-1}(1 - x)^{n-1}\, dx \tag{14}$$

the upper limit x lying between 0 and 1. Evidently the incomplete beta function $B_x(m, n)$ is a function not only of m and n, but also of the upper limit x. In the same symbolism the complete beta function $B(m, n)$ should have the subscript 1, but for brevity and by convention it is generally omitted, just as the subscript ∞ is omitted in writing $\Gamma(n)$ for the complete gamma function.

The student should note that the beta function, complete or incomplete respectively, has one more argument than the gamma function, complete or incomplete.

Interchanging the arguments m and n. By change of variable the above integrals assume a variety of forms [10] and lead to several interesting relations, of which perhaps the most important is that

$$B(m, n) = B(n, m) \tag{15}$$

This is easily proved by replacing x in Eq. 13 by $1 - y$. If $x = 1 - y$, then $dx = -dy$, and

$$B(m, n) = \int_0^1 x^{m-1}(1 - x)^{n-1}\, dx = -\int_1^0 (1 - y)^{m-1} y^{n-1}\, dy$$

$$= \int_0^1 y^{n-1}(1 - y)^{m-1}\, dy = B(n, m)$$

By the same change of variable

$$B_x(m, n) = \int_{1-x}^1 y^{n-1}(1 - y)^{m-1}\, dy$$

$$= \int_0^1 y^{n-1}(1 - y)^{m-1}\, dy - \int_0^{1-x} y^{n-1}(1 - y)^{m-1}\, dy$$

That is,

$$B_x(m, n) = B(n, m) - B_{1-x}(n, m) \tag{16}$$

[10] See, for example, E. B. Wilson, *Advanced Calculus*, Chapter 14 (Ginn and Company, 1912).

Or, in another form,

$$B_x(n, m) = B(m, n) - B_{1-x}(m, n) \qquad (17)$$

Hence in the incomplete beta function $B_x(m, n)$ an interchange of the two arguments alters the value to $B(m, n) - B_{1-x}(m, n)$. That an interchange of the arguments m and n has no effect on the complete

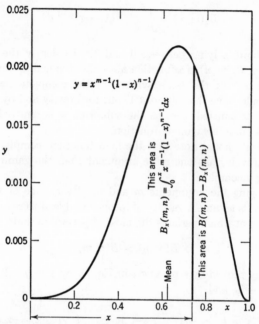

FIG. 60.   The curve $y = x^{m-1}(1 - x)^{n-1}$, this being the integrand in the definition of the beta function.   The curve is drawn for $m = 5$, $n = 3$.   The area under the whole curve from 0 to 1 is the complete beta function $B(m, n)$, and the area between 0 and $x < 1$ is the incomplete beta function $B_x(m, n)$.

beta function $B(m, n)$ follows at once by putting $x = 1$ in Eq. 16 or 17 and noting that $B_0(m, n) = B_0(n, m) = 0$.

These relations are easily recalled by using a plot of the integrand of Eqs. 13 and 14, such as Fig. 60. The effect of interchanging m and n is to produce a mirror image of the figure, which can be visualized as the curve that would be obtained by moving the origin one unit to the right, reversing the positive direction of x, and then looking at the page through the back of the paper. By visualizing this process, the relations exhibited by Eqs. 16 and 17 become evident.

A recursion formula for the incomplete beta function is given in Exercise 16.

Unless m and n are equal, the curve $y = x^{m-1}(1 - x)^{n-1}$ will be unsymmetrical. By varying the choice of m and n, a great variety of forms can be obtained. The student will profit by carrying out the calculations and plotting the curves suggested in Exercise 2.

Two other useful forms of the beta function. Let $x = z^2/(1 + z^2)$, $1 - x = 1/(1 + z^2)$, $dx = 2z \, dz/(1 + z^2)^2$. By introducing this change of variable into Eq. 13 we find that

$$B(m, n) = 2\int_0^\infty z^{2m-1}(1 + z^2)^{-m-n} \, dz = B(n, m) \tag{18}$$

In particular, if m is replaced by $\frac{1}{2}$ and n by $\frac{1}{2}[n - 1]$, this relation gives

$$B(\tfrac{1}{2}[n - 1], \tfrac{1}{2}) = 2\int_0^\infty (1 + z^2)^{-\frac{1}{2}n} \, dz = \int_{-\infty}^\infty (1 + z^2)^{-\frac{1}{2}n} \, dz \tag{19}$$

which will be found convenient for certain calculations.

Another useful form is obtained by the substitution $x = \sin^2 \varphi$, $dx = 2 \sin \varphi \cos \varphi \, d\varphi$, whereupon the original form shown in Eq. 13 goes into

$$B(m, n) = 2\int_0^{\frac{1}{2}\pi} \sin^{2m-1} \varphi \cos^{2n-1} \varphi \, d\varphi \tag{20}$$

In particular,

$$B(\tfrac{1}{2}, \tfrac{1}{2}) = 2\int_0^{\frac{1}{2}\pi} d\varphi = \pi \tag{21}$$

a relation that will be needed shortly (see Eq. 26).

The relation between the gamma and beta functions. From the second form of the gamma function, Eq. 3,

$$\Gamma(m) \, \Gamma(n) = 4\int_0^\infty u^{2m-1}e^{-u^2} \, du \cdot \int_0^\infty v^{2n-1}e^{-v^2} \, dv$$

$$= 4\int_0^\infty \int_0^\infty u^{2m-1}v^{2n-1}e^{-(u^2+v^2)} \, du \, dv \tag{22}$$

It is permissible to change the product of two integrals to a double integral, and *vice versa*, when neither the limits nor the integrands of either integral depend on the variable in the other.[11]

Now consider the surface

$$z = u^{2m-1}v^{2n-1}e^{-(u^2+v^2)} \tag{23}$$

erected on the uv-plane. $u^{2m-1}v^{2n-1}e^{-(u^2+v^2)} \, du \, dv$ may be considered the volume of an elementary rectangular prism erected on the base

[11] E. B. Wilson, *Advanced Calculus*, Art. 143 (Ginn and Company, 1912).

$du\ dv$ and having altitude z, as shown in Fig. 61. The integral written in Eq. 22 calls for the summation of the volumes of all these elementary prisms, by which process one finds the total volume under the surface generated by Eq. 23. The summation (integration) can be carried out by introducing polar coordinates in the uv-plane. Let

$$u = r \cos \varphi, \quad v = r \sin \varphi$$

Then the element of area $du\ dv$ in Eq. 22 is to be replaced by $r\ dr\ d\varphi$. A quadrant of the surface described by Eq. 23 can be mapped out by allowing φ to vary from 0 to $\frac{1}{2}\pi$ and r from 0 to ∞. That the integral

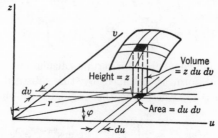

FIG. 61.   The elementary prism of height z and base $du\ dv$, cut from the surface $z = u^{2m-1} v^{2n-1} e^{-(u^2+v^2)}$. The volume under the whole surface is the sum of the volumes of all such prisms.

in Eq. 22 actually exists (i.e., approaches a definite value as the upper limits for u and v are made larger and larger) can be demonstrated by showing that the volume under the surface of Eq. 23 outside the radius R can be reduced to any desired value, however small, simply by making R big enough. Eq. 22 then leads to the following equalities:

$$\Gamma(m)\ \Gamma(n) = 4\int_0^\infty r^{2m-1} r^{2n-1} e^{-r^2} r\ dr \int_0^{\frac{1}{2}\pi} \sin^{2n-1} \varphi \cos^{2m-1} \varphi\ d\varphi$$

$$= 2\int_0^\infty r^{2(n+m)-1} e^{-r^2}\ dr \cdot 2\int_0^{\frac{1}{2}\pi} \sin^{2n-1} \varphi \cos^{2m-1} \varphi\ d\varphi$$

$$= \Gamma(n+m) B(n, m)$$

by Eqs. 3 and 20. Thus we have discovered that

$$B(m, n) = \frac{\Gamma(m)\ \Gamma(n)}{\Gamma(m+n)} \tag{24}$$

Another proof of this relation will be found in Exercise 11 at the end.

To find $\bar{\Gamma}(\frac{1}{2})$. At the close of the paragraph on other useful forms of the beta function it was found that

$$B(\tfrac{1}{2}, \tfrac{1}{2}) = \pi \qquad \text{[See Eq. 21]}$$

Now with $m = n = \tfrac{1}{2}$ in Eq. 24 we have

$$B\left(\frac{1}{2}, \frac{1}{2}\right) = \frac{\Gamma(\tfrac{1}{2})\,\Gamma(\tfrac{1}{2})}{\Gamma(1)} \qquad (25)$$

and since $\Gamma(1) = 1$ it follows that

$$\Gamma(\tfrac{1}{2}) = \sqrt{\pi} \qquad (26)$$

Eq. 3 then gives

$$\Gamma(\tfrac{1}{2}) = 2\int_0^\infty e^{-v^2}\,dy = \int_{-\infty}^\infty e^{-v^2}\,dy = \sqrt{\pi} \qquad (27)$$

The same result is derived in Exercise 25. By means of the integral just written, the student should satisfy himself that the normal curve when written in the form

$$y = \frac{1}{\sigma\sqrt{2\pi}}\, e^{-x^2/2\sigma^2} \qquad (28)$$

is *normalized*, i.e., that the area under the curve of y plotted against x is unity, and that it would not be so if the constant factor were other than $1/\sigma\sqrt{2\pi}$.

Tables of the incomplete beta function. The ratio $B_x(m, n)/B(m, n)$ is the fractional part of the total area under the curve $y = x^{m-1}(1-x)^{n-1}$ (Fig. 60) between 0 and the abscissa $x < 1$, and $1 - B_x(m, n)/B(m, n)$ or $B_{1-x}(n, m)/B(m, n)$ is the fractional part remaining between x and 1. This curve, which is the graph of the integrand of the beta function, constitutes another important frequency curve, the Pearson Type I, and by change of variable, the Pearson Types II, VI, and VII also; hence the incomplete beta function constitutes another important probability integral. The immense task of providing such tables was finally realized in 1934 with the publication by Karl Pearson and his staff of the *Tables of the Incomplete Beta Function* [12]—an even greater undertaking than the *Tables of the Incomplete Gamma Function*. A sample page is shown in Fig. 62. The function tabled is

$$I_x(p, q) = \frac{B_x(p, q)}{B(p, q)} \qquad (29)$$

---

[12] *Tables of the Incomplete Beta Function*, issued by the Office of Biometrika, University College, Gower Street, London W.C. 1. Mention should be made of *Tracts for Computers*, No. 7, by H. E. Soper, entitled *The Numerical Evaluation of the Incomplete Beta Function*, which will be found useful in the absence of the *Tables of the Incomplete Beta Function*, or outside their range.

the arguments being p, q, and x. The introduction gives a history of the incomplete beta function and illustrations of its use. The complete beta function $B(p, q)$ is listed in the headings of the columns, so that, if desired, $B_x(p, q)$ can be found by multiplying $I_x(p, q)$ by $B(p, q)$.

TABLES OF THE INCOMPLETE β-FUNCTION

$p = 5$ to 7.5 $q = 5$ $x = .02$ to $.60$ 128

x	$p = 5$	$p = 5.5$	$p = 6$	$p = 6.5$	$p = 7$	$p = 7.5$
$B(p,q) =$	$.1587\ 3016 \times \frac{1}{10^4}$	$.1108\ 4890 \times \frac{1}{10^4}$	$.7936\ 5079 \times \frac{1}{10^5}$	$.5806\ 3711 \times \frac{1}{10^5}$	$.4329\ 0043 \times \frac{1}{10^5}$	$.3281\ 8619 \times \frac{1}{10^5}$
.02	.0000 004	.0000 001				
.03	.0000 028	.0000 006	.0000 001			
.04	.0000 113	.0000 029	.0000 007	.0000 002		
.05	.0000 332	.0000 096	.0000 028	.0000 008	.0000 002	.0000 001
.06	.0000 798	.0000 254	.0000 079	.0000 024	.0000 007	.0000 002
.07	.0001 666	.0000 572	.0000 193-	.0000 064	.0000 021	.0000 007
.08	.0003 136	.0001 149	.0000 415-	.0000 147	.0000 052	.0000 018
.09	.0005 453	.0002 119	.0000 810	.0000 305+	.0000 114	.0000 042
.10	.0008 909	.0003 646	.0001 469	.0000 584	.0000 229	.0000 089
.11	.0013 838	.0005 936	.0002 507	.0001 044	.0000 429	.0000 175-
.12	.0020 615-	.0009 233	.0004 069	.0001 769	.0000 760	.0000 323
.13	.0029 649	.0013 808	.0006 332	.0002 804	.0001 279	.0000 565+
.14	.0041 384	.0019 986	.0009 505-	.0004 459	.0002 066	.0000 947
.15	.0056 287	.0028 117	.0013 832	.0006 713	.0003 219	.0001 527
.16	.0074 847	.0038 587	.0019 593	.0009 815+	.0004 858	.0002 379
.17	.0097 568	.0051 808	.0027 098	.0013 985-	.0007 131	.0003 598
.18	.0124 962	.0068 224	.0036 694	.0019 475-	.0010 214	.0005 300
.19	.0157 541	.0088 297	.0048 757	.0026 570	.0014 309	.0007 625-
.20	.0195 814	.0112 506	.0063 694	.0035 589	.0019 654	.0010 739
.21	.0240 280	.0141 343	.0081 935+	.0046 883	.0026 515-	.0014 839
.22	.0291 417	.0175 304	.0103 936	.0060 831	.0035 193	.0020 149
.23	.0349 682	.0214 888	.0130 167	.0077 943	.0046 020	.0026 926
.24	.0415 503	.0260 588	.0161 116	.0098 356	.0059 361	.0035 460
.25	.0489 273	.0312 883	.0197 277	.0122 827	.0075 612	.0046 073
.26	.0571 345-	.0372 238	.0239 148	.0151 734	.0095 196	.0059 122
.27	.0662 028	.0439 094	.0287 224	.0185 569	.0118 563	.0074 993
.28	.0761 583	.0513 861	.0341 994	.0224 834	.0146 187	.0094 105+
.29	.0870 218	.0596 916	.0403 932	.0270 037	.0178 560	.0116 907
.30	.0988 087	.0688 598	.0473 490	.0321 685-	.0216 192	.0143 873

FIG. 62.   A sample page from Karl Pearson's *Tables of the Incomplete Beta Function*.

Besides supplying tables for integrals under certain frequency curves, the incomplete beta function sums terms of the point binomial without approximation, as is demonstrated in the next section.

To sum any number of terms of the point binomial without approximation. Let S denote the sum of the terms of the point binomial $(q + p)^n$

as far as the term in p^t, which is to define

$$S = q^n + nq^{n-1}p + \binom{n}{2} q^{n-2}p^2 + \cdots + \binom{n}{t} q^{n-t}p^t$$

$$= \sum_{r=0}^{t} \binom{n}{r} q^{n-r}p^r = \sum_{0}^{t} \frac{n!}{(n-r)!\,r!} q^{n-r}p^r \qquad (30)$$

q and p are to be complementary; i.e., $q = 1 - p$, whence $dq/dp = -1$. Then, if S be differentiated with respect to p, the result is

$$\frac{dS}{dp} = -\sum_{0}^{t} \frac{n!}{(n-r-1)!\,r!} q^{n-r-1}p^r + \sum_{1}^{t} \frac{n!}{(n-r)!\,(r-1)!} q^{n-r}p^{r-1}$$

$$= \qquad\qquad " \qquad\qquad + \sum_{s=0}^{t-1} \frac{n!}{(n-s-1)!\,s!} q^{n-s-1}p^s$$

$$= -\frac{n!}{(n-t-1)!\,t!} q^{n-t-1}p^t \qquad (31)$$

This simple result is produced by the cancellation of plus and minus terms in the two summations; for every term in the first sum, except for $r = t$, there is a term of opposite sign in the second sum.

Now, when $p = 0$, $S = 1$, whence from the last equation,

$$\int_{1}^{S} dS = -\frac{n!}{(n-t-1)!\,t!} \int_{0}^{p} x^t(1-x)^{n-t-1}\,dx \qquad (32)$$

which gives

$$S = 1 - \frac{n!}{(n-t-1)!\,t!} \int_{0}^{p} x^t(1-x)^{n-t-1}\,dx \qquad (33)$$

Moreover, when $p = 1$, $q = 0$ and $S = 0$ if $t < n$; hence the factorials must satisfy the relation

$$\int_{0}^{1} x^t(1-x)^{n-t-1}\,dx = \frac{(n-t-1)!\,t!}{n!} \qquad (34)$$

This relation is equivalent to Eq. 24, but the evaluation of the factorials in terms of the beta function comes in here as a by-product. It follows at once that

$$S = 1 - \frac{\displaystyle\int_{0}^{p} x^t(1-x)^{n-t-1}\,dx}{\displaystyle\int_{0}^{1} x^t(1-x)^{n-t-1}\,dx} \qquad (35)$$

Thus S turns out to be unity diminished by the ratio of an incomplete to a complete beta function, and no approximation has been introduced.[13] Unfortunately, the use of the incomplete beta function, even with tables, is often laborious.   Indeed, Eq. 35 can and has often been used as a means of evaluating the integral $\int_0^p x^t(1 - x)^{n-t-1}\, dx$ by summing terms of the binomial; such was in fact the use that Bayes made of it in 1763 (*Trans. Roy. Soc.* for the year 1763, p. 396).   Laplace later derived the same result with the opposite purpose, viz., to sum terms of the binomial (*Théorie analytique*, 1823: p. 151).

To sum any number of terms of the Poisson exponential limit without approximation.   Let S denote the sum of the terms of the Poisson exponential limit as far as the term in m^t.   Then by definition

$$S = e^{-m}\left\{1 + m + \frac{m^2}{2!} + \cdots + \frac{m^t}{t!}\right\} \tag{36}$$

Differentiation with respect to m gives

$$\frac{dS}{dm} = -S + e^{-m}\left\{1 + \frac{2m}{2!} + \frac{3m^2}{3!} + \cdots + \frac{tm^{t-1}}{t!}\right\}$$

$$= -e^{-m}\frac{m^t}{t!} \tag{37}$$

all the terms in the braces, except for the last, being canceled by opposite terms in S.   Now when $m = 0$, $S = 1$, so

$$\int_1^S dS = -\frac{1}{t!}\int_0^m x^t e^{-x}\, dx$$

$$S = 1 - \frac{1}{t!}\int_0^m x^t e^{-x}\, dx \tag{38}$$

Moreover, when $m = \infty$, $S = 0$, and it follows that

$$\int_0^\infty x^t e^{-x}\, dx = t! \tag{39}$$

[13] This method of deriving Eq. 35 was first published by William Lazarus of Hamburg, *J. Inst. Actuaries*, vol. 15, 1870: pp. 245-57, page 251 in particular.   Lazarus credits this derivation to one Dr. Landi of Trieste.   The result contained in Eq. 35 was first found by Bayes, *Trans. Roy. Soc.*, 1763: page 396.   This paper is available in a booklet entitled *Facsimiles of Two Papers by Bayes*, with commentaries by E. C. Molina and W. Edwards Deming (Graduate School of the Department of Agriculture, 1940).   I am indebted to my mentor Dr. G. J. Lidstone of Edinburgh for calling my attention to Lazarus's paper.

This relation is equivalent to Eq. 8, but the identification of the complete gamma function with the factorial comes in here as a by-product. It follows at once that

$$S = 1 - \frac{\displaystyle\int_0^m x^t e^{-x}\, dx}{\displaystyle\int_0^\infty x^t e^{-x}\, dx} \tag{40}$$

Thus, the incomplete sum of the Poisson exponential limit is expressible in terms of an incomplete gamma function. Turned around, the incomplete gamma function is expressible as a sum of terms of the Poisson limit. Eq. 40 was first published by E. C. Molina in 1915 (*Amer. Math. Monthly*, vol. 22, 1915: p. 223).

EXERCISES ON THE GAMMA AND BETA FUNCTIONS

NOTE. The following exercises illustrate the foregoing text material and prepare a number of results that are needed in later chapters.

Exercise 1. Plot the curve $y = x^{n-1}e^{-x}$ for $n = 1, 2, 3, 6, 10$ between $x = 0$ and $x = 20$. The areas under these curves to infinity are respectively $\Gamma(1) = 1$, $\Gamma(2) = 1$, $\Gamma(3) = 2$, $\Gamma(6) = 5!$, $\Gamma(10) = 9!$. Note that as $x \to \infty$ the slopes of all the curves become zero. If $n = 1$, the curve starts from the point $(0, 1)$ with slope -1; if $n = 2$, the curve starts from the origin with slope $+1$. If $n > 2$, the curve starts from the origin with a horizontal tangent, making contact of order $n - 2$. At infinity all the curves make contact of infinitely high order with the x-axis.

Exercise 2. *a.* Plot the curve $y = x^{m-1}(1 - x)^{n-1}$ for $m, n = 1, 2$; $2, 1; 2, 3; 3, 2; 3, 3; \frac{1}{2}, 2; 2, \frac{1}{2}$ between $x = 0$ and $x = 1$.

b. Determine the numerical values of the areas under these curves, and check roughly with the graphs. Note that when $n = m$, the curves are symmetrical, and when $n \neq m$ the curves are lopsided; if $n < m$, the bulk of the area is thrown to the right, and if $n > m$, the bulk of the area is thrown to the left. The order of contact with the x axis is $m - 2$ at $x = 0$ and $n - 2$ at $x = 1$.

Exercise 3. Given $\Gamma(\tfrac{1}{2}) = \sqrt{\pi}$; find by the recursion formula the values of $\Gamma(\tfrac{3}{2})$, $\Gamma(\tfrac{5}{2})$, $\Gamma(\tfrac{7}{2})$, $\Gamma(\tfrac{9}{2})$. Show that

$$\left(n + \frac{1}{2}\right)! = \Gamma\left(n + \frac{3}{2}\right) = \pi^{\frac{1}{2}} \frac{1 \cdot 3 \cdot 5 \cdots (2n + 1)}{2^{n+1}}$$

Exercise 4. *a.* Show that the recursion formula (Eq. 5) holds for the Gauss infinite product $\pi(n)$, defined on page 471, and that it holds regardless of whether n be positive, negative, or complex.

Solution

$$\pi(n) = \operatorname*{Lim}_{m \to \infty} \frac{m!\, m^n}{(n+1)(n+2)\cdots(n+m)} \qquad \text{[P. 471]}$$

$$= \operatorname*{Lim}_{m \to \infty} m\, \frac{n}{n+m}\, \frac{m!\, m^{n-1}}{n(n+1)(n+2)\cdots(n-1+m)}$$

$$= \operatorname*{Lim}_{m \to \infty} m\, \frac{n}{n+m}\, \pi(n-1)$$

$$= \operatorname*{Lim}_{m \to \infty} \frac{n}{1+\dfrac{n}{m}}\, \pi(n-1)$$

$$= n\pi(n-1) \quad \text{in the limit, as } m \to \infty$$

This is equivalent to the recursion formula

$$\Gamma(n+1) = n\Gamma(n) \qquad \text{[Eq. 5]}$$

if $\pi(n)$ is identified as $\Gamma(n+1)$.

b. Show that the Gauss infinite product $\pi(n)$ equals $n!$ if n is a positive integer.

Solution

$$\pi(n) = \operatorname*{Lim}_{m \to \infty} \frac{m!\, m^n}{(n+1)(n+2)\cdots(n+m)}$$

$$= \operatorname*{Lim}_{m \to \infty} \frac{m!\, m^n}{(n+1)(n+2)\cdots(n+m)}\, \frac{n!}{n!}$$

$$= \operatorname*{Lim}_{m \to \infty} \frac{n!\, m!\, m^n}{(m+n)!}$$

$$= \operatorname*{Lim}_{m \to \infty} \frac{n!\, m^n}{(m+1)(m+2)\cdots(m+n)}$$

$$= \operatorname*{Lim}_{m \to \infty} \frac{n!}{\left(1+\dfrac{1}{m}\right)\left(1+\dfrac{2}{m}\right)\cdots\left(1+\dfrac{n}{m}\right)}$$

$$= n! \quad \text{in the limit as } m \to \infty$$

Exercise 5. *a.* Show that as n approaches a negative integer $-a$ from the positive side, $\Gamma(n)$ approaches $-\infty$ if a is odd and approaches $+\infty$ if a is even.

b. Show that as n approaches a negative integer $-a$ from the negative side, $\Gamma(n)$ approaches $+\infty$ if a is odd and approaches $-\infty$ if a is even.

Solution (Arnold Frank)

a. Let

$$n = -\frac{s}{s+1}a \qquad \begin{array}{l}(a > 0)\\(s > 0)\end{array}$$

Then $n \to -a$ from the positive side as $s \to +\infty$.
Write $\Gamma(n)$ for $\pi(n - 1)$; then

$$\Gamma(n) = \operatorname*{Lim}_{m \to \infty} \frac{m!\, m^{n-1}}{n(n+1)(n+2)\cdots(n+m-1)}$$

$$= \operatorname*{Lim}_{m \to \infty} \frac{m!\, m^{n-1}}{\left[-\dfrac{s}{s+1}a\right]\left[-\dfrac{s}{s+1}a+1\right]\left[-\dfrac{s}{s+1}a+2\right]\cdots\left[-\dfrac{s}{s+1}a+(m-1)\right]}$$

$$= \operatorname*{Lim}_{m \to \infty} \frac{m!\, m^{n-1}(s+1)^m}{[-sa][-s(a-1)+1][-s(a-2)+2]\cdots[-s(a-\overline{m-1})+\overline{m-1}]}$$

As $s \to \infty$, $\Gamma(n) \to +\infty$ or $-\infty$. The first a factors in the denominator are all negative and the rest are positive; hence the sign of $\Gamma(n)$ is negative if a is odd and positive if a is even.

b. Let

$$n = -\frac{s+1}{s}a \qquad \begin{array}{l}(a > 0)\\(s > 0)\end{array}$$

Then $n \to -a$ from the negative side as $s \to +\infty$, and

$$\Gamma(n) = \operatorname*{Lim}_{m \to \infty} \frac{m!\, m^{n-1}}{\left[-\dfrac{s+1}{s}a\right]\left[-\dfrac{s+1}{s}a+1\right]\left[-\dfrac{s+1}{s}a+2\right]\cdots\left[-\dfrac{s+1}{s}a+(m-1)\right]}$$

$$= \operatorname*{Lim}_{m \to \infty} \frac{m!\, m^{n-1}s^m}{[s(-a)-a][s(1-a)-a][s(2-a)-a]\cdots[s(\overline{m-1}-a)-a]}$$

As $s \to \infty$, $\Gamma(n) \to +\infty$ or $-\infty$. The first $a + 1$ factors in the denominator are all negative and the rest are positive; hence the sign of $\Gamma(n)$ is positive if a is odd and negative if a is even.   *Q.E.D.*

Exercise 6.   Show by the recursion formula that $\Gamma(-\frac{1}{4}) = -4\Gamma(\frac{3}{4})$; $\Gamma(-\frac{1}{8}) = -8\Gamma(\frac{7}{8})$, $\Gamma(-\frac{1}{32}) = -32\Gamma(\frac{31}{32})$, $\cdots$; $\Gamma(-\frac{3}{4}) = -(\frac{4}{3})\Gamma(\frac{1}{4})$, $\Gamma(-\frac{7}{8}) = -(\frac{8}{7})\Gamma(\frac{1}{8})$, $\Gamma(-\frac{31}{32}) = -(\frac{32}{31})\Gamma(\frac{1}{32})$, $\cdots$. This illustrates how $\Gamma(n)$ approaches $-\infty$ as n approaches 0 from the negative side or -1 from the positive side.

Exercise 7.   Show that if $n!$ be identified with $\Gamma(n + 1)$, then $0! = 1$.

Exercise 8.   Show that

$$\int_0^\infty e^{-bx-a^2x^2}\, dx = \frac{e^{b^2/4a^2}}{2\,|a|}\sqrt{\pi} - \Gamma_{b^2/4a^2}(\tfrac{1}{2}) \quad (b \geqq 0)$$

$$\int_0^\infty e^{bx-a^2x^2}\, dx = \frac{e^{b^2/4a^2}}{2\,|a|}\sqrt{\pi} + \Gamma_{b^2/4a^2}(\tfrac{1}{2}) \quad (b \geqq 0)$$

By addition

$$2\int_0^\infty e^{-a^2x^2} \cosh bx \, dx = \int_0^\infty e^{bx-a^2x^2} \, dx + \int_0^\infty e^{-bx-a^2x^2} \, dx$$

$$= \frac{e^{b^2/4a^2}}{|a|} \sqrt{\pi} \quad (b \gtrless 0)$$

Given: $\cosh x = \frac{1}{2}(e^x + e^{-x})$.

Exercise 9.

$$\int_0^\infty t^{n-1}e^{-t^n} \, dt = \int_0^\infty t^{2n-1}e^{-t^n} \, dt = \frac{1}{n}$$

$$\int_0^\infty t^{m-1}e^{-t^n} \, dt = \left(\frac{1}{n}\right)\Gamma\left(\frac{m}{n}\right), \quad \text{which becomes } \frac{\pi^{\frac{1}{2}}}{n} \text{ if } n = 2m$$

(Kramp, *Analyse des réfractions*, Strasbourg, 1799.)

Exercise 10.

$$\Gamma\left(\frac{n-1}{2}\right) = \frac{n-3}{2} \cdot \frac{n-5}{2} \cdot \frac{n-7}{2} \cdots \frac{7}{2} \cdot \frac{5}{2} \cdot \frac{3}{2} \cdot \frac{1}{2} \sqrt{\pi} \quad (n \text{ even})$$

$$= \frac{n-3}{2} \cdot \frac{n-5}{2} \cdot \frac{n-7}{2} \cdots \frac{6}{2} \cdot \frac{4}{2} \cdot \frac{2}{2} \quad (n \text{ odd})$$

$$= \sqrt{\pi} \frac{(n-2)!}{2^{n-2}\left(\dfrac{n-2}{2}\right)!} \quad (n \text{ even})$$

$$= [\tfrac{1}{2}(n-3)]! \quad (n \text{ odd})$$

By means of these verify your answers for Exercise 3.

Exercise 11. By trigonometric substitution, show that the Euler integral $\int_0^1 x^{m-1}(1-x)^{n-1} \, dx$ defining $B(m, n)$ is equivalent to

$$2\int_0^{\frac{1}{2}\pi} \sin^{2m-1}\theta \cos^{2n-1}\theta \, d\theta$$

Integrate this by Wallis's formula (p. 63) and show that

$$B(m, n) = B(n, m) = \frac{(m-1)!(n-1)!}{(m+n-1)!} = \frac{\Gamma(m)\,\Gamma(n)}{\Gamma(m+n)}$$

which is Eq. 24.

Exercise 12.

$$\frac{1}{B\left(\dfrac{n-1}{2},\dfrac{1}{2}\right)} = \frac{\Gamma(\tfrac{1}{2}n)}{\Gamma\left(\dfrac{n-1}{2}\right)\Gamma\left(\dfrac{1}{2}\right)}$$

$$= \frac{1}{\pi}\frac{n-2}{n-3}\frac{n-4}{n-5}\frac{n-6}{n-7}\cdots\frac{6\cdot4\cdot2}{5\cdot3\cdot1} = \frac{2^{n-2}\left(\dfrac{n-2}{2}\,!\right)^2}{\pi(n-2)!}$$
(n even)

$$= \frac{1}{2}\frac{n-2}{n-3}\frac{n-4}{n-5}\frac{n-6}{n-7}\cdots\frac{5\cdot3\cdot1}{4\cdot2} = \frac{(n-2)!}{2^{n-2}\left(\dfrac{n-3}{2}\,!\right)^2}$$
(n odd)

Exercise 13. The number of possible combinations of n articles taken r at a time is $n!/(n-r)!\,r!$, which is often abbreviated $\binom{n}{r}$, as in Eq. 30.

a. Show that if the factorials in this expression be identified with gamma functions, then $\binom{n}{r}$ vanishes when r is any negative integer or any positive integer greater than n (a fact already mentioned in the text).

b. Show that $\binom{n}{n} = \binom{n}{0} = 1$ if $0! = \Gamma(1) = 1$.

c. Show that $\Sigma r \binom{n}{r} p^r q^{n-r} = np$ where the sum is to be taken over all integral values of r.

Exercise 14. Show that $\displaystyle\sum_{r=0}^{n} \frac{1}{B(n-r+1, r+1)} = 2^n(n+1)$, or that $\binom{n}{0} + \binom{n}{1} + \binom{n}{2} + \cdots + \binom{n}{n} = 2^n$. HINT. Expand $(\tfrac{1}{2} + \tfrac{1}{2})^n$ by the binomial theorem and note that the sum of the terms is unity. The summation could as well be taken over all integral values of r, positive and negative; see the preceding exercise.

Exercise 15. Show that $B(m, n) = \displaystyle\int_0^1 \frac{x^{m-1} + x^{n-1}}{(1+x)^{m+n}}\,dx$. HINT. Replace x in $B(m, n) = \displaystyle\int_0^1 x^{m-1}(1-x)^{n-1}\,dx$ by $z/(1+z)$ and then by

$1/(1 + z)$ and get

$$B(m, n) = \int_0^\infty z^{m-1}(1 + z)^{-m-n}\, dz$$

and

$$B(m, n) = \int_0^\infty z^{n-1}(1 + z)^{-m-n}\, dz$$

Add these and obtain

$$
\begin{aligned}
B(m, n) &= \frac{1}{2} \int_0^\infty \frac{z^{m-1} + z^{n-1}}{(1 + z)^{m+n}}\, dz \\
&= \frac{1}{2} \int_0^1 \frac{z^{m-1} + z^{n-1}}{(1 + z)^{m+n}}\, dz + \frac{1}{2} \int_1^\infty \frac{z^{m-1} + z^{n-1}}{(1 + z)^{m+n}}\, dz \\
&= \frac{1}{2} \int_0^1 \frac{z^{m-1} + z^{n-1}}{(1 + z)^{m+n}}\, dz + \frac{1}{2} \int_0^1 \frac{v^{n-1} + v^{m-1}}{(1 + v)^{m+n}}\, dv \quad \begin{array}{l}[z = 1/v \text{ in the} \\ \text{last integral}]\end{array} \\
&= \int_0^1 \frac{x^{m-1} + x^{n-1}}{(1 + x)^{m+n}}\, dx
\end{aligned}
$$

Exercise 16. Prove the following recursion formula for the incomplete beta function

$$I_x(p, q) = xI_x(p - 1, q) + (1 - x)I_x(p, q - 1) \tag{1}$$

Here, as in the *Tables of the Incomplete Beta Function,*

$$I_x(p, q) = \frac{B_x(p, q)}{B(p, q)} = \frac{1}{B(p, q)} \int_0^x x^{p-1}(1 - x)^{q-1}\, dx \tag{2}$$

Solution (T. A. Bancroft [14])

$$
\begin{aligned}
\frac{d}{dx} &\{x^{p-1}(1 - x)^q\} \\
&= (p - 1)x^{p-2}(1 - x)^q - qx^{p-1}(1 - x)^{q-1} \\
&= (p - 1)x^{p-2}(1 - x)^{q-1} - (p + q - 1)x^{p-1}(1 - x)^{q-1} \tag{3}
\end{aligned}
$$

$$\frac{d}{dx}\{x^{p-1}(1 - x)^{q-1}\} = (p - 1)x^{p-2}(1 - x)^{q-1} - (q - 1)x^{p-1}(1 - x)^{q-2} \tag{4}$$

Integration of both sides of Eqs. 3 and 4 between the termini 0 and x gives

$$x^{p-1}(1 - x)^q = (p - 1)B_x(p - 1, q) - (p + q - 1)B_x(p, q) \tag{5}$$

$$x^{p-1}(1 - x)^{q-1} = (p - 1)B_x(p - 1, q) - (q - 1)B_x(p, q - 1) \tag{6}$$

[14] T. A. Bancroft, *J. Amer. Stat. Assoc.,* vol. 40, 1945: p. 529.

Multiply and divide the extreme right-hand term of the upper equation by $B(p, q)$, and the extreme right-hand term of the lower equation by $B(p, q - 1)$. Then divide both equations through by $(p - 1)B(p - 1, q)$. The result is

$$\frac{x^{p-1}(1 - x)^q}{(p - 1)B(p - 1, q)} = I_x(p - 1, q) - I_x(p, q) \tag{7}$$

$$\frac{x^{p-1}(1 - x)^{q-1}}{(p - 1)B(p - 1, q)} = I_x(p - 1, q) - I_x(p, q - 1) \tag{8}$$

Divide corresponding members and get Eq. 1.

Q.E.D.

Exercise 17. In the analysis of variance (last chapter) occurs the frequency curve

$$y \, dF = \frac{K^{\frac{1}{2}K} k^{\frac{1}{2}k} F^{\frac{1}{2}K-1}}{B(\frac{1}{2}K, \frac{1}{2}k)[KF + k]^{\frac{1}{2}(K+k)}} \, dF$$

for the distribution of F, which stands for the ratio of two estimates of σ^2. By transforming to x, show that if $P(F)$ denotes the fractional part of the area lying beyond a given abscissa F, then

$$P(F) = 1 - \frac{1}{B(p, q)} \int_0^x x^{p-1}(1 - x)^{q-1} \, dx$$

$$= 1 - I_x(p, q) = I_{1-x}(q, p)$$

wherein x stands for $KF/(KF + k)$, and $p = \frac{1}{2}K$, $q = \frac{1}{2}k$. Thus the evaluation of $P(F)$ can be made to depend on an incomplete beta function.

Exercise 18. The Type III curve is sometimes written

$$y = y_0(1 + x/a)^p e^{-px/a}$$

in order to put the zero of x at the mode (maximum). Select a point B lying at the distance b from the finite end of the curve, and show that the fractional part of the area lying beyond the point B is

$$\frac{\displaystyle\int_{bp/a}^{\infty} v^p e^{-v} \, dv}{\displaystyle\int_0^{\infty} v^p e^{-v} \, dv} = 1 - I(u, p)$$

if $u = bp/a(1 + p)^{\frac{1}{2}}$ in the notation of the *Tables of the Incomplete Gamma Function* (p. 472).

Exercise 19. Prove that $\displaystyle\int_0^t x^{m-1}(a - x)^{n-1} \, dx = a^{m+n-1}B_{t/a}(m, n)$, $a > 0$.

Exercise 20.   The volume bounded by the xy-plane, the plane $x + y = 1$, and the surface $z = x^{l-1}y^{m-1}$, is $\Gamma(l)\, \Gamma(m)/\Gamma(l + m + 1)$.

Exercise 21. a. The integral $\iiint x^{l-1}y^{m-1}z^{n-1}\, dx\, dy\, dz$, taken over all positive values of x, y, z such that $x + y + z \leq 1$, is equal to $\Gamma(l)\, \Gamma(m)\, \Gamma(n)/\Gamma(l + m + n + 1)$.   (This is a Dirichlet integral.)

b. The integral $\iiint x^{l-1}y^{m-1}z^{n-1}\, dx\, dy\, dz$, taken over all positive values of x, y, z such that $x/a + y/b + z/c \leq h$, is equal to

$$a^l b^m c^n h^{l+m+n}\, \Gamma(l)\, \Gamma(m)\, \Gamma(n)/\Gamma(l + m + n + 1).$$

HINT.   Transform the integral of Part a by putting x, y, $z = ah\xi$, $bh\eta$, $ch\zeta$.

c. The integral $\iiint x^{l-1}y^{m-1}z^{n-1}\, dx\, dy\, dz$, taken over all values of x, y, z lying within the positive octant of the surface $(x/a)^p + (y/b)^q + (z/c)^r \leq h$, is equal to

$$h^{l/p+m/q+n/r} \times \frac{a^l b^m c^n}{pqr} \times \frac{\Gamma(l/p)\, \Gamma(m/q)\, \Gamma(n/r)}{\Gamma(l/p + m/q + n/r + 1)}$$

HINT.   Transform to an integral like that in Part a by putting ξh, ηh, $\zeta h = (x/a)^p$, $(y/b)^q$, $(z/c)^r$.

d. By the results of Part c show that the volume of the ellipsoid $(x/a)^2 + (y/b)^2 + (z/c)^2 = 1$ is $\frac{4}{3}\pi abc$ and hence that the volume of a sphere is $\frac{4}{3}\pi r^3$.   [HINT.   Put $h = 1$; $p = q = r = 2$; $l = m = n = 1$.]

e. By a similar artifice, show that the volume of the hypocycloid $(x/a)^{\frac{2}{3}} + (y/b)^{\frac{2}{3}} + (z/c)^{\frac{2}{3}} = 1$ is $4\pi abc/35$.

f. The center of gravity of the positive octant of the ellipsoid of Part d lies at x, y, $z = 3a/8$, $3b/8$, $3c/8$.   HINT.   In Part c put $l = 2$, $m = n = 1$, $h = 1$, $p = q = r = 2$ to find the x-coordinate of the center of gravity.

Exercise 22. a. The n-fold integral

$$\iint \cdots \int x_1^{m_1-1} x_2^{m_2-1} \cdots x_n^{m_n-1}\, dx_1\, dx_2 \cdots dx_n$$

taken over all positive values of x_1, x_2, $\cdots$, x_n such that $x_1 + x_2 + \cdots x_n + \leq 1$, is equal to $\Gamma(m_1)\Gamma(m_2) \cdots \Gamma(m_n)/\Gamma(m_1 + m_2 + \cdots + m_n + 1)$.   [This is an easy extension of Part a of the preceding exercise.]

b. The volume of the n-dimensional ellipsoid $(x_1/a_1)^2 + (x_2/a_2)^2 + \cdots + (x_n/a_n)^2 = 1$ is $V = a_1 a_2 \cdots a_n \dfrac{\pi^{\frac{1}{2}n}}{\Gamma(\frac{1}{2}n + 1)}$, and the volume of the n-dimensional sphere $x_1^2 + x_2^2 + \cdots + x_n^2 = r^2$ is $\dfrac{\pi^{\frac{1}{2}n} r^n}{\Gamma(\frac{1}{2}n + 1)}$.

HINT. The volume V of the n-dimensional ellipsoid is by definition 2^n times the integral of $dx_1 \, dx_2 \, \cdots \, dx_n$ taken over all positive values of $x_1, x_2, \cdots, x_n$ such that $(x_1/a_1)^2 + (x_2/a_2)^2 + \cdots + (x_n/a_n)^2 \leq 1$. By the change of variable $(x_i/a_i)^2 = \xi_i$ and $dx_i = \frac{1}{2} a_i \xi_i^{-\frac{1}{2}} \, d\xi_i$, the volume V is seen to be 2^n times the integral of $\dfrac{a_1 a_2 \cdots a_n}{2^n (\xi_1 \xi_2 \cdots \xi_n)^{\frac{1}{2}}} \, d\xi_1 \, d\xi_2 \cdots d\xi_n$, taken over all positive values of $\xi_1, \xi_2, \cdots, \xi_n$ such that $\xi_1 + \xi_2 + \cdots + \xi_n \leq 1$. The result of Part a with $m_1 = m_2 = \cdots = m_n = \frac{1}{2}$ applies at once, giving $V = a_1 a_2 \cdots a_n \dfrac{\pi^{\frac{1}{2}n}}{\Gamma(\frac{1}{2}n + 1)}$ as written above.

Exercise 23. a. The volume of the thin spherical shell lying between the pair of n-dimensional spheres $x_1^2 + x_2^2 + \cdots + x_n^2 = (r \pm \frac{1}{2}dr)^2$ is $\dfrac{2\pi^{\frac{1}{2}n}}{\Gamma(\frac{1}{2}n)} r^{n-1} \, dr$. [HINT. Simply differentiate the volume of the sphere just found.] This result was used by Helmert in developing the distribution of the standard deviations in samples of n drawn from a normal population. See Czuber's *Beobachtungsfehler* (Teubner, 1891), pages 147–50. The student should observe that this result reduces to $4\pi r^2 \, dr$ for the volume of a spherical shell in three dimensions, and to $2\pi r \, dr$ for the area of a circular ring in two dimensions.

b. The volume of the thin shell lying between the pair of n-dimensional ellipsoids $(x_1/a_1)^2 + (x_2/a_2)^2 + \cdots + (x_n/a_n)^2 = (s \pm \frac{1}{2} \, ds)^2$ is

$$a_1 a_2 \cdots a_n \frac{2\pi^{\frac{1}{2}n}}{\Gamma(\frac{1}{2}n)} s^{n-1} \, ds.$$

This result is useful in developing the distribution of the standard deviations in samples of n drawn from a normal population (next chapter).

Exercise 24. The area of a sphere of radius r is $\{2\pi^{\frac{1}{2}n}/\Gamma(\frac{1}{2}n)\} r^{n-1}$. HINT. Simply take off the dr from the result of Exercise 23. The student should observe that in three dimensions this result reduces to $4\pi r^2$ for the area of a sphere, and in two dimensions to $2\pi r$ for the circumference of a circle.

Exercise 25. Euler in 1768 showed that

$$\int_0^\infty \frac{y^{n-1}}{1 + y} \, dy = \frac{\pi}{\sin n\pi} \quad \text{if} \quad 0 < n < 1$$

Use this result, and put $x = y/(1 + y)$ in Eq. 13. Then:

a. Prove that

$$\Gamma(n) \, \Gamma(1 - n) = B(n, 1 - n) = \pi/\sin n\pi \quad \text{if} \quad 0 < n < 1$$

b. Set $n = \frac{1}{2}$ and prove that $\Gamma(\frac{1}{2}) = \sqrt{\pi}$, whence

$$\int_0^\infty e^{-x^2} \, dx = \frac{1}{2}\sqrt{\pi}$$

c. Prove that

$$\Gamma\left(\frac{1}{2}\right) = \frac{\pi}{\Gamma(\frac{1}{2})}$$

$$\Gamma\left(\frac{3}{4}\right) = \frac{\sqrt{2}\,\pi}{\Gamma(\frac{1}{4})}$$

$$\Gamma\left(\frac{5}{6}\right) = \frac{2\pi}{\Gamma(\frac{1}{6})}$$

Etc.

d. Prove that

$$\Gamma\left(\frac{1}{n}\right) \Gamma\left(\frac{2}{n}\right) \Gamma\left(\frac{3}{n}\right) \cdots \Gamma\left(\frac{n-1}{n}\right) = \frac{(2\pi)^{\frac{1}{2}(n-1)}}{\sqrt{n}}$$

HINT.   Set A equal to the left-hand side.   Then

$$A^2 = \Gamma\left(\frac{1}{n}\right) \Gamma\left(\frac{n-1}{n}\right) \cdot \Gamma\left(\frac{2}{n}\right) \Gamma\left(\frac{n-2}{n}\right) \cdot \quad \cdots \quad \cdot \Gamma\left(\frac{n-1}{n}\right) \Gamma\left(\frac{1}{n}\right)$$

$$= \frac{\pi}{\sin\dfrac{\pi}{n}} \frac{\pi}{\sin\dfrac{2\pi}{n}} \cdots \frac{\pi}{\sin\dfrac{(n-1)\pi}{n}}$$

by Part a.   Then let $\theta \to 0$ in the following trigonometric identity (Hobson, *Plane Trigonometry*):

$$\sin n\theta / \sin \theta = 2^{n-1} \sin\left(\theta + \pi/n\right) \sin\left(\theta + 2\pi/n\right) \cdots \sin\left(\theta + \overline{n-1}\,\pi/n\right)$$

with the result that $A^2 = (2\pi)^{n-1}/n$, which is equivalent to the theorem stated.

Exercise 26.   Show that the relation between the gamma and the beta functions (Eq. 24) can be found without recourse to integration in polar coordinates.

HINT (called to my attention by Mr. Arnold Frank).   Evaluate

$$I = \int_0^\infty \int_0^\infty e^{-x(y+1)} x^{m+n-1} y^{n-1} \, dx \, dy$$

first with respect to x, and then with respect to y, to find

$$I = \Gamma(m+n)B(m, n) \qquad [\text{Let } z = x(y+1)]$$

Now evaluate the same integral, first with respect to y, and then with respect to x, to find

$$I = \Gamma(m)\,\Gamma(n) \qquad\qquad \text{[Let } z = xy]$$

The order of integration is immaterial, whereupon these two results can be set equal, and the relation between the gamma and the beta functions follows. That the order of integration is immaterial can be justified by a theorem due to de la Vallée Poussin. This theorem states that the values of certain infinite multiple integrals are independent of the order of integration.

Exercise 27. Expansion of the incomplete gamma function in series.

a. Expand e^{-x} in powers of x and get

$$\Gamma_x(n) = \int_0^x x^{n-1} e^{-x}\, dx$$

$$= \frac{x^n}{n}\left\{ 1 - \frac{n}{n+1}\,\frac{x}{1!} + \frac{n}{n+2}\,\frac{x^2}{2!} - \frac{n}{n+3}\,\frac{x^3}{3!} + \cdots \right\}$$

This series is convergent for all values of x, but is not convenient for numerical calculation unless x is small.

b. By integrating by parts show that

$$\int_0^x x^{n-1} e^{-x}\, dx = \frac{x^n e^{-x}}{n}\left\{ 1 + \frac{x}{n+1} + \frac{x^2}{(n+1)(n+2)} \right.$$

$$\left. + \frac{x^3}{(n+1)(n+2)(n+3)} + \cdots \right\}$$

This series also converges for all values of x, but is quicker calculated for moderate values of x than the preceding series, especially if n is large.

c. Again, by integrating by parts, develop the asymptotic series

$$\int_0^x x^{n-1} e^{-x}\, dx = \Gamma(n) - x^{n-1} e^{-x}\left\{ 1 + \frac{n-1}{x} + \frac{(n-1)(n-2)}{x^2} \right.$$

$$\left. + \frac{(n-1)(n-2)(n-3)}{x^3} + \cdots \right\}$$

If n is an integer, this series will terminate. If n is not an integer, the series will be infinite and moreover will diverge for any x no matter how large. This can be seen from the ratio test, since the ratio of the sth term to the $(s-1)$th term is $(n-s)/x$. Since $n > 0$, the series will eventually oscillate, and oscillate infinitely for any x. But no matter where curtailed, the error committed will be less than the next term (Laplace, *Théorie analytique*, pp. 174–5 in the 3d ed., 1820). There will always be an optimum term (Bayes, 1763), and the series carried just short of the optimum term will give the best possible representation for those values of x and n. When x is large, a very few terms will suffice for accurate calculations, especially if n is not large.

Series b and c can also be derived by the D operator used inversely, where $D = d/dx$ and $1/D = \int$. See also Exercise 43 for other series.

Exercise 28.

$$\text{Show that} \int_x^\infty x^n e^{-x}\, dx = x^n e^{-x} v, \text{ where}$$

$$v = \cfrac{1}{1 - ny \cfrac{1}{1 + y \cfrac{1}{1 - (n-1)y \cfrac{1}{1 + 2y \cfrac{1}{1 - (n-2)y \cfrac{1}{1 + 3y \cfrac{1}{1 + \cdots}}}}}}}$$

$(y = 1/x)$

HINT.

1. Show that v satisfies the differential equation

$$x\frac{dv}{dx} = (x - n)v - x \tag{1}$$

2. Show that a solution of the general Riccati equation

$$x\frac{dv}{dx} = (x - a)v - x + bv^2 \tag{2}$$

$$v = \frac{1}{1 + kyv_1} = \frac{x}{x + kv_1} \quad \left(y = \frac{1}{x}\right) \tag{3}$$

provided v_1 satisfies the differential equation

$$x\frac{dv_1}{dx} = (x + a + 1)v_1 - \frac{x(b - a)}{k} + kv_1^2 \tag{4}$$

The choice of 1 for the numerator puts $v = 1$ if $x = \infty$, which is required, as is seen by the series in Exercise 27c.

3. This is a Riccati equation like Eq. 2 if $k = b - a$. Hence a solution of Eq. 2 is

$$v = \frac{1}{1 + [b - a]yv_1} \tag{5}$$

wherein v_1 is such that it satisfies the differential equation

$$x\frac{dv_1}{dx} = (x + a + 1) - x + (b - a)v_1^2 \tag{6}$$

4. It is evident, then, that

$$y_1 = \frac{1}{1 + [b + 1]yv_2} \tag{7}$$

if v_2 satisfies

$$\frac{dv_2}{dx} = (x - a)v_2 - x + (b + 1)v_2{}^2 \tag{8}$$

This is like Eq. 2, but with b replaced by $b + 1$. Hence we may say that

$$v = \frac{1}{1 + [b - a]yv_1}$$

$$v_1 = \frac{1}{1 + [b + 1]yv_2}$$

$$v_2 = \frac{1}{1 + [b + 1 - a]yv_3}$$

$$v_3 = \frac{1}{1 + [b + 2]yv_4}$$

$$v_4 = \frac{1}{1 + [b + 2 - a]yv_5}$$

etc., and the required result is established.

Exercise 29. The mean of the curve $y = x^{n-1}e^{-x}$ (Fig. 57, p. 468) between 0 and ∞ is at n, its mode (maximum) is at $n - 1$, its second moment coefficient about the y-axis is $n(n + 1)$, and its standard deviation is $n^{\frac{1}{2}}$.

Exercise 30. The mean of the curve $y = x^{m-1}(1 - x)^{n-1}$ lying between 0 and 1 is at $m/(m + n)$, its mode is at $(m - 1)/(m + n - 2)$, and its standard deviation is $(mn)^{\frac{1}{2}}/(m + n)(m + n + 1)^{\frac{1}{2}}$.

Exercise 31. Show by Eq. 19 that if the area under the curve

$$y \, dz = C(1 + z^2)^{-\frac{1}{2}n} \, dz$$

from $-\infty$ to $+\infty$ is unity, i.e., if the curve is "normalized," C must have the value $1/B[\frac{1}{2}(n - 1), \frac{1}{2}]$. This curve is the distribution of Student's z, n being the sample size.

Exercise 32. Prove that the standard deviation of the distribution of Student's z is $1/\sqrt{(n - 3)}$. (Student's distribution of z is the curve used in the preceding exercise.) What happens to the standard deviation of this curve if n is 3 or less?

Exercise 33. Helmert's curve for the distribution of the standard deviation s in samples of n drawn from a normal universe is

$$y \, ds = C s^{n-2} e^{-ns^2/2\sigma^2} \, ds \qquad \text{[P. 511]}$$

If the area under the curve from $s = 0$ to $s = \infty$ is unity, the value of C must be $\dfrac{n^{\frac{1}{2}(n-1)}}{\Gamma[\frac{1}{2}(n - 1)]2^{\frac{1}{2}(n-3)}\sigma^{n-1}}$, which is the normalizing factor.

HINT. In this exercise and the next nine it will be found convenient to use the change of variable $ns^2/2\sigma^2 = x$ and $2\,ds/s = dx/x$, and to write Eq. 2 in the form

$$\Gamma_x(n) = \int_0^x x^n e^{-x} \frac{dx}{x}$$

Exercise 34. Show that if $P(s)$ is defined by the equation

$$P(s) = \frac{\displaystyle\int_s^\infty \left(\frac{s}{\sigma}\right)^{n-2} e^{-ns^2/2\sigma^2}\,ds}{\displaystyle\int_0^\infty \left(\frac{s}{\sigma}\right)^{n-2} e^{-ns^2/2\sigma^2}\,ds}$$

then

$$P(s) = 1 - \frac{\Gamma_x[\frac{1}{2}(n-1)]}{\Gamma[\frac{1}{2}(n-1)]}$$

The integral $P(s)$ is the area under the distribution of s from the abscissa s to infinity, and the evaluation of $P(s)$ is here seen to depend on an incomplete gamma function. See Exercise 42 for the identification of $P(s)$ with $P(\chi)$, which is the corresponding integral under the distribution of χ^2.

Exercise 35. Show that the mean of the curve in the preceding exercise is

$$Es = \sigma \sqrt{\frac{2\pi}{n}} \frac{1}{B[\frac{1}{2}(n-1), \frac{1}{2}]}$$

The symbol E denotes expectation or mathematical average. Es is the theoretical average standard deviation in samples of n drawn from a normal universe of standard deviation σ.

Exercise 36. Show that for Helmert's distribution of s (Ex. 33)

$$Es^2 = \int_0^\infty s^2 y\,ds = \frac{n-1}{n}\sigma^2$$

Es^2 is the theoretical average of the square of the standard deviation in samples of n drawn from a normal universe or any other universe of standard deviation σ.

Exercise 37. Helmert's curve for the distribution of the root-mean-square error δ in samples of n drawn from a normal universe is

$$y\,d\delta = C\left(\frac{\delta}{\sigma}\right)^{n-1} e^{-n\delta^2/2\sigma^2}\,d\delta \qquad\qquad \text{[P. 507]}$$

If the area under the curve from $\delta = 0$ to $\delta = \infty$ is unity, the value of C must be $\dfrac{n^{\frac{1}{2}n}}{\Gamma(\frac{1}{2}n)2^{\frac{1}{2}(n-2)}\sigma}$, which is the normalizing factor.

Exercise 38. For the curve in the preceding exercise show that

$$E\delta = \int_0^\infty \delta y \, d\delta = \sigma \sqrt{\frac{2\pi}{n}} \frac{1}{B(\frac{1}{2}n, \frac{1}{2})}$$

$$E\delta^2 = \int_0^\infty \delta^2 y \, d\delta = \sigma^2$$

$E\delta$ is the theoretical average root-mean-square error in samples of n drawn from a normal universe, and $E\delta^2$ is the theoretical average mean square error for a normal universe or any other universe.

Exercise 39. Without actually performing the integrations or reducing the integrals to gamma functions, but rather by making use of the results in Exercises 36 and 38, show that the mean of the distribution of s^2 must be $\dfrac{n-1}{n}\sigma^2$ and that the mean of the distribution of δ^2 must be σ^2.

Exercise 40. When there are k degrees of freedom the distribution of χ^2 is

$$y \, d\chi^2 = C(\chi^2)^{\frac{1}{2}(k-2)}e^{-\frac{1}{2}\chi^2} \, d\chi^2$$

If the area under the curve from $\chi^2 = 0$ to $\chi^2 = \infty$ is to be unity, show that C must have the value $1/\Gamma(\frac{1}{2}k)2^{\frac{1}{2}k}$.

Exercise 41. The mean of the distribution of χ^2 is at k, and its standard deviation is $\sqrt{2k}$.

Exercise 42. With $P(s)$ defined as in Exercise 34 and with

$$P(\chi) = \frac{1}{\Gamma(\frac{1}{2}k)2^{\frac{1}{2}k}} \int_{\chi^2}^\infty (\chi^2)^{\frac{1}{2}(k-2)}e^{-\frac{1}{2}\chi^2} \, d\chi^2$$

show that if $ns^2/\sigma^2 = \chi^2$ and $n - 1 = k$, then $P(s) = P(\chi)$. The results of this example show that instead of finding $P(s)$ from the *Tables of the Incomplete Gamma Function*, one may find $P(s)$ from tables of χ^2 which give $P(\chi)$ as a function of χ^2 and k.

Exercise 43. Derive these series for the gamma function:

$$\int_0^t x^{n-1}e^{-x} \, dx = \sum_{r=0}^\infty (-1)^r \frac{t^{r+n}}{r!(r+n)}$$

$$= e^{-t} \sum_{r=0}^\infty \frac{t^{r+n}}{n(n+1)\cdots(r+n)}$$

The first comes by expanding e^{-x} and integrating. The second comes by integrating by parts, with repetition.

Exercise 44.

$$\ln \Gamma(n + 1) = (n + \tfrac{1}{2}) \ln n - n + \ln \sqrt{2\pi}$$

$$+ \int_0^\infty \left\{ \frac{1}{e^t - 1} - \frac{1}{t} + \frac{1}{2} \right\} e^{-nt} \frac{dt}{t}$$

Exercise 45. Let x and y be two positive variates whose cumulative distribution functions are the incomplete gamma functions $\Gamma_x(m)$ and

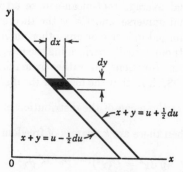

FIG. 63. Showing the element of area in the calculation of the distribution of $x + y$. The space between the u-lines is filled up by integrating y from 0 to u.

$\Gamma_y(n)$. Show that the probability that their sum $x + y$ lies in the interval $u \pm \tfrac{1}{2} du$ is

$$P(u)\, du = \frac{1}{\Gamma(m + n)} u^{m+n-1} e^{-u}\, du$$

wherefore the sum $x + y$ has the cumulative distribution $\Gamma_{x+y}(m + n)$, being the same form as the distributions of x and y separately. Thus (e.g.) the sum of two chi-square values with m and n degrees of freedom also has the chi-square distribution but with $m + n$ degrees of freedom. Extension to three or more chi-square values is obvious.

Solution

Let

$$u = x + y$$

Then u is the variate whose distribution is sought. The area of the blackened element is $dy\, dx$ (base by altitude), but it is also $dy\, du$ because $dx = du$ along a horizontal line. Some students may prefer more formality and set

$$\left. \begin{array}{r} u = x + y \\ y = y \end{array} \right\}$$

whereupon

$$dx\, dy = J\left(\frac{x, y}{u, y}\right) du\, dy = \begin{vmatrix} 1 & -1 \\ 0 & 1 \end{vmatrix} du\, dy = du\, dy$$

which is the same. Anyhow, the distribution of $x + y$ is obtained by writing down the probability of a sample point falling into the blackened element and then integrating this probability throughout the gap between the two parallel 45° lines

$$x + y = u \pm \tfrac{1}{2}\, du \quad \text{(Shown in Fig. 63)}$$

The gap is filled by integrating y from 0 to u (or v or y/u from 0 to 1 near the end of the mathematical steps which follow). First of all, it must be observed that if the cumulative distribution functions of x and y are the incomplete gamma functions $\Gamma_x(m)$ and $\Gamma_y(n)$, then their distributions must be of the form of the integrand of the gamma function depicted in Fig. 57, wherefore the joint probability that the first variate lies in the interval $x \pm \tfrac{1}{2}\, dx$ and the latter in the interval $y \pm \tfrac{1}{2}\, dy$ must be

$$P(x, y)\, dx\, dy = \frac{1}{\Gamma(m)}\, x^{m-1} e^{-x}\, dx\, \frac{1}{\Gamma(n)}\, y^{n-1} e^{-y}\, dy$$

Then

$$P(u)\, du = \int_{y=0}^{u} P(x, y)\, dx\, dy$$

$$= \frac{1}{\Gamma(m)\,\Gamma(n)} \int_{y=0}^{u} x^{m-1} e^{-x} y^{n-1} e^{-y}\, dx\, dy$$

$$= \frac{1}{\Gamma(m)\,\Gamma(n)} \int_{y=0}^{u} (u-y)^{m-1} e^{y-u} y^{n-1} e^{-y}\, dy\, du$$

$$= \frac{e^{-u}\, du}{\Gamma(m)\,\Gamma(n)} \int_{y=0}^{u} (u-y)^{m-1} y^{n-1}\, dy \qquad \left[\text{Put } v = \frac{y}{u} \right]$$

$$= \frac{1}{\Gamma(m)\,\Gamma(n)}\, u^{m+n-1} e^{-u}\, du \int_{v=0}^{1} (1-v)^{m-1} v^{n-1}\, dv$$

$$= \frac{1}{\Gamma(m)\,\Gamma(n)}\, u^{m+n-1} e^{-u}\, du\, B(m, n)$$

$$= \frac{1}{\Gamma(m+n)}\, u^{m+n-1} e^{-u}\, du \qquad\qquad Q.E.D.$$

Exercise 46. In the kinetic theory of gases let $w(r)\, dr$ be the probability that the nearest neighboring molecule lies within the spherical shell $r \pm \tfrac{1}{2}\, dr$.

a. Find $w(r)$.

Answer:
$$w(r)\, dr = \left[1 - \int_0^r w(r)\, dr \right] 4\pi r^2 n\, dr$$

n being the number of particles per unit volume, supposed constant. Differentiate and get

$$\frac{d}{dr}\, \frac{w(r)}{4\pi r^2 n} = -w(r)$$

whence

$$w(r) = 4\pi r^2 n e^{-4\pi n r^3/3}$$

b. Find the average distance between molecules.

Answer:

$$Er = \int_0^\infty r\,w(r)\,dr = \frac{\Gamma(\frac{4}{3})}{(\frac{4}{3}\pi n)^{\frac{1}{3}}} = \frac{0.55396}{n^{\frac{1}{3}}}$$

Reference: S. Chandrasekhar, "Stochastic problems in physics and astronomy," *Rev. Modern Physics*, vol. 15, 1943: pp. 1–89. This article contains a complete treatment of the random walk.

CHAPTER 15. DISTRIBUTION OF THE VARIANCE IN SAMPLES FROM A NORMAL UNIVERSE [1]

The subject of this paper was treated in full, for the first and only time, by Dr. Robert Smith, in the two editions of his *Harmonics* (Cambridge, 1749; London, 1759). The results are the same in both editions, but the improvements of the second edition add considerably to the learned obscurity in which the subject is involved. . . . Dr. Young certainly did not understand Smith's theory. . . . I have my doubts whether Robison had read more of Smith's theory than its results. For myself, I made out what ought to have been the theory from the formulae, and was then successful in mastering Smith's explanations. . . . Dr. Young pronounces Smith's work "a large and obscure volume. . . ." If Dr. Young had said that the work was largely obscure, he would have been correct. . . .—Augustus de Morgan, "On the beats of imperfect consonances," *Trans. Cambridge Phil. Soc.*, vol. x, 1854: pp. 129–41.

Introduction to normal distribution-theory. In the early chapters great dependence was placed on the "method of expected values." By these methods the means and variances of the probability-distributions of $\bar{x}$ and s^2 were found for any universe that possesses a distribution. It should be noted that these methods did not find the *shapes* of the distributions of $\bar{x}$ and s, except for samples from a 2-celled universe. An empirical approach was introduced on pages 439–44 in Chapter 13 to find the distributions of $\bar{x}$ and s, but the labor mounts rapidly with increasing sample-size and the number of cells in the universe and discourages further pursuit along these lines. Moreover, the distributions of $\bar{x}$ and s that were obtained were specific, i.e., they did not contain n and hence do not permit generalizations.

Satisfactory approximations for the probability-distributions of samples from a discrete universe can often be obtained by using a continuous curve fitted approximately to the discrete points of the universe. Calculations are thereby often simplified. Sometimes the integrations for use of the characteristic function (preceding chapter) can be performed, and the sample-size (n) retained as a parameter whereby it is possible to see what happens to the shapes of the distributions as n increases or decreases. A further assumption will be made in this chapter, viz., that the approximating curve is normal, and we shall be content to find the distribution of the mean and variance of samples. As data must always be recorded in discrete numbers, even for continuous phenomena (if any exist), no real universe obtained in practice can be continuous; hence none can be normal. It is a fact, though, that many of the results

[1] Certain parts of this chapter are borrowed heavily from Deming and Birge's "Statistical Theory of Errors," *Rev. Modern Physics*, vol. 6, 1934: pp. 119–61; later published, with supplements, by the Graduate School, Department of Agriculture, Washington.

to be derived here will have wide validity, and the student should be reminded that in Chapters 4, 5, and 10 he derived a number of results that are valid for any universe possessing 2d and 4th moments: these results will reappear here.

The statistician must continually be on guard lest he assume too much generality for normal theory. Unfortunately it is not always easy to see what the effect of nonnormality will be. Moreover, there is the necessity for being specific about any departure from normality: departures from normality can take place in an infinity of ways; hence to deal effectively with the problem of nonnormality it is necessary to deal with general types of departures, as has been done in Chapters 4, 5, and 10. The main thing to keep in mind is that although "tests of significance" based on normal theory may still indicate significance, one should not be too boastful about the exactness of any probabilities involved. Thus, the "5 percent level of t" (next chapter) is 5 percent only for a normal universe, but not for the universes actually met in practice, and the "1 percent level of t" probably in practice departs still further away from the normal 1 percent level, yet perhaps not far enough to affect administrative decisions. One great advantage of the use of 3-sigma limits is that they encompass practically all (99 percent or better) of almost any universe, whereas the probabilities associated with 2-sigma limits depend more critically on the shape of the distribution.

Remark 1. The normal curve admits the occurrence of a small but finite proportion of very large departures from the mean. In practice, however, there are always limits beyond which actual measurements can not go. Thus a galvanometer reading, barring blunders, can not fall beyond the scale. The outside diameter of a piece-part turned on a lathe can not exceed the diameter of the stock. The distribution of weights of marketable potatoes has a lower limit of something like 1 ounce and an upper limit of some amount beyond which a single potato simply does not grow, anywhere. But in spite of the fact that the normal curve does not have a definite cut-off like the distributions just described, the results to be derived from it are nevertheless useful with a few exceptions which will be noted from time to time.

Remark 2. In this and the following chapters on normal distribution theory, the sampling will be done with replacement. One reason is that the applications of distribution-theory are mostly in analysis of causes, wherein the formulas for samples from an infinite universe provide the correct theory, as was learned in Chapter 7.

Remark 3. In applying statistical theory, the main consideration is not what the shape of the universe is, but whether there is any universe at all. No universe can be assumed, nor the statistical theory of this book applied, unless the observations show statistical control. In this state the samples when cumulated over a suitable interval of time give a distribution of a particular shape, and this shape is reproduced hour after hour, day after day, so long as the process remains in statistical control—i.e., exhibits the properties of randomness. In a state of control, n observations may be regarded as a sample from the universe of whatever shape it is. A big enough sample, or enough small samples, enables the statistician to make meaningful and useful predictions *about future samples*. This is as much as statistical theory can do.

In this chapter, statistical control will be assumed. Very often the experimenter, instead of rushing in to apply the theory of this chapter, should be more concerned about attaining statistical control and asking himself whether any predictions at all (the only purpose of his experiment), by statistical theory or otherwise, can be made.

Remark 4. The students of a class may wish to perform some experimental sampling with a normal bowl. It is sufficient to take 200 poker chips [2] and on each to write a number, the proportions being as shown in the table below. It is important to have a smooth bowl, big enough to permit thorough mixing, and to perform the drawing with care so that no physical peculiarity of any chip has any effect on its chance of being drawn.

Mark	Number
0	48
$\pm 0.6\,\sigma$	40
$\pm 1.2\,\sigma$	23
$\pm 1.8\,\sigma$	10
$\pm 2.4\,\sigma$	2
$\pm 3.0\,\sigma$	1

Definitions of errors and residuals. The following definitions will be needed, some of which have already appeared. Fig. 64 further illustrates the relations between the various symbols. The symbols in the left-

Symbol	Definition	Name
ϵ_i	$= x_i - \mu = v_i - u$	Error in x_i
v_i	$= x_i - \bar{x}$	Residual in x_i
u	$= \bar{x} - \mu$	Error in the mean, $\bar{x}$
δ^2	$= \dfrac{1}{n} \Sigma\, (x_i - \mu)^2$	
	$= \dfrac{1}{n} \Sigma\, \epsilon_i{}^2$	
	$= s^2 + u^2$	Mean square error
s^2	$= \dfrac{1}{n} \Sigma\, (x_i - \bar{x})^2$	
	$= \dfrac{1}{n} \Sigma\, v_i{}^2$	Variance or mean square residual

hand column are all random variables, arising from the fact that x_i is a random variable, being the label on a chip drawn from a bowl. The n drawings $x_1, x_2, \cdots, x_n$ constitute the sample.

[2] Professor Holbrook Working uses metal-ringed paper tags, which can usually be purchased cheap at "dime stores."

The distribution of the root-mean-square error. The algebraic form of the normal curve is

$$y \, d\epsilon = \frac{1}{\sigma\sqrt{2\pi}} e^{-\epsilon^2/2\sigma^2} \, d\epsilon \tag{1}$$

Here, as elsewhere in this part of the book, probability curves will be written in differential form, and y will be used indiscriminately for the ordinates of all of them. The differential $d\epsilon$ specifies what sort of frequency curve y is the ordinate of. The entire expression, differentials and all,

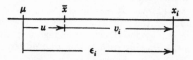

FIG. 64. An observation falls at x_i, and the average of a sample of n such observations falls at $\bar{x}$. The figure shows the relations between the error ϵ_i, the residual v_i, and the error u of the mean $\bar{x}$ of the sample. Here $\mu < \bar{x} < x_i$, hence ϵ_i, v_i, and u are positive, as the arrows indicate.

then represents the probability of an error lying in the interval $\epsilon \pm \frac{1}{2} d\epsilon$. The total area under the curve from $\epsilon = -\infty$ to $\epsilon = +\infty$ is unity, which means that an error *must* lie somewhere.

Suppose that a sample of n observations is drawn from the normal universe. The errors (ϵ) are independent random samples and it follows from Eq. 1 that if n intervals be named, such as

$$\left.\begin{array}{l} \epsilon_1 \pm \frac{1}{2} d\epsilon_1 \\ \epsilon_2 \pm \frac{1}{2} d\epsilon_2 \\ \quad \cdot \\ \quad \cdot \\ \quad \cdot \\ \epsilon_n \pm \frac{1}{2} d\epsilon_n \end{array}\right\} \quad \begin{array}{l} [\epsilon_1, \epsilon_2, \cdots, \epsilon_n \text{ may all be dif-} \\ \text{ferent, some positive and} \\ \text{some negative; likewise } d\epsilon_1, \\ d\epsilon_2, \text{ etc., may all be different} \\ \text{in magnitude]} \end{array}$$

then the probability that the n errors in the sample fall into these intervals is

$$y \, d\epsilon_1 \, d\epsilon_2 \cdots d\epsilon_n = \frac{1}{(\sigma\sqrt{2\pi})^n} e^{-(\epsilon_1^2 + \epsilon_2^2 + \ldots + \epsilon_n^2)/2\sigma^2} \, d\epsilon_1 \, d\epsilon_2 \cdots d\epsilon_n$$

$$= \frac{1}{(\sigma\sqrt{2\pi})^n} e^{-n\delta^2/2\sigma^2} \, d\epsilon_1 \, d\epsilon_2 \cdots d\epsilon_n \tag{2}$$

Here, as in earlier chapters, the mean square error of the sample is

$$\delta^2 = \frac{1}{n}(\epsilon_1^2 + \epsilon_2^2 + \cdots + \epsilon_n^2) \quad \text{[See also p. 332]} \tag{3}$$

It is δ^2 whose distribution is to be found.

To derive the distribution of δ or of δ^2 the procedure will be to find the proportion of samples whose mean square error lies anywhere in the interval $(\delta \pm \frac{1}{2}\delta)^2$. First it is to be noted that errors equal to ϵ_1, ϵ_2, $\cdots$, ϵ_n produce a certain mean square error δ^2. But other sets of errors may also produce the same mean square error. In fact, any set of errors that correspond to a point falling inside the spherical shell of radii $(\delta \pm \frac{1}{2} d\delta) \sqrt{n}$ will have a mean square error differing from δ^2 by only a

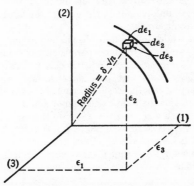

Fig. 65. Showing the three errors ϵ_1, ϵ_2, and ϵ_3 laid off along rectangular axes. The point ϵ_1, ϵ_2, ϵ_3 is at a distance (radius) $\delta\sqrt{3}$ from the origin by definition of δ. A parallelepiped having edges $d\epsilon_1$, $d\epsilon_2$, and $d\epsilon_3$ is erected about the point. The right-hand side of Eq. 2 is the probability that the errors in a sample will fall into this parallelepiped. A spherical shell of radius $(\delta \pm \frac{1}{2} d\delta)\sqrt{3}$ is erected with the origin as center. Eq. 11 gives the volume of this shell. Any set of errors which corresponds to a point lying within this shell has the same value of δ (to within the amount $\frac{1}{2} d\delta$) as the point shown in the figure, and herein lies the key to the procedure for finding the distribution of δ, given in Eq. 12.

small amount; specifically, their mean square error must fall between the values $n(\delta - \frac{1}{2} d\delta)^2$ and $n(\delta + \frac{1}{2} d\delta)^2$. This can be seen by writing the equation of the shell either in the form

$$n(\delta - \tfrac{1}{2} d\delta)^2 < \epsilon_1{}^2 + \epsilon_2{}^2 + \cdots + \epsilon_n{}^2 < n(\delta + \tfrac{1}{2} d\delta)^2 \qquad (4)$$

or its equivalent

$$\epsilon_1{}^2 + \epsilon_2{}^2 + \cdots + \epsilon_n{}^2 = n(\delta \pm \tfrac{1}{2} d\delta)^2 \qquad (5)$$

The probability that the n errors in a sample of n will fall into the particular parallelepiped of edges $d\epsilon_1$, $d\epsilon_2$, $\cdots$, $d\epsilon_n$ (illustrated in three dimensions in Fig. 65) is precisely the probability written as Eq. 2. It follows by summation that the probability that n errors will fall somewhere within the spherical shell of Eq. 4 or 5 is the integral of Eq. 2 taken throughout this shell.

Remark 1. The student whose calculus is shaky need not be frightened by the principles involved in this integration, even though he is unable to perform the calculations by calculus. The principles are more important than the actual manipulation of integral signs, which, after all, are merely short-cuts. Fortunately, the principles are simple. For example, to approximate the required integral of Eq. 2 one might proceed by the following steps: *i.* fill the spherical shell with parallelepipeds; *ii.* evaluate the right-hand side of Eq. 2 for every parallelepiped; *iii.* add up the results. The sum so obtained is the integral required, or an approximation thereto. The smaller the parallelepipeds and the more of them, the better the approximation—also, unfortunately, the greater the labor. By calculus we easily obtain the result *exactly*.

Remark 2. The reader should note that in performing Step ii above, the only part of Eq. 2 that varies from one parallelepiped to another is the edges $d\epsilon_1$, $d\epsilon_2$, $\cdots$, $d\epsilon_n$. These edges need not be equal in any one parallelepiped, and they may differ from one parallelepiped to another. The edges are wholly arbitrary in length, except of course that they are presumably very short. The rest of the right-hand side of Eq. 2 is to be considered a constant because the sum $\epsilon_1^2 + \epsilon_2^2 + \cdots \epsilon_n^2$ in the exponent is constrained to lie in the narrow interval or shell $n(\delta \pm \frac{1}{2} d\delta)^2$ of Eq. 5.

Remark 3. For assistance we may focus attention for a moment on Fig. 65, which is drawn for three dimensions, corresponding to a sample-size $n = 3$. By recalling the rectangular equation of a sphere of radius r, namely,

$$x^2 + y^2 + z^2 = r^2 \tag{6}$$

it can be seen that if δ be held constant and if the three errors ϵ_1, ϵ_2, ϵ_3 be allowed to vary in such manner that

$$\epsilon_1^2 + \epsilon_2^2 + \epsilon_3^2 = 3\delta^2 \tag{7}$$

then the three errors in one sample after another will all lie on a sphere of radius $\delta\sqrt{3}$. Their sum of squares will be constantly equal to $3\delta^2$, and their mean square will be δ^2. If, instead, the three errors are constrained to satisfy the inequalities

$$3(\delta - \tfrac{1}{2} d\delta)^2 < \epsilon_1^2 + \epsilon_2^2 + \epsilon_3^2 < 3(\delta + \tfrac{1}{2} d\delta)^2 \tag{8}$$

then the errors lie within the spherical shell of radii $(\delta \pm \frac{1}{2} d\delta)\sqrt{3}$.

Remark 4. Let dV_1, dV_2, dV_3, $\cdots$ be the volumes of the parallelepipeds that fill the spherical shell. Then the calculations described in Remark 1 lead to the following sum:

$$\frac{1}{(\sigma\sqrt{2\pi})^n} e^{-n\delta^2/2\sigma^2} (dV_1 + dV_2 + dV_3 + \cdots)$$

which is simply Eq. 2 written for one parallelepiped after another, and summed. The ellipsis ($\cdots$) indicates that the entire shell is to be filled. Fortunately, as stated in Remark 2, the multiplier outside is constant, and the parenthesis is merely the volume of the spherical shell (8). In three dimensions this volume is $4\pi r^2 dr$, as one will recall from solid geometry, this volume being merely the area of a sphere of radius r, multiplied by the thickness dr of the shell. Here the radius r is $\delta\sqrt{n}$, n being 3. The integration performed in three dimensions thus gives

$$y \, d\delta = \frac{1}{(\sigma\sqrt{2\pi})^n} e^{-n\delta^2/2\sigma^2} 4\pi r^2 \, dr \quad (r = \delta\sqrt{n})$$

$$= \frac{2n^{\frac{3}{2}n}}{\sqrt{2\pi}\sigma} \left(\frac{\delta}{\sigma}\right)^{n-1} e^{-n\delta^2/2\sigma^2} \, d\delta \tag{9}$$

in which n is to be set equal to 3. This is the required distribution of δ in three dimensions.

To perform the required integration in n dimensions it is only necessary to keep in mind the steps of Remark 1 and the scholium of Remark 2, whereby the exponential on the right-hand side of Eq. 2 differs only negligibly from $e^{-n\delta^2/2\sigma^2}$, while the n errors wander anywhere within the spherical shell defined by Eq. 4 or 5. Thus, the integral of the right-hand side of Eq. 2 taken throughout this spherical shell must be merely

$$\frac{1}{(\sigma\sqrt{2\pi})^n} e^{-n\delta^2/2\sigma^2} \times \text{the volume of the spherical shell} \tag{10}$$

But this is precisely the probability that n errors will have their mean square between $(\delta - \frac{1}{2} \, d\delta)^2$ and $(\delta + \frac{1}{2} \, d\delta)^2$, or their root-mean-square between $\delta - \frac{1}{2} \, d\delta$ and $\delta + \frac{1}{2} \, d\delta$. Moreover, the volume of the spherical shell of radius $r \pm \frac{1}{2} \, dr$ is known from Exercise 23 on p. 491 to be

$$dV = \frac{2\pi^{\frac{1}{2}n}}{\Gamma(\frac{1}{2}n)} r^{n-1} \, dr \tag{11}$$

Here r is to be replaced by $\delta\sqrt{n}$, and it follows that the distribution of the root-mean-square error must be

$$y \, d\delta = \frac{1}{(\sigma\sqrt{2\pi})^n} e^{-n\delta^2/2\sigma^2} \frac{2\pi^{\frac{1}{2}n}}{\Gamma(\frac{1}{2}n)} r^{n-1} \, dr$$

$$= \frac{n^{\frac{1}{2}n}}{\Gamma(\frac{1}{2}n)2^{\frac{1}{2}(n-2)}\sigma} \left(\frac{\delta}{\sigma}\right)^{n-1} e^{-n\delta^2/2\sigma^2} \, d\delta \tag{12}$$

This is the distribution of δ in samples of n.

Remark 5. Note that when $n = 3$, $\Gamma(\frac{1}{2}n) = \frac{1}{2}\Gamma(\frac{1}{2}) = \frac{1}{2}\sqrt{\pi}$, and the result just given for samples of n reduces to Eq. 9 for samples of 3.

Remark 6. The distribution of δ for samples of n was first derived by Helmert [3] in 1875.

---

[3] F. R. Helmert, *Schlömilch's Zeitschrift für Math. and Physik*, vol. 20, 1875: pp. 300–3. Also, *ibid.*, vol. 21, 1876: pp. 192–218. Helmert's derivation is reproduced (with credit) by E. Czuber in his *Beobachtungsfehler* (Teubner, 1891), pp. 147–50.

Remark 7. It is not necessary to resort to knowledge of the volume of an n-dimensional sphere to complete the integration of Eq. 2 throughout the shell. By symmetry and appeal to the dimensions of the problem, it can be asserted that the volume of the shell in n dimensions must be $Cr^{n-1} dr$, wherein r is $\delta\sqrt{n}$ and C is a constant whose value must give unity for the probability that δ lies between 0 and ∞. Thus, the integration of Eq. 2 throughout the shell must produce

$$y\,d\delta = C\delta^{n-1}e^{-n\delta^2/2\sigma^2}\sqrt{n}\,d\delta \tag{13}$$

and C must satisfy the equation

$$C\int_0^\infty \delta^{n-1}e^{-n\delta^2/2\sigma^2}\,d\delta = 1 \tag{14}$$

As on page 496, one may put $n\delta^2/2\sigma^2 = t$ and $2\,d\delta/\delta = dt/t$, whereupon this integral reduces to a gamma function and gives

$$C\int_0^\infty \delta^{n-1}e^{-n\delta^2/2\sigma^2}\,d\delta = C\left(\frac{2\sigma^2}{n}\right)^{\frac12 n}\frac{1}{2}\int_0^\infty t^{\frac12 n}e^{-t}\frac{dt}{t} = 1$$

I.e.,

$$n^{-\frac12 n}2^{(n-2)/2}\sigma^n\,\Gamma(\tfrac12 n)C = 1 \tag{15}$$

whence

$$C = \frac{n^{\frac12 n}}{\Gamma(\tfrac12 n)2^{(n-2)/2}\sigma^n} \quad\text{[See Ex. 33, p. 495]} \tag{16}$$

This value of C used in Eq. 13 gives precisely the same result for the distribution of δ as was found in Eq. 12.

Distribution of the sample mean. As before we now write

$$\left.\begin{aligned}
\epsilon_1 &= v_1 - u \\
\epsilon_2 &= v_2 - u \\
&\;\;\vdots \\
\epsilon_n &= v_n - u = -v_1 - v_2 - \cdots - v_{n-1} - u
\end{aligned}\right\} \tag{17}$$

By squaring each member and adding, it is seen that

$$\Sigma\,\epsilon_i^2 = \Sigma\,v_i^2 + nu^2 \tag{18}$$

whence division through by n gives

$$\delta^2 = s^2 + u^2 \quad\text{[As on p. 332]} \tag{19}$$

The exponential in Eq. 2 may now be converted from errors (ϵ_i) to residuals (v_i), but $u = \bar{x} - \mu$ also now makes its appearance.

It remains to convert the product of differentials $d\epsilon_1 d\epsilon_2 \cdots d\epsilon_n$. This can be done either mathematically or geometrically. The latter method will detain us a shorter time, so we shall follow it out and defer the mathematical conversion (see Ex. 8 at the end of the chapter). It is only necessary to note that the n observations are balanced at $\bar{x}$, which is at a distance u from μ (Fig. 64). Now if x_n is moved to the right a short distance, $\bar{x}$ also moves to the right, but only $1/n$th as much. Thus, $du = (1/n) dv_n$, or $dv_n = n\, du$. But, interval for interval, $d\epsilon_1 = dv_1$, $d\epsilon_2 = dv_2$, $\cdots$, $d\epsilon_n = dv_n$; hence it follows that

$$d\epsilon_1 d\epsilon_2 \cdots d\epsilon_n \text{ is to be replaced by } n\, du\, dv_1\, dv_2 \cdots dv_{n-1} \qquad (20)$$

and the probability expressed by Eq. 2 can now be written as

$$y\, du\, dv_1\, dv_2 \cdots dv_{n-1} = \frac{\sqrt{n}}{\sigma\sqrt{2\pi}} e^{-nu^2/2\sigma^2}\, du$$

$$\times \frac{\sqrt{n}}{(\sigma\sqrt{2\pi})^{n-1}} e^{-ns^2/2\sigma^2}\, dv_1\, dv_2 \cdots dv_{n-1} \qquad (21)$$

which is to be interpreted as the probability that $n - 1$ of the residuals fall in the intervals $v_1 \pm \frac{1}{2} dv_1$, $v_2 \pm \frac{1}{2} dv_2$, $\cdots$, $v_{n-1} \pm \frac{1}{2} dv_{n-1}$, and that the errors in the mean $\bar{x}$ fall in the interval $u \pm \frac{1}{2} du$.

The two factors just written make a clean separation between u and the $n - 1$ residuals $v_1, v_2, \cdots, v_{n-1}$. The letter u is contained only in the left-hand factor; hence this factor gives the distribution of u, which can be written

$$y\, du = \frac{\sqrt{n}}{\sigma\sqrt{2\pi}} e^{-nu^2/2\sigma^2}\, du \qquad \begin{array}{l}\text{[This is the dis-}\\ \text{tribution of the}\\ \text{error } u \text{ in the}\\ \text{mean]}\end{array} \qquad (22)$$

Then, because $u = \bar{x} - \mu$ (Fig. 64, p. 504) and $du = d\bar{x}$, this equation is equivalent to

$$y\, d\bar{x} = \frac{\sqrt{n}}{\sigma\sqrt{2\pi}} e^{-n(\bar{x}-\mu)^2/2\sigma^2}\, d\bar{x} \qquad \begin{array}{l}\text{[This is the dis-}\\ \text{tribution of the}\\ \text{mean } \bar{x}]\end{array} \qquad (23)$$

Both these distributions are obviously normal curves of standard deviation $\sigma/\sqrt{n}$ because they are precisely in the form of Eq. 1 but with $\sigma/\sqrt{n}$ in place of σ. It is thus evident that for samples drawn with replacement from a normal universe the distribution of the means of samples is also normal and with standard deviation reduced by the factor $1/\sqrt{n}$. The reduction-factor $1/\sqrt{n}$ was derived early in Chapter 4 for samples from any universe that possesses a standard deviation. The reader should also recall assertions to the effect that the distribution of the mean $\bar{x}$ approaches the normal with increasing sample-size. Here, the universe is normal and the distribution of the mean is also exactly normal for all sample-sizes. Fig. 66 illustrates the distribution of $\bar{x}$ for $n = 6$.

The distribution of the standard deviation. In deriving the distribution of the root-mean-square error δ, each error (ϵ_i) was always measured from the mean μ of the universe, which remained fixed for sample after sample. But in practice we are more often concerned with *residuals* (v_i) and the root-mean-square residual s, called the standard deviation.

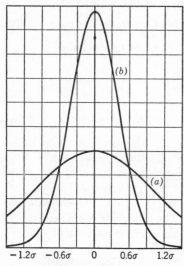

The residuals are always measured from the mean $\bar{x}$ of the sample, as depicted in Fig. 64. But in the next sample there will be a new mean, and the residuals in that sample will be measured from it. Since the mean $\bar{x}$ is continually on the move from sample to sample, the problem of finding the distribution of s presents some features not seen in the development of the distribution of δ. One is the ellipsoidal shell; another is Helmert's transformation (Exs. 9–11) which warps the ellipsoidal shell into a spherical shell. Another feature is the relation $dv_n = n\,du$, already derived geometrically, and to be derived also through the Jacobian (Ex. 8). A by-product was the distribution of the mean, as shown by Eqs. 22 and 23.

Fig. 66. Showing (a) a normal universe of individual values having standard deviation σ, and (b) the theoretical distribution of the means of samples of 6. The standard deviation of the latter curve is the smaller by the factor $1/\sqrt{6}$. Because the universe is normal, the distribution of means is also exactly normal for all sample-sizes.

We return now to the right-hand factor of Eq. 21, which gives the probability that the $n-1$ independent residuals in a sample of n will lie in the $n-1$ intervals $v_1 \pm \frac{1}{2}\,dv_1$, $v_2 \pm \frac{1}{2}\,dv_2$, $\cdots$, $v_{n-1} \pm \frac{1}{2}\,dv_{n-1}$. The variance of this sample is s^2 or $(1/n)(v_1{}^2 + v_2{}^2 + \cdots + v_{n-1}{}^2 + v_n{}^2)$, and it may be obvious that the same value of s^2 can be produced in many other ways, such as by decreasing v_1 and increasing v_2 to keep $v_1{}^2 + v_2{}^2$ the same. To find the distribution of s or of s^2 the procedure will be similar to that which was followed in finding the distribution of δ, viz., to find the proportion of samples that have variance anywhere within the interval $(s \pm \frac{1}{2}\,ds)^2$. This proportion is expressed by the volume of the shell defined by the inequalities

$$n(s - \tfrac{1}{2}\,ds)^2 < v_1{}^2 + v_2{}^2 + \cdots + v_{n-1}{}^2 + v_n{}^2 < n(s + \tfrac{1}{2}\,ds)^2 \quad (24)$$

This is not a spherical shell, but ellipsoidal, as can be seen by substituting

$-v_1 - v_2 - \cdots - v_{n-1}$ for v_n; the result is

$$n(s - \tfrac{1}{2} ds)^2 < 2(v_1{}^2 + v_2{}^2 + \cdots + v_{n-1}{}^2 + v_1 v_2 + v_1 v_3$$

plus all other cross-products up to $v_{n-2} v_{n-1}) < n(s + \tfrac{1}{2} ds)^2$ (25)

The presence of the cross-products makes this an ellipsoidal shell.

Remark 1. With three residuals (two dimensions) the last pair of inequalities would be

$$n(s - \tfrac{1}{2} ds)^2 < 2(v_1{}^2 + v_2{}^2 + v_1 v_2) < n(s + \tfrac{1}{2} ds)^2 \tag{26}$$

These are to be compared with the familiar equation

$$x^2 + y^2 + xy = r^2 \tag{27}$$

which is an ellipse whose axes make angles of 45° with the rectangular x- and y-axes. The presence of the cross-product term xy identifies it as an ellipse and not a circle.

The easiest way to obtain the volume of this shell is to recall the geometrical device mentioned in Remark 7 on page 508. Because of symmetry, and because we are now working in a space of $n - 1$ dimensions, the volume of the shell must be $Cs^{n-2} ds$, where C is a constant whose value must give unity for the probability that s lies between 0 and ∞. Thus, the integration of the right-hand factor of Eq. 21 must produce

$$y \, ds = Cs^{n-2} e^{-ns^2/2\sigma^2} \, ds \tag{28}$$

the constant already present having been absorbed into C. This will be the distribution of s as soon as C is determined. C must satisfy the equation

$$C \int_0^\infty s^{n-2} e^{-ns^2/2\sigma^2} \, ds = 1 \tag{29}$$

As on page 496 one may put $ns^2/2\sigma^2 = t$ and $2 \, ds/s = dt/t$, whereupon this integral reduces to a gamma function and gives

$$C = \frac{n^{\frac{1}{2}(n-1)}}{\Gamma(\tfrac{1}{2}n - \tfrac{1}{2})2^{\frac{1}{2}(n-3)}\sigma^{n-1}} \qquad \begin{array}{l}\text{[See Ex. 33 on} \\ \text{p. 495]}\end{array} \tag{30}$$

This value of C used in Eq. 28 gives

$$y \, ds = \frac{n^{\frac{1}{2}(n-1)}}{\Gamma(\tfrac{1}{2}n - \tfrac{1}{2})2^{\frac{1}{2}(n-3)}\sigma}\left(\frac{s}{\sigma}\right)^{n-2} e^{-ns^2/2\sigma^2} \, ds \tag{31}$$

for the distribution of s in samples of n.

Remark 2. This equation will often be referred to as "Helmert's distribution of s," after its discoverer.[4]

[4] F. R. Helmert, *Astronomische Nachrichtung*, vol. 88, No. 2096, 1876: p. 122. Helmert's derivation is reproduced (with credit) by E. Czuber in his *Beobachtungsfehler* (Teubner, 1891), pp. 159–63.

Curves showing Helmert's distribution of s as a function of n are shown in Fig. 67. They are decidedly skewed when n is small, but as n increases they lose their skewness and become normal about the abscissa

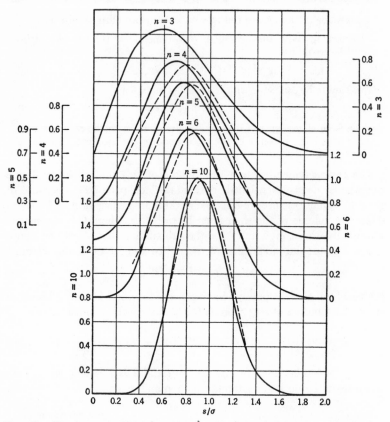

FIG. 67. Showing some Helmert curves for the theoretical frequency distribution of the standard deviation s in samples of n drawn from a normal universe of standard deviation σ (Eq. 31). The dashed curves are normal curves, each having the same mean and standard deviation as the corresponding Helmert curve. It will be observed that with increasing sample-size (n), the Helmert curve is better and better approximated by the normal curve. The mean and standard deviation of any of these curves are given by Eqs. 41, 53, and 64.

$s = \sigma$ with standard deviation approaching $\sigma/\sqrt{2n}$ (Ex. 2, p. 526). In Fig. 67 each solid curve is the graph of the Helmert equation for the sample-size marked, while the corresponding broken curve is a normal curve of standard deviation $\sigma/\sqrt{2n}$, so placed that its center (peak) comes at the mean Es of the Helmert curve. The approaching coincidence of the solid and broken curves with increasing n shows how the Helmert curves approach normality.

Remark 3. In theory, the Helmert curve runs to infinity (Fig. 67), although contact with the s-axis is high and large values of s are thus very improbable. It is the tails in the normal universe that give rise to the small but finite probabilities of large values of s. A universe with a definite cut-off in each tail, as any actual bowl of chips must have, will when sampled repeatedly give forth a distribution of s having also a definite cut-off. For an illustration see Tables 1, 2, and 3 of Chapter 13 (pp. 440–2).

The u, s probability surface. Instead of thinking about the u and s curves separately, it is instructive to think of them simultaneously in the form of the u, s probability surface. This surface can be derived from Eqs. 17; in fact, the work is all done. We need only multiply the equations for the u- and s-curves together to find the simultaneous distribution of u and s. This can be done because u and s are independent, a fact that is evident from Eq. 21, wherein the letter u occurs only in the first factor, which in turn is free of v_i. The multiplication gives

$$y \, du \, ds = \left[\frac{\sqrt{n}}{\sigma \sqrt{2\pi}} e^{-nu^2/2\sigma^2} \, du \right] \left[\frac{n^{\frac{1}{2}(n-1)}}{\Gamma\left(\dfrac{n-1}{2}\right) 2^{\frac{1}{2}(n-3)} \sigma} \left(\frac{s}{\sigma}\right)^{n-2} e^{-ns^2/2\sigma^2} \, ds \right]$$

(32)

which is the equation of the u, s probability surface. This equation gives the volume of an elemental prism whose base measures du one way and ds the other, this volume being the probability of the error in $\bar{x}$ lying in the interval $u \pm \frac{1}{2} \, du$ and of the standard deviation of the sample lying in the interval $s \pm \frac{1}{2} \, ds$.

To construct the surface, or to see how it would be done, give n a certain value, say 6; then with u held constant at first one value and then another, look up values of the first brace in tables of ordinates of the normal curve, for several values of u such as 0, $\pm.5\,\sigma$, $\pm\sigma$, $\pm1.5\,\sigma$, etc. Then compute or look up the second brace for several values of s such as 0, $.5\,\sigma$, σ, $1.5\,\sigma$, $2\,\sigma$. The differentials du and ds are of course to be canceled out, and values of y computed as the product of the two braces.

Wire models of the surface were made by a former pupil of mine, a Mr. Clarence G. Colcord of Washington, now deceased, who also performed most of the calculations therefor. Photographs of his models are shown in Fig. 68.

Fig. 68 should be studied carefully. Note first what happens when n is increased from 6 to 25; the surface is obviously much more concentrated over its limiting position ($u = 0$, $s = \sigma$) for the larger value of n, disclosing the gain in information accruing from large samples. Second, note that the u-curves (the curves running in the s-direction, so-called because u is constant along any one of them) are Helmert curves, and that the skewness so evident when $n = 6$ has practically disappeared when $n = 25$. Third, note that the s-curves are normal curves and hence symmetrical for all sizes of sample large and small, and that they are always centered over the axis of zero error, $u = 0$. The standard deviations of both sets of curves decrease by the factor $1/\sqrt{n}$ as n increases, thus accounting for the increased height and concentration of the surface for the larger value of n.

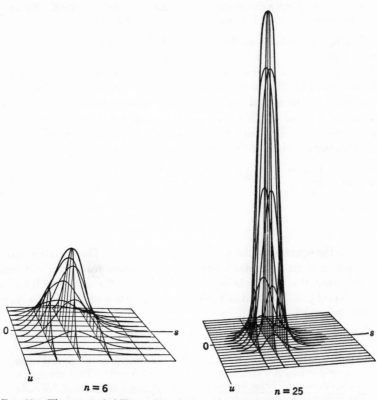

$n = 6$ $n = 25$

Fig. 68. The u, s probability surface, for $n = 6$ and $n = 25$. (Photographed from wire models made for the author by Clarence G. Colcord.)

The distribution of chi-square. By a simple transformation the distributions of both δ and s can be converted into the distribution of χ^2. Let

$$\chi^2 = \frac{n\delta^2}{\sigma^2} \quad \text{and} \quad n = k \tag{33}$$

Then

$$d\chi^2 = \frac{2n\delta}{\sigma^2}\,d\delta = \frac{2k\delta}{\sigma^2}\,d\delta \tag{34}$$

and by substitution, Eq. 12 for the distribution of δ is converted into the distribution of χ^2 which is

$$y \, d\chi^2 = \frac{1}{\Gamma(\frac{1}{2}k)2^{\frac{1}{2}k}} (\chi^2)^{\frac{1}{2}k-1} e^{-\frac{1}{2}\chi^2} \, d\chi^2 \tag{35}$$

k is the number of degrees of freedom (p. 541). In the distribution of δ, k is the same as the sample-size. It is easy to remember the exponent of χ^2 if one recalls that, dimensionally, χ occurs to the kth power, including the differential. By letting

$$\chi^2 = \frac{ns^2}{\sigma^2} \quad \text{and} \quad k = n - 1 \tag{36}$$

Helmert's distribution of s (Eq. 31) also goes into the same distribution of χ^2 but with $k = n - 1$, one degree of freedom having been lost because of Eqs. 17 connecting the residuals.

Many other important distributions can also be converted into the χ^2-distribution by proper transformations. Its properties, once learned, can be kept in mind for the interpretation of other distributions.

Remark 1. By reference to Chapter 14 the student will recognize the distribution of $\frac{1}{2}\chi^2$ to be a Type III curve. Thus, if $x = \frac{1}{2}\chi^2$, the distribution of χ^2 as given in Eq. 35 goes into the curve

$$y \, dx = \frac{1}{\Gamma(\frac{1}{2}k)} x^{\frac{1}{2}k-1} e^{-x} \, dx \qquad \text{[Pictured in Fig. 57, p. 468]}$$

which was used as the integrand of the gamma function. Thus, probabilities that are to be calculated as areas under curves convertible to the χ^2-distribution are easily converted also into incomplete gamma functions and evaluated through tables.

Remark 2. Eqs. 33 and 36, converting δ and s into χ^2, may be written as

$$\chi^2 = \frac{n\delta^2}{\sigma^2} = \left(\frac{\epsilon_1}{\sigma}\right)^2 + \left(\frac{\epsilon_2}{\sigma}\right)^2 + \cdots + \left(\frac{\epsilon_n}{\sigma}\right)^2 \tag{37}$$

and

$$\chi^2 = \frac{ns^2}{\sigma^2} = \left(\frac{v_1}{\sigma}\right)^2 + \left(\frac{v_2}{\sigma}\right)^2 + \cdots + \left(\frac{v_n}{\sigma}\right)^2 \tag{38}$$

Thus χ^2 is the sum of the squares of the standardized errors in one case, and the sum of the squares of the standardized residuals in the other, an error or residual being *standardized* when it is measured in units of σ. χ^2 is thus independent of the unit of measurement (pounds, ohms, number of stores or people, or any other unit).

Remark 3. In curve fitting, if χ^2 be put equal to the sum of the squares of the standardized residuals in both x- and y-coordinates, then χ^2 has the same distribution written above; cf. the author's *Statistical Adjustment of Data*, page 141.

Remark 4. Historical note. The distribution of χ^2 was first published by P. Pizzetti in an article entitled, "I fondamenti matematici per la critica dei resultati sperimentali," *Atti della Regia Universita di Genova*, vol. xi, 1892: pp. 113–333.

Figs. 69 and 70 show a number of χ^2-curves in two different forms, viz., as the distribution of χ^2 and the distribution of χ^2/k. In the second form the bulk of the curve stays on the paper. It can be proved that with increasing k both sets of curves become more and more normal. The mean and standard deviation of the χ^2-distribution have already been

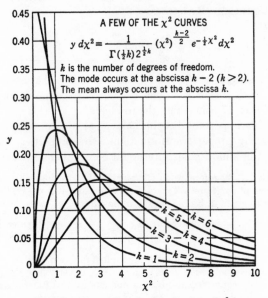

Fig. 69.   Illustrating the distribution of χ^2.

worked out; the mean is always at k, and the standard deviation is $\sqrt{2k}$ (Ex. 41, p. 497); hence with increasing k the curves of Fig. 69 flatten out as the mean of the curves moves to the right. In the distribution of χ^2/k, however, the mean is always at 1 for any value of k, and the standard deviation is $\sqrt{2/k}$, which diminishes with k as the curves of Fig. 70 illustrate.

Probability under the χ^2-curve is defined as

$$P(\chi) \quad \text{or} \quad P(\chi^2) = \frac{1}{\Gamma(\frac{1}{2}k)} \int_{x=\frac{1}{2}\chi^2}^{\infty} x^{\frac{1}{2}k-1} e^{-x} \, dx \tag{39}$$

which is the area under the far tail. The curves of Fig. 71 show pictorially the area $P(\chi^2)$, and they show that beyond a given abscissa χ^2, the (shaded) area $P(\chi^2)$ increases as k increases. Fisher in his *Statistical*

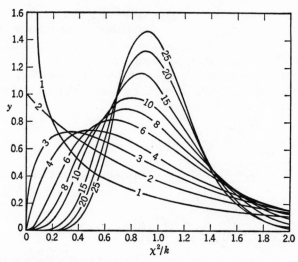

Fig. 70. Illustrating the distribution of χ^2/k. The equation of the curves is $y\,dx = \dfrac{m^m}{\Gamma(m)}\,x^{m-1}e^{-mx}\,dx$, where $m = \frac{1}{2}k$ and $x = \chi^2/k$. The mean of each of the curves is unity. The mode lies at $\dfrac{k-2}{k}$ if $k > 2$. The figure shows that as k increases, the bulk of the area is confined more and more to the vicinity of the mean; that is, χ^2 is confined more and more to a small range about k.

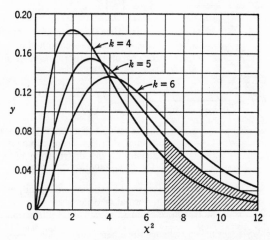

Fig. 71. The meaning of $P(\chi^2)$. The curves show the χ^2 distribution for $k = 4$, 5, 6. By definition, $P(\chi^2)$ for any given value of k is that fraction of the area lying under the curve beyond the designated abscissa χ^2 to infinity, as indicated by the shading beyond $\chi^2 = 7$. It is clear from the three curves that $P(\chi^2)$ increases with k.

Methods for Research Workers gave tables of χ^2 for which $P(\chi^2)$ takes the values 0.01, 0.02, 0.05, and other important probability "levels." His table has been widely copied and is well known. Kendall's chart (Fig. 72) will be found handy for many uses.

The mean and standard deviation of the δ- and s-curves. The following results will be recalled from the exercises at the end of the chapter on the gamma and beta functions.

$$E\delta = \frac{n^{\frac{1}{2}n}}{\Gamma(\frac{1}{2}n)2^{\frac{1}{2}(n-2)}\sigma} \int_0^\infty \delta \left(\frac{\delta}{\sigma}\right)^{n-1} e^{-n\delta^2/2\sigma^2}\, d\delta$$

$$= \sqrt{\frac{2\pi}{n}}\, \frac{1}{B(\frac{1}{2}n, \frac{1}{2})}\, \sigma \qquad \text{[This is the mean of the distribution of } \delta; \text{ see Ex. 38, p. 497]} \tag{40}$$

$$Es = \frac{n^{\frac{1}{2}(n-1)}}{\Gamma(\frac{1}{2}n - \frac{1}{2})2^{\frac{1}{2}(n-3)}\sigma} \int_0^\infty s \left(\frac{s}{\sigma}\right)^{n-2} e^{-ns^2/2\sigma^2}\, ds$$

$$= \sqrt{\frac{2\pi}{n}}\, \frac{1}{B(\frac{1}{2}n - \frac{1}{2}, \frac{1}{2})}\, \sigma$$

$$= c_2\, \sigma$$

[This is the mean of the distribution of s, the curves shown in Fig. 67; see Ex. 35, p. 496. The symbol c_2 as here defined is used in the statistical control of quality as will be seen later. Values of c_2 for various sample-sizes are shown on p. 570.] (41)

$$E\delta^2 = \frac{n^{\frac{1}{2}n}}{\Gamma(\frac{1}{2}n)2^{\frac{1}{2}(n-2)}\sigma} \int_0^\infty \delta^2 \left(\frac{\delta}{\sigma}\right)^{n-1} e^{-n\delta^2/2\sigma^2}\, d\delta$$

$$= \sigma^2 \qquad \text{[This is either the mean of the distribution of } \delta^2 \text{ or the second moment coefficient of the distribution of } \delta; \text{ see Ex. 38, p. 497]} \tag{42}$$

$$Es^2 = \frac{n^{\frac{1}{2}(n-1)}}{\Gamma(\frac{1}{2}n - \frac{1}{2})2^{\frac{1}{2}(n-3)}\sigma} \int_0^\infty s^2 \left(\frac{s}{\sigma}\right)^{n-2} e^{-ns^2/2\sigma^2}\, ds$$

$$= \frac{n-1}{n}\sigma^2 \qquad \text{[This is either the mean of the distribution of } s^2 \text{ or the second moment coefficient of the distribution of } s; \text{ see Ex. 36, p. 496]} \tag{43}$$

Remark 1. The last two results, as the student will recall, are true for samples drawn from any universe, normal or otherwise.

Remark 2. From the transformation $\chi^2 = ns^2/\sigma^2$ it follows that the mean value of ns^2/σ^2 is k or $n - 1$, whence the mean value of s^2 is $\frac{n-1}{n}\sigma^2$. Also, the standard deviation of the distribution of ns^2/σ^2 must be $\sqrt{2(n-1)}$, whence the standard deviation of the distribution of s^2 is $\sigma^2\sqrt{[2(n-1)/n^2]}$, another result for the table on page 530.

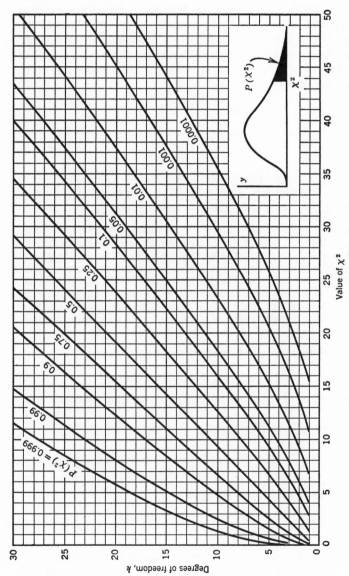

FIG. 72. The Kendall chart for the chi-test. [Reproduced, with the kind permission of Professor Kendall and his publishers, from Yule and Kendall's *Introduction to the Theory of Statistics* (Chas. Griffin, 1937) and also from Kendall's *The Advanced Theory of Statistics* (Chas. Griffin, 1948), vol. I, page 446.] The curve shown in the small inset has been added to remind the reader of the definition of $P(\chi^2)$.

Series for the means, standard deviations, and coefficients of variation of δ and s. The results thus far derived will be placed in the table on page 530, which will be useful for reference. Before making it up, we shall pause to derive two further results. It will be observed that the means of the δ- and s-curves involve beta functions, and it is desirable to express these functions in series so that they can be readily approximated. To arrive at the required series we shall make use of the asymptotic expansion

$$\log n! = \log \Gamma(n+1) = \log\sqrt{2\pi} + (n + \tfrac{1}{2})\log n - n \log e$$

$$+ \left\{ \frac{B_1}{1 \cdot 2n} - \frac{B_2}{3 \cdot 4n^3} + \frac{B_3}{5 \cdot 6n^5} - + \cdots \right\} \log e \quad (44)$$

which is known as the Stirling approximation, although it may also be regarded as an application of the Euler-Maclaurin sum.[5] The numbers B_1, B_2, B_3, B_4, $\cdots$ are the Bernoulli numbers $\frac{1}{6}$, $\frac{1}{30}$, $\frac{1}{42}$, $\frac{1}{30}$, etc.

Remark. This series is asymptotic. It eventually diverges if carried far enough, no matter how large n be.[6] However, an excellent approximation to $\log \Gamma(n+1)$ is obtained by using only the earliest terms of the series. The approximation improves in fact until a term is reached beyond which the next term is always bigger than the preceding term and the series begins to diverge. No matter where curtailed, the error in computing $\log \Gamma(n+1)$ is less than the first term neglected. It follows that the approximation obtained by using any given number of terms improves as n increases. For our purpose here the series need be carried only through the term in B_1.

By recalling that $\Gamma(n+1) = n\Gamma(n)$ and that therefore

$$\log \Gamma(n) = \log \Gamma(n+1) - \log n \quad (45)$$

it follows at once that from the Stirling approximation as written above

$$\log \Gamma(n) = \log \sqrt{2\pi} + (n - \tfrac{1}{2})\log n - n \log e$$

$$+ \left\{ \frac{B_1}{1 \cdot 2n} - + \text{ as before} \right\} \log e \quad (46)$$

In natural logarithms this simplifies to

$$\ln \Gamma(n) = \ln \sqrt{2\pi} + \left(n - \frac{1}{2}\right)\ln n - n + \left\{ \frac{B_1}{1 \cdot 2n} - + \text{ as before} \right\} \quad (47)$$

[5] Whittaker and Robinson, *Calculus of Observations* (Blackie & Son, 1932), Art. 67, p. 135. Whittaker and Watson, *Modern Analysis* (Cambridge), p. 128 in the edition of 1927.

[6] This property was first noticed by Bayes, *Phil. Trans. Royal Soc.*, 1763: pp. 269–71. Bayes's note is available in a publication entitled *Facsimile of Two Papers by Bayes* (The Graduate School, Department of Agriculture, 1941).

Then

$$\ln \Gamma\left(\frac{1}{2}n\right) = \ln \sqrt{2\pi} + \frac{1}{2}(n-1)\ln\frac{1}{2}n - \frac{1}{2}n + \frac{1}{6n} - \cdots$$

$$\ln \Gamma(\tfrac{1}{2}n - \tfrac{1}{2}) = \ln \sqrt{2\pi} + \tfrac{1}{2}(n-2)\ln(\tfrac{1}{2}n - \tfrac{1}{2}) - \tfrac{1}{2}(n-1)$$

$$+ \frac{1}{6(n-1)} - \cdots \qquad \text{[Replace } n \text{ by } n-1 \text{ in the preceding series]}$$

Therefore

$$\ln \frac{\Gamma(\tfrac{1}{2}n)}{\Gamma(\tfrac{1}{2}n - \tfrac{1}{2})}$$

$$= \frac{1}{2}(n-1)\ln\frac{1}{2}n - \frac{1}{2}(n-2)\ln\left(\frac{1}{2}n - \frac{1}{2}\right) - \frac{1}{2} - \frac{1}{6n^2} - \cdots$$

$$= \frac{1}{2}(n-1)\ln\frac{1}{2}n - \frac{1}{2}(n-2)\left[\ln\frac{1}{2}n + \ln\left(1 - \frac{1}{n}\right)\right] - \frac{1}{2} - \frac{1}{6n^2} - \cdots$$

$$= \frac{1}{2}\ln\frac{1}{2}n - \frac{1}{2}(n-2)\ln\left(1 - \frac{1}{n}\right) - \frac{1}{2} - \frac{1}{6n^2} - \cdots$$

$$= \frac{1}{2}\ln\frac{1}{2}n - \frac{1}{2}(n-2)\left(-\frac{1}{n} - \frac{1}{2n^2} - \frac{1}{3n^3} - \cdots\right) - \frac{1}{2} - \frac{1}{6n^2} - \cdots$$

$$= \ln \sqrt{\frac{n}{2}} - \frac{3}{4n} - \frac{1}{2n^2} - \cdots \qquad (48)$$

which gives

$$\ln \sqrt{\frac{2\pi}{n}} \frac{1}{B(\tfrac{1}{2}n - \tfrac{1}{2}, \tfrac{1}{2})} = \ln \sqrt{\frac{2}{n}} + \ln \frac{\Gamma(\tfrac{1}{2}n)}{\Gamma(\tfrac{1}{2}n - \tfrac{1}{2})}$$

$$= -\frac{3}{4n} - \frac{1}{2n^2} - \cdots \qquad (49)$$

wherefore

$$c_2 = \sqrt{\frac{2\pi}{n}} \frac{1}{B(\tfrac{1}{2}n - \tfrac{1}{2}, \tfrac{1}{2})} = e^{-\frac{3}{4n} - \frac{1}{2n^2} - \cdots}$$

$$= 1 - \left(\frac{3}{4n} + \frac{1}{2n^2}\right) + \frac{1}{2!}\left(\frac{3}{4n} + \frac{1}{2n^2}\right)^2 - \cdots$$

$$\left[\text{Recall } e^x = 1 + x + \frac{x^2}{2!} + \cdots\right]$$

$$= 1 - \frac{3}{4n} - \frac{7}{32n^2} - \cdots \qquad (50)$$

As an exercise the student may wish to show in a similar manner, or by replacing n by $n + 1$ in Eq. 48, that

$$\ln \frac{\Gamma(\tfrac{1}{2}n + \tfrac{1}{2})}{\Gamma(\tfrac{1}{2}n)} = \ln \sqrt{\frac{n}{2}} - \frac{1}{4n} + \frac{0}{4n^2} - \cdots$$

and consequently that

$$\sqrt{\frac{2\pi}{n}} \, \frac{1}{B(\tfrac{1}{2}n, \tfrac{1}{2})} = 1 - \frac{1}{4n} + \frac{1}{32n^2} - \cdots \tag{51}$$

It thus follows from Eqs. 40 and 41 that the means of the δ- and s-curves may be written

$$E\delta = \sigma \left(1 - \frac{1}{4n} + \frac{1}{32n^2} - \cdots \right) \tag{52}$$

and

$$Es = \sigma \left(1 - \frac{3}{4n} - \frac{7}{32n^2} - \cdots \right) \tag{53}$$

These equations not only tell us that the means of the δ- and s-curves approach σ in the limit as the sample-size (n) increases, but they also tell us *how rapidly the means approach their limits*, and what error is involved in writing $E\delta = \sigma$ or $Es = \sigma$ for some finite value of n.

Biased and unbiased estimates of σ. Suppose s/c_2 be calculated for each of m samples. Then by Eq. 41 if the universe is normal the theoretical value of the average of s/c_2 will be σ, wherefore s/c_2 is said to be an *unbiased* estimate of σ. Deming and Birge called it the *mean* estimate of σ because the factor c_2 determines the mean of the Helmert curve. The mean estimate is traditionally used in quality control. Values of c_2 are shown for various sample-sizes on page 570; these are strictly valid only for a normal universe.

Now suppose $ns^2/(n-1)$ is calculated for each of the m samples. Then for any universe, normal or otherwise, the theoretical value of the average of $ns^2/(n-1)$ is σ^2 itself (Ch. 10); hence $ns^2/(n-1)$ is said to be an *unbiased* estimate of σ^2. This is the Gauss estimate.

Peculiarly enough, although s/c_2 is an unbiased estimate of σ, the theoretical average value of $(s/c_2)^2$ is not σ^2; hence $(s/c_2)^2$ is a biased estimate of σ^2. Nevertheless $(s/c_2)^2$ is a *consistent* estimate of σ^2, the adjective introduced by Fisher to describe an estimate that is unbiased in the limit for large samples (reference on p. 143).

Similarly, although $ns^2/(n-1)$ is an unbiased estimate of σ^2, the theoretical average of $\sqrt{[ns^2/(n-1)]}$ is not σ; hence $\sqrt{[ns^2/(n-1)]}$ is a biased estimate of σ.

Nevertheless, $\sqrt{[ns^2/(n-1)]}$ is a *consistent* estimate of σ. The Gauss estimate of σ^2 possesses the advantage of being unbiased for any universe, whereas the factor $1/c_2$ for the mean estimate changes when the universe changes. Another advantage of the Gauss estimate is the ease with which the variances of several samples may be pooled to form an estimate of σ^2 (last chapter).

Exercise. Prove that $(s/c_2)^2$ is a consistent estimate of σ^2 for a normal universe, and that $\sqrt{[ns^2/(n-1)]}$ is a consistent estimate of σ for *any* universe.

The maximum likelihood estimate of σ. In theory, the standard deviation s in repeated random samples from a normal universe follows the Helmert curve. Suppose we ask: At a fixed position s on the horizontal scale, and for a given value of n, what value of σ will give the tallest ordinate? Or what value of σ is *most favorable* or *optimum* to the occurrence of the standard deviation s that was observed in a particular sample? The answer is found by differentiating y in the Helmert curve with respect to σ and setting this derivative equal to 0 to find what value of σ makes y a maximum. Thus,

$$y = C \frac{1}{\sigma^{n-1}} e^{-ns^2/2\sigma^2}$$

[From the Helmert distribution of s, Eq. 31, p. 511. s, not being a function of σ, is absorbed into the constant C] $\quad$ (54)

$$\frac{dy}{d\sigma} = y\left[-\frac{n-1}{\sigma} + \frac{ns^2}{\sigma^3}\right] \tag{55}$$

which vanishes if

$$\sigma^2 = \frac{n}{n-1}s^2 \tag{56}$$

This value of σ^2 is the *maximum likelihood estimate* corresponding to an observed standard deviation s in a particular sample. Like s, this estimate is of course a random variable. Peculiarly enough, the maximum likelihood estimate of σ^2 turns out to be identical with the unbiased estimate of σ^2 given earlier. This coincidence is not encountered when sampling from universes other than normal. Three Helmert curves, one with σ equal to the maximum likelihood value and the other two with σ just slightly above and below the maximum likelihood value, are shown in Fig. 73.

If the principle of maximum likelihood be applied to the u, s probability surface in order to find what pair of values of μ and σ maximize the probability of the observed $\bar{x}$ and s, the results are somewhat different. Thus,

$$y = \left[\frac{\sqrt{n}}{\sigma\sqrt{2\pi}} e^{-n(\bar{x}-\mu)^2/2\sigma^2} \right] \left[\frac{n^{\frac{1}{2}(n-1)}}{\Gamma\left(\dfrac{n-1}{2}\right) 2^{\frac{1}{2}(n-3)}\sigma} \left(\frac{s}{\sigma}\right)^{n-2} e^{-ns^2/2\sigma^2} \right] \quad (57)$$

Now treat this as a function of μ and σ, while $\bar{x}$ and s are constants.

$$\frac{\partial y}{\partial \mu} = y\frac{n}{\sigma^2}(\bar{x}-\mu) \quad (58)$$

$$\frac{\partial y}{\partial \sigma} = y\left\{\frac{n}{\sigma} - \frac{ns^2}{\sigma^3}\right\} \quad (59)$$

These derivatives vanish if

$$\left.\begin{array}{l} \mu = \bar{x} \\ \sigma = s \end{array}\right\} \quad (60)$$

Thus $\bar{x}$ and s are the maximum likelihood estimates of μ and σ when both are estimated together. As we know already, $\bar{x}$ is an unbiased estimate of μ, while s is slightly although consistently biased, as $Es = c_2\,\sigma$ and c_2 approaches unity as n increases (Eq. 50).

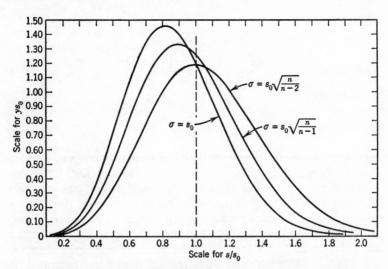

Fig. 73. Illustrating the meaning of likelihood in the Helmert curve, which is the distribution of s in samples of n from a normal universe. A sample of n is drawn, and its standard deviation is observed to be s_0. By placing $\sigma = s_0\sqrt{n/(n-1)}$ the ordinate at $s = s_0$ is made a maximum. Curves drawn with σ slightly smaller and slightly larger than optimum are seen to have lesser ordinates at $s = s_0$.

Remark. Other estimates can be made, but one further illustration will suffice. Let s be multiplied by a factor f such that if fs be computed for a large number of samples, half the values of fs would theoretically be less than σ, and half of them greater. fs is then the *median estimate* of σ. The factor f for any sample-size n satisfies the equation

$$\int_0^{1/f} y \, ds = \tfrac{1}{2} \tag{61}$$

wherein y is the ordinate of a Helmert curve (Eq. 31, p. 511). For purposes of calculation, this integral would be transformed into the incomplete gamma function

$$\frac{\Gamma_{n/2f^2}(\tfrac{1}{2}n - 1)}{\Gamma(\tfrac{1}{2}n - 1)} = \frac{1}{2} \tag{62}$$

whence $n/2f^2$ or χ^2, and hence f, can be found from the *Tables of the Incomplete Gamma Function* or from χ^2-tables.

Exercise 1. Let $h = 1/\sigma\sqrt{2}$ and $h' = 1/s\sqrt{2}$. h is commonly found in books on least squares and is called the *precision index*. Starting with Helmert's equation for s (Eq. 31, p. 511):

a. Show that the distribution of $h' = 1/s\sqrt{2}$ is

$$y \, dh' = \frac{n^{\frac{1}{2}(n-1)}}{\Gamma\left(\dfrac{n-1}{2}\right) 2^{\frac{1}{2}(n-3)} h} \left(\frac{h}{h'}\right)^{n-2} e^{-nh^2/2\sigma^2} \, dh'$$

b. Show that the modal value of h' is h regardless of n (H. M. Feldman, *Annals Math. Stat.*, vol. 3, 1932: pp. 20–31).

c. By setting $dy/dh = 0$, find that the optimum or maximum likelihood estimate of h for an observed h' is

$$\frac{1}{h} = \frac{1}{h'} \sqrt{\frac{n}{n-1}}$$

which is equivalent to the relation

$$\sigma = s \sqrt{\frac{n}{n-1}}$$

exactly as found otherwise.

EXERCISES LEADING TO THE TABLE OF MEANS, STANDARD DEVIATIONS, AND COEFFICIENTS OF VARIATION, FOUND ON PAGE 530

Exercise 2. Find the variance and the standard deviation of Helmert's distribution of s.

Solution

$$\sigma_s{}^2 = Es^2 - (Es)^2 \qquad\qquad\qquad \text{[Ch. 3]}$$

$$= \sigma^2 \left\{ \frac{n-1}{n} - \frac{2\pi}{n} \frac{1}{[B(\frac{1}{2}n - \frac{1}{2}, \frac{1}{2})]^2} \right\}$$

$$= \sigma^2 \left\{ 1 - \frac{1}{n} - \left(1 - \frac{3}{4n} - \frac{7}{32n^2} - \cdots \right)^2 \right\} \qquad \text{[From Eq. 50]}$$

$$= \sigma^2 \left\{ 1 - \frac{1}{n} - \left(1 - \frac{3}{2n} - \frac{7}{16n^2} + \frac{9}{16n^2} + \cdots \right) \right\}$$

$$= \frac{\sigma^2}{2n} \left\{ 1 - \frac{1}{4n} - \frac{3}{8n^2} - \cdots \right\} \tag{63}$$

Therefore

$$\sigma_s = \frac{\sigma}{\sqrt{2n}} \left\{ 1 - \frac{1}{8n} - \frac{25}{128n^2} + \cdots \right\} \tag{64}$$

By replacing n by $k + 1$ one finds also that

$$\sigma_s = \frac{\sigma}{\sqrt{2k}} \left\{ 1 - \frac{5}{8k} + \frac{47}{128k^2} - \cdots \right\} \qquad \begin{array}{l}[k = \text{degrees of} \\ \text{freedom}]\end{array} \tag{65}$$

The terms in $1/n^2$ and $1/k^2$ will compel the student to go back and find one more term in Eqs. 48, 49, 50, and 53, as some may wish to do. The term $-1/8n$ in the series for the standard deviation of s will be found to disagree with the series given in a famed paper by Romanovsky on p. 12 of *Metron*, vol. 5, No. 4, 1925: pp. 3–46. Romanovsky's slip has inadvertently been carried into textbooks, for example, Kenny's *Mathematics of Statistics*, vol. 2 (Van Nostrand, 1939), p. 136.

Exercise 3. Find the standard deviation of Helmert's distribution of s^2.

Answer:

$$\text{S.D. } s^2 = \sigma^2 \sqrt{\frac{2}{n}\left(1 - \frac{1}{n}\right)} \tag{66}$$

Solution

The variance of the distribution of χ^2 is $2k$, as is known from Exercise 41 in Chapter 14 (p. 497), and because the transformation $ns^2/\sigma^2 = \chi^2$ leads to the distribution of χ^2 with $k = n - 1$ degrees of freedom it is only necessary to set

$$\text{Var } \frac{ns^2}{\sigma^2} = 2k = 2(n-1)$$

whence

$$\text{Var } s^2 = \left(\frac{\sigma^2}{n}\right)^2 \times \text{Var } \frac{ns^2}{\sigma^2} = \frac{2(n-1)}{n^2}\sigma^4 \qquad Q.E.D.$$

Remark 1. The standard deviation of s^2 for samples drawn from any universe whatever was found on page 338 in terms of β_2, and the student should observe that if β_2 is there put equal to 3 for a normal universe, the earlier result reduces to this one exactly.

Remark 2. Note that the standard deviation of s^2 is roughly double the standard deviation of s (p. 133).

Exercise 4. Find the variance and the standard deviation of Helmert's distribution of δ.

Solution

$$\text{Var } \delta = E\delta^2 - (E\delta)^2 \qquad \text{[Definition]}$$

$$= \sigma^2 \left\{ 1 - \frac{2\pi}{n} \frac{1}{[B(\frac{1}{2}n, \frac{1}{2})]^2} \right\}$$

$$= \sigma^2 \left\{ 1 - \left(1 - \frac{1}{4n} + \frac{1}{32n^2} - \cdots \right)^2 \right\}$$

$$= \frac{\sigma^2}{2n} \left\{ 1 - \frac{1}{4n} - \cdots \right\} \qquad (67)$$

Therefore

$$\sigma_\delta = \frac{\sigma}{\sqrt{2n}} \left\{ 1 - \frac{1}{8n} - \cdots \right\} \qquad (68)$$

Exercise 5. Find the standard deviation of Helmert's distribution of δ^2.

Answer:

$$\text{S.D. } \delta^2 = \sigma^2 \sqrt{\frac{2}{n}}$$

Solution

The transformation $n\delta^2/\sigma^2 = \chi^2$ leads to the distribution of χ^2 with $k = n$ degrees of freedom, so it is only necessary to set

$$\text{Var } \frac{n\delta^2}{\sigma^2} = \text{Var } \chi^2 = 2k = 2n$$

whence

$$\text{Var } \delta^2 = \left(\frac{\sigma^2}{n} \right)^2 \times \text{Var } \frac{n\delta^2}{\sigma^2} = \frac{2\sigma^4}{n} \qquad (69)$$

which is the solution to the problem.

Remark 1. The variance of δ^2 for samples drawn from any universe whatever was found on page 338, and the student should observe that if β_2 is there put equal to 3 for a normal universe, the earlier result reduces to this one exactly.

Remark 2. Note that the standard deviation of δ^2 is roughly double the standard deviation of δ (cf. Exercise 3f, p. 133).

Exercise 6. Show that in the Helmert distribution of s

a.
$$\text{C.V. } s = \frac{1}{\sqrt{2n}}\left\{1 + \frac{5}{8n} + \cdots\right\}$$

$$= \frac{1}{\sqrt{2k}}\left\{1 + \frac{1}{8k} + \cdots\right\} \tag{70}$$

b.
$$\text{C.V. } s^2 = \sqrt{\frac{2}{k}} \quad \text{exactly} \qquad \begin{bmatrix} k = \text{degrees of} \\ \text{freedom} \end{bmatrix} \tag{71}$$

NOTE. The reader should bear in mind the fact that these equations are true only for samples from a normal universe.

Solution

a.
$$\text{Var } s = Es^2 - (Es)^2 \qquad \text{[Definition]}$$

$$Es = c_2\,\sigma \qquad \text{[P. 521]}$$

$$(\text{C.V. } s)^2 = \frac{\text{Var } s}{(Es)^2} = \frac{Es^2}{(Es)^2} - 1$$

$$= \frac{n-1}{n}\frac{1}{c_2{}^2} - 1$$

$$= \left(1 - \frac{1}{n}\right)e^{\frac{3}{2n}+\frac{1}{n^2}+\cdots} - 1 \qquad \text{[From Eq. 50, p. 521]}$$

$$= \left(1 - \frac{1}{n}\right)\left(1 + \frac{3}{2n} + \frac{1}{n^2} + \frac{9}{8n^2} + \cdots\right) - 1 \qquad \begin{bmatrix} \text{Because} \\ e^x = 1 + x + \\ x^2/2! + \cdots \end{bmatrix}$$

$$= 1 + \frac{1}{2n} + \frac{5}{8n^2} + \cdots - 1$$

$$= \frac{1}{2n}\left[1 + \frac{5}{4n} + \cdots\right]$$

Thus the
$$\text{C.V. } s = \frac{1}{\sqrt{2n}}\left[1 + \frac{5}{8n} + \cdots\right] \qquad Q.E.D.$$

The series in k is derived by setting $k = n - 1$.

b.
$$(\text{C.V. } s^2)^2 = \frac{\text{Var } s^2}{(Es^2)^2} \qquad \begin{bmatrix} \text{Var } s^2 \text{ was found in Ex. 3;} \\ Es^2 = (n-1)\sigma^2/n \end{bmatrix}$$

$$= \frac{\dfrac{2(n-1)}{n^2}}{\left(\dfrac{n-1}{n}\right)^2} = \frac{2}{n-1} = \frac{2}{k} \tag{72}$$

Remark 1. The solutions in terms of k are also valid for the pooled standard deviations of several samples, k being adjusted to equal the number of degrees of freedom in the distribution of s (*vide* Ch. 17, p. 541).

Remark 2. The coefficient of variation of s^2 for samples drawn from any universe whatever was found on page 338 in terms of β_2, and the student should observe that if β_2 is there put equal to 3 for a normal universe, the earlier result reduces to this one exactly.

Remark 3. Note that the coefficient of variation of s^2 is roughly double the coefficient of variation of s (cf. p. 133).

Exercise 7. Show that in the Helmert distribution of δ

a.
$$\text{C.V. } \delta = \frac{1}{\sqrt{2n}} \left\{ 1 + \frac{1}{8n} + \cdots \right\} \tag{73}$$

b.
$$\text{C.V. } \delta^2 = \sqrt{\frac{2}{n}} \quad \text{exactly} \tag{74}$$

Solution

a.
$$\text{Var } \delta = \frac{\sigma^2}{2n} \left\{ 1 - \frac{1}{4n} - \cdots \right\} \qquad \text{[Eq. 67]}$$

$$E\delta = \sigma \left\{ 1 - \frac{1}{4n} + \frac{1}{32n^2} - \cdots \right\} \qquad \text{[Eq. 52, p. 522]}$$

$$\therefore \ (\text{C.V. } \delta)^2 = \frac{\text{Var } \delta}{(E\delta)^2}$$

$$= \frac{1}{2n} \left\{ 1 - \frac{1}{4n} + \cdots \right\} \left\{ 1 - \frac{1}{4n} + \cdots \right\}^{-2}$$

$$= \frac{1}{2n} \left\{ 1 + \frac{1}{4n} + \cdots \right\}$$

and the

$$\text{C.V. } \delta = \frac{1}{\sqrt{2n}} \left\{ 1 + \frac{1}{8n} + \cdots \right\}$$

as required.

b.
$$(\text{C.V. } \delta^2)^2 = \frac{\text{Var } \delta^2}{(E\delta^2)^2} \qquad \begin{array}{l} \text{[Var } \delta^2 = 2\sigma^4/n \text{ from} \\ \text{Ex. 5; } E\delta^2 = \sigma^2 \text{]} \end{array}$$

$$= \frac{2}{n} \text{ exactly}$$

Table of means, standard deviations, and coefficients of variation of several important distributions. The table on page 530 can now be

TABLE OF CHARACTERISTICS OF THE DISTRIBUTIONS OF χ^2, δ, δ^2, s, AND s^2 FOR SAMPLES DRAWN WITH REPLACEMENT FROM A NORMAL UNIVERSE

Variate	Mean	Mode	Standard deviation	Coefficient of variation
χ^2	k	$k-2$	$\sqrt{2k}$	$\sqrt{\dfrac{2}{k}}$
$\dfrac{\chi^2}{k}$	1	$1-\dfrac{2}{k}$	$\sqrt{\dfrac{2}{k}}$	$\sqrt{\dfrac{2}{k}}$
δ (Note a)	$\sqrt{\dfrac{2\pi}{n}}\dfrac{1}{B(\frac{1}{2}n,\frac{1}{2})}\sigma$ $= \sigma\left\{1-\dfrac{1}{4n}+\dfrac{1}{32n^2}+\cdots\right\}$ (Note b)	$\sigma\sqrt{\dfrac{n-1}{n}}$ $= \sigma\left\{1-\dfrac{1}{2n}-\dfrac{1}{8n^2}-\cdots\right\}$	$\dfrac{\sigma}{\sqrt{2n}}\left\{1-\dfrac{1}{8n}+\cdots\right\}$ (Note e)	$\dfrac{1}{\sqrt{2n}}\left\{1+\dfrac{1}{8n}+\cdots\right\}$
δ^2 (Note a)	σ^2	$\dfrac{n-2}{n}\sigma^2 = \sigma^2\left\{1-\dfrac{2}{n}\right\}$	$\sigma^2\sqrt{\dfrac{2}{n}}$ (Note e)	$\sqrt{\dfrac{2}{n}}$ (Note e)
s	$\sqrt{\dfrac{2\pi}{n}}\dfrac{1}{B\left(\frac{n-1}{2},\frac{1}{2}\right)}\sigma$ $= \sigma\left\{1-\dfrac{3}{4n}-\dfrac{7}{32n^2}+\cdots\right\}$ $= c_2\sigma$ (Note c)	$\sigma\sqrt{\dfrac{n-2}{n}}$ $= \sigma\left\{1-\dfrac{1}{n}-\dfrac{1}{2n^2}-\cdots\right\}$	$\dfrac{\sigma}{\sqrt{2n}}\left\{1-\dfrac{1}{8n}+\cdots\right\}$ or $= \dfrac{\sigma}{\sqrt{2k}}\left\{1-\dfrac{5}{8k}+\cdots\right\}$ (Note e)	$\dfrac{1}{\sqrt{2n}}\left\{1+\dfrac{5}{8n}+\cdots\right\}$ (Note f)
s^2	$\dfrac{n-1}{n}\sigma^2 = \sigma^2\left\{1-\dfrac{1}{n}\right\}$ (Note d)	$\dfrac{n-3}{n}\sigma^2 = \sigma^2\left\{1-\dfrac{3}{n}\right\}$	$\sigma^2\sqrt{\dfrac{2}{n}\left(1-\dfrac{1}{n}\right)}$ (Note e) $= \sigma^2\sqrt{\dfrac{2}{k+1}\dfrac{k}{k+1}}$	$\sqrt{\dfrac{2}{k}}$ (Note e) (Note g)

a. For δ and δ^2 all expressions are the same in k as in n.
b. This expression holds whether sampling with or without replacement, and regardless of the shape of the universe.
c. The symbol c_2 is used in quality control and is tabulated on page 570.
d. This expression holds regardless of the shape of the universe.
e. Results that are valid regardless of the shape of the universe will be found on pages 338 and 340.
f. This will also be the coefficient of variation of any estimate of σ which is formed as a multiple of s.
g. " " " " " " " " " " σ^2 " " " " " " s^2.

filled in by recapitulation of results. The letter n denotes the size of sample, and k denotes the number of degrees of freedom in the distribution. It is to be noted that the results that are expressed in terms of k are valid not only for statistics computed from a single sample, but also for statistics computed by pooling several samples of equal or unequal sizes, as will be done in later chapters.

It should be recalled in using the table that any estimate of σ that is a multiple of s has the same coefficient of variation as s itself, and likewise any estimate of σ^2 that is a multiple of s^2 has the same coefficient of variation as s^2 (p. 338). The coefficients of variation of the distributions of s and s^2 are undisturbed if, following Fisher, one uses the symbol n for the number of degrees of freedom rather than for sample-size and redefines the symbol s so that ns^2 remains the second moment of the sample.

ADDITIONAL EXERCISES

Exercise 8. By definition

$$\left.\begin{aligned}
\epsilon_1 &= v_1 - u \\
\epsilon_2 &= v_2 - u \\
&\;\;\vdots \\
\epsilon_n &= v_n - u = -v_1 - v_2 - \cdots - v_{n-1} - u
\end{aligned}\right\} \quad \text{[As in Eqs. 17]}$$

If J denotes the Jacobian of the left-hand variables with respect to those on the right, show that $J = n$, as was found on page 509 by geometrical considerations.

Solution

$$J = \begin{vmatrix}
\dfrac{\partial \epsilon_1}{\partial v_1} & \dfrac{\partial \epsilon_1}{\partial v_2} & \dfrac{\partial \epsilon_1}{\partial v_3} & \cdots & \dfrac{\partial \epsilon_1}{\partial v_{n-1}} & \dfrac{\partial \epsilon_1}{\partial u} \\[2ex]
\dfrac{\partial \epsilon_2}{\partial v_1} & \dfrac{\partial \epsilon_2}{\partial v_2} & \dfrac{\partial \epsilon_2}{\partial v_3} & & \dfrac{\partial \epsilon_2}{\partial v_{n-1}} & \dfrac{\partial \epsilon_2}{\partial u} \\[2ex]
\dfrac{\partial \epsilon_3}{\partial v_1} & \dfrac{\partial \epsilon_3}{\partial v_2} & \dfrac{\partial \epsilon_3}{\partial v_3} & & \dfrac{\partial \epsilon_3}{\partial v_{n-1}} & \dfrac{\partial \epsilon_3}{\partial u} \\[1ex]
\vdots & & & & & \vdots \\[1ex]
\dfrac{\partial \epsilon_{n-1}}{\partial v_1} & \dfrac{\partial \epsilon_{n-1}}{\partial v_2} & \dfrac{\partial \epsilon_{n-1}}{\partial v_3} & & \dfrac{\partial \epsilon_{n-1}}{\partial v_{n-1}} & \dfrac{\partial \epsilon_{n-1}}{\partial u} \\[2ex]
\dfrac{\partial \epsilon_n}{\partial v_1} & \dfrac{\partial \epsilon_n}{\partial v_2} & \dfrac{\partial \epsilon_n}{\partial v_3} & \cdots & \dfrac{\partial \epsilon_n}{\partial v_{n-1}} & \dfrac{\partial \epsilon_n}{\partial u}
\end{vmatrix} \quad \begin{array}{l} (n \text{ rows and } n \\ \quad \text{columns)} \end{array}$$

Now $\partial\epsilon_1/\partial v_1 = 1$, but $\partial\epsilon_1/\partial v_2 = 0$ because, in the top row of Eqs. 17, $\epsilon_1 = v_1 - u$ and does not involve v_2 at all. Similarly $\partial\epsilon_1/\partial v_3 = 0$, etc. So

$$
J = \begin{vmatrix}
1 & 0 & 0 & \cdots & 0 & 1 \\
0 & 1 & 0 & & 0 & 1 \\
0 & 0 & 1 & & 0 & 1 \\
\vdots & & & & & \vdots \\
0 & 0 & 0 & & 1 & 1 \\
-1 & -1 & -1 & \cdots & -1 & 1
\end{vmatrix}
$$

Evaluate this determinant by adding all the rows together for a new bottom row, thus obtaining

$$
J = \begin{vmatrix}
1 & 0 & 0 & \cdots & 0 & 1 \\
0 & 1 & 0 & & 0 & 1 \\
0 & 0 & 1 & & 0 & 1 \\
\vdots & & & & & \vdots \\
0 & 0 & 0 & & 1 & 1 \\
0 & 0 & 0 & \cdots & 0 & n
\end{vmatrix}
$$

which is obviously equal to n. Therefore $d\epsilon_1\, d\epsilon_2 \cdots d\epsilon_n$ in Eq. 2 is to be replaced by $n\, dv_1\, dv_2 \cdots dv_{n-1}\, du$, as was done on page 509.

The next three exercises follow Helmert's derivation of the distribution of s (1876). References were given on page 511.

Exercise 9 (Helmert's transformation). Put

$$
t_1 = \sqrt{\tfrac{2}{1}}\left(v_1 + \tfrac{1}{2}v_2 + \tfrac{1}{2}v_3 + \tfrac{1}{2}v_4 + \cdots + \tfrac{1}{2}v_{n-1}\right)
$$

$$
t_2 = \sqrt{\tfrac{3}{2}}\left(v_2 + \tfrac{1}{3}v_3 + \tfrac{1}{3}v_4 + \cdots + \tfrac{1}{3}v_{n-1}\right)
$$

$$
t_3 = \sqrt{\tfrac{4}{3}}\left(v_3 + \tfrac{1}{4}v_4 + \cdots + \tfrac{1}{4}v_{n-1}\right)
$$

$$
\bullet \qquad\qquad\qquad\qquad\qquad \bullet
$$

$$
\bullet \qquad\qquad\qquad\qquad\qquad \bullet
$$

$$
\bullet \qquad\qquad\qquad\qquad\qquad \bullet
$$

$$
t_{n-1} = \sqrt{\frac{n}{n-1}}\, v_{n-1}
$$

Then prove that the Jacobian of the transformation is

$$J\left(\frac{t}{v}\right) \equiv \begin{vmatrix} \dfrac{\partial t_1}{\partial v_1} & \dfrac{\partial t_1}{\partial v_2} & \dfrac{\partial t_1}{\partial v_3} & \cdots & \dfrac{\partial t_1}{\partial v_{n-1}} \\[2mm] \dfrac{\partial t_2}{\partial v_1} & \dfrac{\partial t_2}{\partial v_2} & \dfrac{\partial t_2}{\partial v_3} & \cdots & \dfrac{\partial t_2}{\partial v_{n-1}} \\[2mm] \dfrac{\partial t_3}{\partial v_1} & \dfrac{\partial t_3}{\partial v_2} & \dfrac{\partial t_3}{\partial v_3} & \cdots & \dfrac{\partial t_3}{\partial v_{n-1}} \\[2mm] \vdots & \vdots & \vdots & & \vdots \\[2mm] \dfrac{\partial t_{n-1}}{\partial v_1} & \dfrac{\partial t_{n-1}}{\partial v_2} & \dfrac{\partial t_{n-1}}{\partial v_3} & \cdots & \dfrac{\partial t_{n-1}}{\partial v_{n-1}} \end{vmatrix}$$

$$= \sqrt{\frac{2 \cdot 3 \cdot 4 \cdots n}{1 \cdot 2 \cdot 3 \cdots n-1}} = \sqrt{n}$$

whence

$$J\left(\frac{v}{t}\right) = \frac{1}{\sqrt{n}}$$

and it follows that $\sqrt{n}\, dv_1\, dv_2 \cdots dv_{n-1}$ in Eq. 21 is to be replaced by $dt_1\, dt_2 \cdots dt_{n-1}$.

Exercise 10. With t_i defined as in the last exercise, prove that

$$\begin{aligned} t_1{}^2 + t_2{}^2 + \cdots + t_{n-1}{}^2 = 2(&v_1v_1 + v_1v_2 + v_1v_3 + \cdots + v_1v_{n-1} \\ &+ v_2v_2 + v_2v_3 + \cdots + v_2v_{n-1} \\ &+ v_3v_3 + \cdots + v_3v_{n-1} \\ & \quad\vdots \\ & \quad + v_{n-1}v_{n-1}) \end{aligned}$$

$$= v_1{}^2 + v_2{}^2 + \cdots + v_{n-1}{}^2$$
$$\quad + (-v_1 - v_2 - \cdots - v_{n-1})^2$$
$$= v_1{}^2 + v_2{}^2 + \cdots + v_{n-1}{}^2 + v_n{}^2$$
$$= ns^2 \qquad\qquad \text{[By definition of } s\text{]}$$

Thus Helmert's transformation removes the cross-products.

Exercise 11. From the results of the last two exercises the integration of the right-hand side of Eq. 21 throughout the *ellipsoidal* shell

$$n(s - \tfrac{1}{2} ds)^2 < v_1{}^2 + v_2{}^2 + \cdots + v_n{}^2 < n(s + \tfrac{1}{2} ds)^2 \quad \text{[See Ineq. 24]}$$

in the *v*-space of $n - 1$ dimensions may be expressed as the integration of

$$\frac{1}{\sigma(\sqrt{2\pi}\,)^{n-1}} e^{-ns^2/2\sigma^2} \, dt_1 \, dt_2 \cdots dt_{n-1}$$

throughout the *spherical* shell

$$n(s - \tfrac{1}{2} ds)^2 < t_1{}^2 + t_2{}^2 + \cdots + t_{n-1}{}^2 < n(s + \tfrac{1}{2} ds)^2$$

in the *t*-space of $n - 1$ dimensions. With reference to the volume of a spherical shell of radius r and thickness dr in a space of n dimensions, which is

$$dV = \frac{2\pi^{\frac{1}{2}n}}{\Gamma(\frac{1}{2}n)} r^{n-1} \, dr \qquad \text{[Pp. 491 or 507]}$$

it can be seen that with $r = \sqrt{n} \cdot s$ the required integration in $n - 1$ dimensions must give

$$
\begin{aligned}
y \, ds &= \frac{1}{(\sigma\sqrt{2\pi}\,)^{n-1}} e^{-ns^2/2\sigma^2} \frac{2\pi^{\frac{1}{2}(n-1)}(\sqrt{n} \cdot s)^{n-2}}{\Gamma(\frac{1}{2}n - 1)} \sqrt{n} \, ds \\
&= \frac{n^{\frac{1}{2}(n-1)}}{\Gamma(\frac{1}{2}n - 1)2^{\frac{1}{2}(n-3)}\sigma} \left(\frac{s}{\sigma}\right)^{n-2} e^{-ns^2/2\sigma^2} \, ds
\end{aligned}
$$

which is Helmert's distribution of s as given in Eq. 31.

Exercise 12. Fisher in his *Statistical Methods for Research Workers* (Oliver and Boyd) cleverly introduces the function $\sqrt{2\chi^2} - \sqrt{2k - 1}$ to be used beyond the range of k (his n) provided by his Table III, with the statement that the chi-test may be made by assuming this function to be normally distributed about 0 with unit variance. Prove that this is so.

Solution [7] (*A. George Carlton*)

Let

$$
\begin{aligned}
\phi(t) &= \int_0^\infty e^{ti(\sqrt{2\chi^2} - \sqrt{2k-1})} \frac{1}{\Gamma(\frac{1}{2}k)2^{\frac{1}{2}k}} (\chi^2)^{\frac{1}{2}k-1} e^{-\frac{1}{2}\chi^2} \, d\chi^2 \\
&= \frac{e^{-ti\sqrt{2k-1}}}{\Gamma(\frac{1}{2}k)} \int_0^\infty v^{\frac{1}{2}k-1} e^{2tiv^{\frac{1}{2}} - v} \, dv, \quad v = \tfrac{1}{2}\chi^2
\end{aligned}
$$

---

[7] This solution makes use of the characteristic function, a topic treated in texts more mathematical such as Wetherburn. Wilks, Hoel, Kendall, Cramér.

$$= \quad " \quad \int_0^\infty v^{\frac{1}{2}k-1} e^{-v}\, dv \left\{ 1 + 2itv^{\frac{1}{2}} + 2(it)^2 v + \cdots \right\}$$

$$= \quad " \quad \left\{ \Gamma(\tfrac{1}{2}k) + 2it\,\Gamma(\tfrac{1}{2}k + \tfrac{1}{2}) + 2(it)^2\,\Gamma(\tfrac{1}{2}k + 1) + \cdots \right\}$$

$$= e^{-it\sqrt{2k-1}} \left\{ 1 + 2it\gamma + 2(it)^2 \tfrac{1}{2}k + \cdots \right\}$$

$$= \left\{ 1 - it\sqrt{2k-1} + (it)^2 \frac{2k-1}{2!} + \cdots \right\} \left\{ 1 + 2it\gamma + (it)^2 k + \cdots \right\}$$

$$= 1 + it(2\gamma - \sqrt{2k-1}) + \frac{(it)^2}{2!}(4k - 4\gamma\sqrt{2k-1} - 1) + \cdots$$

wherein

$$\gamma = \frac{\Gamma(\tfrac{1}{2}k + \tfrac{1}{2})}{\Gamma(\tfrac{1}{2}k)}$$

The coefficient of it is the mean, which is $2\gamma - \sqrt{2k-1}$. The coefficient of $(it)^2/2!$ is the 2d moment coefficient about the origin, which is $4k - 4\gamma\sqrt{2k-1} - 1$. The desired results are seen by expanding these quantities in powers of $1/k$. To make use of a series already derived, we put $k = n - 1$. Then

$$\gamma \equiv \frac{\Gamma(\tfrac{1}{2}k + \tfrac{1}{2})}{\Gamma(\tfrac{1}{2}k)} = \sqrt{\pi}\, \frac{\Gamma(\tfrac{1}{2}n)}{\Gamma\left(\tfrac{1}{2}n - \tfrac{1}{2}\right)\Gamma\left(\tfrac{1}{2}\right)} = \sqrt{\frac{n}{2}}\,\sqrt{\frac{2\pi}{n}}\, \frac{1}{B\left(\dfrac{n-1}{2}, \dfrac{1}{2}\right)}$$

$$= \sqrt{\frac{1}{2}n} \left\{ 1 - \frac{3}{4n} - \frac{7}{32n^2} - \cdots \right\} \qquad \text{[Series, p. 521]}$$

$$= \sqrt{\frac{1}{2}(k+1)} \left\{ 1 - \frac{3}{4(k+1)} - \frac{7}{32(k+1)^2} - \cdots \right\}$$

$$= \sqrt{\frac{1}{2}k} \left(1 + \frac{1}{k}\right)^{\frac{1}{2}} \left\{ 1 - \frac{3}{4k}\left(1 - \frac{1}{k} - \cdots\right) - \frac{7}{32k^2} - \cdots \right\}$$

$$= \sqrt{\frac{1}{2}k} \left\{ 1 + \frac{1}{2k} - \frac{1}{8k^2} - \cdots \right\} \left\{ 1 - \frac{3}{4k} + \frac{3}{4k^2} - \frac{7}{32k^2} - \cdots \right\}$$

$$= \sqrt{\frac{1}{2}k} \left\{ 1 - \frac{1}{4k} + \frac{1}{32k^2} + \cdots \right\}$$

Also

$$\sqrt{2k-1} = \sqrt{2k}\left(1 - \frac{1}{2k}\right)^{\frac{1}{2}} = \sqrt{2k}\left\{ 1 - \frac{1}{4k} - \frac{1}{32k^2} - \cdots \right\}$$

So the mean lies at

$$\mu = 2\gamma - \sqrt{2k-1}$$

$$= \sqrt{2k}\left\{ 1 - \frac{1}{4k} + \frac{1}{32k^2} + \cdots \right\} - \sqrt{2k}\left\{ 1 - \frac{1}{4k} - \frac{1}{32k^2} - \cdots \right\}$$

$$= 0 + \frac{0}{k^{\frac{1}{2}}} + \frac{\sqrt{2}}{16k^{\frac{3}{2}}} + \cdots$$

Thus the mean μ approaches 0 rapidly with increasing k. If $k = 30$, μ lies at 0.00054, and if $k > 30$, μ is still closer to 0. Now for the standard deviation:

$$\text{2d moment} = 4k - 4\sqrt{2k-1}\,\gamma - 1$$

$$= 4k - 1 - 2\sqrt{2k-1}\sqrt{2k}\left\{1 - \frac{1}{4k} + \frac{1}{32k^2} + \cdots\right\} \quad [\text{P. }522]$$

$$= 4k - 1 - 2\cdot 2k\left(1 - \frac{1}{2k}\right)^{\frac{1}{2}}\left\{1 - \frac{1}{4k} + \frac{1}{32k^2} + \cdots\right\}$$

$$= 4k - 1 - 4k\left\{1 - \frac{1}{4k} - \frac{1}{32k^2} - \cdots\right\}\left\{1 - \frac{1}{4k} + \frac{1}{32k^2} + \cdots\right\}$$

$$= 4k - 1 - 4k\left\{1 - \frac{1}{2k} + \frac{1}{16k^2} + \cdots\right\}$$

$$= 1 - \frac{1}{4k} + \cdots \to 1 \text{ as } k \to \infty$$

Now this is the 2d moment coefficient about the origin. But the mean differs from the origin only by the amount $\sqrt{2}/16k^{\frac{3}{2}} + \cdots$ (as was proved), which affects the variance only in the term in $1/k^3$; consequently the variance of the curve is $1 - 1/4k + \cdots$, and the standard deviation is the square root of this or

$$1 - \frac{1}{8k} + \cdots \to 1 \text{ as } k \to \infty$$

This is close to unity if k is large. If $k = 30$, for example, the standard deviation is less than unity by only $\frac{1}{240}$, and for larger values of k it is still closer to unity.

CHAPTER 16. TESTS FOR HYPOTHESES IN NORMAL THEORY

Many paths lie before me; a wise man is known from a fool by his choice.—Ibsen, *Peer Gynt*.

There is no knowledge of external reality without the anticipation of future experience. . . . what the concept denotes has always some temporal spread. . . . There is no knowledge without interpretation. . . . Thus if there is to be any knowledge at all, some knowledge must be a priori.—C. I. Lewis, *Mind and The World-Order* (Scribners, 1929), p. 157.

The distribution of Student's z. It will be recalled from Chapter 4 that for samples drawn with replacement from any universe having a standard deviation σ

$$\sigma_{\bar{x}}^2 = \frac{\sigma^2}{n} \qquad (1)$$

In the use of this equation, there is not only the difficulty of ascertaining whether randomness exists, and thus whether the formula is applicable at all, but also of determining σ within useful limits, which brings up the problem of estimating σ. Gauss [1] long ago introduced the unbiased estimate

$$\sigma''^2 = \frac{n}{n-1} s^2 \qquad \text{[Eq. 10, p. 333]}$$

and from his time onward the usual practice had been to assume that, for want of a more accurate test, the ratio t of the error in $\bar{x}$ to its estimated standard deviation, viz.,

$$t = \frac{\bar{x} - \mu}{(\sigma''/\sqrt{n})} = \frac{\bar{x} - \mu}{s} \sqrt{n-1} \qquad (2)$$

is normal about 0 with unit standard deviation. Student,[2] in 1908, in a paper now famous, became concerned about the sampling fluctuations in σ'' and the consequent warping away from normality of this distribution. He had little doubt of the validity of the usual practice for large samples but sought to find what sample-size can be considered large, and what happens in small samples. Helmert's earlier work,[3] giving the distribu-

---

[1] Gauss, *Theoria Combinationis Observationum Erroribus Minimis Obnoxiae*, pars posterior (Göttingen, 1823; vol. 4 of his *Werke*), Art. 38.

[2] Student, "On the probable error of a mean," *Biometrika*, vol. vi, 1908: pp. 1–25; *ibid.*, vol. xi, 1915–17: pp. 414–7. Birge once called my attention to the fact that Student nowhere in these papers mentioned the probable error of the mean. These papers are now available in *Student's Collected Papers*, edited by E. S. Pearson and John Wishart (Office of Biometrika, University College, London, W.C. 1; 1942).

[3] References are given in the preceding chapter.

tion of s^2, even though printed and used in textbooks [4] on the continent as early as 1891, seems to have been unknown in London. Student accordingly began at the beginning. He arrived first at the distribution of s^2 by discovering what its first four moments must be, and he ingeniously recognized these moments as belonging to the Pearson Type III curve, which he thereupon declared to be the distribution of s^2 beyond reasonable doubt. His next step was the distribution of u/s (which Fisher later on converted into the more useful distribution of t). His success laid the foundation for modern theoretical statistics.

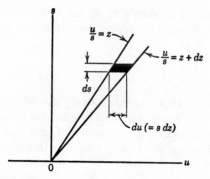

Fig. 74. A view of the u, s frequency surface, looking down on the us-plane. The ratio of u to s is constant along any line through the origin. The area of the element shown in black is $du\ ds$ (base $\times$ altitude), or $s\ ds\ dz$.

The symbol u denotes $\bar{x} - \mu$, and to derive Student's distribution of u/s we return to the u, s probability surface of page 513 and look down upon it, seeing the u- and s-axes laid out as shown in Fig. 74. Every sample puts a point on the u, s plane, and the u, s probability surface shows the theoretical density of these points. The line $u/s =$ constant is obviously a line (really a plane) radiating from the origin. If z denotes the ratio of u/s, then z is constant all along the line: any two samples falling on the line have the same value of z, although they have different values of u and s. If the line is turned through a small angle, then the ratio u/s is changed into $z + dz$. To find the distribution of u/s we merely ask what fraction of the volume under the u, s probability surface is contained in the wedge defined by the inequalities

$$z < \frac{u}{s} < z + dz \tag{3}$$

The elementary area shown black in Fig. 74 is $du\ ds$ (base $\times$ altitude). But along any horizontal line, $ds = 0$ and $du = s\ dz$; wherefore the area

[4] For example, Czuber, *Beobachtungsfehler* (Teubner, 1891).

$du\,ds = s\,ds\,dz$ (see also the exercise on p. 541). Then suppose that y denotes the height of the u, s surface at the point u, s; then $y\,du\,ds$ or $y\,s\,ds\,dz$ is the volume of a prism of height y erected on the blackened elementary area of Fig. 74. This volume is the probability of a sample point falling into the blackened area, hence of having u/s lying between z and $z + dz$, and with standard deviation lying between s and $s + ds$. Then the probability that a sample point will fall *anywhere* in the wedge (3) will merely be these elementary probabilities summed (integrated) throughout the entire wedge. This summation can be obtained by returning to Eq. 32 on page 513 and proceeding to integrate. The first step is to multiply the two factors together and to replace $u^2 + s^2$ by $s^2(1 + z^2)$ in the exponential; also to replace $du\,ds$ by $s\,ds\,dz$. In this way there is obtained

$$y\,dz = \frac{n^{\frac{1}{2}n}}{\sqrt{2\pi}\,\Gamma\left(\dfrac{n-1}{2}\right)2^{\frac{1}{2}(n-3)}\sigma^2}\,dz\int_0^\infty \left(\frac{s}{\sigma}\right)^{n-2} e^{-\frac{ns^2}{2\sigma^2}(1+z^2)}\,s\,ds \qquad (4)$$

Now let

$$v = \frac{ns^2}{2\sigma^2}\,(1+z^2)$$

whence

$$\frac{dv}{v} = \frac{2\,ds}{s} \qquad\qquad (5)$$

It is to be noted that z and dz are constants for a particular wedge, whence by substitution

$$y\,dz = \frac{dz}{\sqrt{\pi}\,\Gamma\left(\dfrac{n-1}{2}\right)(1+z^2)^{\frac{1}{2}n}}\int_0^\infty v^{\frac{1}{2}n-1}e^{-v}\,dv$$

$$= \frac{\Gamma(\frac{1}{2}n)}{\sqrt{\pi}\,\Gamma\left(\dfrac{n-1}{2}\right)}\,\frac{dz}{(1+z^2)^{\frac{1}{2}n}}$$

$$= \frac{1}{B\left(\dfrac{n-1}{2},\dfrac{1}{2}\right)}\,\frac{dz}{(1+z^2)^{\frac{1}{2}n}} \qquad\qquad (6)$$

This is Student's distribution of z, i.e., of u/s, in samples of n drawn with replacement from a normal universe. It is obviously symmetrical about

$z = 0$. Its variance is

$$\sigma_z{}^2 = \frac{1}{B\left(\dfrac{n-1}{2},\dfrac{1}{2}\right)} \int_0^\infty z^2(1+z^2)^{-\frac{1}{2}n}\, dz \qquad \left[\text{Let} \quad \begin{aligned} z &= \tan\theta \\ 1 + z^2 &= \sec^2\theta \\ dz &= \sec^2\theta\, d\theta \end{aligned}\right.$$

$$= \frac{1}{B\left(\dfrac{n-1}{2},\dfrac{1}{2}\right)} \int_0^{\frac{1}{2}\pi} \sin^2\theta \cos^{n-4}\theta\, d\theta$$

$$= \frac{1}{B\left(\dfrac{n-1}{2},\dfrac{1}{2}\right)} \frac{1}{2} B\left(\frac{3}{2}, \frac{n-3}{2}\right) = \frac{1}{n-3} \tag{7}$$

as was found in Exercise 32, p. 495. If $n = 2$ or 3, the integral does not exist, i.e., the variance is infinite. With increasing n the curves become more and more normal and concentrated about $z = 0$ as illustrated in Fig. 75.

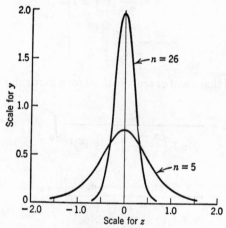

FIG. 75. The distribution of Student's z or u/s. The equation of the curve is

$$y\, dz = \frac{1}{B\left(\dfrac{n-1}{2},\dfrac{1}{2}\right)} \frac{1}{(1+z^2)^{n/2}}\, dz$$

As the sample-size n increases, the standard deviation decreases as the curves approach normality. Two curves are shown, one for $n = 5$ and one for $n = 26$. These curves correspond to the t-curves in the following figure. (Drawn for the author by Mr. Pete James.)

Exercise. Determine algebraically that the element of area $du\,ds$ in Fig. 74 (p. 538) is to be replaced by $s\,ds\,dz$ in the development of Student's z.

Solution

Let

$$s = s \left.\vphantom{\begin{matrix}s\\u\end{matrix}}\right\}$$
$$u = sz$$

Then

$$J\left(\frac{u,\,s}{z,\,s}\right) = \begin{vmatrix} \dfrac{\partial u}{\partial z} & \dfrac{\partial u}{\partial s} \\[2mm] \dfrac{\partial s}{\partial z} & \dfrac{\partial s}{\partial s} \end{vmatrix} = \begin{vmatrix} s & z \\ 0 & 1 \end{vmatrix} = s$$

whence $du\,ds = s\,ds\,dz$.

Fisher's t. Fisher defined t as given in Eq. 2, which is equivalent to putting

$$t = z\sqrt{k} = \frac{u}{(\sigma''/\sqrt{n})} \tag{8}$$

wherein $\sigma'' = s\sqrt{n/(n-1)}$ and $k = n-1$, being the number of degrees of freedom [5] in the estimate of σ. By substituting $z = t/\sqrt{k}$ into the distribution of z, one obtains the equation

$$y\,dt = \frac{1}{\sqrt{k}\,B\left(\dfrac{1}{2}k,\,\dfrac{1}{2}\right)} \frac{dt}{\left(1 + \dfrac{t^2}{k}\right)^{\frac{1}{2}(k+1)}} \tag{9}$$

for the distribution of t. The standard deviation of this curve is

$$\sigma_t = \sqrt{\frac{n-1}{n-3}} = \sqrt{\frac{k}{k-2}} \to 1 \quad \text{as} \quad k \to \infty \tag{10}$$

As t increases, the t-curve approaches the normal curve of unit standard deviation. The advantage of t over z is that the t-curve is easier to visualize, as it barely changes shape as the sample-size increases, as is seen in Fig. 76. For sample-sizes of 25 or 30 or more, the t-curve will not differ appreciably from the normal curve of unit standard deviation, except far out in the tails. For small samples the t-curve is flatter than the normal curve, containing more area in the tails and hence requiring larger values of t for significance than would be required if t were a normal deviate.

[5] The number of degrees of freedom, as has been explained, is the divisor required under the 2d moment (ns^2) to get an unbiased estimate of σ^2. Thus $ns^2/(n-1)$ is an unbiased estimate of σ^2, and $n-1$ is the number of degrees of freedom in the estimate. Or, this number is simply the number k in the related χ^2-distribution (pp. 514–5).

Exercise 1. Prove that if n is large compared with m, then

a.
$$\frac{\Gamma(m+n)}{\Gamma(n)} = n^m \left\{ 1 + \frac{m(m-1)}{2n} \right.$$
$$\left. + \frac{m(m-1)(m-2)(3m-1)}{24n^2} + \cdots \right\}$$

b.
$$\frac{1}{B(m,n)} = \frac{n^m}{\Gamma(m)} \{\text{same series}\}$$

For the use of this asymptotic series in calculating ordinates of the z- and t-curves, see Exercise 4 ahead.

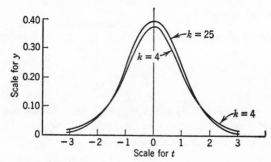

FIG. 76.   The distribution of t, the equation for which is

$$y \, dt = \frac{1}{\sqrt{k}\, B(\tfrac{1}{2}k, \tfrac{1}{2})} \frac{1}{(1 + t^2/k)^{(k+1)/2}} \, dt$$

These curves are nearly similar for all values of k, but they approach the normal curve of unit standard deviation as k increases. Two curves are shown here, one for $k = 4$ and one for $k = 25$. The normal curve of unit standard deviation is not shown because it could hardly be distinguished from the t-curve for $k = 25$. These curves correspond to the z-curves in the preceding figure. (Drawn for the author by Mr. Pete James.)

Solution

Return to the Stirling approximation given on page 520, whence

$$\ln \frac{\Gamma(m+n)}{\Gamma(n)} = \ln \Gamma(m+n) - \ln \Gamma(n)$$

$$= \left(m + n - \frac{1}{2} \right) \ln (m+n) - (m+n) + \left\{ \frac{1}{6(m+n)} - + \cdots \right\}$$

$$- \left(n - \frac{1}{2} \right) \ln n + n - \left\{ \frac{1}{6n} - + \cdots \right\}$$

$$= \left(m + n - \frac{1}{2} \right) \ln \left(1 + \frac{m}{n} \right) + m \ln n - m$$

$$+ \left\{ \frac{1}{6(m+n)} - \frac{1}{6n} + \cdots \right\} \qquad \begin{array}{l} \text{[Recall } \ln (1 + x) \\ = x - \dfrac{x^2}{2} + \dfrac{x^3}{3} - + \cdots \text{]} \end{array}$$

$$= \left(m + n - \frac{1}{2}\right)\left(\frac{m}{n} - \frac{m^2}{2n^2} + \frac{m^3}{3n^3} - \cdots\right)$$

$$+ m \ln n - m - \frac{m}{6n^2} - \cdots$$

$$= \ln n^m + \frac{m(m-1)}{2n} - \frac{m(m-1)(2m-1)}{12n^2} + \cdots$$

$$\therefore \frac{\Gamma(m+n)}{\Gamma(n)} = n^m e^{\frac{m(m-1)}{2n} - \frac{m(m-1)(2m-1)}{12n^2} + \cdots}$$

The required results follow by evaluating the exponential through application of the series

$$e^x = 1 + x + \frac{x^2}{2!} + \frac{x^3}{3!} + \cdots$$

Exercise 2. If m is an integer, the numerator over n^r in the series of the preceding exercise must contain $m(m-1)(m-2) \cdots (m-r)$ as a factor.

HINT. If m is an integer,

$$\frac{\Gamma(m+n)}{\Gamma(n)} = (m+n-1)(m+n-2) \cdots n \qquad \text{[From the recursion formula]}$$

which must equal n^m multiplied by the series in Part a of the preceding exercise. But this can be so only if the series terminates with the term in $1/n^{m-1}$; all following terms must be zero and so must contain $m - r$ as a factor.

Exercise 3. Show that as k increases, the distribution of t approaches the normal curve of unit standard deviation.

Solution

$$y = \frac{1}{\sqrt{k}} \frac{1}{B(\frac{1}{2}k, \frac{1}{2})} \left(1 + \frac{t^2}{k}\right)^{-\frac{1}{2}(k+1)}$$

$$\ln\left(1 + \frac{t^2}{k}\right)^{-\frac{1}{2}(k+1)} = -\frac{1}{2}(k+1) \ln\left(1 + \frac{t^2}{k}\right)$$

$$= -\frac{1}{2}(k+1) \left\{\frac{t^2}{k} - \frac{t^4}{2k^2} + -\cdots\right\}$$

$$= -\frac{1}{2}t^2 + \frac{t^4 - 2t^2}{4k} + \cdots$$

$$\therefore \left(1 + \frac{t^2}{k}\right)^{-\frac{1}{2}(k+1)} = e^{-\frac{1}{2}t^2 + \frac{t^4 - 2t^2}{4k} + \cdots}$$

$$= e^{-\frac{1}{2}t^2} \left\{1 + \frac{t^4 - 2t^2}{4k} + \cdots\right\}$$

Also, from Exercise 1, with $m = \frac{1}{2}$ and $n = \frac{1}{2}k$,

$$\frac{1}{\sqrt{k}}\frac{1}{B(\frac{1}{2}k, \frac{1}{2})} = \frac{(\frac{1}{2}k)^{\frac{1}{2}}}{\sqrt{k}\,\Gamma(\frac{1}{2})}\left\{1 - \frac{1}{4k} + \frac{1}{32k^2} - \cdots\right\}$$

$$= \frac{1}{\sqrt{2\pi}}\left\{1 - \frac{1}{4k} + \frac{1}{32k^2} - \cdots\right\} \quad [\text{Recall } \Gamma(\tfrac{1}{2}) = \sqrt{\pi}, \text{ p. 479}]$$

So

$$y = \frac{1}{\sqrt{2\pi}}\,e^{-\frac{1}{2}t^2}\left\{1 - \frac{1}{4k} + \cdots\right\}\left\{1 + \frac{t^4 - 2t^2}{4k} + \cdots\right\}$$

$$\rightarrow \frac{1}{\sqrt{2\pi}}\,e^{-\frac{1}{2}t^2} \quad \text{as} \quad k \rightarrow \infty$$

The limit is thus the normal curve of unit standard deviation, as required. The correction terms in the braces give a rough measure of the skewness and flatness of the curve. In the center (at $t = 0$) the t-curve falls below the normal curve as is shown by the term $-1/4k$ in the first brace. Also, near the center the term $(t^4 - 2t^2)/4k$ in the second brace is negative, which shortens the ordinate (y) still further. However, this latter term becomes positive beyond $t = \pm\sqrt{2}$ and acts in the other direction. The series can not be expected to give accurate values of y for large values of t unless k is very large.

Exercise 4 (Calculation of the ordinates of the z- and t-curves in Figs. 75 and 76). The student should satisfy himself of the following calculations.

a. For $n = 5$ and $k = n - 1 = 4$,

$$y_z = \frac{1}{B(2, \frac{1}{2})}(1 + z^2)^{-\frac{5}{2}} = 0.75(1 + z^2)^{-\frac{5}{2}}$$

$$y_t = \frac{1}{\sqrt{k}}y_z = \frac{1}{2}y_z \text{ at abscissa } t = z\sqrt{k} = 2z$$

b. For $n = 26$ and $k = 25$,

$$y_z = \frac{1}{B(12\frac{1}{2}, \frac{1}{2})}(1 + z^2)^{-13} = 1.9749(1 + z^2)^{-13}$$

$$y_t = \frac{1}{\sqrt{k}}y_z = \frac{1}{5}y_z \text{ at abscissa } t = z\sqrt{k} = 5z$$

The numerical factor $1/B(12\frac{1}{2}, \frac{1}{2})$ is calculated by the series in Exercise 1 which gives

$$\frac{1}{\sqrt{k}}\frac{1}{B(\frac{1}{2}k, \frac{1}{2})} = \frac{(\frac{1}{2}k)^{\frac{1}{2}}}{\sqrt{k}\,\Gamma(\frac{1}{2})}\left\{1 - \frac{1}{4k} + \frac{1}{32k^2} - \cdots\right\}$$

$$= \frac{1}{\sqrt{2\pi}}\{\text{same series}\}$$

$$= 0.398\,942\{1 - 0.01 + 0.00005 - \cdots\}$$

$$= 0.394\,973$$

The factor for the z-curve is $\sqrt{25}$ or 5 times as great, being 1.9749.

Exercise 5. Define $\sigma'' = s \sqrt{n/(n-1)}$ as an estimate of σ, in conformity with earlier parts of the book. Prove that the

$$\text{r.m.s.e. } \sigma'' = \frac{\sigma}{\sqrt{2k}} \left\{ 1 - \frac{1}{16k} + \cdots \right\}$$

wherein $k = n - 1$ and r.m.s.e. σ'' stands for the root-mean-square error in σ''.

Tests of hypotheses concerning the universe. The only excuse for taking a survey is to enable a rational decision to be made on some problem that has arisen and on which a decision, right or wrong, will be made. It is really not the sample but the universe whence the sample came that is important, yet the decision must be based on the sample. If the mean μ of the universe lies below (e.g.) 0.02 the decision should be one way; if μ lies above (e.g.) 0.04 the decision should be another way. In between, there may be a region of indecision. No sample short of 100 percent can definitely fix the value of μ, but by probability theory it will be possible to calculate the risk involved in following a rule of decision based on samples of less than 100 percent. The risk can be adjusted by varying the size of sample: the bigger the sample, the less the risk. But the cost of the sample increases with its size [6] and will eclipse the risk of a wrong decision if the sample is too big. Proper adjustment of the cost of sampling to the risk involved is one of the main problems that confronts the statistician (Chs. 1 and 2).

In Chapter 9 a study was made of statistical rules of decision, based on binomial theory. The rules were such that certain risks were controlled to predetermined levels. Each article in the sample could be classified unequivocally as accepted or rejected—as good or bad, in other words. Now, more theory has been developed, by which it will be possible to tackle problems in which each sampling unit may take any one of several values, and not just two. For instance, the articles of a sample can be measured: people and families and farms can be classified not merely yes or no, with or without a refrigerator or pigs, owners or renters, but by age-class, number of children, number of hours worked, income-class, etc. Something further than the binomial theory in Chapter 9 is required.

A peculiarity of binomial sampling is that an estimate of the mean p of the universe furnishes also an estimate of σ: this is because $\sigma^2 = pq$. But when there are three or more classes the mean μ and the standard deviation σ must be estimated separately. These estimates are independent in normal theory (the independence of $\bar{x}$ and s has already been remarked upon in the preceding chapter), and nearly independent in

---

[6] It is important to note that the cost of a sample depends also on the method of delineating and drawing the primary and secondary units.

some non-normal theory, although no proof can be given by the author.

As an example, suppose that a distributor has under consideration the idea of installing a plant for packaging and distribution in a certain city in order to cut costs, give better service, and increase his business. The question is whether the potential market in the surrounding area is sufficient. A survey with a well-designed questionnaire may help him decide. It has been decided that if in the surrounding region (which would require careful definition) there are 100,000 or more families that

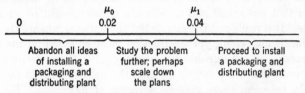

Fig. 77. Illustrating the risks in testing a hypothesis as a basis for action. If 4 percent of the families residing within a given region are potential customers, a packaging and distributing plant in the chief city of the region would pay profit. However, if the number of potential customers is below 2 percent, the plant would not pay and its installation would bring financial disaster. A sample of the population is questioned in an attempt to ascertain the number of potential customers. The sample contains the proportion $\bar{x}$. The problem is to determine a rule of decision by which the risks of a wrong decision on the basis of the sample can be calculated.

bought his or a competitor's product during the past year, or give other evidence of interest in his wares, the new plant should yield a neat profit. If the number is 50,000 or fewer, it will not. Between 50,000 and 100,000 the matter must undergo further study or postponement. A sample-survey shows an estimated 90,000. Owing to sampling errors the "true" number could still be 100,000 or over. What is the risk of proceeding with his plans; rather, what is the probability that the "true" number is 100,000? It should be remembered that, because of sampling errors, there would still be a risk if the sample showed 110,000 families. (The numbers 50,000 and 100,000 correspond in the accompanying Fig. 77 to $\mu = 0.02$ and 0.04, the latter figures being the proportions of the total families.)

As another illustration, a manufacturer tries out a new and costlier process, hoping thereby to improve the quality of a particular article. This quality might be hardness, average period of wear, or some other desired characteristic. Let us say that the old quality average is μ_1 and that the new and costlier process is hardly worth while unless the quality is raised to a new level μ_0. The market will pay only (e.g.) 5¢ more per article at the higher quality-level, while the additional cost of manufacture is 4¢ per article. There is thus a differential of 1¢, but the question is whether a new level is really being attained. As a practical matter the manufacturer would also be interested in the shape of the distribution of quality, and not merely its mean, but within the scope of this book we can give consideration only to the comparison of the means μ_0 and μ_1. By taking samples of the product the manufacturer wishes to know whether the new and desired level is really being attained. The testing of each article in the sample costs 50¢, and the question is how big a sample to take.

Such questions bring up an important class of statistical problems known as the *testing of hypotheses*. In the first place there will always be two alternative hypotheses: never only one. In other words, a hypothesis H_0 is never tested unless there is an alternative hypothesis H_1 that could be true, and which if true would call for a decision different from the decision that would be made if H_0 were known to be true. If H_0 had no alternative, no test would be necessary; in fact no sample would be necessary, because it could not help matters any. It is the existence of alternative hypotheses and alternative decisions and the risks involved that bring up the need for information and the question of how to use it.

In the second place, in making decisions, one after another, anyone, even a statistician, will occasionally commit one or another of three kinds of mistakes:

i. He will occasionally reject H_0 and accept H_1 when H_0 is true. (Let α be the relative frequency of this error.)

ii. He will occasionally accept H_0 and reject H_1 when H_1 is true. (Let β be the relative frequency of this error.)

iii. He will occasionally fail to make a decision one way or the other when one of the alternatives H_0 and H_1 is true. (This possibility is precluded when the region of no decision is not admitted.)

Through a statistical plan it is possible to control the relative frequency of these mistakes and to achieve a net economic minimum of their consequences. Consideration of costs and risks enables an intelligent sampling plan to be drawn up. If the sample is already completed, the risks in making one decision or another can still be calculated though not controlled. It is impossible without large samples to reduce all three mistakes to 0, although it is possible to reduce any one alone to 0. But in reducing one of them to 0 we increase the others. A rule of decision, based on a statistical plan, will consist of:

i. A choice of the magnitudes of α and β (10 percent, 5 percent, 1 percent, or smaller). The choice will depend on the cost of sampling and the cost of making mistakes in accepting or rejecting the two hypotheses.

ii. A choice of the regions of rejection and acceptance.

One portion of the sample-space will be so delineated that if a sample-point falls therein, the hypothesis H_0 is rejected and H_1 is accepted, the action being governed accordingly. There will also be a region for accepting H_0 and rejecting H_1. The rest of the sample-space (if any) is the *region of no decision*, in which the sample is not adequate to distinguish, with the desired protection, between H_0 and H_1: a second sample is then required if the difficulty is to be resolved. The regions for rejection and acceptance should be so placed in the sample-space that

i. H_0, when true, is rejected only with probability α, predetermined.

ii. H_1, when true, is accepted as often as possible, and oftener than when it is not true.

iii. H_1, when true, is rejected only with probability β, predetermined.

iv. H_0, when true, is accepted as often as possible, and oftener than when it is not true.

We have already seen in Chapters 8 and 9 and in the examples cited there how several different risks are encountered in practice. A further extremely important example was cited by Neyman in his Washington lectures, 1938.[7] It involved tests of varieties of sugar-beets. Many new varieties are produced by a breeder: which ones should be saved and propagated? If the breeder saves only those that in a small sample show 6 percent or better than standard, he may and will occasionally discard a variety that is really good but which unfortunately through accident of seed, soil, and physical environment gave a low sugar-content in the test. On the other hand, he can not save and propagate all his new varieties, for some of them are really not worth propagating and would impair the breeder's reputation. The statistical problem is how to devise a rule that will give him minimum net losses.

The region of no decision was encountered in Chapter 8; in double- and sequential-sampling plans if a sample-point fell in the region of no decision, another sample was required. In contrast, it will be recalled that the single-sampling plans admitted no region of no decision: as the name implies, a definite decision to accept or reject the lot is arrived at on the basis of a single sample.

Double-sampling plans could doubtless be practicable and economical in social, economic, and market surveys. A small sample would be

[7] J. Neyman, *Lectures and Conferences in Mathematical Statistics* (The Graduate School, Department of Agriculture, Washington, 1938; now out of print), pp. 60–88.

taken: if the results are decisive, no further investigation is required. But if the results are not decisive, a further sample would be taken. The results of the first sample would nearly always provide the necessary estimates of variance and of β_2 (Ch. 10) by which the second sample could be very economically planned and made neither too large nor too small to provide definite and sufficient evidence for a rational decision.[8]

This is a very different kind of planning from that in which, through lack of a statistical plan, a sample fails to provide significant results, leaving no time or money for a second sample. In the double-sampling plan, both samples would be planned simultaneously, and the field-force would be engaged ahead of time to be ready to move into the second sample if required.

The theory of double sampling in social, economic, and market surveys will not be attempted here.

Remark. In practice, the real difficulty is to frame the questions of a survey so as to get quantitative answers, and to design the sample in such form that a statistical test is possible: otherwise the usefulness of a survey may be seriously impaired. In fact, proper planning demands as a pre-requisite a chart something like Fig. 77 on page 546 and a decision in regard to the risks α and β that can be tolerated (next section; also Ch. 8).

It may be trite to remark that the bias in any characteristic that is to be tested must be held to some minimum level, for otherwise the test will be meaningless (Ch. 2).

It will be seen that in the u-test and chi-test, it is necessary to know σ. In agricultural experimentation, in industrial sampling, and in physical science, it is often possible to keep records of a series of experiments by which σ can be estimated pretty closely and by which, of foremost importance, the series can be tested for randomness to see whether σ remains constant within the allowable limits of statistical fluctuations. As will be seen, knowledge of σ permits use of the u-test, which is preferable to the t-test, particularly in small samples.

The u-test. The us-plane is a "sample-space"; any sample has coordinates u and s and puts a point in the us-plane. The u, s probability surface (Fig. 68, p. 514) shows the "expected" density of sample-points; hence it can be used for computing the probability (volume) lying inside or outside or beyond any region or curve or line in the us-plane that might be designated as a region of acceptance or rejection.[9]

[8] This idea was explored by William G. Madow at a statistical meeting in Washington in 1945: his subject was "On the selection of a sample in repeated steps."

[9] Other tests are possible, not involving the us-plane. For example, a test that involves the number of pluses and minuses in a sample, or the number of ups and downs, does not involve u or s. See for example Allen Wallis and Geoffrey Moore, "A significance test for time series analysis," *J. Amer. Stat. Assoc.*, vol. 36, 1941: pp. 401–9.

A reference on statistical tests that has been most helpful to the author, particularly for emphasis on the planning of experiments, is Part II of the book by the Statistical Research Group, *Techniques of Statistical Analysis* (McGraw-Hill, 1947).

Only the s-coordinate can be computed: the u-coordinate is unknown because $u = \bar{x} - \mu$ and in practice μ is unknown. However, a hypothesis H_0 regarding μ_0 provides a u-coordinate equal to $\bar{x} - \mu_0$ and the sample-point may then be plotted on the u, s plane for testing purposes. The alternative hypothesis μ_1 may be measured off from μ_0 at a distance $\mu_1 - \mu_0$. Or, if desired, axes $\bar{x}$ and s may be used, on which $\bar{x} = \mu_0$ and $\bar{x} = \mu_1$ constitute two contours (vertical lines on the us-plane).

Now because u and s are independent it is simpler to deal with the u-axis by itself when testing a hypothesis regarding μ alone (the $\bar{x}$-axis would do as well, of course). The situation might be that shown in Fig. 78 in which the two alternative hypotheses are these:

$$H_0: \text{ viz., that } \quad \mu \lessgtr \mu_0 = 0.02$$

$$H_1: \text{ viz., that } \quad \mu \gtrless \mu_1 = 0.04$$

These are both multiple hypotheses, in the sense that there are many possible values of μ beyond 0.04 and many below 0.02 (theoretically, an infinite number). To simplify the calculations the hypotheses may at first be made *simple* or single-valued, say $H_0 = 0.04$ and $H_1 = 0.02$.

To rehearse for a moment the aims in choosing a region for rejecting H_0, we recall that if α is to be 0.01, the region of rejection could be 1 percent of the area in the left-hand tail of a normal curve centered at μ_0, or 1 percent in the right-hand tail, or 1 percent in the middle or anywhere else. For the situation described above, one should certainly choose the region of rejection in the right-hand tail, not the left nor the middle, because in rejecting H_0 we accept H_1, and we wish to accept H_1 correctly as often as possible. The probability of accepting H_1 correctly increases toward the right because if $y_0 \, d\bar{x}$ is the probability of the mean of the sample falling in the interval $\bar{x} \pm \frac{1}{2} \, d\bar{x}$ on the supposition that μ_0 is the mean of the universe, and $y_1 \, d\bar{x}$ is the probability on the supposition that μ_1 is the mean of the universe, then

$$y_0 \, d\bar{x} = \frac{\sqrt{n}}{\sigma \sqrt{2\pi}} e^{-n(\bar{x}-\mu_0)^2/2\sigma^2} \, d\bar{x} \qquad \begin{array}{c}\text{[From Eq. 23,}\\ \text{p. 509]}\end{array} \quad (11)$$

$$y_1 \, d\bar{x} = \quad `` \quad e^{-n(\bar{x}-\mu_1)^2/2\sigma^2} \, d\bar{x} \qquad (12)$$

whence

$$\frac{y_1}{y_0} = e^{\frac{n}{\sigma^2}[\mu_1 - \mu_0][\bar{x} - \frac{1}{2}(\mu_1 + \mu_0)]} \qquad (13)$$

This ratio increases toward the right (recall $\mu_1 > \mu_0$), and thus any piece of area of a given size is most effective in guarding against errors of the second kind (falsely accepting H_0), if placed in the extreme right-hand tail of the curve. If α is desired to be 0.01, then the region for rejecting H_0 will be the area in the right-hand tail under a normal curve of standard deviation $\sigma/\sqrt{n}$ centered at μ_0, beyond the abscissa ξ which lies at a distance 2.33 $\sigma/\sqrt{n}$ to the right of μ_0. (The figure 2.33 comes from a table of the normal integral; see the table on page 552.) Likewise, if β is desired to be 0.05, the region for accepting H_0 and rejecting H_1 will be the left-hand tail under a normal curve of standard deviation $\sigma/\sqrt{n}$ centered at μ_1, beyond (to the left of) the abscissa ξ', which lies at a distance $1.64\sigma/\sqrt{n}$ to the left of μ_1. Between ξ and ξ' is an interval of no decision.

Sometimes (as in the example of the sugar-beets) the interval of no decision is inadmissible, as when a definite decision must be made—accept either H_0 or H_1: in this case, if α is predetermined, β can not be

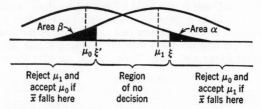

FIG. 78. Showing the relation between α and β in terms of the two alternative hypotheses. α is the probability of falsely rejecting H_0 when it is true, and β is the probability of falsely accepting H_0 when the alternative H_1 is true. α and β can be predetermined at (e.g.) 0.01 and 0.05. If the region of no decision is not allowed, ξ' moves up to ξ, and β increases while α remains fixed. Under these circumstances as α is made smaller, β is necessarily made larger, and the only way to decrease β is to take a bigger sample.

except by adjusting the size of sample. The right-hand tail for rejecting H_0 and accepting H_1 will then give the smallest possible value of β for a given size of sample and will thus be *most powerful*. Moreover, it will be most powerful for all alternatives $\mu_1 > \mu_0$; hence it is *uniformly most powerful*.

If $\mu_1 < \mu_0$, the left-hand tail would be a uniformly most powerful region for rejecting μ_0.

Fig. 78 shows the regions for accepting and rejecting the hypothesis when the u-test is used, and the areas under the normal curves by which α and β can be calculated. An extremely important set of curves showing β as a function of $\xi - \mu_0$ for $\alpha = 0.01$ and $\alpha = 0.05$ (no interval of no decision) was published by Neyman, Iwaszkiewicz, and Kolodziejczyk [10] in the solution of problems met in the breeding of sugar-beets.

[10] Neyman, Iwaszkiewicz, and Kolodziejczyk, Supplement to the *J. Roy. Stat. Soc.*, vol. ii, 107–80, 1935; pp. 133 and 134 in particular.

Let P_u be the area in the tail of a normal curve all the way to the right of any abscissa $u = \bar{x} - \mu_0$. Then

$$P_u = \frac{1}{\sqrt{2\pi}} \int_t^\infty e^{-\frac{1}{2}t^2}\, dt \qquad [t = u/(\sigma/\sqrt{n})] \quad (14)$$

For calculating α, place $u = \xi - \mu_0$; then $P_u = \alpha$. For calculating β, place $u = \mu_1 - \xi'$; then $P_u = \beta$. The numerical value of the integral can be found from any of numerous tables of the normal integral, or from the table given here.

Remark 1. The quantity u appearing in the table below and in Fig. 79 is a *standardized error* and is said to be in *standard units*, as it is measured in terms of its standard error.

Remark 2. Caution should be exercised in looking up probabilities in a table. There are many varieties of tables, differing in the particular portion of the area tabled, and in the scale. Some tables show probabilities for both left and right tails; others for the right-hand half of the internal area. The user of a table, to be sure that he understands the figures that he is reading, should compare them with the following simple table (which is in fact all that I ever use). Fig. 79 defines the quantity P_u in this table.

NORMAL INTEGRAL

u (in standard units)	P_u	$2P_u$
0	0.5	1.00
0.6745	.25	0.50
1	.16	.32
1.28	.10	.20
1.64	.05	.10
1.96	.025	.05
2	.02	.04
2.33	.01	.02
3	Practically 0	

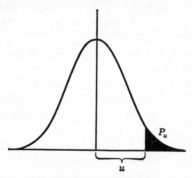

FIG. 79. Showing the area P_u which is defined as the area in one tail beyond the standardized error u.

The t-test. It will be noted that calculation of P_u requires knowledge of σ. If σ must be estimated from the sample, what we have is the t-test, not the u-test. A straight line through the origin in the us-plane gives a z-contour (or a t-contour it could be called, because t as well as z is constant along it). The volume of the u, s probability

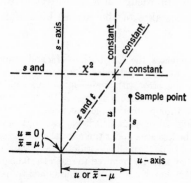

FIG. 80. Showing three contours in the us-plane. On this plane curves or lines will be drawn to designate regions for accepting or rejecting a hypothesis. For example, the area to one side of the u-contour (or the t-contour, or the s-contour) might be designated as one of these regions. A sample-point falls somewhere on the plane, and depending on where it falls—

  i. H_0 will be rejected and H_1 accepted,

or

  ii. H_0 will be accepted and H_1 rejected,

or

  iii. No decision will be reached, and another sample is necessary.

Sometimes the third possibility is not admitted, as when a definite decision must be made now.

surface lying to the right of this contour is equal to the area P_t under the distribution of t (Fig. 76, p. 542) beyond the abscissa t, where

$$t = \frac{u}{s}\sqrt{n-1} = \frac{u}{(\sigma''/\sqrt{n})} \qquad \text{[As on p. 541]}$$

and

$$P_t = \frac{1}{\sqrt{k}\, B\left(\frac{1}{2}k, \frac{1}{2}\right)} \int_t^\infty \frac{1}{\left(1 + \dfrac{t^2}{k}\right)^{\frac{1}{2}(k+1)}}\, dt \qquad (15)$$

This integral has been tabulated by Fisher in his *Statistical Methods for Research Workers*, and the student is expected to become familiar with his treatment. In this table, t is read out for various levels of P_t and various degrees of freedom. As with his table of normal probabilities, both tails are included, so a bit of caution is necessary. Student's tables of z in 1908, 1915, and 1925 were tables of P_t against t in another form.

The Nekrassoff nomograph may be used in place of the t-table.

In the absence of knowledge of σ, the t-test gives the best possible protection against mistakes of the second kind (falsely accepting H_0 when H_1 is true).

Remark. As n increases, the distributions of u and t approach coincidence (p. 542). This is because the estimate of σ made from a sample is subject to less and less sampling error with increasing n. It should be noted, though, that the approach to coincidence is slowest in the tails, and it is the area in the tails that is used for protection against falsely accepting H_0 when it is not true. From Fig. 76 on page 542 it is clear that beyond any positive abscissa there will be less area under the normal curve than under a t-curve. Hence if t is assumed to be normally distributed (as is valid only in large samples), the value of t apparently required to reach the 0.05 or the 0.01 level of probability is smaller than if t is (correctly) assumed to have the t-distribution. Thus, instances have been pointed out in the literature of incorrect allegations of "significance," arising through use of normal probabilities in place of t-probabilities.

It is easy, however, to be misled on the importance of the distinction between the u- and t-curves. The difference $P_t - P_u$ is much more important than the ratio $P_t : P_u$, for if P_u is very small, say 0.005, it matters not if for a small sample, P_t is twice or thrice as large: any action based on a single small sample will still be the same.

In acceptance sampling, where many successive samples are taken, and action taken on each, there is opportunity for extreme refinement in probabilities, with economic gains. First of all, repeated samples give opportunity to attain and maintain statistical control so that probability-calculations really apply. Second, there is opportunity to learn the shape of the distribution that the samples are being drawn from, and to make special mathematical calculations leading to other types of tests and discoveries of assignable causes of variability.

The student may wish to show that $P_t - P_u$ is always positive.

The s-test or chi-test. This is a test of a proposed standard deviation σ_0 for the universe. This test is important when a change in precision is expected or indicated. A new process (as in manufacturing) may have been introduced which (it is hoped) should lower the standard deviation of some particular quality-characteristic from σ_1 to $\sigma_0 < \sigma_1$. A sample of n articles shows standard deviation s, which is lower than σ_1. But should the new process be adopted? It is more costly than the old one and not worth while unless $\sigma_1 - \sigma_0 \geqq 5$. If the process is in statistical control and if normal theory suffices, then the risks α and β in accepting the hypothesis that the variability has been lowered to σ_0 can be calcu-

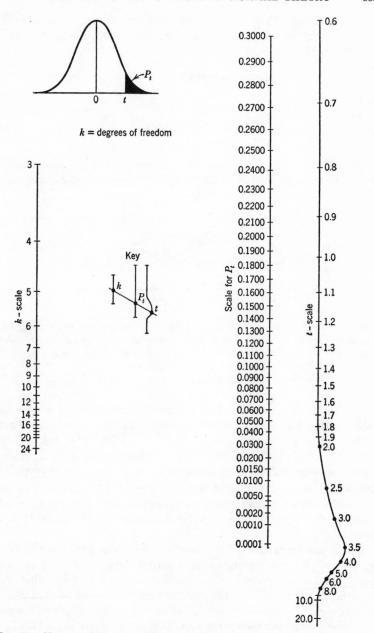

Fig. 81. Nomograph for P_t. (Developed and published by V. A. Nekrassoff in *Metron*, vol. 8, No. 3, 1930. Reproduced with the kind permission of the Bell Telephone Laboratories, Inc.)

lated as follows. First, draw a rough diagram like Fig. 82 and decide on the region for rejecting σ_0. Assuming Helmert's distribution of s we are able to write at once,

$$\alpha, \text{ which is } P(\chi^2) \text{ for } \chi^2 = \frac{ns^2}{\sigma_0^2} \quad [\sigma_0 = \sigma_1 - 5] \quad (16)$$

and

$$\beta, \text{ which is } 1 - P(\chi^2) \text{ for } \chi^2 = \frac{ns^2}{\sigma_1^2} \quad (17)$$

For a given value of σ it is not difficult to compute the risks α and β in Fig. 82 corresponding to an observed standard deviation s in a sample

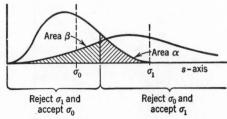

FIG. 82. Showing the region for rejecting σ_0 and for calculating the risks α and β. The curves are assumed to be Helmert's distribution of s, which is strictly valid only in normal theory. (No region of no decision is shown here, as it is assumed that the decision must be definite: adopt the new process or hold to the old one. Compare with Fig. 78, which contains a region of no decision.)

of n. The computation could be carried out with tables of χ^2 or with tables of the incomplete gamma function (Ch. 14). Kendall's chart with $k = n - 1$ will be found excellent for reading off α but will be found insensitive to β (Fig. 72, p. 519). When n is 25 or 30 or bigger, the calculations will be good enough if χ^2 is assumed to be normally distributed about k with standard deviation $\sqrt{2k}$ (cf. the table on page 530). Frequently the problem occurs the other way around: what must be the sample-size in order that neither α nor β shall exceed 0.05?

Confidence limits for σ. The adjective *fiducial* was introduced in 1930 by Fisher [11] for the description of a certain interesting relation that exists between a constant of the universe and the corresponding parameter of a sample when the sampling distribution of the latter depends only on the former. Such is the case with σ and s. Thus, if a set of n observations has been taken and the standard deviation is found to be s, we may arbitrarily put $P_s = 0.95$, using the observed value of s

[11] R. A. Fisher, *Proc. Camb. Phil. Soc.*, vol. 26, 1930: pp. 528–35.

for the lower limit of integration in

$$P_s = \int_s^\infty y \, ds \qquad \begin{array}{l} \text{[}y \text{ being the ordinate} \\ \text{from Helmert's equa-} \\ \text{tion for the distribu-} \\ \text{tion of } s, \text{ p. 511]} \end{array} \qquad (18)$$

to find the value of σ required in the integral. This is the same thing as drawing the s-contour of Fig. 80 at a distance from the u-axis equal to the observed standard deviation s, and then arbitrarily selecting for σ that value which will put 95 percent of the volume of the u, s probability surface above the contour and the remaining 5 percent below it.

Now it so happens that in the incomplete gamma function to which Eq. 18 reduces, s and σ occur only in the ratio $s:\sigma$. This ratio will of course be a function of P_s for a given value of n. If, then, for $P_s = 0.95$ this ratio be denoted by $1/f_{95}$, Eq. 18 gives

$$P(\chi^2) = 0.95 \qquad \text{[P. 497]} \quad (19)$$

where

$$\chi^2 = \frac{n}{f_{95}{}^2} \qquad (20)$$

from which the numerical evaluation of f_{95} for different values of n can be accomplished.

Values for f_{95} are shown in Table 1 for n running from 2 to 25. These values for f_{95} provide a fiducially reciprocal relation between the standard deviation of the universe and the standard deviation of the sample. If for every observed standard deviation s one calculates $\bar{\sigma}_{95} = sf_{95}$, then it is a fact that the inequality

$$\sigma \leq \bar{\sigma}_{95} \qquad (21)$$

will hold in 95 samples out of 100, on the average. Thus, $\bar{\sigma}_{95}$ is the upper 95 percent confidence limit for σ, corresponding to the observed standard deviation s (cf. Ch. 9).

In repetition of the ideas expressed in Chapter 9 one could say that if for every random sample in his experience with random samples the Ineq. 21 be written on a card and these cards classified true or false, the "expected" height of the pile labeled "true" will be 19 times that of the pile labeled "false." This is so even if σ varies from one sample to another.

It is essential to understand that there must be no selection of values of s; the rule for calculating $\bar{\sigma}_{95}$ must be followed consistently for all values of s in one's entire sampling experience, whether or not σ varies from one sample to another, and whether or not any such variation is known to

TABLE 1. FIDUCIAL FACTORS BETWEEN s AND σ

Multiplying factors for finding the 50 and the 5 percent fiducial values of σ, corresponding to an observed standard deviation s in a sample of size n. f_{95} is defined as the ratio of the 5 percent fiducial value of σ to the observed value of s. f_{95} is obtained by setting P_s or P_{χ^2} equal to 0.95, by which

$$\frac{\Gamma_x[\tfrac{1}{2}(n-1)]}{\Gamma[\tfrac{1}{2}(n-1)]} = 0.05$$

wherein the subscript $x = n/2f_{95}^2 = \tfrac{1}{2}x_{95}^2$.

n	f_{50}	f_{95}
2	2.096 716	22.552 803
3	1.471 068	5.353 057
4	1.300 244	3.371 735
5	1.220 476	2.652 372
6	1.174 244	2.288 667
7	1.144 059	2.068 899
8	1.122 797	1.921 235
9	1.107 009	1.815 807
10	1.094 821	1.734 191
11	1.085 127	1.670 828
12	1.077 232	1.619 586
13	1.070 678	1.577 196
14	1.065 150	1.541 478
15	1.060 424	1.510 922
16	1.056 338	1.484 443
17	1.052 769	1.461 245
18	1.049 626	1.440 730
19	1.046 836	1.422 439
20	1.044 343	1.406 011
21	1.042 102	1.391 165
22	1.040 077	1.377 670
23	1.038 238	1.365 341
24	1.036 560	1.354 027
25	1.035 023	1.343 599

exist. If $\bar{\sigma}_{95}$ were calculated only whenever s fell within some previously selected range, no such statement would hold.[12]

It would be well to turn here to a simple diagram for showing the fiducial relation between s and σ. In Fig. 83 the horizontal line at ordinate σ is shaded proportional to the density of samples following a

[12] This was first pointed out to the author in 1934 by Dr. T. Koopmans, now of Chicago, then of Amsterdam; also by Professor Egon Pearson in 1934.

Helmert distribution of s, for a chosen value of n. Five percent of the values of s lie to the left of B and 95 percent to the right. Now suppose that in a sample of n the standard deviation is found to be s. Let the distance s be laid off from O on the horizontal axis. A vertical line through s then cuts OA at D, and the ordinate of D is $\bar\sigma_{95} = sf_{95}$.

Whenever s falls to the right of C, as pictured, and as is "expected" in 95 percent of the samples, then $\bar\sigma_{95} > \sigma$. But if s falls to the left of

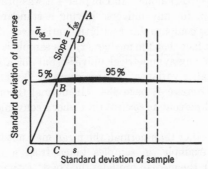

Fig. 83. Diagram for finding $\bar\sigma_{95}$. In Neyman's terminology, OA forms a *confidence belt* with the s-axis of *confidence coefficient* 0.95. The slope of OA depends only on n.

C, as will happen in 5 percent of the samples, then $\bar\sigma_{95} < \sigma$. These statements hold no matter what σ may be, known or unknown, and whether σ be constant or not from one sample to another.

Odds other than 19:1 may be used if desired. The student may recall the "median" estimate of σ from the preceding chapter (p. 525): this is none other than the 50 percent fiducial value of σ, because if sf_{50} were calculated for every one of a series of samples, half the values of sf_{50} would be greater and half less than the standard deviations of the universe whence the samples came. Values of f_{50} are shown in Table 1.

More on fixed- and varying-interval predictions. So far in this chapter the discussion has centered around problems in the testing of hypotheses, which are in practice problems of devising rules of decision. This subject was commenced in Chapter 9 with binomial theory, and continued here with normal theory. There is also the kindred problem of estimation, which is the problem of determining, from the sample and perhaps from other information as well, an approximation to some characteristic of the universe (proportion or total unemployed, proportion or total defective, the standard deviation of the number unemployed per household), along with some kind of probability limits, with the purpose of providing either (a) a basis for action, or (b) increased knowledge of the universe.

Statistical literature is replete with mathematical research on estimation, and in spite of the brilliance displayed since 1908, some of the most astounding results have been recent. It is impossible here to do justice to the problem except to state some of the principles from a practical man's viewpoint, so we shall move into it at once. For a normal bowl, the mean $\bar{x}$ of a random sample is the "best" estimate of the mean μ, in the sense of having the smallest variance of any function of the sample $(x_1, x_2, \cdots, x_n)$ alone. In Chapter 4 it was proved that $E\bar{x} = \mu$ and Var $\bar{x} = \sigma^2/n$ for any universe having a standard deviation, but Chapter 4 did not compare Var $\bar{x}$ with the variance of any other possible estimate. It is a fact that the median $(\dot{x})$ of a sample is also an unbiased estimate of μ for any symmetrical universe, but for a normal bowl the median has a higher variance; in large samples Var $\bar{x}$:Var $\dot{x} = 0.64$ as was proved in Exercise 4, page 434. In normal theory the median is therefore only 64 percent as *efficient* (in Fisher's convenient terminology) as the mean.[13]

For universes other than normal, the mean may not have the smallest variance. For example, in samples taken from a Cauchy universe $y = 2/\pi(1 + x^2)$, the median provides the "best" estimate:[14] the mean $\bar{x}$ is here no better than a single observation for estimating μ. (This is no contradiction of the law Var $\bar{x} = \sigma^2/n$ because σ does not exist: see p. 60). For estimating the mean of a rectangular universe, the average of the largest and smallest observations in a sample is much more efficient (in large samples) than the mean of the sample: as a matter of fact the variance of the mean is the larger by the ratio $6n/(n + 1)(n + 2)$, which actually tends to 0 with increasing sample-size.[15]

For a proper study of the subject of efficiency from the standpoint of variance and mean square error, cost disregarded, the student is urged to pursue Harald Cramér's *Mathematical Methods of Statistics* (Princeton, 1946), particularly pp. 478–80. Under certain mathematical restrictions (continuity of function and existence of moments) Cramér shows that there always exists a lower bound on the mean square error of an estimate of a parameter in a distribution. This lower bound is the right-hand

[13] The article by Mosteller cited on page 291 should be consulted for other easily calculated functions, more efficient than the median. It should be noted that when costs of calculating and sampling are taken into consideration, the measurement of efficiency by comparing variances is not adequate: rather (as has been done in Washington for several years) efficiency should be measured by the ratio of total costs (sample plus computation) for equal variances.

[14] R. A. Fisher, "On the mathematical foundations of theoretical statistics," *Phil. Trans. Royal Soc.*, vol. A222, 1921–22: pp. 309–68.

[15] A. George Carlton, "Estimating the parameters of a rectangular universe," *Annals Math. Statistics*, vol. xvii, 1946: pp. 355–8

side of Cramér's inequality

$$E(\alpha^* - \alpha)^2 \geqq \frac{\left(1 + \dfrac{db}{d\alpha}\right)^2}{n \displaystyle\int_{-\infty}^{\infty} \left(\frac{\partial \ln f}{\partial \alpha}\right)^2 f(x; \alpha)\, dx} \qquad (22)$$

Here α is the parameter, α^* an estimate, and b the bias of this estimate. $f(x; \alpha)$ is the distribution function of x. Without this inequality one would never know whether a particular estimate is really the most efficient one that is possible: he could only compare the efficiency of one estimate with the efficiency of another, using (e.g.) Fisher's measure of efficiency as the reciprocal of variance.

To return to the normal bowl, what is wanted in the problem of estimation is a pair of limits affixed to $\bar{x}$ having operational meaning in the probability sense. The u and t integrals provide devices for computing fixed and varying intervals of prediction, which are used in problems of estimation as in Chapter 9. If $u_{2.5}$ and $t_{2.5}$ are so selected that they render the u and t integrals both equal to 0.025, i.e., if

$$P_u = \frac{1}{\sqrt{2\pi}} \int_{\frac{u_{2.5}}{\sigma/\sqrt{n}}}^{\infty} e^{-\frac{1}{2}t^2}\, dt = 0.025 \qquad (23)$$

and

$$P_t = \frac{1}{\sqrt{k}\, B\left(\dfrac{1}{2}k, \dfrac{1}{2}\right)} \int_{t_{2.5}}^{\infty} \frac{1}{\left(1 + \dfrac{t^2}{k}\right)^{\frac{1}{2}(k+1)}}\, dt = 0.025 \qquad [k = n - 1] \quad (24)$$

then because $u = \bar{x} - \mu$ and $t = \sqrt{n-1}\,(\bar{x} - \mu)/s$ the inequalities

$$\mu - u_{2.5} < \bar{x} < \mu + u_{2.5} \qquad \begin{array}{c}\text{[Fixed ends; vary-}\\ \text{ing center]}\end{array} \quad (25)$$

and

$$\bar{x} - \frac{st_{2.5}}{\sqrt{n-1}} < \mu < \bar{x} + \frac{st_{2.5}}{\sqrt{n-1}} \qquad \begin{array}{c}\text{[Varying ends;}\\ \text{fixed center]}\end{array} \quad (26)$$

will be "expected" to hold in 95 percent of a series of random samples. The upper inequality gives a *fixed interval* of width $\mu \pm u_{2.5}$ within which 95 percent of the sample means ($\bar{x}$) are "expected" to fall. The lower inequality gives a series of *varying intervals* or *confidence intervals* $\bar{x} \pm st_{2.5}/\sqrt{n-1}$, an "expected" 95 percent of which will contain the mean μ of the universe.

For large samples the varying intervals settle down: their lengths vary only slightly from one sample to another, and their centers ($\bar{x}$)

do not jump so much as they do for small samples. This assertion follows from the fact that i. the variance s^2, in large samples, in practice hovers pretty close to its "expected" value $(n - 1)\sigma^2/n$ (Chs. 10 and 13); and ii. the mean $\bar{x}$ hovers pretty close to its "expected" value μ. Moreover, the distribution of t approaches the distribution of $u/(\sigma/\sqrt{n})$, which is normal about 0 with unit standard deviation. Thus, for large samples, the varying intervals $\bar{x} \pm st_{2.5}/\sqrt{n - 1}$ when laid off one after another will be found to form a distribution that is concentrated close to the fixed interval $\mu \pm u_{2.5}$.

Probability-levels other than 95 percent can be used and the Ineqs. 25 and 26 recalculated by taking a new u and a new t from the tables or charts to correspond with the probability-level desired.

In my experience, the calculation of intervals falls into two classes. There are social and economic surveys which are so large that the fixed interval can be calculated for any desired probability, and there need be no talk of varying intervals. For such surveys, either the simple 1-sigma limits $\bar{x} \pm \sigma_{\bar{x}}$ (or merely $\sigma_{\bar{x}}$, the standard error), or the 3-sigma limits $\bar{x} \pm 3\sigma_{\bar{x}}$, will usually suffice. The standard error is understood by people who have acquired some mastery over statistical theory; knowing the standard error they can calculate any probabilities they wish. The 3-sigma limits correspond to near certainty, and this is what administrators need or think they need when they can not take chances and demand upper or lower limits or both. In some instances 2-sigma limits are more suited to the risks involved (see also p. 299).

Then there are small surveys and experiments with no background in which only the varying or confidence interval can be computed. If it is a frequently recurring survey or experiment, with decisions to be taken on each one, the proper procedure is to work out a plan of decision which will involve the risks α and β with the t-test as given in the earlier part of this chapter. If the survey is not to be repeated, the statistical test is at best only part of the evidence that is used in coming to a decision.

Experimental illustration of the two kinds of prediction. For an experimental illustration it is desirable to put P_u and P_t in Eqs. 14 and 15 both equal to some fairly large value, like $\frac{1}{4}$, so that a fairly short series of experiments will contain enough "expected" satisfactions of the Ineqs. 25 and 26 to permit theory and practice to be compared. Deming and Birge showed a series of 100 samples of 4 each, drawn by Shewhart from a normal bowl of 0 mean and unit standard deviation.[16] The fixed limits $u \pm 0.6745\,\sigma/\sqrt{n}$ were plotted; these are "expected" to contain

[16] These drawings are listed in Shewhart's *Economic Control of Quality* (Van Nostrand, 1931), Table D, p. 454.

half of the 100 sample means ($\bar{x}$). Also, for each sample, $\bar{x}$ and the varying limits $\bar{x} \pm 0.442s$ were plotted; half of these are "expected" to overlap the mean of the bowl (0). The results are shown in Fig. 84,

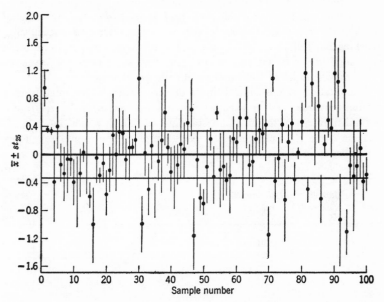

FIG. 84. $\bar{x} \pm st_{25}$ is laid off for each of 100 samples. The mean of the universe is $\mu = 0$, and its standard deviation is unity. The horizontal lines at distances $\pm 0.67\sigma/\sqrt{4}$ from the mean μ define a fixed interval predicting the inclusion of 50 percent of the means of future samples: actually, in 52 out of 100 samples the means were so included. Theory predicted that 50 percent of the vertical segments (varying intervals) would overlap the mean μ: actually, in 51 out of the 100 samples the mean was so included. This figure was published by Deming and Birge in 1934, but they attributed the idea to a figure that Shewhart had exhibited at a joint meeting of the American Mathematical Society and Section K of the American Association for the Advancement of Science in Atlantic City, 27 December 1932. Meanwhile the same chart had appeared as Fig. 2 in Supplement A of the *A.S.T.M. Manual on the Presentation of Data*, 1933, published by the American Society for Testing Materials, 1916 Race St., Philadelphia 3. The A.S.T.M. chart shows also the 90 percent confidence intervals, $\bar{x} \pm st_{45}$.

in which it will be observed that the fixed limits (pair of horizontal lines) include 52 of the 100 means, and that 51 of the varying intervals include the mean of the bowl (0). It should be noted that some of the varying intervals are extremely short (the 2d and 3d, for example—hardly visible) while some are very long (the 4th, 30th, 96th).

This figure illustrates the failing of small samples. The only possible use of statistical theory is for making predictions regarding future samples, but of what use would any of the varying intervals be for predicting where and within what limits half of the next 100 points will fall? None, or very little. It is necessary to have a large sample if one desires to reduce the risks of making decisions and predictions.

Shewhart, on page 62 of his *Statistical Method from the Viewpoint of Quality Control* (The Graduate School, Department of Agriculture, Washington, 1939) demonstrates how futile it is to attempt to estimate the magnitude of σ from samples of 4, but that samples of 100 are excellent, while samples of 1000 give practically perfect results.

Computation of the control limits or error band for samples from a normal universe. Let m successive samples of size n be drawn from a normal universe of mean μ and standard deviation σ, and let $\bar{x}$, s, s^2, and the range R be calculated for each sample. The results are shown symbolically in the accompanying table.

Sample number	Individual values	$\bar{x}$	s	s^2	R
1	x_{11}				
	x_{12}				
	.				
	.				
	.				
	x_{1n}				
		$\bar{x}_1$	s_1	$s_1{}^2$	R_1
2	x_{21}				
	x_{22}				
	.				
	.				
	.				
	x_{2n}				
		$\bar{x}_2$	s_2	$s_2{}^2$	R_2
.					
.					
.					
m	x_{m1}				
	x_{m2}				
	.				
	.				
	.				
	x_{mn}				
		$\bar{x}_m$	s_m	$s_m{}^2$	R_m
Average	$\bar{\bar{x}}$	$\bar{\bar{x}}$	$\bar{s}$	$\bar{s^2}$	$\bar{R}$
"Expected" average	μ	μ	Es	Es^2	ER

The following theoretical relations exist for a normal universe:

$$Es = \sqrt{\frac{2\pi}{n}} \frac{\sigma}{B(\tfrac{1}{2}n - 1, \tfrac{1}{2})} \qquad \text{[P. 496]}$$

$$= \sigma \left\{ 1 - \frac{3}{4n} - \frac{7}{32n^2} - \cdots \right\} \qquad \text{[P. 522]}$$

$$= c_2 \sigma \qquad \text{[P. 521]} \quad (27)$$

$$ER = d_2 \sigma \qquad \begin{array}{l}\text{[The constant } d_2 \text{ is not calculated here.} \\ \text{Numerical values are shown in} \\ \text{Table 2, p. 570.]}\end{array} \quad (28)$$

The symbols c_2 and d_2, also A_1, A_2, B_3, B_4, D_3, and D_4, to be seen presently, were initiated by Shewhart in his original writings on quality-control. Their numerical values are shown in the table on page 570. As n increases, c_2 approaches unity from below, as is obvious from the series above. When $n \geq 16$, the error in calculating c_2 by the simple series $1 - 3/4n$ is less than 1 part in 1000.

Although the factors c_2 and d_2 are valid only for a normal universe, they are nevertheless useful in practice. The factor d_2 especially appears to withstand considerable departures from normality.

In work in quality-control, where the aim is to find assignable causes of variability in quality (inches, pounds, hardness, or other characteristics of quality) it is useful to adopt and test the hypothesis that the quality is varying like numbers being drawn from a bowl in the ideal-bowl experiment. If they do so behave, the manufacturing process (including the sampling and measuring) is said to be in statistical control. The alternative hypothesis H_1 is that, occasionally at least, the level (μ) of quality slips away to one side or the other below or above its average value, and slips far enough to cause concern. It is as if someone changed bowls occasionally. The hypothesis H_2 is similar with reference to the dispersion (σ).

To test the hypothesis H_0, control-charts are plotted. There will be a chart for $\bar{x}$ and one for either R or s (but not both). Points are plotted to show $\bar{x}$ and either R or s for 25 or more successive small samples, in the order in which they occur. There will be central lines at the level $\bar{\bar{x}}$ on the $\bar{x}$-chart, and a central line at the level $\bar{s}$ or $\bar{R}$ on the s- or R-chart.

Control limits are placed above and below the central lines. If H_1 is true, points on the $\bar{x}$-chart will be sent outside the control limits oftener than if H_0 is true. And if H_2 is true, points on the s- or R-chart will likewise go beyond the control limits oftener than if H_0 is true. If a point falls beyond the $\bar{x}$-control limits, an assignable cause of variability is indicated, and the hypothesis H_0 that statistical control exists is rejected, and the alternative hypothesis H_1 is accepted that the level

(μ) of the bowl is not constant but is occasionally displaced above or below its usual range by an assignable cause of variability. Likewise, if a point on the R-chart falls beyond the control limits on that chart (and similarly for the s-chart), the hypothesis H_2 is accepted. On the other hand, if the points all fall within the control limits, and if no suspicious runs, trends, or other patterns can be discerned, H_0 is accepted, and the process is said to exhibit *statistical control* or *stability*.

The spacing of the control limits is important. If too widely spaced, they will not detect assignable causes of variability frequently enough, and if too narrow they will give the alarm too often when there is no assignable cause to be found. In practice, 3-sigma limits have been found to be the correct spacing as judged by the criterion that the control-chart procedure should strike a balance between these two mistakes:

i. Looking for assignable causes too often when they do not exist;

ii. Failing to look for assignable causes on some occasions when they actually do exist.

The theory of the 3-sigma limits is as follows: *i.* the means of samples of n are distributed with standard deviation $\sigma/\sqrt{n}$; *ii.* practically all of any ordinary distribution is included within the compass ± 3 standard deviations, and any point beyond 3 sigma is taken as an indication of an assignable cause. For the $\bar{x}$-chart, the position of the 3-sigma control limits will be

$$\bar{x} \pm 3\,\frac{\sigma}{\sqrt{n}}$$

σ is estimated from the average value of R or the average value of s. Thus,

$$\sigma' = \frac{\bar{R}}{d_2} \quad \text{if the } R\text{-chart is used}$$

$$\sigma' = \frac{\bar{s}}{c_2} \quad \text{if the } s\text{-chart is used}$$

The 3-sigma control limits for the $\bar{x}$-chart are thus computed as

$$\bar{x} \pm \frac{3}{d_2\sqrt{n}}\,\bar{R} \quad \text{if the } R\text{-chart is used}$$

$$\bar{x} \pm \frac{3}{c_2\sqrt{n}}\,\bar{s} \quad \text{if the } s\text{-chart is used}$$

$\bar{R}$ and $\bar{s}$, it should be noted, are the central lines on the R- and s-charts.

For convenience, let

$$A_1 = \frac{3}{c_2\sqrt{n}} \tag{29}$$

$$A_2 = \frac{3}{d_2\sqrt{n}} \tag{30}$$

A_1 and A_2 are shown in the table on page 570 for various sizes of sample.

The purpose of quality-control is to find assignable causes and (if practicable) to eliminate them one by one until statistical control is exhibited. The distribution of the particular quality-characteristic under consideration is then definite, predictable, and stable.

If statistical control is exhibited, the control limits may be projected a short distance into the future, say for the next 5, 10, or 20 samples. The points for these future samples should be plotted as soon as possible, so that if any point falls outside the control limits there is still a good chance of discovering what change (assignable cause) in the manufacturing process is responsible.

Similar remarks apply to the charts for R or s. The $\bar{x}$-chart is used to discover assignable causes that affect the level of $\bar{x}$; the R- or s-chart is used to discover assignable causes that affect the dispersion of quality. Examples will be found in the references at the end of Chapter 8 (p. 283).

Remark 1. It is customary to start a control-chart for $\bar{x}$ on the basis of 25 samples, but fewer may be used if there is a paucity of good data. Why specify 25 samples? This is the same question as asking how accurately the control limits need to be fixed. Control limits are one form of fixed-interval prediction. The coefficient of variation of width of the interval between them is the same as the coefficient of variation of σ' or of s itself. Each sample of n contributes $n - 1$ degrees of freedom to the distribution of s. If there are m points on which to base the calculations, there are $m(n - 1)$ degrees of freedom altogether, and this is the value of k to be used in the formula

$$\text{C.V. } s^2 = \sqrt{\frac{2}{k}} \qquad \text{[If the universe may be considered normal, p. 340]}$$

or

$$\text{C.V. } s^2 = \sqrt{\frac{\beta_2 - 1}{k} - \frac{\beta_2 - 3}{k^2}\left[1 + \frac{1}{k + 1}\right]} \qquad \begin{array}{l}\text{[To be used if the universe} \\ \text{is far from normal; see} \quad (31) \\ \text{p. 338. } \beta_2 = \mu_4/\sigma^4]\end{array}$$

For a rough idea of the variability in the setting of the control limits the assumption of normality will be made, which is probably not bad in most practice when a state of control has been reached. On this assumption the coefficient of variation in s and hence in the spacing of the control limits for 25 samples of 5 will be determined by the relation

$$\text{C.V. } s = \frac{1}{2}\sqrt{\frac{2}{m(n - 1)}} = \frac{1}{2}\sqrt{\frac{2}{25(5 - 1)}} = \frac{1}{2}\sqrt{\frac{2}{100}}$$

$$= \tfrac{1}{14} \text{ or 7 percent} \tag{32}$$

Now at the 3-sigma limit, what variation in probability corresponds to a coefficient of variation of 7 percent in an estimate of $\sigma_{\bar{x}}$? A rough answer is found by calculating the area of a vertical strip under a normal curve, the strip being centered at $3\,\sigma$, its height being the height of the normal curve at that point and its half-width being $0.07\,\sigma$. Explicitly, the variation in probability will be roughly

$$P_{3\sigma} \pm 3 \times 0.07 \times \frac{1}{\sqrt{2\pi}}\, e^{-\frac{1}{2}\cdot 3^2} = P_{3\sigma} \pm 3 \times 0.07 \times 0.0044$$
$$= P_{3\sigma} \pm 0.0009 \qquad (33)$$

where $P_{3\sigma}$ is the area under the normal probability curve beyond the 3-sigma limit on one side (about 0.00135). Strictly, this variability is not equal on each side of the 3-sigma limit, but is more like 0.0012 on the inner side and 0.0006 on the outer side, as the student may wish to verify either from a table of the normal probability integral or by direct calculation. Thus, assuming a normal distribution for $\bar{x}$, the coefficient of variation in estimating σ from the average standard deviation ($\bar{s}$) calculated from 25 samples of 5 corresponds to a play between

$$0.0014 + 0.0012 = 0.0026$$
and
$$0.0014 - 0.0006 = 0.0008$$

in the probability beyond a 3-sigma limit for $\bar{x}$. Added to this is the sampling fluctuation of $\bar{x}$ itself, which will have a standard deviation of $\sigma/\sqrt{mn}$ or $\sigma/\sqrt{125} = 0.09\,\sigma$ in this example which about doubles the variability just calculated. While this is some considerable play, it is not too much, and it is to be remembered that 75 more points would only trim this variability to half its value. It is therefore recommended that the control-chart be started with data from 25 points, or (as previously stated) with even fewer if there is a paucity of data.

Now, how about the control limits for the s-chart? Again, we may say that practically all of the m standard deviations ($s_1, s_2, \cdots, s_m$) will fall within $Es \pm 3\sigma_s$, so the 3-sigma control limits for the s-chart will be

$$Es \pm 3\sigma_s = Es \left\{ 1 \pm 3\,\frac{\sigma_s}{Es} \right\}$$

$$= Es \left\{ 1 \pm 3\,\frac{[Es^2 - (Es)^2]^{\frac{1}{2}}}{Es} \right\}$$

$$= B_4 Es \quad \text{for the upper control limit} \Big\}$$
$$= B_3 Es \quad \text{for the lower control limit} \Big\} \qquad (34)$$

B_4 is defined with the $+$ sign, B_3 with the $-$ sign, except that B_3 is never to be negative. Series for B_3 and B_4 will now be developed.

$$\left.\begin{array}{c} B_4 \\ B_3 \end{array}\right\} = Es \left\{ 1 \pm 3 \left[\frac{Es^2}{(Es)^2} - 1 \right]^{\frac{1}{2}} \right\}$$

$$= Es \left\{ 1 \pm 3 \left[\frac{n-1}{n} \frac{1}{c_2{}^2} - 1 \right]^{\frac{1}{2}} \right\}$$

$$\doteq Es \left\{ 1 \pm 3 \left[\frac{n-1}{n} e^{\frac{3}{2n} + \frac{1}{n^2} + \cdots} - 1 \right]^{\frac{1}{2}} \right\} \qquad \text{[Eq. 50, p. 521]}$$

$$\doteq Es \left\{ 1 \pm 3 \left[\left(1 - \frac{1}{n} \right) \left(1 + \frac{3}{2n} + \frac{17}{8n^2} + \cdots \right) - 1 \right]^{\frac{1}{2}} \right\}$$

$$= Es \left\{ 1 \pm 3 \left[\frac{1}{2n} - \frac{5}{8n^2} - \cdots \right]^{\frac{1}{2}} \right\}$$

$$= Es \left\{ 1 \pm \frac{3}{\sqrt{2n}} \left[1 - \frac{10}{8n} - \cdots \right]^{\frac{1}{2}} \right\}$$

$$= Es \left\{ 1 \pm \frac{3}{\sqrt{2n}} \left[1 - \frac{5}{8n} - \cdots \right] \right\} \qquad (35)$$

As an exercise the student may wish to carry one more term in the series for B_3 and B_4, finding that

$$\left.\begin{array}{c} B_4 s \\ B_3 s \end{array}\right\} = Es \pm 3\sigma_s = Es \left\{ 1 \pm \frac{3}{\sqrt{2n}} \left[1 - \frac{5}{8n} - \frac{21}{64n^2} - \cdots \right] \right\} \qquad (36)$$

Numerical values of B_3 and B_4 are in Table 2.

Calculation of the corresponding factors D_3 and D_4 for the control limits for R will not be attempted here.

Remark 2. Charts for s and R are never used simultaneously. It is difficult to say which is better, s or R, for discovering assignable causes in dispersion or for fixing the control limits for $\bar{x}$. The range, R, besides being simple to compute (it can be written down mentally), is also very efficient in discovering assignable causes of change of dispersion. Very little theoretical work has been done by which the efficiencies of s- and R-charts can be compared. Much depends on the state of control and the shape of the universe. If the universe were normal, the range would be less efficient for estimating σ for fixing the spread of the control limits for $\bar{x}$, and considerably so for samples above 10 or 12, although a sample of (e.g.) 25 can be broken up into 5 samples of 5 and an average range taken and corrected for size of sample (cf. the references on quality-control). However, estimation of σ is not one of the problems until control has been attained, in which state any method of estimation, efficient or not, is good enough, as there will then be a superfluity of data.

Remark 3. If more power is needed to detect assignable causes (lack of control) than is provided by the 3-sigma limits, as when it is extremely costly not to detect assignable causes, and when it is not costly to look for trouble occasionally when there is none, limits arbitrarily narrower than 3 sigma may be used, such as 2 or $2\frac{1}{2}$ sigma, or "probability limits" may be assigned if there is any basis for arriving at appropriate values of α and β.

Remark 4. Control-charts direct attention to time-trends of the points and to the extreme values of $\bar{x}$, R, s, or p (the p-chart, for fraction defective, the simplest chart of all, is not discussed here). The analysis of variance (Ch. 17), on the other hand, directs attention to the general spread of the points, through a comparison of the variance of the plotted points ($\bar{x}$, R, s, p) with the variance as estimated internally.[17]

TABLE 2. FACTORS FOR COMPUTING 3-SIGMA CONTROL LIMITS

Number of observations in sample	Chart for averages			Chart for standard deviations					Chart for ranges					Number of observations in sample
	Factors for control limits			Factor for central line	Factors for control limits				Factor for central line	Factors for control limits				
n	A	A_1	A_2	c_2	B_1	B_2	B_3	B_4	d_2	D_1	D_2	D_3	D_4	n
2	2.121	3.759	1.880	.5642	0	2.064	0	3.658	1.128	0	3.686	0	3.268	2
3	1.732	2.394	1.023	.7236	0	1.948	0	2.692	1.693	0	4.358	0	2.574	3
4	1.500	1.880	0.729	.7979	0	1.859	0	2.330	2.059	0	4.698	0	2.282	4
5	1.342	1.596	.577	.8407	0	1.789	0	2.128	2.326	0	4.918	0	2.114	5
6	1.225	1.410	.483	.8686	.003	1.735	.003	1.997	2.534	0	5.078	0	2.004	6
7	1.134	1.277	.419	.8882	.086	1.690	.097	1.903	2.704	0.205	5.203	.076	1.924	7
8	1.061	1.175	.373	.9027	.153	1.653	.169	1.831	2.847	0.387	5.307	.136	1.864	8
9	1.000	1.094	.337	.9139	.207	1.621	.227	1.774	2.970	0.546	5.394	.184	1.816	9
10	0.949	1.028	.308	.9227	.252	1.594	.273	1.727	3.078	0.687	5.469	.223	1.777	10
11	.905	0.973	.285	.9300	.290	1.570	.312	1.688	3.173	0.812	5.534	.256	1.744	11
12	.866	.925	.266	.9359	.324	1.548	.346	1.654	3.258	0.925	5.593	.284	1.717	12
13	.832	.884	.249	.9410	.353	1.529	.375	1.625	3.336	1.026	5.646	.308	1.692	13
14	.802	.848	.235	.9453	.378	1.512	.400	1.599	3.407	1.121	5.693	.329	1.671	14
15	.775	.817	.223	.9490	.401	1.497	.423	1.577	3.472	1.207	5.737	.348	1.652	15
16	.750	.788	....	.9523	.422	1.483	.443	1.557	....	....	....	....	....	16
17	.728	.762	....	.9551	.441	1.470	.462	1.539	....	....	....	....	....	17
18	.707	.738	....	.9577	.458	1.458	.478	1.522	....	....	....	....	....	18
19	.688	.717	....	.9599	.473	1.447	.493	1.507	....	....	....	....	....	19
20	.671	.698	....	.9619	.488	1.436	.507	1.493	....	....	....	....	....	20
21	.655	.680	....	.9638	.501	1.427	.520	1.481	....	....	....	....	....	21
22	.639	.662	....	.9655	.513	1.418	.531	1.469	....	....	....	....	....	22
23	.626	.647	....	.9670	.525	1.409	.543	1.457	....	....	....	....	....	23
24	.612	.632	....	.9684	.535	1.401	.552	1.447	....	....	....	....	....	24
25	.600	.619	....	.9697	.545	1.394	.562	1.438	....	....	....	....	....	25

Table 2 is taken from Appendix 1 to the American War Standard, Z1.3–1942, which is in turn a copy, with minor corrections, of Table 1

[17] Henry Scheffé, "The relation of control charts to analysis of variance and chi-square tests," *J. Am. Stat. Assoc.*, vol. 42, 1947: pp. 425–31.

in Supplement B of the *Manual on Presentation of Data,* published by the American Society for Testing Materials.   Permission to use this table has been granted by the American Society for Testing Materials and the American Standards Association.

FORMULAS FOR USE OF TABLE 2

Purpose of chart	Chart for	Central line	3-sigma control limits
For analyzing past inspection data for control ($\bar{x}$, $\bar{s}$, $\bar{R}$ are average values for the data being analyzed)	Averages	$\bar{x}$	$\bar{x} \pm A_1\bar{s}$, or $\bar{x} \pm A_2\bar{R}$
	Standard deviations	$\bar{s}$	$B_3\bar{s}$ and $B_4\bar{s}$
	Ranges	$\bar{R}$	$D_3\bar{R}$ and $D_4\bar{R}$
For controlling quality during production ($\bar{x}'$, σ', R' are selected standard values; $R' = d_2\sigma'$ for samples of size n)	Averages	$\bar{x}'$	$\bar{x}' \pm A\sigma'$, or $\bar{x}' \pm A_2R'$
	Standard deviations	$c_2\sigma'$	$B_1\sigma'$ and $B_2\sigma'$
	Ranges	$d_2\sigma'$, or R'	$D_1\sigma'$ and $D_2\sigma'$, or D_3R' and D_4R'

CHAPTER 17. THE DISTRIBUTION OF THE EXTERNAL AND INTERNAL VARIANCES

> But this whole business is too large to deal with at the tail-end of a letter.—Screwtape to Wormwood, C. S. Lewis, *The Screwtape Letters* (Macmillan, 1943).

Some reasons for introducing these variances. Notation. In the mechanism for the formation of clusters (Ch. 5) two estimates of σ were encountered, one being formed from the external variances of the clusters, the other from their internal variances. The distribution of the ratio of the two estimates is important in comparing models for the formation of clusters, and hence in the design of samples drawn from primary units that have been formed from these models. The comparison of the two estimates of the same variance is equally important in deciding whether a particular system of stratification is effective or necessary. But the main use of the ratio of the two estimates is in the design of experiments and the analysis of variance, the importance of which can not be overstated.

Let m samples be drawn with replacement from a normal universe of N sampling units (chips) whose mean is μ and standard deviation is σ. Each sample will have a mean $(\bar{x}_j)$ and there will be a general mean $\bar{\bar{x}}$.

Sample number	Size of sample	Mean	Variance	Residual of the mean $\bar{x}_j - \bar{\bar{x}}$	Error of the mean $\bar{x}_j - \mu$
1	n_1	$\bar{x}_1$	s_1^2	v_1	u_1
2	n_2	$\bar{x}_2$	s_2^2	v_2	u_2
3	n_3	$\bar{x}_3$	s_3^2	v_3	u_3
.					
.					
.					
m	n_m	$\bar{x}_m$	s_m^2	v_m	u_m
Wtd. av.	$\bar{n}$	$\bar{\bar{x}}$	s_w^2	0	u

The notation will be closely consistent with that encountered earlier. The accompanying figure will help. Further, let

$$n = n_1 + n_2 + \cdots + n_m, \quad \text{the total sample} \tag{1}$$

$$\bar{\bar{x}} = \frac{n_1\bar{x}_1 + n_2\bar{x}_2 + \cdots + n_m\bar{x}_m}{n_1 + n_2 + \cdots + n_m}, \quad \text{the weighted mean of} \atop \text{the samples} \tag{2}$$

$$s_w{}^2 = \frac{n_1 s_1{}^2 + n_2 s_2{}^2 + \cdots + n_m s_m{}^2}{n_1 + n_2 + \cdots + n_m}, \quad \text{the average internal variance} \quad (3)$$

$$s_b{}^2 = \frac{n_1 v_1{}^2 + n_2 v_2{}^2 + \cdots + n_m v_m{}^2}{n_1 + n_2 + \cdots + n_m}, \quad \text{the external variance} \quad (4)$$

$$\bar{n} = \frac{n}{m}, \quad \text{the average size of sample} \quad (5)$$

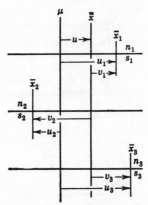

FIG. 85. Showing the relations between the various symbols. As the figure is drawn, v_1, v_3, u_1, u_3, and u are positive; v_2 and u_2 are negative.

Unbiased estimates of the variance of the universe, formed from the internal variances of the samples. No new principles are introduced here; as a matter of fact the development has already been carried out in Chapter 10 and need not be repeated here. From page 333 it is seen that for the jth sample, being drawn with replacement,

$$E \frac{n_j s_j{}^2}{n_j - 1} = \sigma^2 \qquad \text{[Ch. 10, Eq. 10, p. 333]} \quad (6)$$

or

$$E n_j s_j{}^2 = (n_j - 1)\sigma^2 \qquad \text{[As } n_j \text{ is constant]} \quad (7)$$

Addition of such equations over all m samples gives

$$\sum_1^m E n_j s_j{}^2 = (n - m)\sigma^2 \quad (8)$$

By Eq. 3, the left-hand side is $E n s_w{}^2$ wherefore

$$E s_w{}^2 = \frac{n - m}{n}\sigma^2 = \frac{\bar{n} - 1}{\bar{n}}\sigma^2 \quad (9)$$

and an unbiased estimate of σ^2 formed from the average internal variance $s_w{}^2$ is

$$\sigma''^2 = \frac{ns_w{}^2}{n - m} = \frac{\bar{n}}{\bar{n} - 1} s_w{}^2 \tag{10}$$

As a matter of fact, each sample gives its own estimate

$$\sigma_j''^2 = \frac{n_j}{n_j - 1} s_j{}^2 \tag{11}$$

The estimate σ''^2 is obtained by "pooling" the internal variances. Both of the last equations should be compared with Eq. 11 on page 333.

In the estimate σ''^2, $n - m$ is the number of degrees of freedom: it is the denominator by which the total internal 2d moment $ns_w{}^2$ is divided to get an unbiased estimate of σ^2. Each sample subtracts one degree of freedom.

By reasoning along the line of Exercise 3 on page 338 it can be shown that

$$\text{C.V. } \sigma'' = \frac{1}{2} \sqrt{\frac{\beta_2 - 1}{n - m}} \tag{12}$$

wherein $\beta_2 = \mu_4/\sigma^4$ as heretofore.

Unbiased estimate of the variance of the universe, formed from the external variance of the samples. First define the error in $\bar{x}$ as

$$u = \bar{\bar{x}} - \mu \qquad \text{[Illustrated in Fig. 85]} \tag{13}$$

Then write

$$\bar{x}_j - \mu = (\bar{x}_j - \bar{\bar{x}}) + (\bar{\bar{x}} - \mu)$$

whence

$$n_j(\bar{x}_j - \mu)^2 = n_j(\bar{x}_j - \bar{\bar{x}})^2 + n_j(\bar{\bar{x}} - \mu)^2 + 2(\bar{\bar{x}} - \mu)n_j(\bar{x}_j - \bar{\bar{x}})$$

Summation over all m samples gives

$$\sum_1^m n_j(\bar{x}_j - \mu)^2 = \sum_1^m n_j(\bar{x}_j - \bar{\bar{x}})^2 + (\bar{\bar{x}} - \mu)^2 \sum_1^m n_j + 0$$

I.e.,

$$\sum_1^m n_j u_j{}^2 = \sum_1^m n_j v_j{}^2 + nu^2$$

$$= ns_b{}^2 + nu^2 \tag{14}$$

Now as $Eu_j{}^2 = \text{Var } u_j = \sigma^2/n_j$ and $Eu^2 = \text{Var } u = \sigma^2/n$ (Ch. 4), the "expected" value of each term gives

$$m\sigma^2 = nEs_b{}^2 + \sigma^2$$

whence

$$Es_b{}^2 = \frac{m-1}{n}\sigma^2 \tag{15}$$

Thus an unbiased estimate of σ^2 formed from the external variance is

$$\sigma'^2 = \frac{n}{m-1}s_b{}^2 \qquad \begin{array}{l}\text{[The \textit{external} esti-}\\ \text{mate of } \sigma^2]\end{array} \tag{16}$$

The number of degrees of freedom in the external estimate just written is $m-1$. For the precision of this estimate we may call upon theory already developed in Chapters 4 and 10, whence it may be said that if the sample-sizes n_j were all equal to $\bar{n}$ then the

$$\text{C.V. }\sigma' = \frac{1}{2}\sqrt{\frac{\beta_2(\bar{x})-1}{m-1}} \qquad \text{[Ch. 10, p. 338]}$$

$$= \frac{1}{2}\sqrt{\frac{1}{m-1}\left[2+\frac{1}{\bar{n}}(\beta_2-3)\right]} \qquad \text{[Ch. 4, p. 104]} \tag{17}$$

wherein β_2 refers to the initial universe, and $\beta_2(\bar{x})$ refers to the distribution of $\bar{x}$. If the sample-sizes are unequal but not too far different, this equation will still be useful if $\bar{n}$ is taken as the average sample-size. Clearly, as $\bar{n}$ increases, the bracketed quantity under the last root sign approaches the normal value for $\beta_2 - 1$, viz., 2.

Both σ'' and σ' are estimates of σ. They are random variables, and in any one set of samples can not be expected to agree with σ nor with each other. If, contrary to hypothesis, the m samples are drawn from universes of the same standard deviation but whose means are not all equal, σ'' will not be affected, but σ' will on the average be increased. Hence a large value of $\sigma':\sigma''$ leads to suspicion that some of the samples were drawn from different universes. If the universes are all normal and of the same standard deviation σ, the probability-distribution of $\sigma':\sigma''$ leads to the z-distribution which is to be treated further on. The sampling fluctuations of this ratio constitute the foundation for the analysis of variance, originated by Fisher, for testing whether there is statistical evidence that some of the samples came from different universes. Consideration of the ratio $\sigma':\sigma''$ is also important in the choice of sampling unit (Ch. 5). Without waiting for refined results it may be crudely stated that if the two estimates σ''^2 and σ'^2 are separated by 3 estimated standard deviations of their difference, a violation of the hypothesis that the m samples all came from the same universe is clearly indicated. This circumstance is fulfilled by the inequality

$$F \geq 1 + 3\sqrt{\frac{1}{m-1}\left[2+\frac{m}{n}(\beta_2-3)\right] + \frac{\beta_2-1}{n-m}} \tag{18}$$

wherein F denotes $(\sigma':\sigma'')^2$. The radical encloses the sum of the variances of σ'^2 and σ''^2, under the assumption that all m samples came from the same universe. A lower degree of significance is obtained by using the factor 2 in place of 3.

Exercise. If the universe is normal, this crude test reduces to the inequality

$$F \geq 1 + 3 \sqrt{\frac{2}{K} + \frac{2}{k}} \qquad (18a)$$

wherein $K = m - 1$ and $k = n - m$. However, the more exact F- or z-test would then be preferred (*vide infra*).

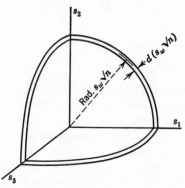

FIG. 86.   Illustrating the shell formed by confining $s_1, s_2, \cdots, s_m$ to the inequality $n(s_w - \frac{1}{2} ds_w)^2 < n_1 s_1^2 + n_2 s_2^2 + \cdots + n_m s_m^2 < n(s_w + \frac{1}{2} ds_w)^2$.

Remark. In addition to comparing σ' and σ'', it is highly important that the individual means $\bar{x}_1, \bar{x}_2, \cdots, \bar{x}_m$ be compared with each other, and the individual internal estimates $\sigma_1'', \sigma_2'', \cdots, \sigma_m''$ also, and perhaps plotted, as on a control-chart (preceding chapter). A particularly high or low value of $\bar{x}_j$ or σ_j'' can often be identified with special circumstances that lead to the discovery of causes of variability, and increased knowledge of the underlying sources of the data.

The distribution of the internal variance.   From Helmert's distribution of the standard deviation s (p. 511), the probability that the m standard deviations of the aforementioned samples will fall in the intervals $s_1 \pm \frac{1}{2} ds_1, s_2 \pm \frac{1}{2} ds_2, \cdots, s_m \pm \frac{1}{2} ds_m$ will be

$$C s_1^{n_1 - 2} s_2^{n_2 - 2} \cdots s_m^{n_m - 2} e^{-n s_w^2 / 2\sigma^2} \, ds_1 \, ds_2 \cdots ds_m$$

These m intervals define a parallelepiped of volume $ds_1 \, ds_2 \cdots ds_m$, and the expression just written is the probability of s_w falling within it. But the same range of $s_w \pm \frac{1}{2} ds_w$ as defined by this parallelepiped can be defined in many other ways; it is only necessary that $s_1, s_2, \cdots, s_m$ satisfy the inequality

$$n(s_w - \tfrac{1}{2} ds_w)^2 < n_1 s_1{}^2 + n_2 s_2{}^2 + \cdots + n_m s_m{}^2 < n(s_w + \tfrac{1}{2} ds_w)^2 \quad (19)$$

Throughout this shell the factor $e^{-ns_w{}^2/2\sigma^2}$ is constant to within a differential; hence to find the total probability for the interval $s_w \pm \tfrac{1}{2} ds_w$ it is only necessary to integrate

$$s_1{}^{n_1-2} s_2{}^{n_2-2} \cdots s_m{}^{n_m-2} \, ds_1 \, ds_2 \cdots ds_m$$

in the shell defined by the above inequality (illustrated in Fig. 86). The result will be $s_w{}^{n-m-1} \, ds_w$ multiplied by some constant whose value can be determined when needed. (The exponent $n - m - 1$ is arrived at instantly by adding the exponents of $s_1, s_2, \cdots, s_m, ds_1, ds_2, \cdots, ds_m$.) The distribution of s_w must therefore be

$$y \, ds_w = C' s_w{}^{n-m-1} e^{-ns_w{}^2/2\sigma^2} \, ds \quad (20)$$

The constant C' must be such that $\int_0^\infty y \, ds = 1$, and by returning to the exercises in the gamma function on pages 495–6 its value may be found and inserted into the last equation which then gives

$$y \, ds_w = \frac{n^{\frac{1}{2}(n-m)}}{\Gamma\left(\dfrac{n-m}{2}\right) 2^{\frac{1}{2}(n-m-2)}\sigma} \left(\frac{s_w}{\sigma}\right)^{n-m-1} e^{-ns_w{}^2/2\sigma^2} \, ds_w \quad (21)$$

for the distribution of s_w. This result may be seen instantly by making use of Exercise 45 on page 498: $n_i s_i{}^2/\sigma^2$ is distributed as χ^2 with $n_i - 1$ degrees of freedom, so $ns_w{}^2/\sigma^2 = \Sigma \, n_i s_i{}^2/\sigma^2$ must also be distributed as χ^2 with $\Sigma \, (n_i - 1)$ or $n - m$ degrees of freedom.

Exercise 1. Define σ''^2 as in Eq. 10 for samples drawn with replacement, viz.,

$$\sigma''^2 = \frac{n}{n-m} s_w{}^2$$

and prove that

a. the distribution of σ''^2 is

$$y \, d\sigma''^2 = \frac{k^{\frac{1}{2}k}}{\Gamma(\frac{1}{2}k) 2^{\frac{1}{2}k}\sigma^2} \left(\frac{\sigma''^2}{\sigma^2}\right)^{\frac{1}{2}(k-2)} e^{-k\sigma''^2/2\sigma^2} \, d\sigma''^2$$

b. the distribution of σ'' is

$$y \, d\sigma'' = \frac{k^{\frac{1}{2}k}}{\Gamma(\frac{1}{2}k) 2^{\frac{1}{2}(k-2)}\sigma} \left(\frac{\sigma''}{\sigma}\right)^{(k-1)} e^{-k\sigma''^2/2\sigma^2} \, d\sigma''$$

wherein $k = n - m$, the number of degrees of freedom in the estimate σ''^2 or σ''. These results are exactly true only for a normal universe, although $E\sigma''^2 = \sigma^2$ for any universe, as is known from Chapter 10.

HINT. Put

$$\sigma''^2 = \frac{n}{n-m} s_w{}^2 = \frac{n}{k} s_w{}^2$$

$$\frac{d\sigma''^2}{\sigma''^2} = \frac{2\, ds_w}{s_w}$$

$$\frac{d\sigma''}{\sigma''} = \frac{ds_w}{s_w}$$

Then substitute into Eq. 21.

Exercise 2. Put $\chi^2 = k\sigma''^2/\sigma^2 = ns_w{}^2/\sigma^2$. Prove that the distribution of χ^2 for samples drawn from a normal universe is

$$y\, d\chi^2 = \frac{1}{\Gamma(\frac{1}{2}k)2^{\frac{1}{2}k}}\,(\chi^2)^{\frac{1}{2}(k-2)}e^{-\frac{1}{2}\chi^2}\, d\chi^2$$

where again

$$k = n - m$$

The student will recall from page 497 that $E\chi^2 = k$, which is only saying that $E\sigma''^2 = \sigma^2$.

Exercise 3. From the distribution of either σ''^2 or χ^2

a. Prove that $E\sigma''^2 = \sigma^2$, a result that is true for any universe.

b. Prove that $\mathrm{Var}\,\sigma''^2 = 2\sigma^4/k$, a result that is strictly true only for a normal universe. For a non-normal universe the C.V. σ'' is given by Eq. 12, which of course reduces to the result just written for $\mathrm{Var}\,\sigma''^2$ when β_2 has its normal value of 3.

Exercise 4. The standard error of σ'' is

$$\frac{\sigma}{\sqrt{2k}}\left\{1 - \frac{1}{16k} + \cdots\right\}$$

provided the universe is normal.

The distribution of the external variance. Define anew

$$\chi^2 = \frac{1}{\sigma^2}\sum_1^m n_j v_j{}^2 = \frac{ns_b{}^2}{\sigma^2} = \frac{K\sigma'^2}{\sigma^2} \qquad \begin{bmatrix} v_j = \bar{x}_j - \bar{x}; \\ K = m - 1 \end{bmatrix} \quad (22)$$

To find the distribution of χ^2, first write the distribution of u_j as

$$y\, du_j = \frac{\sqrt{n_j}}{\sigma\sqrt{2\pi}}\, e^{-n_j u_j{}^2/2\sigma^2}\, du_j \tag{23}$$

whence

$$P = \frac{\sqrt{n_1 n_2 \cdots n_m}}{(\sigma\sqrt{2\pi})^m} e^{-\frac{1}{2\sigma^2}(n_1 u_1{}^2 + \ldots + n_m u_m{}^2)} du_1 \, du_2 \cdots du_m \qquad (24)$$

is the probability that the m errors in $\bar{x}_j$ will fall into the m ranges $u_j \pm \frac{1}{2} du_j$ $(j = 1, 2, \cdots, m)$. Next, this probability is to be written in terms of χ^2 and u^2 by the defining identities

$$\left.\begin{aligned}
u_1 &= v_1 + u \\
u_2 &= v_2 + u \\
&\;\;\vdots \\
u_{m-1} &= v_{m-1} + u \\
u_m &= v_m + u \\
&= -\frac{n_1 v_1 + n_2 v_2 + \cdots + n_{m-1} v_{m-1}}{n_m} + u
\end{aligned}\right\} \qquad (25)$$

The last row expresses the fact that $\Sigma n_j v_j = 0$, which in turn arises from the definition of $\bar{x}$, from which the residuals (v_j) are all measured. The Jacobian of this transformation is

$$J\left(\frac{u_1,\, u_2,\, \cdots,\, u_{m-1},\, u_m}{v_1,\, v_2,\, \cdots,\, v_{m-1},\, u}\right) = \begin{vmatrix}
\dfrac{\partial u_1}{\partial v_1} & \dfrac{\partial u_1}{\partial v_2} & \cdots & \dfrac{\partial u_1}{\partial u} \\[2mm]
\dfrac{\partial u_2}{\partial v_1} & \dfrac{\partial u_2}{\partial v_2} & & \dfrac{\partial u_2}{\partial u} \\[2mm]
\vdots & & & \vdots \\[2mm]
\dfrac{\partial u_m}{\partial v_1} & \dfrac{\partial u_m}{\partial v_2} & \cdots & \dfrac{\partial u_m}{\partial u}
\end{vmatrix} \qquad (26)$$

$$= \begin{vmatrix}
1 & 0 & 0 & \cdots & 1 \\
0 & 1 & 0 & \cdots & 1 \\
0 & 0 & 1 & \cdots & 1 \\
\vdots & & & & \vdots \\
-\dfrac{n_1}{n_m} & -\dfrac{n_2}{n_m} & -\dfrac{n_3}{n_m} & \cdots & 1
\end{vmatrix}$$

$$
= \begin{vmatrix}
1 & 0 & 0 & \cdots & 0 \\
0 & 1 & 0 & \cdots & 0 \\
0 & 0 & 1 & \cdots & 0 \\
\vdots & & & & \vdots \\
-\dfrac{n_1}{n_m} & -\dfrac{n_2}{n_m} & -\dfrac{n_3}{n_m} & \cdots & \dfrac{n_1 + n_2 + \cdots + n_m}{n_m}
\end{vmatrix}
$$

$$
= \frac{n}{n_m} \tag{27}
$$

The mth column of the last determinant comes from the preceding determinant by adding therein the negatives of all the other columns to the mth column for a new mth column.

Moreover, the exponent in the probability P is

$$
\begin{aligned}
\Sigma\, n_j u_j^2 &= \Sigma\, n_j(\bar{x}_j - \mu)^2 \\
&= \Sigma\, n_j[(\bar{x}_j - \bar{x}) + (\bar{x} - \mu)]^2 = \Sigma\, n_j[v_j + u]^2 \\
&= \Sigma\, n_j v_j^2 + 2u\, \Sigma\, n_j v_j + nu^2 \\
&= \sigma^2 \chi^2 + nu^2 \tag{28}
\end{aligned}
$$

Therefore, in terms of v_j and u,

$$
P = \frac{\sqrt{n_1 n_2 \cdots n_m}}{(\sigma\sqrt{2\pi}\,)^m} \frac{n}{n_m} e^{-\frac{1}{2}\chi^2 - \frac{nu^2}{2\sigma^2}}\, dv_1\, dv_2 \cdots dv_{m-1}\, du
$$

$$
= \frac{1}{(\sigma\sqrt{2\pi}\,)^{m-1}} \sqrt{\frac{n_1 n_2 \cdots n_{m-1} n}{n_m}}\, e^{-\frac{1}{2}\chi^2}\, dv_1\, dv_2 \cdots dv_{m-1}
$$

$$
\times \frac{\sqrt{n}}{\sigma\sqrt{2\pi}}\, e^{-nu^2/2\sigma^2}\, du \tag{29}
$$

The second factor shows that u (the error in $\bar{x}$) is distributed normally about 0 with standard deviation $\sigma/\sqrt{n}$—a fact already known. The integral of the second factor from $u = -\infty$ to $u = +\infty$ gives unity, leaving only the first factor, containing χ^2 and the parallelepiped $dv_1\, dv_2 \cdots dv_{m-1}$. In this parallelepiped χ^2 is confined to a narrow range, say $\chi^2 \pm \frac{1}{2} d\chi^2$. But the same χ^2 is produced by all the parallelepipeds in the shell

$$
[\sigma(\chi - \tfrac{1}{2}d\chi)]^2 < \Sigma\, n_j v_j^2 < [\sigma(\chi + \tfrac{1}{2}d\chi)]^2 \tag{30}
$$

so we need the integral of P throughout this shell. As

$$v_m = -\frac{1}{n_m}(v_1 + v_2 + \cdots + v_{m-1})$$

it follows that

$$\Sigma\, n_j v_j{}^2 = n_1\left(1 + \frac{n_1}{n_m}\right)v_1{}^2 + n_2\left(1 + \frac{n_2}{n_m}\right)v_2{}^2 + \cdots$$

$$+ n_{m-1}\left(1 + \frac{n_{m-1}}{n_m}\right)v_{m-1}{}^2$$

$$+ \frac{1}{n_m}\sum_i\sum_j n_i n_j v_i v_j \qquad \begin{matrix}[i,\ j\ =\ 1,\ 2,\ \cdots,\\ m-1;\, j \neq i]\end{matrix} \quad (31)$$

Because of the cross-products, this is an *ellipsoidal shell*. To rid this of the cross-products, introduce new variables defined by the transformation by which

$$t_1 = \sqrt{n_1\frac{n_1+n_m}{n_m}}\left[v_1 + \frac{n_2}{n_1+n_m}v_2 + \frac{n_3}{n_1+n_m}v_3 + \cdots + \frac{n_{m-1}}{n_1+n_m}v_{m-1}\right]$$

$$t_2 = \sqrt{n_2\frac{n_1+n_2+n_m}{n_1+n_m}}\left[v_2 + \frac{n_3}{n_1+n_2+n_m}v_3 + \cdots + \frac{n_{m-1}}{n_1+n_2+n_m}v_{m-1}\right]$$

$$t_3 = \sqrt{n_3\frac{n_1+n_2+n_3+n_m}{n_1+n_2+n_m}}\left[v_3 + \cdots + \frac{n_{m-1}}{n_1+n_2+n_3+n_m}v_{m-1}\right]$$

$$\vdots$$

$$t_{m-1} = \sqrt{n_{m-1}\frac{n}{n-n_{m-1}}}\,v_{m-1}$$

$$\cdots \quad (32)$$

With a little algebraic effort it will be seen that

$$t_1{}^2 + t_2{}^2 + \cdots + t_{m-1}{}^2 = \sum_1^m n_j v_j{}^2 \qquad (33)$$

Thus the cross-products have disappeared and χ^2 is now confined to the *spherical* shell

$$[\sigma(\chi - \tfrac{1}{2}\,d\chi)]^2 < \sum_1^{m-1} t_j{}^2 < [\sigma(\chi + \tfrac{1}{2}\,d\chi)]^2 \qquad (34)$$

The Jacobian of the transformation from v_j to t_j is

$$J\begin{pmatrix}t_1,\ t_2,\ \cdots,\ t_{m-1}\\ v_1,\ v_2,\ \cdots,\ v_{m-1}\end{pmatrix} = \sqrt{\frac{n_1 n_2 \cdots n_{m-1} n}{n_m}} = \frac{1}{J\begin{pmatrix}v_1,\ v_2,\ \cdots,\ v_{m-1}\\ t_1,\ t_2,\ \cdots,\ t_{m-1}\end{pmatrix}} \qquad (35)$$

wherefore in the probability P in Eq. 29

$$\sqrt{\frac{n_1 n_2 \cdots n_{m-1} n}{n_m}} \, dv_1 \, dv_2 \cdots dv_{m-1}$$

is to be replaced by $dt_1 \, dt_2 \cdots dt_{m-1}$.

The volume of the spherical shell in the t-space, being of radius $\sigma\chi$ and thickness $d(\sigma\chi)$ in $m - 1$ dimensions, must be

$$dV = \frac{2\pi^{\frac{1}{2}(m-1)}}{\Gamma[\frac{1}{2}(m-1)]} (\sigma\chi)^{m-2} \, d(\sigma\chi) \quad \text{or} \quad \frac{\pi^{\frac{1}{2}(m-1)}}{\Gamma[\frac{1}{2}(m-1)]} \sigma^{m-1}(\chi^2)^{\frac{1}{2}(m-3)} \, d\chi^2$$

[P. 491]

So when P is integrated over this shell the result is

$$y \, d\chi^2 = \frac{1}{\Gamma(\frac{1}{2}K)2^{\frac{1}{2}K}} (\chi^2)^{\frac{1}{2}(K-2)} e^{-\frac{1}{2}\chi^2} \, d\chi^2 \tag{36}$$

wherein $K = m - 1$. This is the same distribution of χ^2 as was obtained earlier under the definition given by Exercise 2, the only difference being that the number of degrees of freedom is now $K = m - 1$, whereas for σ'' it was $k = n - m$. It follows that the distribution of σ'^2 is the same as the distribution of σ''^2 as given in Exercise 1 except for the degrees of freedom. The C.V. σ' has already been evaluated in Eq. 17.

Some properties of the χ^2-distribution have been mentioned earlier (pp. 516–9).

Exercise 5. Take the case of three samples. Helmert's t-transformation (p. 581) is

$$t_1 = \sqrt{n_1 \frac{n_1 + n_3}{n_3}} \left(v_1 + \frac{n_2}{n_1 + n_3} v_2 \right)$$

$$t_2 = \sqrt{n_2 \frac{n_1 + n_2 + n_3}{n_1 + n_3}} \, v_2$$

By actual multiplication show that

$$t_1{}^2 + t_2{}^2 = n_1 \left(1 + \frac{n_1}{n_3} \right) v_1{}^2 + n_2 \left(1 + \frac{n_2}{n_3} \right) v_2{}^2 + 2 \frac{n_1 n_2}{n_3} v_1 v_2$$

$$= n_1 v_1{}^2 + n_2 v_2{}^2 + n_3 v_3{}^2 = m s_b{}^2$$

$$= \sigma^2 \chi^2$$

Let $t_1{}^2 + t_2{}^2$ be confined to the circular ring in Fig. 87 defined by

$$[\sigma(\chi - \tfrac{1}{2}d\chi)]^2 \leq t_1{}^2 + t_2{}^2 \leq [\sigma(\chi + \tfrac{1}{2}d\chi)]^2$$

or

$$t_1{}^2 + t_2{}^2 = [\sigma(\chi \pm \tfrac{1}{2}d\chi)]^2$$

The area of this ring is $2\pi r\, dr$ where $r = \sigma\chi$; i.e., the area $= 2\pi\sigma^2\chi\, d\chi = \pi\sigma^2\, d\chi^2$, which is the integral of $\sqrt{\dfrac{n_1 n_2 n}{n_3}}\, dv_1\, dv_2$ in Eq. 29. It follows that in two dimensions,

$$y\, d\chi^2 = \frac{1}{(\sigma\sqrt{2\pi}\,)^2}\, e^{-\frac{1}{2}\chi^2}\pi\sigma^2\, d\chi^2$$

$$= \tfrac{1}{2}e^{-\frac{1}{2}\chi^2}\, d\chi^2$$

which also comes directly from Eq. 36 by putting $K = 2$.

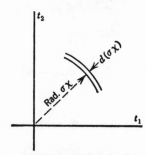

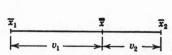

<table>
<tr><td>FIG. 87. Illustrating the spherical shell in the t-space for three samples.</td><td>FIG. 88. Two sample-means. The residual v_2 is positive; the residual v_1 is negative. $\bar{x}$, v_1, and v_2 are random variables.</td></tr>
</table>

In the next four exercises a pair of samples is drawn with replacement from the same normal universe of mean 0 and standard deviation σ. (See Fig. 88.)

$$\bar{x} = \frac{n_1 v_1 + n_2 v_2}{n_1 + n_2}$$

$$n_1 v_1 + n_2 v_2 = 0$$

$$\chi^2 = \frac{1}{\sigma^2}\left(n_1 v_1{}^2 + n_2 v_2{}^2\right)$$

$$m = 2, \quad K = 1 \quad \text{degree of freedom}$$

Exercise 6. The χ^2-distribution for the two means is

$$y\, d\chi^2 = \frac{1}{\sqrt{2\pi}}\, (\chi^2)^{-\frac{1}{2}}e^{-\frac{1}{2}\chi^2}\, d\chi^2$$

or

$$y\, d|\chi| = \frac{2}{\sqrt{2\pi}}\, e^{-\frac{1}{2}\chi^2}\, d|\chi|$$

This is the right-hand half of a normal curve, doubled. Why doubled?

Exercise 7.

a.
$$\bar{x}_2 - \bar{x}_1 = v_2 - v_1 = -\frac{n}{n_2}v_1 = \frac{n}{n_1}v_2$$

b.
$$\chi^2 = \frac{1}{\sigma^2}(n_1 v_1{}^2 + n_2 v_2{}^2) = \frac{(\bar{x}_2 - \bar{x}_1)^2}{\sigma^2\left(\dfrac{1}{n_1} + \dfrac{1}{n_2}\right)}$$

c. The distribution of $|\chi|$ then gives

$$y\, d|\bar{x}_2 - \bar{x}_1| = \frac{2}{\sqrt{2\pi}\,\sigma\sqrt{\dfrac{1}{n_1} + \dfrac{1}{n_2}}}\, e^{-\frac{(\bar{x}_2 - \bar{x}_1)^2}{2\sigma^2\left(\frac{1}{n_1}+\frac{1}{n_2}\right)}}\, d|\bar{x}_2 - \bar{x}_1|$$

d.
$$y\, d(\bar{x}_2 - \bar{x}_1) = \frac{1}{\sqrt{2\pi}\,\sigma\sqrt{\dfrac{1}{n_1} + \dfrac{1}{n_2}}} \qquad `` \qquad d(\bar{x}_2 - \bar{x}_1)$$

Thus $\bar{x}_2 - \bar{x}_1$ is distributed normally about 0 with variance $\sigma^2(1/n_1 + 1/n_2)$.

Exercise 8. The distributions of v_1 and v_2 in a pair of samples are

$$y\, dv_1 = \frac{\sqrt{n_1\left(1 + \dfrac{n_1}{n_2}\right)}}{\sigma\sqrt{2\pi}}\, e^{-\frac{n_1}{2\sigma^2}\left(1+\frac{n_1}{n_2}\right)v_1{}^2}\, dv_1$$

$$y\, dv_2 = \frac{\sqrt{n_2\left(1 + \dfrac{n_2}{n_1}\right)}}{\sigma\sqrt{2\pi}}\, e^{-\frac{n_2}{2\sigma^2}\left(1+\frac{n_2}{n_1}\right)v_2{}^2}\, dv_2$$

Note that the distribution of v_1 approaches that of u_1 as n_2 increases. Recall that $u_1 = \bar{x}_1 - \mu$. It follows that

$$\text{Var } v_1 = \frac{\sigma^2}{n_1\left(1 + \dfrac{n_1}{n_2}\right)} \quad \text{and} \quad \text{Var } v_2 = \frac{\sigma^2}{n_2\left(1 + \dfrac{n_2}{n_1}\right)}$$

Exercise 9. In a single sample of n drawn with replacement, the distribution of any one of the residuals is

$$y\, dv = \frac{\sqrt{n/(n-1)}}{\sigma\sqrt{2\pi}}\, e^{-\frac{nv^2}{2\sigma^2(n-1)}}\, dv$$

which is a normal curve with standard deviation $\sigma\sqrt{\dfrac{n-1}{n}}$. The distribution of the ith residual is thus normal like the distribution of an error (ϵ_i) but with standard deviation $\sigma \sqrt{1 - \dfrac{1}{n}}$ in place of σ.

Exercise 10. A pair of samples from two universes having coincident means but different standard deviations, σ_1 and σ_2. Prove that the distribution of $\bar{x}_2 - \bar{x}_1$ is normal about 0 with variance $\sigma_1^2/n_1 + \sigma_2^2/n_2$.

Solution

A solution may be obtained by the propagation of error, as on page 130. Still another solution follows. Let

$$n_2' = n_2 \left(\frac{\sigma_1}{\sigma_2}\right)^2$$

Then

$$\chi^2 = \frac{n_1 v_1^2}{\sigma_1^2} + \frac{n_2 v_2^2}{\sigma_2^2} = \frac{1}{\sigma_1^2}(n_1 v_1^2 + n_2' v_2^2)$$

$$\bar{x} = \frac{n_1 \bar{x}_1 + n_2' \bar{x}_2}{n_1 + n_2'}$$

$$n_1 v_1 + n_2' v_2 = 0$$

In other words χ^2, $\bar{x}$, v_1, and v_2 all satisfy the same equations that were used to find the distributions of χ^2 and t, except that n_2' replaces n_2. So $\bar{x}_2 - \bar{x}_1$ will be distributed about 0 with variance $\sigma_1^2 \left(\dfrac{1}{n_1} + \dfrac{1}{n_2'}\right) = \dfrac{\sigma_1^2}{n_1} + \dfrac{\sigma_2^2}{n_2}$.

Exercise 11. It is known from the preceding exercise that

$$\text{Var}\,(\bar{x}_2 - \bar{x}_1) = \frac{\sigma_1^2}{n_1} + \frac{\sigma_2^2}{n_2} \qquad [\bar{x}_1 \text{ and } \bar{x}_2 \text{ independent}]$$

Show that estimates of the $\text{Var}\,(\bar{x}_2 - \bar{x}_1)$ may be written as

$$\sigma''^2 \left(\frac{1}{n_1} + \frac{1}{n_2}\right) = \frac{n_1 s_1^2 + n_2 s_2^2}{n_1 + n_2 - 2}\left(\frac{1}{n_1} + \frac{1}{n_2}\right) \to \frac{s_1^2}{n_2} + \frac{s_2^2}{n_1}$$

[If it is believed that the sampled universes have the same standard deviation]

and

$$\frac{s_1^2}{n_1 - 1} + \frac{s_2^2}{n_2 - 1} \to \frac{s_1^2}{n_1} + \frac{s_2^2}{n_2}$$

[If it is believed that the sampled universes have different standard deviations]

Exercise 12. Let a pair of samples be drawn with replacement from two universes of the same standard deviation σ but with means μ_1 and

μ_2 differing by the amount $D = \mu_2 - \mu_1$. Prove that if

$$\bar{x} = \frac{n_1 \bar{x}_1 + n_2 \bar{x}_2}{n_1 + n_2} \qquad \text{[Eq. 2 with } m = 2\text{]}$$

$$\sigma''^2 = \frac{n_1 s_1{}^2 + n_2 s_2{}^2}{n_1 + n_2 - 2} \qquad \text{[Eq. 10 with } m = 2\text{]}$$

$$s_b{}^2 = \frac{1}{n}(n_1 v_1{}^2 + n_2 v_2{}^2) \qquad \text{[Eq. 4]}$$

$$\sigma'^2 = \frac{n}{m-1} s_b{}^2 = n_1 v_1{}^2 + n_2 v_2{}^2 \quad \text{[Eq. 16 with } m = 2\text{]}$$

then

$$E\sigma'^2 = \sigma^2 + \frac{D^2}{\dfrac{1}{n_1} + \dfrac{1}{n_2}}$$

while

$$E\sigma''^2 = \sigma^2 \quad \text{as before}$$

Solution

First, find Ev_1 which will be needed in a moment.

$$v_1 = \bar{x}_1 - \bar{x} \qquad \text{[Definition]}$$

$$= \bar{x}_1 - \frac{n_1 \bar{x}_1 + n_2 \bar{x}_2}{n_1 + n_2} \qquad \text{[By definition of } \bar{x}\text{]}$$

$$= \frac{n_2}{n}(\bar{x}_1 - \bar{x}_2) \quad \text{where} \quad n = n_1 + n_2$$

So

$$Ev_1 = \frac{n_2}{n}(\mu_1 - \mu_2) = \frac{n_2}{n} D$$

Then

$$\sigma'^2 = n_1 v_1{}^2 + n_2 v_2{}^2$$

$$= \frac{(n_1 v_1)^2}{n_1} + \frac{(n_2 v_2)^2}{n_2} \qquad \text{[Because } n_1 v_1 + n_2 v_2 = 0\text{]}$$

$$= \left(\frac{1}{n_1} + \frac{1}{n_2}\right) n_1{}^2 v_1{}^2$$

Hence

$$E\sigma'^2 = \left(\frac{1}{n_1} + \frac{1}{n_2}\right) n_1{}^2 Ev_1{}^2$$

$$= \left(\quad \text{"} \quad \right) n_1{}^2 \left[\text{Var } v_1{}^2 + (Ev_1)^2\right] \qquad \text{[P. 71]}$$

K and k.   Fisher put

$$z = \tfrac{1}{2} \ln F = \tfrac{1}{2} \ln \frac{\sigma'^2}{\sigma''^2} \qquad (48)$$

ortant feature of the z-distribution is its proximity to the nor- with mean at zero and unit standard deviation for large values k (denoted by n_1 and n_2 by Fisher and most other writers). feature is the relative ease of interpolation, particularly against nonic means of K and k.   Fisher provided tables of z for levels of the integral

$$P(z) = \int_z^{\infty} y \, dz \qquad (49)$$

nd original tables have been widely copied.   An abbreviated table of z r $P(z) = 0.01$ is shown here, through the kind permission of Professor Fisher and of his publishers, Messrs. Oliver & Boyd.

Remark 1.   Existing tables of F are mere numerical transformations of Fisher's z.   The advantage of z over F is its ease of interpolation.[3]

Remark 2.   It should be borne in mind that both F and z as tabled apply strictly only for a normal universe.   For a universe that is not normal (more particularly for a universe whose β_2 is much higher than 3) the crude test in Ineq. 18 may be preferable.

Remark 3.   With reference to the fact that integrals under the F-curve in Eq. 47 are expressible as an incomplete beta function, and that the incomplete sum of a binomial is also expressible as an incomplete beta function (p. 489), it may be seen that probability levels for F and hence for z can be found numerically by aid of the Mosteller-Tukey double square-root paper described on pages 306–12.

[3] Fisher gives simple formulas of interpolation along with the tables of z in his *Statistical Methods for Research Workers* (Oliver & Boyd).   Fisher and Yates show more extensive tables and accompanying formulas of interpolation.   A. H. Carter gives a more exact but more complicated formula in his article "Approximation to percentage points on the z-distribution," *Biometrika*, vol. xxxiv, Parts III and IV, December 1947: pp. 352–8.   A remarkable normalization of a function of F was published by Edward Paulson, *Annals Math. Stat.*, vol. xiii, 1942: pp. 233–5.

$$= (\quad " \quad) n_1^2 \left[\frac{\sigma^2}{n_1\left(1 + \frac{n_1}{n_2}\right)} + \left(\frac{n_2}{n_1 + n_2} D\right)^2 \right]$$

$$= \sigma^2 + \frac{D^2}{\frac{1}{n_1} + \frac{1}{n_2}}$$

The distribution of Fisher's t for a pair of samples.   Define

$$t = \frac{|\bar{x}_2 - \bar{x}_1|}{\sigma'' \sqrt{\frac{1}{n_1} + \frac{1}{n_2}}} \qquad (37)$$

$$= \frac{\text{The difference between } \bar{x}_2 \text{ and } \bar{x}_1}{\text{The internal estimate of the standard error of this difference}}$$

This fraction will have the t-distribution of Eq. 9 in Chapter 16, the number of degrees of freedom being $n_1 + n_2 - 2$.   *Proof*: Let

$$u = \bar{x}_2 - \bar{x}_1 \qquad (38)$$

then the simultaneous distribution of σ'' and u will be

$$y \, d\sigma'' \, du = \frac{k^{\frac{1}{2}k}}{\Gamma(\frac{1}{2}k) 2^{\frac{1}{2}(k-2)} \sigma} \left(\frac{\sigma''}{\sigma}\right)^{k-1} e^{-k\sigma''^2/2\sigma^2} \, d\sigma''$$

$$\times \frac{1}{\sigma\sqrt{2\pi} \sqrt{\frac{1}{n_1} + \frac{1}{n_2}}} e^{-\frac{u^2}{2\sigma^2\left(\frac{1}{n_1} + \frac{1}{n_2}\right)}} \, du \qquad (39)$$

This is merely a product of the distribution of σ'' from Exercise 1 on page 576 and the normal distribution of $u = \bar{x}_2 + \bar{x}_1$.   Now write

$$\left. \begin{array}{l} t = \dfrac{u}{\sigma'' \sqrt{\dfrac{1}{n_1} + \dfrac{1}{n_2}}} \\ \sigma'' = \sigma'' \end{array} \right\} \qquad (40)$$

then

$$d\sigma'' \, du = \sigma'' \, d\sigma'' \, dt \sqrt{\frac{1}{n_1} + \frac{1}{n_2}} \qquad (41)$$

Integration of Eq. 39 from $\sigma'' = 0$ to $\sigma'' = \infty$ will give

$$y \, dt = \frac{1}{\sqrt{k} \, B(\frac{1}{2}k, \frac{1}{2})} \left(1 + \frac{t^2}{k}\right)^{-\frac{1}{2}(k+1)} dt \quad [k = n_1 + n_2 - 2] \quad (42)$$

for the distribution of t. (The details are parallel to the integration performed on page 539 for the distribution of $t = u/s$, hence are omitted.) Integrals under the t-distribution may be evaluated with Nekrasoff's nomograph on page 555.

The distribution of Fisher's z. The above exercise is fundamental in the analysis of variance. If the means of two universes are unequal, $E\sigma'^2$ is increased above σ^2 while $E\sigma''^2$ remains fixed. Earlier in the chapter it was remarked that a high value of $\sigma':\sigma''$ may indicate unequal means,

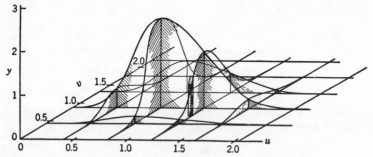

FIG. 89.   The probability surface for the analysis of variance.   Here

$$u = \left(\frac{\sigma'}{\sigma}\right)^2 \quad v = \left(\frac{\sigma''}{\sigma}\right)^2$$

wherein σ' denotes an estimate of σ made from K degrees of freedom (illustrated with $K = 5$), and σ'' denotes an estimate of σ made from k degrees of freedom (illustrated with $k = 8$).   The equation of the surface is given by Eq. 44.

and a crude rule was devised in Ineq. 18 for coming to a quick decision regarding the question. On the assumption of normal universes we are now ready to find the distribution of the ratio $\sigma':\sigma''$, and to discover a more exact test when required.

It will be recalled from Exercise 2 on page 578 and from Eq. 36 that the distribution of $k\sigma''^2/\sigma^2$ is the χ^2-distribution with $k = n - m$ degrees of freedom and that the distribution of $K\sigma'^2/\sigma^2$ is likewise the χ^2-distribution with $K = m - 1$ degrees of freedom. In pursuit of the distribution of the ratio $\sigma':\sigma''$ let

$$u = \left(\frac{\sigma'}{\sigma}\right)^2 \quad \text{and} \quad v = \left(\frac{\sigma''}{\sigma}\right)^2 \tag{43}$$

Then the simultaneous distribution of u and v will be

$$y \, du \, dv = \frac{K^{\frac{1}{2}K}k^{\frac{1}{2}k}u^{\frac{1}{2}K-1}v^{\frac{1}{2}k-1}}{\Gamma(\frac{1}{2}K)\Gamma(\frac{1}{2}k)2^{\frac{1}{2}(K+k)}} e^{-\frac{1}{2}(Ku+kv)} \, du \, dv \tag{44}$$

The symbols u and v here are not to be confused with the same letters used earlier in this chapter for other meanings.

The locus of y against u and v is a surface... Now with Snedecor [1] let

$$F = \left(\frac{\sigma'}{\sigma''}\right)^2 = \frac{u}{v}$$

F is constant along any radial line on the uv-pla... now a wedge by the inequalities

$$F < \frac{u}{v} < F + dF$$

as pictured in Fig. 90. The shaded area is $du \, dv$. But ... along the horizontal, $du = v \, dF$, whence $du \, dv$ in Eq. 44 n...

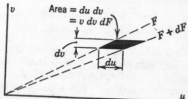

FIG. 90.   Looking down on the uv-surface.   The F-ratio is constant along any line. $u = (\sigma'/\sigma)^2$ and $v = (\sigma''/\sigma)^2$; $F = u/v$.

by $v \, dv \, dF$. The total volume of the surface pictured in Fig. 89, loca... within the wedge pictured in Fig. 90, is therefore

$$y \, dF = \frac{K^{\frac{1}{2}K}k^{\frac{1}{2}k} \, dF}{\Gamma(\frac{1}{2}K)\Gamma(\frac{1}{2}k)2^{\frac{1}{2}(K+k)}} \int_0^\infty (Fv)^{\frac{1}{2}K-1}v^{\frac{1}{2}k-1}e^{-\frac{1}{2}(KF+k)v} \, v \, dv$$

$$= \frac{K^{\frac{1}{2}K}k^{\frac{1}{2}k}F^{\frac{1}{2}K-1} \, dF}{B(\frac{1}{2}K, \frac{1}{2}k)[KF + k]^{\frac{1}{2}(K+k)}}$$

This is the distribution of F. Smith and Duncan [2] show an F... plotted for $K = 4$, $k = 3$. Values of F that cut off 1 percent or 5 ... of the area in the far tail of this curve are referred to as the 1 per... 5 percent "levels" of F: they may be evaluated numerically ... incomplete beta function (cf. Ex. 17, p. 489). The observe... $\sigma':\sigma''$ is "significantly" greater than unity if F exceeds the 1 pe... 5 percent level, whichever level has been chosen in advance ... significance from nonsignificance. Snedecor and others hav... tables of the 5 and 1 percent levels of F against K and k fo...

[1] George W. Snedecor, *Statistical Methods* (The Collegiate Press, 1937)
[2] James G. Smith and Acheson J. Duncan, *Sampling Statistics and* ... (McGraw-Hill, 1945), p. 113.

THE 1 PERCENT POINTS ON THE DISTRIBUTION OF z

$$z = \tfrac{1}{2} \ln \frac{\text{External estimate of } \sigma^2}{\text{Internal estimate of } \sigma^2}$$

k, degrees of freedom in the internal estimate	K, degrees of freedom in the external estimate									
	1	2	3	4	5	6	8	12	24	∞
1	4.1535	4.2585	4.2974	4.3175	4.3297	4.3379	4.3482	4.3585	4.3689	4.3794
2	2.2950	2.2976	2.2984	2.2988	2.2991	2.2992	2.2994	2.2997	2.2999	2.3001
3	1.7649	1.7140	1.6915	1.6786	1.6703	1.6645	1.6569	1.6489	1.6404	1.6314
4	1.5270	1.4452	1.4075	1.3856	1.3711	1.3609	1.3473	1.3327	1.3170	1.3000
5	1.3943	1.2929	1.2449	1.2164	1.1974	1.1838	1.1656	1.1457	1.1239	1.0997
6	1.3103	1.1955	1.1401	1.1068	1.0843	1.0680	1.0460	1.0218	.9948	.9643
7	1.2526	1.1281	1.0672	1.0300	1.0048	.9864	.9614	.9335	.9020	.8658
8	1.2106	1.0787	1.0135	.9734	.9459	.9259	.8983	.8673	.8319	.7904
9	1.1786	1.0411	.9724	.9299	.9006	.8791	.8494	.8157	.7769	.7305
10	1.1535	1.0114	.9399	.8954	.8646	.8419	.8104	.7744	.7324	.6816
11	1.1333	.9874	.9136	.8674	.8354	.8116	.7785	.7405	.6958	.6408
12	1.1166	.9677	.8919	.8443	.8111	.7864	.7520	.7122	.6649	.6061
13	1.1027	.9511	.8737	.8248	.7907	.7652	.7295	.6882	.6386	.5761
14	1.0909	.9370	.8581	.8082	.7732	.7471	.7103	.6675	.6159	.5500
15	1.0807	.9249	.8448	.7939	.7582	.7314	.6937	.6496	.5961	.5269
20	1.0457	.8831	.7985	.7443	.7058	.6768	.6355	.5864	.5253	.4421
25	1.0251	.8585	.7712	.7148	.6747	.6442	.6006	.5481	.4816	.3872
30	1.0116	.8423	.7531	.6954	.6540	.6226	.5773	.5224	.4519	.3481
40	.9949	.8223	.7307	.6712	.6283	.5956	.5481	.4901	.4138	.2952
60	.9784	.8025	.7086	.6472	.6028	.5687	.5189	.4574	.3746	.2352
120	.9622	.7829	.6867	.6234	.5774	.5419	.4897	.4243	.3339	.1612
∞	.9462	.7636	.6651	.5999	.5522	.5152	.4604	.3908	.2913	0

For high values of K and k,

$$z \text{ (1 per cent)} = \frac{2.3263}{\sqrt{h - 1.4}} - 1.235 \left(\frac{1}{K} - \frac{1}{k} \right)$$

approximately, where $\dfrac{2}{h} = \dfrac{1}{K} + \dfrac{1}{k}$. In other parts of the table interpolation is approximately linear against $1/K$ and $1/k$.

This table is reprinted from Table VI of R. A. Fisher: *Statistical Methods for Research Workers*, Oliver & Boyd Ltd., Edinburgh, by permission of the author and publishers.

APPENDIX

A PAGE OF RANDOM NUMBERS

This is a portion of page 22 which was drawn at random from H. Burke Horton's *Random Decimal Digits* (Interstate Commerce Commission, 1949). It is included here to assist students in pursuing the exercises.

1	2	3	4	5	6	7
02946	96520	81881	56247	17623	47441	27821
85697	62000	87957	07258	45054	58410	92081
26734	68426	52067	23123	73700	58730	06111
47829	32353	95941	72169	58374	03905	06865
76603	99339	40571	41186	04981	17531	97372
47526	26522	11045	83565	45639	02485	43905
70100	85732	19741	92951	98832	38188	24080
86819	50200	50889	06493	66638	03619	90906
41614	30074	23403	03656	77580	87772	86877
17930	26194	53836	53692	67125	98175	00912
24649	31845	25736	75231	83808	98997	71829
79899	34061	54308	59358	56462	58166	97302
76801	49594	81002	30397	52728	15101	72070
62567	08480	61873	63162	44873	35302	04511
49723	15275	09399	11211	67352	41526	23497
42658	70183	89417	57676	35370	14915	16569
65080	35569	79392	14937	06081	74957	87787
02906	38119	72407	71427	58478	99297	43519
75153	86376	63852	60557	21211	77299	74967
14192	49525	78844	13664	98964	64425	33536
32059	11548	86264	74406	81496	23996	56872
81716	80301	96704	57214	71361	41989	92589
43315	50483	02950	09611	36341	20326	37489
27510	10769	09921	46721	34183	22856	18724
81782	04769	36716	82519	98272	13969	12429
19975	48346	91029	78902	75689	70722	88553
98356	76855	18769	52843	64204	95212	31320
29708	17814	31556	68610	16574	42305	56300
88014	27583	78167	25057	93552	74363	30951
94491	19238	17396	10592	48907	79840	34607
56957	05072	53948	07850	42569	82391	20435
50915	31924	80621	17495	81618	15125	48087
49631	93771	80200	84622	31413	33756	15218
99683	58162	45516	39761	77600	15175	67415
86017	20264	94618	85979	42009	78616	45210
77339	64605	82583	85011	02955	84348	46436
61714	57933	37342	26000	93611	93346	71212
15232	48027	15832	62924	11509	95853	02747

INDEX

A

Acceptance, region of, 271, 550, 551, 553, 556.
Acceptance number 271.
Accidental variations 17.
Accuracy (unmeasurable) 17.
Russell ACKOFF 9.
Age-sex pyramid 387.
Allocation of sample by cost functions 150, 151, 154, 204, 208; allocation in single-stage sampling 218–26; Neyman sampling and proportionate sampling compared 230–2; allocation in 2-stage stratified sampling 238–40, 379.
Analytic use of samples 247–9, 252; variances under different assumptions 254.
Oskar ANDERSON 301.
AOQ and AOQL 276.
Appraisal of precision, in single-stage sampling 331; in 2-stage sampling 350; of a ratio-estimate 178–82; in stratified sampling 352; of a systematic sample 352–5; numerical examples 369, 385.
Arc sine transformation 132, 306; Mosteller-Tukey double square-root paper—description 306; use in single-sampling plan 273; use in inventory problem 426.
R. C. ARCHIBALD 407.
Area of a sphere 491.
Asymptotic series 493, 544.
Attributes, sampling for, 109.

B

T. A. BANCROFT 488.
G. A. BARNARD 267.
M. S. BARTLETT 420.
Sam BARTON 28.
H. BATEMAN 465.
T. BAYES 482, 493, 520.
Joseph BERKSON 52.
Jacques BERNOULLI 113, 402; Bernoulli numbers 403, 520; Bernoulli sums 403; Bernoulli sampling, Bernoulli series, see Binomial sampling.
Beta function, complete 474; incomplete 475, 476; illustration of table 480; other forms 477.

Bias defined 16, 17; illustrated and compared with sampling error 20; must remain unknown 17; list of biases 26–9; biases of sampling 30, 47, 94, 97; bias of nonresponse 33–6; of the auspices 43; of the interviewer 42; in response 37; dilution bias 147; of a ratio-estimate 178, 180; biased formulas of estimation 242–6, 291.
Biased and unbiased statistical estimates 88–98; of σ^2 333, 522, 537; of the sampling error 333, 341, 347, 350.
Bimodal distribution 405.
Binomial sampling, binomial series, Bernoulli sampling, Bernoulli series 109–13, 115, 399 ff; chart 117; exercises 119–27, 419–20, 426–8, 444–66; plotted 401, 402, 409, 412–7; binomial series summed without approximation 480.
Raymond T. BIRGE 501, 537, 563.
David BLACKWELL 282.
Richard H. BLYTHE 6, 233, 372.
George W. BROWN 425.
Breaking strength 127.
K. A. BROWNLEE 283.
Bulk products, sampling of, 160 ff, 183 ff.
Business, census of, 41.
β_2, definition, 64; for the binomial 121; effect of β_2 on estimates of the sampling error 340; control of β_2 344.

C

c_2 521; table 570.
Werner J. CAHNMAN 15.
Calculated risk 266.
Calculation (numerical) of mean and variance 66–8.
Cards, sampling file of, 88 ff.
A. George CARLTON 534, 560.
H. S. CARSLAW 471.
A. H. CARTER 591.
CAUCHY distribution 60, 434.
S. CHANDRASEKHAR 513.
Roy A. CHAPMAN 134.
Characteristics of a good statistical program 3.
G. CHEVRY 7.
China, sampling units in, 81.

595

A CATALOGUE OF SELECTED DOVER BOOKS
IN ALL FIELDS OF INTEREST

A CATALOGUE OF SELECTED DOVER
BOOKS IN ALL FIELDS OF INTEREST

CONDITIONED REFLEXES, Ivan P. Pavlov. Full translation of most complete statement of Pavlov's work; cerebral damage, conditioned reflex, experiments with dogs, sleep, similar topics of great importance. 430pp. 5⅜ x 8½. 60614-7 Pa. $4.50

NOTES ON NURSING: WHAT IT IS, AND WHAT IT IS NOT, Florence Nightingale. Outspoken writings by founder of modern nursing. When first published (1860) it played an important role in much needed revolution in nursing. Still stimulating. 140pp. 5⅜ x 8½. 22340-X Pa. $3.00

HARTER'S PICTURE ARCHIVE FOR COLLAGE AND ILLUSTRATION, Jim Harter. Over 300 authentic, rare 19th-century engravings selected by noted collagist for artists, designers, decoupeurs, etc. Machines, people, animals, etc., printed one side of page. 25 scene plates for backgrounds. 6 collages by Harter, Satty, Singer, Evans. Introduction. 192pp. 8⅞ x 11¾. 23659-5 Pa. $5.00

MANUAL OF TRADITIONAL WOOD CARVING, edited by Paul N. Hasluck. Possibly the best book in English on the craft of wood carving. Practical instructions, along with 1,146 working drawings and photographic illustrations. Formerly titled *Cassell's Wood Carving*. 576pp. 6½ x 9¼. 23489-4 Pa. $7.95

THE PRINCIPLES AND PRACTICE OF HAND OR SIMPLE TURNING, John Jacob Holtzapffel. Full coverage of basic lathe techniques—history and development, special apparatus, softwood turning, hardwood turning, metal turning. Many projects—billiard ball, works formed within a sphere, egg cups, ash trays, vases, jardiniers, others—included. 1881 edition. 800 illustrations. 592pp. 6⅛ x 9¼. 23365-0 Clothbd. $15.00

THE JOY OF HANDWEAVING, Osma Tod. Only book you need for hand weaving. Fundamentals, threads, weaves, plus numerous projects for small board-loom, two-harness, tapestry, laid-in, four-harness weaving and more. Over 160 illustrations. 2nd revised edition. 352pp. 6½ x 9¼. 23458-4 Pa. $6.00

THE BOOK OF WOOD CARVING, Charles Marshall Sayers. Still finest book for beginning student in wood sculpture. Noted teacher, craftsman discusses fundamentals, technique; gives 34 designs, over 34 projects for panels, bookends, mirrors, etc. "Absolutely first-rate"—E. J. Tangerman. 33 photos. 118pp. 7¾ x 10⅝. 23654-4 Pa. $3.50

THE CURVES OF LIFE, Theodore A. Cook. Examination of shells, leaves, horns, human body, art, etc., in *"the* classic reference on how the golden ratio applies to spirals and helices in nature . . . ."—Martin Gardner. 426 illustrations. Total of 512pp. 5⅜ x 8½.   23701-X Pa. $5.95

AN ILLUSTRATED FLORA OF THE NORTHERN UNITED STATES AND CANADA, Nathaniel L. Britton, Addison Brown. Encyclopedic work covers 4666 species, ferns on up. Everything. Full botanical information, illustration for each. This earlier edition is preferred by many to more recent revisions. 1913 edition. Over 4000 illustrations, total of 2087pp. 6⅛ x 9¼.   22642-5, 22643-3, 22644-1 Pa., Three-vol. set $25.50

MANUAL OF THE GRASSES OF THE UNITED STATES, A. S. Hitchcock, U.S. Dept. of Agriculture. The basic study of American grasses, both indigenous and escapes, cultivated and wild. Over 1400 species. Full descriptions, information. Over 1100 maps, illustrations. Total of 1051pp. 5⅜ x 8½.   22717-0, 22718-9 Pa., Two-vol. set $15.00

THE CACTACEAE,, Nathaniel L. Britton, John N. Rose. Exhaustive, definitive. Every cactus in the world. Full botanical descriptions. Thorough statement of nomenclatures, habitat, detailed finding keys. The one book needed by every cactus enthusiast. Over 1275 illustrations. Total of 1080pp. 8 x 10¼.   21191-6, 21192-4 Clothbd., Two-vol. set $35.00

AMERICAN MEDICINAL PLANTS, Charles F. Millspaugh. Full descriptions, 180 plants covered: history; physical description; methods of preparation with all chemical constituents extracted; all claimed curative or adverse effects. 180 full-page plates. Classification table. 804pp. 6½ x 9¼.
23034-1 Pa. $12.95

A MODERN HERBAL, Margaret Grieve. Much the fullest, most exact, most useful compilation of herbal material. Gigantic alphabetical encyclopedia, from aconite to zedoary, gives botanical information, medical properties, folklore, economic uses, and much else. Indispensable to serious reader. 161 illustrations. 888pp. 6½ x 9¼. (Available in U.S. only)
22798-7, 22799-5 Pa., Two-vol. set $13.00

THE HERBAL or GENERAL HISTORY OF PLANTS, John Gerard. The 1633 edition revised and enlarged by Thomas Johnson. Containing almost 2850 plant descriptions and 2705 superb illustrations, Gerard's *Herbal* is a monumental work, the book all modern English herbals are derived from, the one herbal every serious enthusiast should have in its entirety. Original editions are worth perhaps $750. 1678pp. 8½ x 12¼.
23147-X Clothbd. $50.00

MANUAL OF THE TREES OF NORTH AMERICA, Charles S. Sargent. The basic survey of every native tree and tree-like shrub, 717 species in all. Extremely full descriptions, information on habitat, growth, locales, economics, etc. Necessary to every serious tree lover. Over 100 finding keys. 783 illustrations. Total of 986pp. 5⅜ x 8½.
20277-1, 20278-X Pa., Two-vol. set $11.00

AMERICAN BIRD ENGRAVINGS, Alexander Wilson et al. All 76 plates. from Wilson's *American Ornithology* (1808-14), most important ornithological work before Audubon, plus 27 plates from the supplement (1825-33) by Charles Bonaparte. Over 250 birds portrayed. 8 plates also reproduced in full color. 111pp. 9⅜ x 12½. 23195-X Pa. $6.00

CRUICKSHANK'S PHOTOGRAPHS OF BIRDS OF AMERICA, Allan D. Cruickshank. Great ornithologist, photographer presents 177 closeups, groupings, panoramas, flightings, etc., of about 150 different birds. Expanded *Wings in the Wilderness*. Introduction by Helen G. Cruickshank. 191pp. 8¼ x 11. 23497-5 Pa. $6.00

AMERICAN WILDLIFE AND PLANTS, A. C. Martin, et al. Describes food habits of more than 1000 species of mammals, birds, fish. Special treatment of important food plants. Over 300 illustrations. 500pp. 5⅜ x 8½. 20793-5 Pa. $4.95

THE PEOPLE CALLED SHAKERS, Edward D. Andrews. Lifetime of research, definitive study of Shakers: origins, beliefs, practices, dances, social organization, furniture and crafts, impact on 19th-century USA, present heritage. Indispensable to student of American history, collector. 33 illustrations. 351pp. 5⅜ x 8½. 21081-2 Pa. $4.50

OLD NEW YORK IN EARLY PHOTOGRAPHS, Mary Black. New York City as it was in 1853-1901, through 196 wonderful photographs from N.-Y. Historical Society. Great Blizzard, Lincoln's funeral procession, great buildings. 228pp. 9 x 12. 22907-6 Pa. $8.95

MR. LINCOLN'S CAMERA MAN: MATHEW BRADY, Roy Meredith. Over 300 Brady photos reproduced directly from original negatives, photos. Jackson, Webster, Grant, Lee, Carnegie, Barnum; Lincoln; Battle Smoke, Death of Rebel Sniper, Atlanta Just After Capture. Lively commentary. 368pp. 8⅜ x 11¼. 23021-X Pa. $8.95

TRAVELS OF WILLIAM BARTRAM, William Bartram. From 1773-8, Bartram explored Northern Florida, Georgia, Carolinas, and reported on wild life, plants, Indians, early settlers. Basic account for period, entertaining reading. Edited by Mark Van Doren. 13 illustrations. 141pp. 5⅜ x 8½. 20013-2 Pa. $5.00

THE GENTLEMAN AND CABINET MAKER'S DIRECTOR, Thomas Chippendale. Full reprint, 1762 style book, most influential of all time; chairs, tables, sofas, mirrors, cabinets, etc. 200 plates, plus 24 photographs of surviving pieces. 249pp. 9⅞ x 12¾. 21601-2 Pa. $7.95

AMERICAN CARRIAGES, SLEIGHS, SULKIES AND CARTS, edited by Don H. Berkebile. 168 Victorian illustrations from catalogues, trade journals, fully captioned. Useful for artists. Author is Assoc. Curator, Div. of Transportation of Smithsonian Institution. 168pp. 8½ x 9½.

23328-6 Pa. $5.00

YUCATAN BEFORE AND AFTER THE CONQUEST, Diego de Landa. First English translation of basic book in Maya studies, the only significant account of Yucatan written in the early post-Conquest era. Translated by distinguished Maya scholar William Gates. Appendices, introduction, 4 maps and over 120 illustrations added by translator. 162pp. 5⅜ x 8½.
23622-6 Pa. $3.00

THE MALAY ARCHIPELAGO, Alfred R. Wallace. Spirited travel account by one of founders of modern biology. Touches on zoology, botany, ethnography, geography, and geology. 62 illustrations, maps. 515pp. 5⅜ x 8½.
20187-2 Pa. $6.95

THE DISCOVERY OF THE TOMB OF TUTANKHAMEN, Howard Carter, A. C. Mace. Accompany Carter in the thrill of discovery, as ruined passage suddenly reveals unique, untouched, fabulously rich tomb. Fascinating account, with 106 illustrations. New introduction by J. M. White. Total of 382pp. 5⅜ x 8½. (Available in U.S. only)   23500-9 Pa. $4.00

THE WORLD'S GREATEST SPEECHES, edited by Lewis Copeland and Lawrence W. Lamm. Vast collection of 278 speeches from Greeks up to present. Powerful and effective models; unique look at history. Revised to 1970. Indices. 842pp. 5⅜ x 8½.   20468-5 Pa. $8.95

THE 100 GREATEST ADVERTISEMENTS, Julian Watkins. The priceless ingredient; His master's voice; 99 44/100% pure; over 100 others. How they were written, their impact, etc. Remarkable record. 130 illustrations. 233pp. 7⅞ x 10 3/5.   20540-1 Pa. $5.95

CRUICKSHANK PRINTS FOR HAND COLORING, George Cruickshank. 18 illustrations, one side of a page, on fine-quality paper suitable for water-colors. Caricatures of people in society (c. 1820) full of trenchant wit. Very large format. 32pp. 11 x 16.   23684-6 Pa. $5.00

THIRTY-TWO COLOR POSTCARDS OF TWENTIETH-CENTURY AMERICAN ART, Whitney Museum of American Art. Reproduced in full color in postcard form are 31 art works and one shot of the museum. Calder, Hopper, Rauschenberg, others. Detachable. 16pp. 8¼ x 11.
23629-3 Pa. $3.00

MUSIC OF THE SPHERES: THE MATERIAL UNIVERSE FROM ATOM TO QUASAR SIMPLY EXPLAINED, Guy Murchie. Planets, stars, geology, atoms, radiation, relativity, quantum theory, light, antimatter, similar topics. 319 figures. 664pp. 5⅜ x 8½.
21809-0, 21810-4 Pa., Two-vol. set $11.00

EINSTEIN'S THEORY OF RELATIVITY, Max Born. Finest semi-technical account; covers Einstein, Lorentz, Minkowski, and others, with much detail, much explanation of ideas and math not readily available elsewhere on this level. For student, non-specialist. 376pp. 5⅜ x 8½.
60769-0 Pa. $4.50

THE ANATOMY OF THE HORSE, George Stubbs. Often considered the great masterpiece of animal anatomy. Full reproduction of 1766 edition, plus prospectus; original text and modernized text. 36 plates. Introduction by Eleanor Garvey. 121pp. 11 x 14¾.   23402-9 Pa. $6.00

BRIDGMAN'S LIFE DRAWING, George B. Bridgman. More than 500 illustrative drawings and text teach you to abstract the body into its major masses, use light and shade, proportion; as well as specific areas of anatomy, of which Bridgman is master. 192pp. 6½ x 9¼. (Available in U.S. only)
22710-3 Pa. $3.50

ART NOUVEAU DESIGNS IN COLOR, Alphonse Mucha, Maurice Verneuil, Georges Auriol. Full-color reproduction of *Combinaisons ornementales* (c. 1900) by Art Nouveau masters. Floral, animal, geometric, interlacings, swashes—borders, frames, spots—all incredibly beautiful. 60 plates, hundreds of designs. 9⅜ x 8-1/16.   22885-1 Pa. $4.00

FULL-COLOR FLORAL DESIGNS IN THE ART NOUVEAU STYLE, E. A. Seguy. 166 motifs, on 40 plates, from *Les fleurs et leurs applications decoratives* (1902): borders, circular designs, repeats, allovers, "spots." All in authentic Art Nouveau colors. 48pp. 9⅜ x 12¼.
23439-8 Pa. $5.00

A DIDEROT PICTORIAL ENCYCLOPEDIA OF TRADES AND IN-DUSTRY, edited by Charles C. Gillispie. 485 most interesting plates from the great French Encyclopedia of the 18th century show hundreds of working figures, artifacts, process, land and cityscapes; glassmaking, paper-making, metal extraction, construction, weaving, making furniture, clothing, wigs, dozens of other activities. Plates fully explained. 920pp. 9 x 12.
22284-5, 22285-3 Clothbd., Two-vol. set $40.00

HANDBOOK OF EARLY ADVERTISING ART, Clarence P. Hornung. Largest collection of copyright-free early and antique advertising art ever compiled. Over 6,000 illustrations, from Franklin's time to the 1890's for special effects, novelty. Valuable source, almost inexhaustible.
Pictorial Volume. Agriculture, the zodiac, animals, autos, birds, Christmas, fire engines, flowers, trees, musical instruments, ships, games and sports, much more. Arranged by subject matter and use. 237 plates. 288pp. 9 x 12.
20122-8 Clothbd. $14.50

Typographical Volume. Roman and Gothic faces ranging from 10 point to 300 point, "Barnum," German and Old English faces, script, logotypes, scrolls and flourishes, 1115 ornamental initials, 67 complete alphabets, more. 310 plates. 320pp. 9 x 12.   20123-6 Clothbd. $15.00

CALLIGRAPHY (CALLIGRAPHIA LATINA), J. G. Schwandner. High point of 18th-century ornamental calligraphy. Very ornate initials, scrolls, borders, cherubs, birds, lettered examples. 172pp. 9 x 13.
20475-8 Pa. $7.00

ART FORMS IN NATURE, Ernst Haeckel. Multitude of strangely beautiful natural forms: Radiolaria, Foraminifera, jellyfishes, fungi, turtles, bats, etc. All 100 plates of the 19th-century evolutionist's *Kunstformen der Natur* (1904). 100pp. 9⅜ x 12¼.   22987-4 Pa. $5.00

CHILDREN: A PICTORIAL ARCHIVE FROM NINETEENTH-CENTURY SOURCES, edited by Carol Belanger Grafton. 242 rare, copyright-free wood engravings for artists and designers. Widest such selection available. All illustrations in line. 119pp. 8⅜ x 11¼.
23694-3 Pa. $4.00

WOMEN: A PICTORIAL ARCHIVE FROM NINETEENTH-CENTURY SOURCES, edited by Jim Harter. 391 copyright-free wood engravings for artists and designers selected from rare periodicals. Most extensive such collection available. All illustrations in line. 128pp. 9 x 12.
23703-6 Pa. $4.50

ARABIC ART IN COLOR, Prisse d'Avennes. From the greatest ornamentalists of all time—50 plates in color, rarely seen outside the Near East, rich in suggestion and stimulus. Includes 4 plates on covers. 46pp. 9⅜ x 12¼.   23658-7 Pa. $6.00

AUTHENTIC ALGERIAN CARPET DESIGNS AND MOTIFS, edited by June Beveridge. Algerian carpets are world famous. Dozens of geometrical motifs are charted on grids, color-coded, for weavers, needleworkers, craftsmen, designers. 53 illustrations plus 4 in color. 48pp. 8¼ x 11. (Available in U.S. only)   23650-1 Pa. $1.75

DICTIONARY OF AMERICAN PORTRAITS, edited by Hayward and Blanche Cirker. 4000 important Americans, earliest times to 1905, mostly in clear line. Politicians, writers, soldiers, scientists, inventors, industrialists, Indians, Blacks, women, outlaws, etc. Identificatory information. 756pp. 9¼ x 12¾.   21823-6 Clothbd. $40.00

HOW THE OTHER HALF LIVES, Jacob A. Riis. Journalistic record of filth, degradation, upward drive in New York immigrant slums, shops, around 1900. New edition includes 100 original Riis photos, monuments of early photography. 233pp. 10 x 7⅞.   22012-5 Pa. $7.00

NEW YORK IN THE THIRTIES, Berenice Abbott. Noted photographer's fascinating study of city shows new buildings that have become famous and old sights that have disappeared forever. Insightful commentary. 97 photographs. 97pp. 11⅜ x 10.   22967-X Pa. $5.00

MEN AT WORK, Lewis W. Hine. Famous photographic studies of construction workers, railroad men, factory workers and coal miners. New supplement of 18 photos on Empire State building construction. New introduction by Jonathan L. Doherty. Total of 69 photos. 63pp. 8 x 10¾.
23475-4 Pa. $3.00

THE DEPRESSION YEARS AS PHOTOGRAPHED BY ARTHUR ROTH-STEIN, Arthur Rothstein. First collection devoted entirely to the work of outstanding 1930s photographer: famous dust storm photo, ragged children, unemployed, etc. 120 photographs. Captions. 119pp. 9¼ x 10¾.
23590-4 Pa. $5.00

CAMERA WORK: A PICTORIAL GUIDE, Alfred Stieglitz. All 559 illustrations and plates from the most important periodical in the history of art photography, Camera Work (1903-17). Presented four to a page, reduced in size but still clear, in strict chronological order, with complete captions. Three indexes. Glossary. Bibliography. 176pp. 8⅜ x 11¼.
23591-2 Pa. $6.95

ALVIN LANGDON COBURN, PHOTOGRAPHER, Alvin L. Coburn. Revealing autobiography by one of greatest photographers of 20th century gives insider's version of Photo-Secession, plus comments on his own work. 77 photographs by Coburn. Edited by Helmut and Alison Gernsheim. 160pp. 8⅛ x 11.
23685-4 Pa. $6.00

NEW YORK IN THE FORTIES, Andreas Feininger. 162 brilliant photographs by the well-known photographer, formerly with Life magazine, show commuters, shoppers, Times Square at night, Harlem nightclub, Lower East Side, etc. Introduction and full captions by John von Hartz. 181pp. 9¼ x 10¾.
23585-8 Pa. $6.95

GREAT NEWS PHOTOS AND THE STORIES BEHIND THEM, John Faber. Dramatic volume of 140 great news photos, 1855 through 1976, and revealing stories behind them, with both historical and technical information. Hindenburg disaster, shooting of Oswald, nomination of Jimmy Carter, etc. 160pp. 8¼ x 11.
23667-6 Pa. $5.00

THE ART OF THE CINEMATOGRAPHER, Leonard Maltin. Survey of American cinematography history and anecdotal interviews with 5 masters—Arthur Miller, Hal Mohr, Hal Rosson, Lucien Ballard, and Conrad Hall. Very large selection of behind-the-scenes production photos. 105 photographs. Filmographies. Index. Originally Behind the Camera. 144pp. 8¼ x 11.
23686-2 Pa. $5.00

DESIGNS FOR THE THREE-CORNERED HAT (LE TRICORNE), Pablo Picasso. 32 fabulously rare drawings—including 31 color illustrations of costumes and accessories—for 1919 production of famous ballet. Edited by Parmenia Migel, who has written new introduction. 48pp. 9⅜ x 12¼. (Available in U.S. only)
23709-5 Pa. $5.00

NOTES OF A FILM DIRECTOR, Sergei Eisenstein. Greatest Russian filmmaker explains montage, making of Alexander Nevsky, aesthetics; comments on self, associates, great rivals (Chaplin), similar material. 78 illustrations. 240pp. 5⅜ x 8½.
22392-2 Pa. $4.50

HOLLYWOOD GLAMOUR PORTRAITS, edited by John Kobal. 145 photos capture the stars from 1926-49, the high point in portrait photography. Gable, Harlow, Bogart, Bacall, Hedy Lamarr, Marlene Dietrich, Robert Montgomery, Marlon Brando, Veronica Lake; 94 stars in all. Full background on photographers, technical aspects, much more. Total of 160pp. 8⅜ x 11¼.   23352-9 Pa. $6.00

THE NEW YORK STAGE: FAMOUS PRODUCTIONS IN PHOTO-GRAPHS, edited by Stanley Appelbaum. 148 photographs from Museum of City of New York show 142 plays, 1883-1939. *Peter Pan, The Front Page, Dead End, Our Town,* O'Neill, hundreds of actors and actresses, etc. Full indexes. 154pp. 9½ x 10.   23241-7 Pa. $6.00

DIALOGUES CONCERNING TWO NEW SCIENCES, Galileo Galilei. Encompassing 30 years of experiment and thought, these dialogues deal with geometric demonstrations of fracture of solid bodies, cohesion, leverage, speed of light and sound, pendulums, falling bodies, accelerated motion, etc. 300pp. 5⅜ x 8½.   60099-8 Pa. $4.00

THE GREAT OPERA STARS IN HISTORIC PHOTOGRAPHS, edited by James Camner. 343 portraits from the 1850s to the 1940s: Tamburini, Mario, Caliapin, Jeritza, Melchior, Melba, Patti, Pinza, Schipa, Caruso, Farrar, Steber, Gobbi, and many more—270 performers in all. Index. 199pp. 8⅜ x 11¼.   23575-0 Pa. $7.50

J. S. BACH, Albert Schweitzer. Great full-length study of Bach, life, background to music, music, by foremost modern scholar. Ernest Newman translation. 650 musical examples. Total of 928pp. 5⅜ x 8½. (Available in U.S. only)   21631-4, 21632-2 Pa., Two-vol. set $11.00

COMPLETE PIANO SONATAS, Ludwig van Beethoven. All sonatas in the fine Schenker edition, with fingering, analytical material. One of best modern editions. Total of 615pp. 9 x 12. (Available in U.S. only)
                              23134-8, 23135-6 Pa., Two-vol. set $15.50

KEYBOARD MUSIC, J. S. Bach. Bach-Gesellschaft edition. For harpsichord, piano, other keyboard instruments. English Suites, French Suites, Six Partitas, Goldberg Variations, Two-Part Inventions, Three-Part Sinfonias. 312pp. 8⅛ x 11. (Available in U.S. only)   22360-4 Pa. $6.95

FOUR SYMPHONIES IN FULL SCORE, Franz Schubert. Schubert's four most popular symphonies: No. 4 in C Minor ("Tragic"); No. 5 in B-flat Major; No. 8 in B Minor ("Unfinished"); No. 9 in C Major ("Great"). Breitkopf & Hartel edition. Study score. 261pp. 9⅜ x 12¼.
                                                   23681-1 Pa. $6.50

THE AUTHENTIC GILBERT & SULLIVAN SONGBOOK, W. S. Gilbert, A. S. Sullivan. Largest selection available; 92 songs, uncut, original keys, in piano rendering approved by Sullivan. Favorites and lesser-known fine numbers. Edited with plot synopses by James Spero. 3 illustrations. 399pp. 9 x 12.   23482-7 Pa. $9.95

PRINCIPLES OF ORCHESTRATION, Nikolay Rimsky-Korsakov. Great classical orchestrator provides fundamentals of tonal resonance, progression of parts, voice and orchestra, tutti effects, much else in major document. 330pp. of musical excerpts. 489pp. 6½ x 9¼.　　　21266-1 Pa. $7.50

TRISTAN UND ISOLDE, Richard Wagner. Full orchestral score with complete instrumentation. Do not confuse with piano reduction. Commentary by Felix Mottl, great Wagnerian conductor and scholar. Study score. 655pp. 8⅛ x 11.　　　22915-7 Pa. $13.95

REQUIEM IN FULL SCORE, Giuseppe Verdi. Immensely popular with choral groups and music lovers. Republication of edition published by C. F. Peters, Leipzig, n. d. German frontmaker in English translation. Glossary. Text in Latin. Study score. 204pp. 9⅜ x 12¼.
23682-X Pa. $6.00

COMPLETE CHAMBER MUSIC FOR STRINGS, Felix Mendelssohn. All of Mendelssohn's chamber music: Octet, 2 Quintets, 6 Quartets, and Four Pieces for String Quartet. (Nothing with piano is included). Complete works edition (1874-7). Study score. 283 pp. 9⅜ x 12¼.
23679-X Pa. $7.50

POPULAR SONGS OF NINETEENTH-CENTURY AMERICA, edited by Richard Jackson. 64 most important songs: "Old Oaken Bucket," "Arkansas Traveler," "Yellow Rose of Texas," etc. Authentic original sheet music, full introduction and commentaries. 290pp. 9 x 12.　　　23270-0 Pa. $7.95

COLLECTED PIANO WORKS, Scott Joplin. Edited by Vera Brodsky Lawrence. Practically all of Joplin's piano works—rags, two-steps, marches, waltzes, etc., 51 works in all. Extensive introduction by Rudi Blesh. Total of 345pp. 9 x 12.　　　23106-2 Pa. $14.95

BASIC PRINCIPLES OF CLASSICAL BALLET, Agrippina Vaganova. Great Russian theoretician, teacher explains methods for teaching classical ballet; incorporates best from French, Italian, Russian schools. 118 illustrations. 175pp. 5⅜ x 8½.　　　22036-2 Pa. $2.50

CHINESE CHARACTERS, L. Wieger. Rich analysis of 2300 characters according to traditional systems into primitives. Historical-semantic analysis to phonetics (Classical Mandarin) and radicals. 820pp. 6⅛ x 9¼.
21321-8 Pa. $10.00

EGYPTIAN LANGUAGE: EASY LESSONS IN EGYPTIAN HIERO-GLYPHICS, E. A. Wallis Budge. Foremost Egyptologist offers Egyptian grammar, explanation of hieroglyphics, many reading texts, dictionary of symbols. 246pp. 5 x 7½. (Available in U.S. only)
21394-3 Clothbd. $7.50

AN ETYMOLOGICAL DICTIONARY OF MODERN ENGLISH, Ernest Weekley. Richest, fullest work, by foremost British lexicographer. Detailed word histories. Inexhaustible. Do not confuse this with Concise Etymological Dictionary, which is abridged. Total of 856pp. 6½ x 9¼.
21873-2, 21874-0 Pa., Two-vol. set $12.00

A MAYA GRAMMAR, Alfred M. Tozzer. Practical, useful English-language grammar by the Harvard anthropologist who was one of the three greatest American scholars in the area of Maya culture. Phonetics, grammatical processes, syntax, more. 301pp. 5⅜ x 8½. 23465-7 Pa. $4.00

THE JOURNAL OF HENRY D. THOREAU, edited by Bradford Torrey, F. H. Allen. Complete reprinting of 14 volumes, 1837-61, over two million words; the sourcebooks for *Walden*, etc. Definitive. All original sketches, plus 75 photographs. Introduction by Walter Harding. Total of 1804pp. 8½ x 12¼. 20312-3, 20313-1 Clothbd., Two-vol. set $70.00

CLASSIC GHOST STORIES, Charles Dickens and others. 18 wonderful stories you've wanted to reread: "The Monkey's Paw," "The House and the Brain," "The Upper Berth," "The Signalman," "Dracula's Guest," "The Tapestried Chamber," etc. Dickens, Scott, Mary Shelley, Stoker, etc. 330pp. 5⅜ x 8½. 20735-8 Pa. $4.50

SEVEN SCIENCE FICTION NOVELS, H. G. Wells. Full novels. *First Men in the Moon, Island of Dr. Moreau, War of the Worlds, Food of the Gods, Invisible Man, Time Machine, In the Days of the Comet.* A basic science-fiction library. 1015pp. 5⅜ x 8½. (Available in U.S. only) 20264-X Clothbd. $8.95

ARMADALE, Wilkie Collins. Third great mystery novel by the author of *The Woman in White* and *The Moonstone*. Ingeniously plotted narrative shows an exceptional command of character, incident and mood. Original magazine version with 40 illustrations. 597pp. 5⅜ x 8½. 23429-0 Pa. $6.00

MASTERS OF MYSTERY, H. Douglas Thomson. The first book in English (1931) devoted to history and aesthetics of detective story. Poe, Doyle, LeFanu, Dickens, many others, up to 1930. New introduction and notes by E. F. Bleiler. 288pp. 5⅜ x 8½. (Available in U.S. only) 23606-4 Pa. $4.00

FLATLAND, E. A. Abbott. Science-fiction classic explores life of 2-D being in 3-D world. Read also as introduction to thought about hyperspace. Introduction by Banesh Hoffmann. 16 illustrations. 103pp. 5⅜ x 8½. 20001-9 Pa. $2.00

THREE SUPERNATURAL NOVELS OF THE VICTORIAN PERIOD, edited, with an introduction, by E. F. Bleiler. Reprinted complete and unabridged, three great classics of the supernatural: *The Haunted Hotel* by Wilkie Collins, *The Haunted House at Latchford* by Mrs. J. H. Riddell, and *The Lost Stradivarious* by J. Meade Falkner. 325pp. 5⅜ x 8½. 22571-2 Pa. $4.00

AYESHA: THE RETURN OF "SHE," H. Rider Haggard. Virtuoso sequel featuring the great mythic creation, Ayesha, in an adventure that is fully as good as the first book, *She*. Original magazine version, with 47 original illustrations by Maurice Greiffenhagen. 189pp. 6½ x 9¼. 23649-8 Pa. $3.50

UNCLE SILAS, J. Sheridan LeFanu. Victorian Gothic mystery novel, considered by many best of period, even better than Collins or Dickens. Wonderful psychological terror. Introduction by Frederick Shroyer. 436pp. 5⅜ x 8½. 21715-9 Pa. $6.00

JURGEN, James Branch Cabell. The great erotic fantasy of the 1920's that delighted thousands, shocked thousands more. Full final text, Lane edition with 13 plates by Frank Pape. 346pp. 5⅜ x 8½.
23507-6 Pa. $4.50

THE CLAVERINGS, Anthony Trollope. Major novel, chronicling aspects of British Victorian society, personalities. Reprint of Cornhill serialization, 16 plates by M. Edwards; first reprint of full text. Introduction by Norman Donaldson. 412pp. 5⅜ x 8½. 23464-9 Pa. $5.00

KEPT IN THE DARK, Anthony Trollope. Unusual short novel about Victorian morality and abnormal psychology by the great English author. Probably the first American publication. Frontispiece by Sir John Millais. 92pp. 6½ x 9¼. 23609-9 Pa. $2.50

RALPH THE HEIR, Anthony Trollope. Forgotten tale of illegitimacy, inheritance. Master novel of Trollope's later years. Victorian country estates, clubs, Parliament, fox hunting, world of fully realized characters. Reprint of 1871 edition. 12 illustrations by F. A. Faser. 434pp. of text. 5⅜ x 8½. 23642-0 Pa. $5.00

YEKL and THE IMPORTED BRIDEGROOM AND OTHER STORIES OF THE NEW YORK GHETTO, Abraham Cahan. Film *Hester Street* based on *Yekl* (1896). Novel, other stories among first about Jewish immigrants of N.Y.'s East Side. Highly praised by W. D. Howells—Cahan "a new star of realism." New introduction by Bernard G. Richards. 240pp. 5⅜ x 8½. 22427-9 Pa. $3.50

THE HIGH PLACE, James Branch Cabell. Great fantasy writer's enchanting comedy of disenchantment set in 18th-century France. Considered by some critics to be even better than his famous *Jurgen*. 10 illustrations and numerous vignettes by noted fantasy artist Frank C. Pape. 320pp. 5⅜ x 8½. 23670-6 Pa. $4.00

ALICE'S ADVENTURES UNDER GROUND, Lewis Carroll. Facsimile of ms. Carroll gave Alice Liddell in 1864. Different in many ways from final Alice. Handlettered, illustrated by Carroll. Introduction by Martin Gardner. 128pp. 5⅜ x 8½. 21482-6 Pa. $2.50

FAVORITE ANDREW LANG FAIRY TALE BOOKS IN MANY COLORS, Andrew Lang. The four Lang favorites in a boxed set—the complete *Red, Green, Yellow* and *Blue* Fairy Books. 164 stories; 439 illustrations by Lancelot Speed, Henry Ford and G. P. Jacomb Hood. Total of about 1500pp. 5⅜ x 8½. 23407-X Boxed set, Pa. $15.95

HOUSEHOLD STORIES BY THE BROTHERS GRIMM. All the great Grimm stories: "Rumpelstiltskin," "Snow White," "Hansel and Gretel," etc., with 114 illustrations by Walter Crane. 269pp. 5⅜ x 8½.
21080-4 Pa. $3.50

SLEEPING BEAUTY, illustrated by Arthur Rackham. Perhaps the fullest, most delightful version ever, told by C. S. Evans. Rackham's best work. 49 illustrations. 110pp. 7⅞ x 10¾.
22756-1 Pa. $2.50

AMERICAN FAIRY TALES, L. Frank Baum. Young cowboy lassoes Father Time; dummy in Mr. Floman's department store window comes to life; and 10 other fairy tales. 41 illustrations by N. P. Hall, Harry Kennedy, Ike Morgan, and Ralph Gardner. 209pp. 5⅜ x 8½.
23643-9 Pa. $3.00

THE WONDERFUL WIZARD OF OZ, L. Frank Baum. Facsimile in full color of America's finest children's classic. Introduction by Martin Gardner. 143 illustrations by W. W. Denslow. 267pp. 5⅜ x 8½.
20691-2 Pa. $3.50

THE TALE OF PETER RABBIT, Beatrix Potter. The inimitable Peter's terrifying adventure in Mr. McGregor's garden, with all 27 wonderful, full-color Potter illustrations. 55pp. 4¼ x 5½. (Available in U.S. only)
22827-4 Pa. $1.25

THE STORY OF KING ARTHUR AND HIS KNIGHTS, Howard Pyle. Finest children's version of life of King Arthur. 48 illustrations by Pyle. 131pp. 6⅛ x 9¼.
21445-1 Pa. $4.95

CARUSO'S CARICATURES, Enrico Caruso. Great tenor's remarkable caricatures of self, fellow musicians, composers, others. Toscanini, Puccini, Farrar, etc. Impish, cutting, insightful. 473 illustrations. Preface by M. Sisca. 217pp. 8⅜ x 11¼.
23528-9 Pa. $6.95

PERSONAL NARRATIVE OF A PILGRIMAGE TO ALMADINAH AND MECCAH, Richard Burton. Great travel classic by remarkably colorful personality. Burton, disguised as a Moroccan, visited sacred shrines of Islam, narrowly escaping death. Wonderful observations of Islamic life, customs, personalities. 47 illustrations. Total of 959pp. 5⅜ x 8½.
21217-3, 21218-1 Pa., Two-vol. set $12.00

INCIDENTS OF TRAVEL IN YUCATAN, John L. Stephens. Classic (1843) exploration of jungles of Yucatan, looking for evidences of Maya civilization. Travel adventures, Mexican and Indian culture, etc. Total of 669pp. 5⅜ x 8½.   20926-1, 20927-X Pa., Two-vol. set $7.90

AMERICAN LITERARY AUTOGRAPHS FROM WASHINGTON IRVING TO HENRY JAMES, Herbert Cahoon, et al. Letters, poems, manuscripts of Hawthorne, Thoreau, Twain, Alcott, Whitman, 67 other prominent American authors. Reproductions, full transcripts and commentary. Plus checklist of all American Literary Autographs in The Pierpont Morgan Library. Printed on exceptionally high-quality paper. 136 illustrations. 212pp. 9⅛ x 12¼.
23548-3 Pa. $12.50

AN AUTOBIOGRAPHY, Margaret Sanger. Exciting personal account of hard-fought battle for woman's right to birth control, against prejudice, church, law. Foremost feminist document. 504pp. 5⅜ x 8½.
20470-7 Pa. $5.50

MY BONDAGE AND MY FREEDOM, Frederick Douglass. Born as a slave, Douglass became outspoken force in antislavery movement. The best of Douglass's autobiographies. Graphic description of slave life. Introduction by P. Foner. 464pp. 5⅜ x 8½. 22457-0 Pa. $5.50

LIVING MY LIFE, Emma Goldman. Candid, no holds barred account by foremost American anarchist: her own life, anarchist movement, famous contemporaries, ideas and their impact. Struggles and confrontations in America, plus deportation to U.S.S.R. Shocking inside account of persecution of anarchists under Lenin. 13 plates. Total of 944pp. 5⅜ x 8½.
22543-7, 22544-5 Pa., Two-vol. set $12.00

LETTERS AND NOTES ON THE MANNERS, CUSTOMS AND CONDITIONS OF THE NORTH AMERICAN INDIANS, George Catlin. Classic account of life among Plains Indians: ceremonies, hunt, warfare, etc. Dover edition reproduces for first time all original paintings. 312 plates. 572pp. of text. 6⅛ x 9¼. 22118-0, 22119-9 Pa.. Two-vol. set $12.00

THE MAYA AND THEIR NEIGHBORS, edited by Clarence L. Hay, others. Synoptic view of Maya civilization in broadest sense, together with Northern, Southern neighbors. Integrates much background, valuable detail not elsewhere. Prepared by greatest scholars: Kroeber, Morley, Thompson, Spinden, Vaillant, many others. Sometimes called Tozzer Memorial Volume. 60 illustrations, linguistic map. 634pp. 5⅜ x 8½.
23510-6 Pa. $10.00

HANDBOOK OF THE INDIANS OF CALIFORNIA, A. L. Kroeber. Foremost American anthropologist offers complete ethnographic study of each group. Monumental classic. 459 illustrations, maps. 995pp. 5⅜ x 8½.
23368-5 Pa. $13.00

SHAKTI AND SHAKTA, Arthur Avalon. First book to give clear, cohesive analysis of Shakta doctrine, Shakta ritual and Kundalini Shakti (yoga). Important work by one of world's foremost students of Shaktic and Tantric thought. 732pp. 5⅜ x 8½. (Available in U.S. only)
23645-5 Pa. $7.95

AN INTRODUCTION TO THE STUDY OF THE MAYA HIEROGLYPHS, Syvanus Griswold Morley. Classic study by one of the truly great figures in hieroglyph research. Still the best introduction for the student for reading Maya hieroglyphs. New introduction by J. Eric S. Thompson. 117 illustrations. 284pp. 5⅜ x 8½. 23108-9 Pa. $4.00

A STUDY OF MAYA ART, Herbert J. Spinden. Landmark classic interprets Maya symbolism, estimates styles, covers ceramics, architecture, murals, stone carvings as artforms. Still a basic book in area. New introduction by J. Eric Thompson. Over 750 illustrations. 341pp. 8⅜ x 11¼.
21235-1 Pa. $6.95

GEOMETRY, RELATIVITY AND THE FOURTH DIMENSION, Rudolf Rucker. Exposition of fourth dimension, means of visualization, concepts of relativity as Flatland characters continue adventures. Popular, easily followed yet accurate, profound. 141 illustrations. 133pp. 5⅜ x 8½.
23400-2 Pa. $2.75

THE ORIGIN OF LIFE, A. I. Oparin. Modern classic in biochemistry, the first rigorous examination of possible evolution of life from nitrocarbon compounds. Non-technical, easily followed. Total of 295pp. 5⅜ x 8½.
60213-3 Pa. $4.00

PLANETS, STARS AND GALAXIES, A. E. Fanning. Comprehensive introductory survey: the sun, solar system, stars, galaxies, universe, cosmology; quasars, radio stars, etc. 24pp. of photographs. 189pp. 5⅜ x 8½. (Available in U.S. only) 21680-2 Pa. $3.75

THE THIRTEEN BOOKS OF EUCLID'S ELEMENTS, translated with introduction and commentary by Sir Thomas L. Heath. Definitive edition. Textual and linguistic notes, mathematical analysis, 2500 years of critical commentary. Do not confuse with abridged school editions. Total of 1414pp. 5⅜ x 8½. 60088-2, 60089-0, 60090-4 Pa., Three-vol. set $18.50

Prices subject to change without notice.

Available at your book dealer or write for free catalogue to Dept. GI, Dover Publications, Inc., 31 East Second Street, Mineola, N.Y. 11501. Dover publishes more than 175 books each year on science, elementary and advanced mathematics, biology, music, art, literary history, social sciences and other areas.